KU-318-364

# GENERALIZED FUNCTIONS

**Volume 5**

**Integral Geometry and Representation Theory**


**I. M. GEL'FAND**

**M. I. GRAEV**

and

**N. Ya. VILENKIN**

Academy of Sciences, U.S.S.R.

Translated by

**EUGENE SALETAN**

Department of Physics
Northeastern University
Boston, Massachusetts



1966

ACADEMIC PRESS · New York and London

182178

QA 331

COPYRIGHT © 1966 BY ACADEMIC PRESS INC.

ALL RIGHTS RESERVED.
NO PART OF THIS BOOK MAY BE REPRODUCED IN ANY FORM, BY PHOTOSTAT, MICROFILM, OR ANY OTHER MEANS, WITHOUT WRITTEN PERMISSION FROM THE PUBLISHERS.

ACADEMIC PRESS INC.
111 Fifth Avenue, New York, New York 10003

*United Kingdom Edition published by*
ACADEMIC PRESS INC. (LONDON) LTD.
Berkeley Square House, London W.1

ISBN-0-12-279505-9

QUEEN MARY
COLLEGE
LIBRARY

PRINTED IN THE UNITED STATES OF AMERICA

GENERALIZED FUNCTIONS:
VOLUME 5: INTEGRAL GEOMETRY AND REPRESENTATION THEORY

THIS BOOK IS A TRANSLATION OF
OBOBSHCHENNYE FUNKTSII, VYPUSK 5,
INTEGRAL'NAYA GEOMETRIYA I SVYAZANNYE S NEĬ VOPROSY
TEORII PREDSTAVLENIĬ,
MOSCOW, 1962

# Translator's Note

This English translation of the fifth volume of Professor Gel'fand's series on generalized functions contains all the material of the Russian fifth volume with the exception of its appendix. This appendix, in which generalized functions of a complex variable are discussed, appears as Appendix B of the first volume in the English translation.

The text of the translation does not deviate significantly from the Russian, although some minor typographical errors have been corrected and some equations have been renumbered. The symbol # has been used to indicate the end of the Remarks (set in small type in the Russian).

The subjects discussed in this book are often of interest both to mathematicians and to physicists, and each discipline has its own terminology. An attempt has been made to keep to the mathematicians' terminology, but some confusion is inevitable. I will appreciate suggestions for improvement in terminology and notation.

I wish to express my gratitude to the members of the Department of Mathematics at Northeastern University who have helped with the terminology. I am especially grateful to Professors Flavio Reis and Robert Bonic. I also wish to thank Dr. Eric H. Roffman and Professor E. C. G. Sudershan, who read the manuscript and galley proof and offered many helpful suggestions.

E. J. S.

# Foreword

Originally the material in this book had been intended for some chapters of Volume 4, but it was later decided to devote a separate volume to the theory of representations. This separation was based on a suggestion by G. F. Rybkin, to whom the authors express their deep gratitude, for it is in excellent accord with the aims of the entire undertaking.

The theory of representations is a good example of the use of algebraic and geometric methods in functional analysis, in which transformations are performed not on the points of a space, but on the functions defined on it.

As we proceeded in our study of representation theory, we began to recognize that this theory is based on what we shall call integral geometry [see Gel'fand and Graev(9)]. Essentially, we shall understand integral geometry to involve the transition from functions defined on one set of geometrical objects (for instance on the points of some linear surface) to functions defined on some other set (for instance on the lines generating this surface).* Stated in this way, integral geometry is of the same general nature as classical geometry (Plücker, Klein, and others), in which new homogeneous spaces are formed out of elements taken from an originally given space. In integral geometry, however, we shall deal with such problems in what perhaps may be called their modern aspect: the transition from one space to the other shall be accomplished with the simultaneous transformation of the functions defined on it. This may be compared to the difference between classical and quantum mechanics: the transformations in classical mechanics are point transformations, while those of quantum mechanics are transformations in function space. (See the introduction to Chapter II.)

We have presumed to devote an entire volume to these elegant special problems in order to emphasize particularly this modern point of view relating geometry to functional analysis, as well as to point out the algebraic-geometric approach to functional analysis, an approach still in its earliest development.

In this book we shall not attempt a complete description of the theory of representations, for that would probably take several such volumes. Instead we shall restrict ourselves to the group of two-dimensional

* The term "integral geometry" as we use it here differs from its traditional meaning in which it involves calculating invariant measures on homogeneous spaces.

complex matrices of determinant one, which is of interest for many reasons. First, it is the simplest noncommutative and noncompact group. Further, it is the transformation group of many important spaces. In particular, it is locally isomorphic to the group of Lobachevskian motions, to the group of linear-fractional transformations of the complex plane, and several others. Finally, it is important in physics, for it is locally isomorphic to the proper Lorentz group.

The method we use in this book to develop the representation theory is not the only one possible. We have chosen the most natural approach, one based on the theory of generalized functions and making use of the excellent work of Bruhat(4). In this approach many of the phenomena of representation theory, in particular the relation between finite and infinite dimensional representations, become somewhat easier to understand.

This volume can be read almost independently of the previous ones. We assume only a knowledge of Chapters I and II and some of Chapter III of Volume 1, as well as their extension to the complex domain as discussed in Appendix B of that volume. The authors apologize beforehand for the incompleteness of the present volume. We hope that its underlying point of view will nevertheless be useful for those who are interested in new developments in functional analysis. The book is written to be read in one of two possible ways. Readers interested only in integral geometry may study Chapters I, II, and V, which are concerned only with integral geometry and are independent of the rest of the book. On the other hand readers interested only in representation theory can start with Chapter III, although an outline of the problems discussed is already given in Section 2 of Chapter II.

Chapters I and II were written by Gel'fand and Graev. The rest of the book was written by the three authors together. It contains a rewritten and expanded version of chapters originally written for Volume 4 by Gel'fand and Vilenkin (Chapters III and IV of this volume).

The authors express their deep gratitude to A. A. Kirillov and F. V. Shirokov, who read over the manuscript and made many helpful observations. They are especially grateful to L. I. Kopeykina whose help in the final stages of the manuscript greatly accelerated the publication of the book, and to S. A. Vilenkina for important help in the manuscript stage.

I. M. GEL'FAND
M. I. GRAEV
N. YA. VILENKIN

*1962*

# Contents

## Chapter II

## Chapter VI

## Chapter VII

CHAPTER I

# RADON TRANSFORM OF TEST FUNCTIONS AND GENERALIZED FUNCTIONS ON A REAL AFFINE SPACE

## 1. The Radon Transform on a Real Affine Space

### 1.1. Definition of the Radon Transform

In this section we shall study the relations between the functions $f(x) = f(x_1, ..., x_n)$ on a real affine space and their integrals over all possible hypersurfaces.

We shall start by defining the integrals over a hyperplane. Consider an $n$-dimensional real affine space consisting of points $x = (x_1, ..., x_n)$. We shall always assume the space to be oriented. Further, we shall consider the differential form

$$dx = dx_1 \cdots dx_n$$

to be the volume element in this space, which then defines the integral of a function.

We now wish to define the integral of $f(x)$ over the hyperplane whose equation is

$$(\xi, x) \equiv \xi_1 x_1 + \cdots + \xi_n x_n = p.$$

To do this we must define a volume element on the hyperplane, namely a differential form of degree $n - 1$, and give the orientation of the hyperplane. Thus with the hyperplane we shall associate the differential form defined by[1]

$$d(\xi, x) \cdot \omega = dx_1 \cdots dx_n . \tag{1}$$

It is a simple matter to obtain an expression for $\omega$ in any system of coordinates on the hyperplane. For instance, if the points on the hyper-

[1] Here $d(\xi, x) \cdot \omega$ is the exterior product of the differential forms. (We refer the reader to Volume 1, pp. 214 ff.)

plane are given by the $n - 1$ coordinates $x_1, \ldots, x_{i-1}, x_{i+1}, \ldots, x_n$, we arrive at[2]

$$\omega = \frac{(-1)^{i-1}\, dx_1 \cdots dx_{i-1}\, dx_{i+1} \cdots dx_n}{\xi_i} \tag{2}$$

We then define the integral of $f(x)$ over the $(\xi, x) = p$ hyperplane by

$$\check{f}(\xi, p) = \int_{(\xi,x)=p} f(x)\omega, \tag{3}$$

where $\omega$ is given by Eq. (2) and the orientation of the hyperplane is chosen so that it becomes the boundary of the region $(\xi, x) < p$.

By using the $\delta$ function we may write (3) in the convenient form

$$\check{f}(\xi, p) = \int f(x)\, \delta(p - (\xi, x))\, dx. \tag{4}$$

(The integral is taken over the entire space.[3])

Thus with a function $f(x)$ on an $n$-dimensional real affine space, we may associate another function $\check{f}(\xi, p)$ which is defined on the set of hyperplanes. *We shall call this new function the Radon of transform of $f(x)$.*

Let us specify the class of functions whose Radon transforms we shall be discussing. In general, in order for the integral of (3) to converge for all values of $\xi$ and $p$ we need only require that $f(x)$ be absolutely summable over the entire space. Unless stated otherwise, however, we shall place stronger requirements on $f(x)$. We shall assume, namely, that $f(x)$ is infinitely differentiable rapidly decreasing, as are all of its derivatives.[4] Then the Radon transform of $f(x)$ is easily shown to be an infinitely differentiable function of $\xi$ and $p$.

Note that according to Eq. (4) $\check{f}(\xi, p)$ is an even homogeneous function of $\xi$ and $p$ of degree $-1$. This means that for any real $\alpha \neq 0$,

$$\check{f}(\alpha\xi, \alpha p) = |\alpha|^{-1} \check{f}(\xi, p). \tag{5}$$

[2] Note that $\omega$ depends not only on the hyperplane, but also on the equation by which this hyperplane is given. If, in particular, $(\xi, x) = p$ is replaced by $(\alpha\xi, x) = \alpha p$ for $\alpha \neq 0$, then $\omega$ will be replaced by $\alpha^{-1}\omega$.

[3] See Volume 1, pp. 220 ff., where integration over an arbitrary smooth surface $P(x) = 0$ is discussed and where $\delta(P)$ is defined.

[4] We call $f(x)$ *rapidly decreasing* if for every $k > 0$ we have

$$\lim_{|x|\to\infty} |x|^k f(x) = 0, \qquad \text{where} \quad |x| = (x_1^2 + \cdots + x_n^2)^{\frac{1}{2}}.$$

Henceforth unless otherwise stated a *rapidly decreasing function* will be one whose derivatives also are rapidly decreasing.

Consequently $\check{f}(\xi, p)$ is known so long as it is known for some fixed $p$, say for $p = 1$, and for all $\xi$, and it therefore depends on the same number of variables as does $f(x)$.

An easily visualized interpretation of $\check{f}(\xi, p)$ is the following. Let $f(x)$ be the density with which some finite mass is distributed throughout space. Let $M(\xi, p)$ be the total mass in the region $(\xi, x) < p$. Then

$$M(\xi, p) = \int_{(\xi, x) < p} f(x)\, dx = \int f(x)\theta(p - (\xi, x))\, dx, \tag{6}$$

where, as usual, $\theta(p) = 1$ for $p > 0$ and $\theta(p) = 0$ for $p < 0$. Now we know that $\theta'(p) = \delta(p)$. Thus the derivative of (6) with respect to $p$ gives

$$\frac{\partial M(\xi, p)}{\partial p} = \int f(x)\, \delta(p - (\xi, x))\, dx = \check{f}(\xi, p).$$

Consequently if $f(x)$ is the density with which a finite mass is distributed throughout space, its Radon transform is

$$\check{f}(\xi, p) = \frac{\partial M(\xi, p)}{\partial p},$$

where $M(\xi, p)$ is the mass in the half-space $(\xi, x) < p$.

Let us consider a particularly interesting example of the Radon transform. Specifically, let us find the geometric meaning of the Radon transform of the characteristic function of some bounded region. Let $V$ be a bounded region and let $f(x)$ be its characteristic function; i.e., $f(x) = 1$ for $x \in V$, and $f(x) = 0$ for $x \notin V$. Then the Radon transform $\check{f}(\xi, p)$ of $f(x)$ is given by

$$\check{f}(\xi, p) = \frac{\partial V(\xi, p)}{\partial p},$$

where $V(\xi, p)$ is the volume of that part of $V$ which lies in $(\xi, x) < p$.

We shall treat our space as Euclidean with metric given by $ds^2 = dx^2 + \cdots + dx_n^2$, and we shall give the hyperplanes by normalized equations [i.e., such than $|\xi| \equiv (\xi_1^2 + \cdots + \xi_n^2)^{\frac{1}{2}} = 1$ in the equation $(\xi, x) = p$]. Then $\check{f}(\xi, p)$ is simply the area of the intersection of $V$ and $(\xi, x) = p$. It is thus evident that the Radon transform of the characteristic function of a bounded region $V$ can be found from geometrical considerations, namely by calculating appropriate areas. In particular, the Radon transform of the characteristic function of the ball

$$x_1^2 + x_2^2 + \cdots + x_n^2 \leqslant R^2$$

is given by

$$\check{f}(\xi, p) = \frac{\pi^{\frac{1}{2}(n-1)}}{\Gamma(\frac{1}{2}n + \frac{1}{2})} \frac{1}{|\xi|} \left(R^2 - \frac{p^2}{|\xi|^2}\right)^{\frac{1}{2}(n-1)} \quad \text{for} \quad \frac{p^2}{|\xi|^2} < R^2,$$

$$\check{f}(\xi, p) = 0, \quad \text{for} \quad \frac{p^2}{|\xi|^2} > R^2.$$

In Section 2.4 of this chapter we shall also define the Radon transform of the characteristic function of an unbounded region.

## 1.2. Relation between Radon and Fourier Transforms

Let us find the relation between the Radon transform of $f(x)$ and its Fourier transform

$$\tilde{f}(\xi) = \int f(x) e^{i(\xi, x)} \, dx. \tag{1}$$

This Fourier transform can be written directly in terms of integrals of $f(x)$ over hyperplanes. Specifically, in order to calculate (1) we first integrate over the $(\xi, x) = p$ hyperplane and then integrate the expression so obtained over $p$ for fixed $\xi$. In other words,

$$\tilde{f}(\xi) = \int_{-\infty}^{+\infty} \check{f}(\xi, p) e^{ip} \, dp. \tag{2}$$

This is the desired relation between the Fourier and Radon transforms of $f(x)$.

Equation (2) can be written somewhat differently. We replace $\xi$ by $\alpha\xi$, where $\alpha \neq 0$, and then change variables in the integrand, writing $p = \alpha p_1$. Using the homogeneity of $\check{f}(\xi, p)$, we obtain

$$\tilde{f}(\alpha\xi) = \int_{-\infty}^{+\infty} \check{f}(\xi, p) e^{i\alpha p} \, dp. \tag{3}$$

This shows that the Fourier transform in $n$ dimensions reduces to the Radon transform followed by a one-dimensional Fourier transform.

By taking the inverse Fourier transform of (3) we obtain

$$\check{f}(\xi, p) = \frac{1}{2\pi} \int_{-\infty}^{+\infty} \tilde{f}(\alpha\xi) e^{-i\alpha p} \, d\alpha. \tag{4}$$

Thus we see that in a real affine space the Radon transform of $f(x)$ is closely related to its Fourier transform; specifically, one is obtained from the other by a one-dimensional Fourier transform. The Radon transform has some advantage over the Fourier transform in that it is geometrically more meaningful.

**Remark.** We shall see later (Chapter II, Section 3, and Chapter V) that the analog of the Radon transform can be defined not only on a real affine space, but also on other homogeneous spaces. Now we have seen that the ordinary Fourier transform can be obtained by two successive transformations, namely the Radon transform followed by a one-dimensional Fourier transform. In turns out, however, that although the analog of the Radon transform exists in many homogeneous spaces, the second of these transform is peculiar only to Euclidean spaces. Strictly speaking, the analog of this second transform does exist in other homogeneous spaces, but is related to representations of groups and is constructed differently in different spaces. (The analog of the one-dimensional Fourier transform for real and imaginary Lobachevskian spaces is developed in Chapter VI.) The advantage of the Radon transformation over the Fourier transformation is perhaps best illustrated by this more general view. There is therefore good reason for trying to construct an operational calculus based on the Radon transform. In part this will be done in the present chapter. #

## 1.3. Elementary Properties of the Radon Transform

Let us study the behavior of the Radon transform of a function under transformations of the function. Further, let us also calculate the Radon transform of the convolution of two functions.

Note first the obvious fact that the Radon transform is linear, that is, that

$$(a_1 f_1 + a_2 f_2)^{\vee} = a_1 \check{f}_1 + a_2 \check{f}_2$$

for all functions $f_1$ and $f_2$ and for all numbers $a_1$ and $a_2$. Almost as obviously, we have the following properties.

(a) *Let $A$ be a nonsingular linear transformation of the $x_i$; we write*

$$Ax = (y_1, \ldots, y_n),$$

*where $y_k = \sum_{l=1}^{n} a_{kl} x_l$, $\det| a_{kl} | \neq 0$. Then the Radon transform of*

$$f_A(x) \equiv f(A^{-1}x)$$

*is*

$$\check{f}_A(\xi, p) = | \det A | \check{f}(A'\xi, p), \tag{1}$$

*where $\check{f}(\xi, p)$ is the Radon transform of $f(x)$, and $A'$ is the transpose of $A$.*[5]

[5] The transpose is defined by $(\xi, Ax) = (A'\xi, x)$.

**Proof.** The Radon transform of $f_A(x)$ is givev by

$$\check{f}_A(\xi, p) = \int f(A^{-1}x)\, \delta(p - (\xi, x))\, dx.$$

We now write $x = Ay$. This yields

$$\check{f}_A(\xi, p) = |\det A| \int f(y)\, \delta(p - (\xi, Ay))\, dy$$

$$= |\det A| \int f(y)\, \delta(p - (A'\xi, y))\, dy = |\det A|\, \check{f}(A'\xi, p).$$

(b) *The Radon transforms of* $f_a(x) \equiv f(x + a) = f(x_1 + a_1, \ldots, x_n + a_n)$ *is*

$$\check{f}_a(\xi, p) = \check{f}(\xi, p + (\xi, a)), \tag{2}$$

*where* $\check{f}(\xi, p)$ *is the Radon transform of* $f(x)$.

**Proof.** The Radon transform of $f_a(x)$ is

$$\check{f}_a(\xi, p) = \int f(x + a)\, \delta(p - (\xi, x))\, dx.$$

The change of variables $x = y - a$ yields

$$\check{f}_a(\xi, p) = \int f(y)\, \delta(p + (\xi, a) - (\xi, y))\, dy = \check{f}(\xi, p + (\xi, a)).$$

From (b) we arrive immediately at the following.

(c) *The Radon transform of*

$$\left(a, \frac{\partial}{\partial x}\right) f(x) \equiv \sum_{k=1}^{n} a_k \frac{\partial f}{\partial x_k}$$

(*where the* $a_i$ *are arbitrary numbers*) *is*

$$(a, \xi) \frac{\partial \check{f}(\xi, p)}{\partial p} \equiv \left(\sum_{k=1}^{n} a_k \xi_k\right) \frac{\partial \check{f}(\xi, p)}{\partial p}. \tag{3}$$

A consequence of (c) is that if $P$ is a homogeneous polynomial of degree $k$ with constant coefficients, the Radon transform of $P(\partial/\partial x) f(x)$ is

$$P(\xi) \frac{\partial^k \check{f}(\xi, p)}{\partial p^k}.$$

(d) *The Radon transform* $\check{f}_1(\xi, p)$ *of*

$$f_1(x) \equiv (a, x) f(x) = \Big(\sum_{k=1}^{n} a_k x_k\Big) f(x)$$

*is related to that of* $f(x)$ *by*

$$\frac{\partial \check{f}_1(\xi, p)}{\partial p} = -\Big(a, \frac{\partial}{\partial \xi}\Big) \check{f}(\xi, p). \tag{4}$$

**Proof.** By differentiating

$$\check{f}(\xi, p) = \int f(x)\, \delta(p - (\xi, x))\, dx$$

with respect to the $\xi_i$, we obtain

$$\Big(a, \frac{\partial}{\partial \xi}\Big) \check{f}(\xi, p) = -\int (a, x) f(x)\, \delta'(p - (\xi, x))\, dx = -\frac{\partial \check{f}_1(\xi, p)}{\partial p}.$$

(e) *The Radon transform of the convolution*

$$f(x) = \int f_1(y) f_2(x - y)\, dy$$

*is given by*

$$\check{f}(\xi, p) = \int_{-\infty}^{\infty} \check{f}_1(\xi, t) \check{f}_2(\xi, p - t)\, dt, \tag{5}$$

*where* $\check{f}_1(\xi, t)$ *and* $\check{f}_2(\xi, t)$ *are the Radon transforms of* $f_1(x)$ *and* $f_2(x)$.

**Proof.** By definition, the Radon transform of $f(x)$ is

$$\check{f}(\xi, p) = \int f_1(y) f_2(x - y)\, \delta(p - (\xi, x))\, dy\, dx.$$

We interchange the order of integration and transform from $x$ to the new variable $x - y$. This yields

$$\begin{aligned}\check{f}(\xi, p) &= \int f_1(y) f_2(x)\, \delta(p - (\xi, x) - (\xi, y))\, dx\, dy \\ &= \int f_1(y) \check{f}_2(\xi, p - (\xi, y))\, dy.\end{aligned}$$

The integral over $y$ can be reduced to an integral over the $(\xi, y) = t$ hyperplane followed by an integral over $t$ for fixed $\xi$, which is just the result stated in Eq. (5).

## 1.4. The Inverse Radon Transform

Consider the Radon transform

$$\check{f}(\xi, p) = \int f(x)\, \delta(p - (\xi, x))\, dx \tag{1}$$

of $f(x)$. We wish to obtain a formula expressing $f(x)$ in terms of its integrals over hyperplanes, or in other words to invert Eq. (1).[6] This formula will be found to depend on whether the space has even or odd dimension, so we start with the simpler case of odd dimension.

Let $n$ (an odd integer) be the dimension of the space. Let us differentiate $\check{f}(\xi, p)$ $n - 1$ times with respect to $p$,[7] and let us write

$$\psi(\xi, p) = \check{f}_p^{(n-1)}(\xi, p). \tag{2}$$

We now average $\psi(\xi, p)$ over the set of hyperplanes passing through some fixed point $x$. It will be shown that this average is equal to a constant factor times the value of $f$ at $x$.

We have not yet defined what is meant by averaging $\psi$ over the hyperplanes passing through $x$. Before doing so, however, we note that a hyperplane passing through $x_0$ is defined by an equation of the form $(\xi, x) = (\xi, x_0)$. Thus what we must average over $\xi$ are functions of the form $\psi(\xi, (\xi, x))$. We shall define these averages by the integrals

$$\int_\Gamma \psi(\xi, (\xi, x))\omega(\xi) \tag{3}$$

over any closed surface $\Gamma$ enclosing the point $\xi = 0$ in $\xi$ space, where[8]

$$\omega(\xi) = \sum_{k=1}^{n} (-1)^{k-1} \xi_k\, d\xi_1 \cdots d\xi_{k-1}\, d\xi_{k+1} \cdots d\xi_n\,. \tag{4}$$

It is easily shown that (3) is independent of the choice of $\Gamma$. Indeed, both the function

$$\psi(\xi, (\xi, x)) = \check{f}_p^{(n-1)}(\xi, (\xi, x))$$

[6] The solution of this problem was first obtained by Radon [38]. We obtained a solution also in Volume 1 (Chapter I, Section 3.10) in studying $\int_{|\omega|=1} |(\omega, x)|^\lambda\, d\omega$ where $\lambda$ is a complex number. Here we shall treat this problem somewhat differently.

[7] It is natural to call $\partial/\partial p$ the infinitesimal parallel displacement operator of the plane, since variation of $p$ causes the $(\xi, x) = p$ hyperplane to move so that it remains parallel to itself.

[8] The differential form $\omega(\xi)$ has a simple geometric meaning, namely $\omega(\xi)/n$ is the volume of the cone whose vertex is at $\xi = 0$ and whose base is an element of area on $\Gamma$ (see Volume 1, p. 298).

in the integrand and the differential form $\omega(\xi)$ are homogeneous in $\xi$, the former of degree $-n$, and the latter of degree $n$. Therefore, $\psi(\xi, (\xi, x))\omega(\xi)$ is homogeneous of degree zero, or invariant under replacement of $\xi$ by $\alpha\xi$, where $\alpha > 0$. In other words, this expression remains constant along any ray passing through the origin in $\xi$ space, or the integral of Eq. (3) is invariant under deformations of $\Gamma$.[9]

We now wish to prove the following result.

*Let $\check{f}(\xi, p)$ be the Radon transform of $f(x)$ in a space of odd dimension.*[10] *Then the inverse Radon transform formula is*

$$\int_\Gamma \check{f}_p^{(n-1)}(\xi, (\xi, x))\omega(\xi) = cf(x), \tag{5}$$

*where the integral is taken over any hypersurface $\Gamma$ enclosing the origin in $\xi$ space, and $\omega(\xi)$ is given by Eq.* (4). We shall show later that $c = (-1)^{\frac{1}{2}(n-1)}2(2\pi)^{n-1}$.

**Proof.** We first prove the validity of (5) for $x = 0$, or

$$\int_\Gamma \check{f}_p^{(n-1)}(\xi, 0)\omega(\xi) = cf(0). \tag{6}$$

This integral defines a continuous functional on the space of infinitely differentiable rapidly decreasing functions which we may write in the form

$$(F, f) = \int_\Gamma \check{f}_p^{(n-1)}(\xi, 0)\omega(\xi). \tag{7}$$

What we wish to prove is that $F(x) = c\delta(x)$. We first show that $F$ satisfies the condition

$$(F, f(A^{-1}x)) = (F, f(x)), \tag{8}$$

where $A$ may be any nonsingular linear transformation. This follows from the fact established in Section 1.3 that the Radon transform of $f(A^{-1}x)$ is $|\det A|\, \check{f}(A'\xi, p)$, where $A'$ is the transpose of $A$. Thus

$$(F, f(A^{-1}x)) = \int_\Gamma |\det A|\, \check{f}_p^{(n-1)}(A'\xi, 0)\omega(\xi).$$

[9] Since $\psi(\xi, (\xi, x))\omega(\xi)$ is constant on each ray passing through the origin in $\xi$ space, it can be considered a differential form defined in the ray space. Thus (3) can be considered an integral over the set of rays diverging from the origin.

[10] Recall that $f(x)$ is an infinitely differentiable and rapidly decreasing function. This inverse Radon transform formula is, however, valid also for much weaker conditions on $f(x)$, but we shall not go into them here.

Now obviously by replacing $\xi$ by $A'^{-1}\xi$, we arrive at the original equation (7). This then proves condition (8).

Let us average the left-hand side of (8) over the set of orthogonal transformations $A$, i.e., over all transformations which leave invariant $|x|^2 = x_1^2 + \cdots + x_n^2$. This gives

$$(F, f_1(x)) = (F, f(x)), \tag{9}$$

where $f_1(x)$ is the average of $f(x)$ over the sphere $|x|^2 = r^2$. Thus in calculating $(F, f)$ we may replace $f$ by its averages over such spheres. In other words, $F$ can be treated as a functional on functions on the half-line $0 \leqslant r < \infty$.

Note that according to Eq. (8) $(F, f(\alpha x)) = (F, f(x))$, so that on the half-line $F$ is homogeneous of degree $-1$. It is easily shown that up to a constant factor the only homogeneous generalized function of degree $-1$ is the $\delta$ function (cf. Volume 1, Chapter I, Section 3.11). This proves Eq. (6), namely $F = c\,\delta(x)$.

The expression for $f(x)$ at any other point $x_0$ can now be obtained from (6). For this we need only apply (6) to $f_1(x) = f(x + x_0)$. Recall that the Radon transform of $f_1(x)$ is $\check{f}(\xi, p + (\xi, x_0))$ (see Section 1.3). On so doing, we obtain the result

$$\int_\Gamma \check{f}_p^{(n-1)}(\xi, (\xi, x_0))\omega(\xi) = cf(x_0). \tag{10}$$

**Remark.** In deriving (10) we have not actually used the fact that we are working in a space of odd dimension. But in a space of even dimension (10) is a trivial result, since the integral simply vanishes. This is because for $n$ even $\psi(\xi) = \check{f}_p^{(n-1)}(\xi, (\xi, x_0))$ is an odd function of $\xi$ [that is, $\psi(-\xi) = -\psi(\xi)$].[11] #

Let us now calculate the constant $c$ in (6) for odd $n$ by applying (6) to the test function

$$f(x) = \exp\{-x_1^2 - \cdots - x_n^2\} = \exp\{-|x|^2\}.$$

The Radon transform of this function is

$$\check{f}(\xi, p) = \int \exp\{-|x|^2\}\,\delta(p - (\xi, x))\,dx.$$

An orthogonal transformation on the $x_i$ in this integral leads to

$$\check{f}(\xi, p) = \int \exp\{-|x|^2\}\,\delta(p - |\xi|\,x_1)\,dx.$$

[11] This is because $\check{f}_p^{(n-1)}(\alpha\xi, \alpha p) = \alpha^{-n} \operatorname{sgn} \alpha \check{f}_p^{(n-1)}(\xi, p)$.

Then we easily obtain

$$\check{f}(\xi, p) = \frac{\pi^{\frac{1}{2}(n-1)}}{|\xi|} \exp\left(-\frac{p^2}{|\xi|^2}\right).$$

Expanding the exponential in a power series, we have

$$\check{f}_p^{(n-1)}(\xi, 0) = (-1)^{\frac{1}{2}(n-1)} \frac{(n-1)!}{(\frac{1}{2}n - \frac{1}{2})!} \frac{\pi^{\frac{1}{2}(n-1)}}{|\xi|^n}.$$

Now we insert the particular $f(x)$ and the expression for $\check{f}_p^{(n-1)}(\xi, 0)$ into (6), from which we obtain

$$c = (-\pi)^{\frac{1}{2}(n-1)} \frac{(n-1)!}{(\frac{1}{2}n - \frac{1}{2})!} \int_\Gamma \frac{\omega(\xi)}{|\xi|^n} = (-\pi)^{\frac{1}{2}(n-1)} \frac{(n-1)!}{(\frac{1}{2}n - \frac{1}{2})!} \Omega_n,$$

where $\Omega_n$ is the surface area of the unit sphere. Since

$$\Omega_n = \frac{2\pi^{\frac{1}{2}n}}{\Gamma(\frac{1}{2}n)} = \frac{2(2\pi)^{\frac{1}{2}(n-1)} 2^{\frac{1}{2}(n-1)} (\frac{1}{2}n - \frac{1}{2})!}{(n-1)!},$$

we arrive finally at

$$c = (-1)^{\frac{1}{2}(n-1)} 2(2\pi)^{n-1}.$$

Our final result is therefore the following. In a space of odd dimension $n$, the inverse Radon transform formula is

$$f(x) = \frac{(-1)^{\frac{1}{2}(n-1)}}{2(2\pi)^{n-1}} \int_\Gamma \check{f}_p^{(n-1)}(\xi, (\xi, x)) \omega(\xi), \tag{11}$$

where $\omega(\xi)$ is given by (4) and the integral is over any hypersurface which encloses the origin in $\xi$ space.

It can be shown that *if $n$ is even the inverse Radon transform formula is*

$$f(x) = \frac{(-1)^{\frac{1}{2}n}(n-1)!}{(2\pi)^n} \int_\Gamma \left[\int_{-\infty}^{+\infty} \check{f}(\xi, p)\{p - (\xi, x)\}^{-n}\, dp\right] \omega(\xi), \tag{12}$$

*where the integral over $p$ is understood in terms of its regularization.*[12]

[12] Specifically,

$$\int_{-\infty}^{+\infty} p^{-n}\varphi(p)\, dp = \int_0^\infty p^{-n} \Big\{ \varphi(p) + \varphi(-p) - 2\left[\varphi(0) + \frac{p^2}{2!}\varphi''(0) + \cdots + \frac{p^{n-2}}{(n-2)!}\varphi^{(n-2)}(0)\right] \Big\}\, dp$$

(see Volume 1, p. 335).

There is a fundamental difference between these inversion formulas for odd and even dimension. In the case of odd dimension, the inversion formula is local in that the value of $f$ at some point $x$ depends only on the integrals of $f$ over hyperplanes passing through $x$ and over hyperplanes infinitesimally close to these. In the case of even dimension, as is seen from (12), the value of $f$ at $x$ depends on its integrals over all possible hyperplanes.

The derivation of (12) is quite similar to that of (11). Specifically, by using the same considerations as were used to derive the analogous result for odd dimension, it can be shown that

$$\int_\Gamma \left[\int_{-\infty}^{+\infty} \check{f}(\xi, p)p^{-n}\, dp\right] \omega(\xi) = cf(0). \tag{13}$$

It is easily verified that this is a trivial result for odd dimension, for which the integral vanishes. For even dimension, however, $c \neq 0$; again by using the test function $f(x) = \exp(-|x|^2)$, we may deduce that

$$c = \frac{(-1)^{\frac{1}{2}n}}{(n-1)!}(2\pi)^n.$$

### 1.5. Analog of Plancherel's Theorem for the Radon Transform

Let $f(x)$ and $g(x)$ be two infinitely differentiable rapidly decreasing functions in an $n$-dimensional affine space, and let $\check{f}(\xi, p)$ and $\check{g}(\xi, p)$ be their Radon transforms. We shall prove that *if $n$ is odd, then*

$$\int f(x)\bar{g}(x)\, dx = \frac{(-1)^{\frac{1}{2}(n-1)}}{2(2\pi)^{n-1}} \int_\Gamma \left[\int_{-\infty}^{\infty} \check{f}(\xi, p)\bar{\check{g}}_p^{(n-1)}(\xi, p)\, dp\right] \omega(\xi), \tag{1}$$

*where, as before,*

$$\omega(\xi) = \sum_{k=1}^{n} (-1)^{k-1}\xi_k\, d\xi_1 \cdots d\xi_{k-1}\, d\xi_{k+1} \dots d\xi_n$$

*and $\Gamma$ is any hypersurface enclosing the origin in $\xi$ space.* This is the analog of Plancherel's theorem for the Radon transform. This result can also be written (by integrating by parts with respect to $p$) in the form

$$\int f(x)\bar{g}(x)\, dx = \frac{1}{2(2\pi)^{n-1}} \int_\Gamma \left[\int_{-\infty}^{\infty} \check{f}_p^{(m)}(\xi, p)\bar{\check{g}}_p^{(m)}(\xi, p)\, dp\right] \omega(\xi), \tag{1'}$$

where $m = \frac{1}{2}(n-1)$.

If $n$ is even, Eq. (1) becomes[13]

$$\int f(x)\bar{g}(x)\,dx = \frac{(-1)^{\frac{1}{2}n}(n-1)!}{(2\pi)^n}\int_\Gamma \left[\iint \check{f}(\xi, p_1)\bar{\check{g}}(\xi, p_2)(p_1 - p_2)^{-n}\,dp_1\,dp_2\right]\omega(\xi). \tag{2}$$

We shall prove this result for odd $n$; the proof for even $n$ is analogous. Consider the convolution of $f(x)$ and $g^*(x) = \bar{g}(-x)$, namely

$$F(x) = \int f(y)\bar{g}(y - x)\,dy.$$

According to Section 1.3, the Radon transform of $F(x)$ is

$$\check{F}(\xi, p) = \int_{-\infty}^{\infty} \check{f}(\xi, t)\bar{\check{g}}(\xi, t - p)\,dt.$$

We now use the inversion formula to write $F(x)$ in terms of its Radon transform. We have

$$\begin{aligned} F(0) &= \frac{(-1)^{\frac{1}{2}(n-1)}}{2(2\pi)^{n-1}}\int \check{F}_p^{(n-1)}(\xi, 0)\omega(\xi) \\ &= \frac{(-1)^{\frac{1}{2}(n-1)}}{2(2\pi)^{n-1}}\int_\Gamma \left[\int_{-\infty}^{\infty} \check{f}(\xi, t)\bar{\check{g}}_p^{(n-1)}(\xi, t)\,dt\right]\omega(\xi). \end{aligned}$$

But

$$F(0) = \int f(y)\bar{g}(y)\,dy.$$

When we insert this expression for $F(0)$ in the preceding equation and replace $t$ by $p$, we arrive at (1).

Let us rewrite the right-hand side of (1) by integrating by parts $\frac{1}{2}(n-1)$ times with respect to $p$. Since the derivatives of $f$ and $g$ are rapidly decreasing functions, $\check{f}(\xi, p)$ and $\check{g}(\xi, p)$ and their derivatives with respect to $p$ converge to zero as $p \to \pm\infty$, or as the $(\xi, x) = p$ hyperplanes move off to infinity. Equation (1′) follows immediately from these considerations.

[13] The integral over $p_1$ and $p_2$ in the square brackets is to be understood in the sense of its regularization. Specifically, by a change of variables, we write it in the form

$$\int_{-\infty}^{\infty}\left[\int_{-\infty}^{\infty} \check{f}(\xi, p + p_2)\,\bar{\check{g}}(\xi, p_2)\,dp_2\right] p^{-n}\,dp.$$

which reduces the problem to regularizing an integral over $p$. This regularization was described in footnote 12.

Note that (1′) has been obtained for rapidly decreasing $f$ and $g$. But such functions form an everywhere dense set in the space of square integrable functions, i.e., those for which

$$\int |f(x)|^2 \, dx < \infty.$$

Thus Eq. (1′) is valid also for this larger set.

An interesting consequence of Plancherel's theorem is obtained when $g(x)$ is the characteristic function of some bounded region $V$ in a space of odd dimension [i.e., $g(x) = 1$ for $x \in V$ and $g(x) = 0$ for $x \notin V$]. Let us assume the space to be Euclidean. Then as we have shown, $\check{g}(\xi, p)$ for $|\xi| = 1$ is the area of the intersection of $V$ and the $(\xi, x) = p$ hyperplane. Plancherel's theorem is an expression for the integral of $f(x)$ over $V$ in terms of its integrals $\check{f}(\xi, p)$ over the $(\xi, x) = p$ hyperplanes, namely

$$\int_V f(x)\, dx = \frac{(-1)^{\frac{1}{2}(n-1)}}{2(2\pi)^{n-1}} \int_\Gamma \left[\int_{-\infty}^{+\infty} \check{f}(\xi, p) \frac{\partial^{n-1} S(\xi, p)}{\partial p^{n-1}}\, dp\right] \omega(\xi). \tag{3}$$

Here $S(\xi, p)$ is the area of the intersection of $V$ with the appropriate hyperplane, and $\Gamma$ is the unit sphere.

In Section 2.4 this formula will be generalized to unbounded $V$.

Now let *both* $f(x)$ and $g(x)$ be the characteristic function of $V$. This leads to an expression for the volume $v$ of $V$ in a space of odd dimension in terms of the intersection $S(\xi, p)$ of $V$ with all possible hyperplanes. From (1′) we then have

$$v = \frac{1}{2(2\pi)^{n-1}} \int_\Gamma \left[\int_{-\infty}^{\infty} \left\{\frac{\partial^m S(\xi, p)}{\partial p^m}\right\}^2 dp\right] \omega(\xi), \tag{4}$$

where $m = \frac{1}{2}(n - 1)$.

We shall not discuss the somewhat trickier case of even $n$.

**Remark.** Plancherel's formula can be used to extend the (odd $n$) mapping $f(x) \to f_p^{(m)}(\xi, p)$, with $m = \frac{1}{2}(n - 1)$ to an isometric mapping of square integrable functions into the space of functions $\psi(\xi, p)$ that satisfy the homogeneity condition

$$\psi(\alpha\xi, \alpha p) = \alpha^{-m-1} \operatorname{sgn} \alpha\, \psi(\xi, p) \tag{5}$$

and such that

$$\int_\Gamma \left[\int_{-\infty}^{+\infty} |\psi(\xi, p)|^2\, dp\right] \omega(\xi) < \infty. \tag{6}$$

It can be shown that this extended mapping is *onto*, but we shall not go into the proof here. We may add that both Plancherel's theorem and the extended mapping can be obtained by writting the Radon transform in terms of the Fourier transform and using the known properties of the latter. This is suggested as an exercise for the reader. #

## 1.6. Analog of the Paley-Wiener Theorem for the Radon Transform[14]

In this section we shall find the necessary and sufficient conditions for a function $\check{f}(\xi, p)$ to be the Radon transform of some infinitely differentiable rapidly decreasing function. These conditions may be formulated uniformly for spaces of all dimensions, although the proofs differ somewhat for even and odd dimension. As before, we shall restrict our considerations for simplicity to spaces of odd dimension. It is particularly interesting that in addition to expected requirements involving its differentiability and its rate of decrease, it is found that $\check{f}(\xi, p)$ must satisfy additional algebraic relations (see Condition 4).[15]

To proceed, let $f(x)$ be an infinitely differentiable rapidly decreasing function on a space of odd dimension. We wish to find the necessary conditions satisfied by its Radon transform

$$\check{f}(\xi, p) = \int_{(\xi, x) = p} f(x)\omega = \int f(x)\, \delta(p - (\xi, x))\, dx. \tag{1}$$

The first condition has already been established.

*Condition 1.* The function $\check{f}(\xi, p)$ has the homogeneity property

$$\check{f}(\alpha\xi, \alpha p) = |\alpha|^{-1} \check{f}(\xi, p) \tag{2}$$

for any $\alpha \neq 0$.

In addition, from the definition of the Radon transform and the conditions imposed on $f(x)$, the following two results are easily established.

*Condition 2.* The function $\check{f}(\xi, p)$ is infinitely differentiable with respect to $\xi$ and with respect to $p$ for $\xi \neq 0$.

[14] The classical Paley–Wiener theorem concerns the Fourier transforms of functions with compact support on the line. In this book we shall mean by analogs of this theorem those theorems which describe functions obtained from "sufficiently good" functions by any of the transforms of integral geometry.

[15] These algebraic relations are associated with the occurrence of finite-dimensional representations of the group of affine transformations (including parallel transfer). Similar relations in the case of the motions of Lobachevskian space are related to the occurrence of finite-dimensional representations of this group also (see the Paley–Wiener theorem for the Lorentz group in Chapter IV, Section 5).

*Condition 3.* Asymptotically as $|p| \to \infty$ and for any $k > 0$, we have

$$|\check{f}(\xi, p)| = o(|p|^{-k}) \tag{3}$$

uniformly in $\xi$ for $\xi$ running through a bounded closed region not containing the origin. This equation holds also for each derivative of $\check{f}(\xi, p)$ to any order with respect to $\xi$ or $p$.

Finally, it is easily shown that $\check{f}(\xi, p)$ has the following additional property.

*Condition 4.* For any nonnegative integer $k$, the integral

$$\int_{-\infty}^{\infty} \check{f}(\xi, p) p^k \, dp \tag{4}$$

is a polynomial in $\xi$, homogeneous of degree $k$.

Indeed replace $\check{f}(\xi, p)$ in (4) by its expression in terms of $f(x)$, obtaining

$$\int_{-\infty}^{\infty} \check{f}(\xi, p) p^k \, dp = \int f(x)(\xi, x)^k \, dx.$$

The assertion then follows immediately.

These then are necessary conditions that must be satisfied by the Radon transform of an infinitely differentiable rapidly decreasing function. We shall now show that they are also sufficient. Specifically, we shall show that every function $\check{f}(\xi, p)$ satisfying Conditions 1–4 is the Radon transform of some infinitely differentiable rapidly decreasing function $f(x)$.

We start by writing $f(x)$ in terms of $\check{f}(\xi, p)$ by means of the inversion formula[16]

$$f(x) = \frac{(-1)^{\frac{1}{2}(n-1)}}{2(2\pi)^{n-1}} \int_{\Gamma} \check{f}_p^{(n-1)}(\xi, (\xi, x)) \omega(\xi). \tag{5}$$

We must show that $f(x)$ is an infinitely differentiable rapidly decreasing function and that its Radon transform is $\check{f}(\xi, p)$.

That $f(x)$ is infinitely differentiable follows from Condition 2. We assert that it is also rapidly decreasing, i.e., that for any $x_0 \neq 0$ the function $f(tx_0)$ decreases rapidly as a function of $t$ as $t \to \infty$. For the proof we may, without loss of generality, make the choice

$$x_0 = (1, 0, ..., 0).$$

[16] It follows from Condition 1 that this integral is independent of the choice of $\Gamma$.

Further, we choose the surface of integration in Eq. (5) to be the pair of hyperplanes $\xi_n = \pm 1$. Since the function inside the integral sign in (5) is not altered when $\xi$ is replaced by $-\xi$ we may replace the integral by twice the integral over the $\xi_n = 1$ hyperplane. Then $f(tx_0)$ will be given by

$$\begin{aligned} f(tx_0) &= \int \psi(\xi_1, ..., \xi_{n-1}, 1; t\xi_1)\, d\xi_1 \cdots d\xi_{n-1} \\ &= t^{-1} \int \psi(pt^{-1}, \xi_2, ..., \xi_{n-1}, 1; p)\, dp\, d\xi_2 \cdots d\xi_{n-1}, \end{aligned} \tag{6}$$

where we have written

$$\psi(\xi, p) = \frac{(-1)^{\frac{1}{2}(n-1)}}{(2\pi)^{n-1}} \check{f}_p^{(n-1)}(\xi, p). \tag{7}$$

Now the integrand in this expression is an infinitely differentiable function of $t^{-1}$ in the neighborhood of $t^{-1} = 0$. It can therefore be expanded in an asymptotic Taylor's series in powers of $t^{-1}$, to yield

$$\psi(pt^{-1}, \xi_2, ..., \xi_{n-1}, 1; p) \sim \sum_{k=0}^{\infty} \frac{t^{-k}p^k}{k!} \psi_{\xi_1}^{(k)}(0, \xi_2, ..., \xi_{n-1}, 1; p). \tag{8}$$

We now integrate this series term by term[17] over $\xi_2, ..., \xi_{n-1}$ and $p$ to obtain an asymptotic series for $f(tx_0)$ as $t \to \infty$, namely

$$f(tx_0) \sim \sum_{k=0}^{\infty} \frac{t^{-k-1}}{k!} \int p^k \psi_{\xi_1}^{(k)}(0, \xi_2, ..., \xi_{n-1}; p)\, dp\, d\xi_2 \cdots d\xi_{n-1}. \tag{9}$$

Now all the terms of this asymptotic series vanish, as will now be shown, so that $f(tx_0)$ decreases more rapidly than any negative power of $t$ as $t \to \infty$. Indeed, recall that according to Condition 4

$$\int_{-\infty}^{+\infty} \check{f}(\xi, p) p^m\, dp$$

is a polynomial in $\xi$ of degree $m$. Integration by parts then shows that

$$\int_{-\infty}^{+\infty} \psi(\xi, p) p^k\, dp = \frac{(-1)^{\frac{1}{2}(n-1)}}{(2\pi)^{n-1}} \int \check{f}_p^{(n-1)}(\xi, p) p^k\, dp$$

[17] That (8) may indeed be integrated term by term may be verified by calculating the remainder. This is a tedious, but standard task which we shall omit.

is a polynomial in $\xi$ of degree $k - n + 1$ (for $k < n - 1$ the integral vavishes). Hence

$$\int_{-\infty}^{+\infty} \psi_{\xi_1}^{(k)}(\xi, p) p^k \, dp = \frac{\partial^k}{\partial \xi_1^k} \int_{-\infty}^{+\infty} \psi(\xi, p) p^k \, dp = 0.$$

Thus all the terms of (9) vanish. This shows that $f(x)$ is a rapidly decreasing function. Similar considerations can be used to obtain the same result for any derivative of $f(x)$.

We have thus shown that according to the definition in terms of $\check{f}(\xi, p)$, that is, according to Eq. (5), $f(x)$ is an infinitely differentiable rapidly decreasing function. What remains is to shown that the Radon transform of $f(x)$ is $\check{f}(\xi, p)$. Thus let $\check{g}(\xi, p)$ be the Radon transform of $f(x)$. Then the inversion formula gives

$$f(x) = \frac{(-1)^{\frac{1}{2}(n-1)}}{2(2\pi)^{n-1}} \int_{\Gamma} \check{g}_p^{(n-1)}(\xi, (\xi, x)) \omega(\xi). \tag{10}$$

It follows then that

$$\check{F}(\xi, p) = \check{f}(\xi, p) - \check{g}(\xi, p)$$

has the property that

$$\int_{\Gamma} \check{F}_p^{(n-1)}(\xi, (\xi, x)) \omega(\xi) \equiv 0. \tag{11}$$

What we must show is that $\check{F}(\xi, p) = 0$. To do this we introduce the new function $\Phi(\xi) = \check{F}_p^{(n-1)}(\xi, 1)$ and show that $\Phi(\xi) \equiv 0$.

Let us choose $\Gamma$ in (11) to be the two hyperplanes

$$(\xi, x) = \pm 1$$

in $\xi$ space. Now the function under the integral sign in (11) does not change sign when $\xi$ is replaced by $-\xi$, so that the integral may be written

$$2 \int_{(\xi, x)=1} \Phi(\xi) \omega \equiv 0.$$

This means that the Radon transform of $\Phi(\xi)$ vanishes identically, and therefore so does $\Phi$.[18]

[18] Note that $\Phi(\xi)$ is infinitely differentiable for $\xi \neq 0$. It is easily shown that if we define $\Phi(0) = 0$, it will be infinitely differentiable also at $\xi = 0$. Further, for this function and any of its derivatives $D\Phi(\xi)$ we may write

$$| D\Phi(\xi) | = O(| \xi |^{-n}).$$

It follows that the Radon transform of $\Phi(\xi)$ exists, and that $\Phi(\xi)$ itself is uniquely defined by the inversion formula in terms of its Radon transform.

We have thus shown that $\check{F}_p^{(n-1)}(\xi, 1) \equiv 0$. But then $\check{F}_p^{(n-1)}(\xi, p) \equiv 0$ and therefore also $\check{F}(\xi, p) = 0$. This completes the proof: any function $\check{f}(\xi, p)$ satisfying Conditions 1–4 is the Radon transform of an infinitely differentiable rapidly decreasing function $f(x)$.

## 1.7. Asymptotic Behavior of Fourier Transforms of Characteristic Functions of Regions

Let us turn aside briefly to consider a certain interesting problem of analysis. We wish to find the asymptotic behavior as $|\xi| \to \infty$ of the integral

$$\varphi(\xi) = \int_V e^{i(\xi,x)}\, dx, \tag{1}$$

over some bounded region $V$ in $n$ dimensions. Assume that $V$ is bounded by a convex surface which is $\frac{1}{2}(n+3)$ times differentiable and centrally symmetric about the origin. Assume also that at each point on the surface the product of the principal radii of curvature is nonzero. Note that Eq. (1) is the Fourier transform of the characteristic function of $V$.

We shall solve this problem, namely obtain the asymptotic behavior of (1), from simple geometric considerations by going over from the Fourier to the Radon transform of the characteristic function.

Let $S(\xi, p)$ be the Radon transform of the characteristic function of $V$. For $|\xi| = 1$, $S(\xi, p)$ is the area of the intersection of $V$ and the $(\xi, x) = p$ hyperplane. We may express $\varphi(\xi)$ in terms of $S(\xi, p)$ by the equation

$$\varphi(r\xi) = \int_{-a(\xi)}^{a(\xi)} S(\xi, p) e^{ipr}\, dp, \tag{2}$$

where $2a(\xi)$ is the diameter of $V$ along the $\xi$ vector (Fig. 1). What we

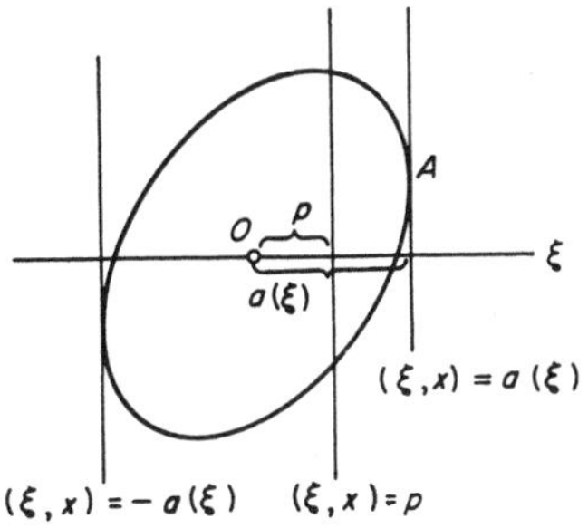


FIG. 1.

must do is analyze the behavior of $S(\xi, p)$ in the neighborhood of $p = \pm a(\xi)$. Let the hyperplane $(\xi, x) = a(\xi)$ be tangent to the boundary of $V$ at some point $A$. We choose a new coordinate system to describe the surface, in which the coordinate axes are the normal to the surface and to the principal curvature vectors at $A$. Then in the neighborhood of $A$ the equation of the surface is

$$a - p = \sum_{i=1}^{n-1} \frac{x_i^2}{2\rho_i} + \cdots, \qquad a = a(\xi),$$

where the $\rho_i$ are the principal curvature vectors at $A$ (we have omitted terms of higher order in the $x_i$).

Now as $p \to a(\xi)$, the shape of the intersection of $V$ and the $(\xi, x) = p$ hyperplane approaches the ellipsoid

$$\sum_{i=1}^{n-1} \frac{x_i^2}{2\rho_i} = a - p$$

to second order in the $x_i$. The volume of this ellipsoid is

$$\frac{(2\pi)^{\frac{1}{2}(n-1)}}{\Gamma(\frac{1}{2}n + \frac{1}{2})} (\rho_1 \cdots \rho_{n-1})^{\frac{1}{2}} (a - p)^{\frac{1}{2}(n-1)}.$$

Thus

$$S(\xi, p) = \frac{(2\pi)^{\frac{1}{2}(n-1)}}{\Gamma(\frac{1}{2}n + \frac{1}{2})} (\rho_1 \cdots \rho_{n-1})^{\frac{1}{2}} (a - p)^{\frac{1}{2}(n-1)} + (a - p)^{\frac{1}{2}n} S_1(\xi, p),$$

where $S_1(\xi, p)$ has $\frac{1}{2}(n + 3)$ continuous derivatives with respect to $p$ for $|p| \leqslant a(\xi)$. Our assumption that $V$ is symmetric implies that $S(\xi, p)$ is an even function of $p$, so that we may write

$$S(\xi, p) = \frac{\pi^{\frac{1}{2}(n-1)}}{\Gamma(\frac{1}{2}n + \frac{1}{2})} \frac{(\rho_1 \cdots \rho_{n-1})^{\frac{1}{2}}}{a^{\frac{1}{2}(n-1)}} (a^2 - p^2)^{\frac{1}{2}(n-1)} + (a^2 - p^2)^{\frac{1}{2}n} S_2(\xi, p),$$

where $S_2(\xi, p)$ also has $\frac{1}{2}(n + 3)$ continuous derivatives with respect to $p$ for $|p| \leqslant a(\xi)$.

Let us now go over from the Radon transform $S(\xi, p)$ to the Fourier transform $\varphi$ by using Eq. (2). This yields

$$\varphi(r\xi) = \frac{(\rho_1 \cdots \rho_{n-1})^{\frac{1}{2}} a^{\frac{1}{2}} \cdot (2\pi)^{\frac{1}{2}n}}{r^{\frac{1}{2}n}} J_{\frac{1}{2}n}(ar) + \int_{-a}^{a} (a^2 - p^2)^{\frac{1}{2}n} S_2(\xi, p) e^{ipr}\, dp, \quad (3)$$

where $J_{\frac{1}{2}n}(ar)$ is the Bessel function of order $\frac{1}{2}n$.[19] Now as $r \to \infty$ we have the asymptotic expression[20]

$$J_{\frac{1}{2}n}(ar) = (2/\pi ar)^{\frac{1}{2}} \cos[ar - (\tfrac{1}{4}n + \tfrac{1}{4})\pi][1 + O(1/r)]$$

for the Bessel function, with the condition $|\cos[ar - (\frac{1}{4}n+\frac{1}{4})\pi]| \geqslant \delta > 0$. On the other hand, the second term in Eq. (3) can be integrated by parts, and it is then easily seen that for large $r$ it behaves according to

$$\int_{-a}^{a} (a^2 - p^2)^{\frac{1}{2}n} S_2(\xi, p) e^{ipr}\, dp = O(r^{\frac{1}{2}(n+2)}).$$

We now insert both of these expressions into (3). Then for $\varphi(\xi) = \int_V e^{i(\xi,x)}\, dx$, we obtain the asymptotic expression

$$\varphi(r\xi) = 2(2\pi)^{\frac{1}{2}(n-1)}(\rho_1 \cdots \rho_{n-1})^{\frac{1}{2}} \frac{\cos[ar - \frac{1}{4}(n+1)\pi]}{r^{\frac{1}{2}(n+1)}} \left[1 + O\left(\frac{1}{r^{\frac{1}{2}}}\right)\right]$$

under the condition $|\cos[ar - \frac{1}{4}(n+1)\pi]| \geqslant \delta > 0$ (recall that $2a$ is the diameter of $V$ along the vector $\xi$).

## 2. The Radon Transform of Generalized Functions

We have been considering Radon transforms of rapidly decreasing functions, and have defined them in terms of integrals over hyperplanes. It is clear that this definition can be extended to all summable functions, in particular to characteristic functions of bounded regions. In this section we shall go on to define the Radon transforms of generalized functions. In addition, the Radon transforms of some nonsummable functions (e.g., characteristic functions of unbounded regions) will be defined. This will then generalize to nonsummable functions the concept of the integral over a hyperplane.

The Radon transform of the characteristic function of an unbounded region is of particular interest. Let us try to explain its meaning in some detail, at least for the case of a space of odd dimension. Let $S(\xi, p)$ be the Radon transform of the characteristic function of some bounded region $V$ (e.g., of an ellipsoid, of a bounded convex region, or of something similar). We have seen that $S(\xi, p)$ is then the area of the intersection

[19] We have made use of Poisson's integral representation of the Bessel function. See formula **8.411**(10) in Gradshteyn, I. S., and Ryzhik, I. M., "Table of Integrals, Series, and Products," Academic Press, New York, 1966, p. 953.

[20] *Ibid.*, Formula **8.451** (1), p. 961.

of this region with the $(\xi, x) = p$ hyperplane, and have obtained the following formula. Let $F$ be the characteristic function of $V$. Then

$$\int F(x)f(x)\,dx = \frac{(-1)^{\frac{1}{2}(n-1)}}{2(2\pi)^{n-1}} \int_{\Gamma} \left[\int_{-\infty}^{\infty} S(\xi, p)\check{f}_p^{(n-1)}(\xi, p)\,dp\right] \omega(\xi), \qquad (1)$$

where $\check{f}(\xi, p)$ is the Radon transform of $f(x)$. Since the integral on the left-hand side is $\int_V f(x)\,dx$, Eq. (1) is a solution of the following problem: to calculate the integral of $f(x)$ over the bounded region $V$ if what is known is the integral of $f(x)$ over every hyperplane. We could, on the other hand, have formulated the same problem for an unbounded region $V$. We shall see in Section 2.4 that its solution will again be given by Eq. (1), where $S(\xi, p)$ is the Radon transform of the characteristic function of $V$. Thus when we calculate the Radon transform for such $V$, we will have solved the following problem: to calculate the integral of $f(x)$ over the unbounded region $V$ if what is known is the integral of $f(x)$ over every hyperplane. It is natural then to consider $S(\xi, p)$ the generalization, to unbounded regions, of the concept of area of an intersection.

The next few sections will be devoted mainly to calculating Radon transforms of some generalized functions. In particular, in Sections 2.2 and 2.3 we do this for generalized functions concentrated at a point, on an interval, on a ray, and on a line. In Sections 2.5, 2.6, and 2.8 we shall calculate the Radon transforms of the characteristic functions of some unbounded regions, specifically the interior of one sheet of a circular cone, the interior of one sheet of a hyperboloid, and of an octant. In all the examples it is not only the final formulas which are of importance, but also the methods by which they are obtained.

## 2.1. Definition of the Radon Transform for Generalized Functions

We shall define the Radon transform for generalized functions so that on test functions it coincides with our previous definition. The definition will be based on Plancherel's theorem for the Radon transform, which we obtained in Section 1.5. Since the form of the theorem depends on whether the space is of even or odd dimension, we shall have to treat these two cases separately. Actually we shall consider only the simpler case of odd dimension.[21]

Plancherel's theorem for the case of odd dimension $n$ can be written

$$\int F(x)f(x)\,dx = \frac{(-1)^{\frac{1}{2}(n-1)}}{2(2\pi)^{n-1}} \int_{\Gamma} \left[\int_{-\infty}^{\infty} \check{F}(\xi, p)\check{f}_p^{(n-1)}(\xi, p)\,dp\right] \omega(\xi), \qquad (1)$$

[21] All the definitions and the reasoning of Sections 2.1 and 2.2 for odd dimension are easily extended to the case of even dimension.

where $\check{F}$ and $\check{f}$ are the Radon transforms of test functions $F$ and $f$. Recall that the integral on the right-hand side is over any surface which encloses the origin in $\xi$ space, and that

$$\omega(\xi) = \sum_{k=1}^{n} (-1)^{k-1} \xi_k \, d\xi_1 \cdots d\xi_{k-1} \, d\xi_{k+1} \cdots d\xi_n \, .$$

Let the $(n - 1)$st derivative of $\check{f}(\xi, p)$ with respect to $p$ be written

$$\psi(\xi, p) = \check{f}_p^{(n-1)}(\xi, p). \tag{2}$$

Then the integral on the right-hand side of (1) is

$$\int_\Gamma \left[\int_{-\infty}^{\infty} \check{F}(\xi, p)\psi(\xi, p)\, dp\right] \omega(\xi) = (\check{F}, \psi), \tag{3}$$

which means that $\check{F}$ defines a functional on the space of the $\psi$ functions. Then Plancherel's theorem itself can be written

$$(F, f) = \frac{(-1)^{\frac{1}{2}(n-1)}}{2(2\pi)^{n-1}} (\check{F}, \psi). \tag{4}$$

Thus this formula defines the Radon transform $\check{F}$ of $F$ as a functional on the space of $\psi$ functions.

We are now able to define the *Radon transform of a generalized function* $F$. Let $F$ be a generalized function defined on the test functions $f(x)$. Consider the space of $\psi(\xi, p)$ functions given by Eq. (2), where $\check{f}(\xi, p)$ is the Radon transform of a test function. We shall call the Radon transform of the generalized function $F$ the functional $\check{F}$ defined by Eq. (4) on the space of test functions $\psi$.

The test function space we choose will be $S$, namely the space of infinitely differentiable rapidly decreasing functions. We have already discussed in Section 1.6 the Radon transforms $\check{f}(\xi, p)$ of these functions, and therefore also their derivatives $\psi(\xi, p) = \check{f}_p^{(n-1)}(\xi, p)$. Specifically, the $\psi$ functions are characterized by the following conditions.

*Condition 1.* Every $\psi(\xi, p)$ is an even function of $\xi$ and $p$, homogeneous of degree $-n$; in other words,

$$\psi(\alpha\xi, \alpha p) = |\alpha|^{-n}\psi(\xi, p)$$

for any $\alpha \neq 0$.

*Condition 2.* Every $\psi(\xi, p)$ is infinitely differentiable with respect to $\xi$ and $p$ for $\xi \neq 0$.

*Condition 3.* For every $k \geqslant 0$, as $|p| \to \infty$ we have

$$|\psi(\xi, p)| = o(|p|^{-k})$$

uniformly in $\xi$ for $\xi$ in any bounded closed region which does not contain the origin. The same is true of the derivatives of $\psi$.

*Condition 4.* For any integer $k \geqslant 0$, the integral

$$\int_{-\infty}^{\infty} \psi(\xi, p)p^k \, dp$$

is a homogeneous polynomial in $\xi$ of degree $k - n + 1$ (for $k < n - 1$ the integral vanishes).

Conditions 1, 2, and 3 are direct consequences of the analogous conditions for $f$ which we derived in Section 1.6. Condition 4 for $k \geqslant n - 1$ is obtained by integrating by parts in the similar condition for $f$; for $k < n - 1$, the result follows by taking the $(n - 1)$st derivative.

Thus the Radon transform of a generalized function $F$ is a functional on the test functions $\psi(\xi, p)$ satisfying Conditions 1–4. We may now extend this functional $\check{F}$ in one of various ways to the space of all functions $\psi(\xi, p)$ satisfying only Conditions 1–3, namely homogeneity, infinite differentiability, and rapid decrease in $p$. This space is topologized in a natural way. In this way the Radon transform of a generalized function is found to be a generalized function in the usual sense; nevertheless it is not uniquely defined.

Then the question arises as to which are the "unessential" functions $\check{F}(\xi, p)$ to which corresponds the functional equal to zero on the subspace of test functions fulfilling Condition 4. In other words, for which generalized functions $\check{F}(\xi, p)$ is it true that

$$(\check{F}, \psi) \equiv \int_{\Gamma} \left[\int_{-\infty}^{\infty} \check{F}(\xi, p)\psi(\xi, p) \, dp\right] \omega(\xi) = 0$$

for $\psi(\xi, p)$ fulfilling Condition 4? It is found that *the subspace of unessential functions is generated by functions of the form*

$$p^k a_{-k-1}(\xi), \tag{5}$$

*where for $k < n - 1$ the function $a_{-k-1}(\xi)$ is arbitrary except that it must have the homogeneity property*

$$a_{-k-1}(\alpha\xi) = \alpha^{-k} \,|\, \alpha \,|^{-1} a_{-k-1}(\xi) \tag{6}$$

*for any $\alpha \neq 0$, while for $k \geqslant n - 1$ it must satisfy in addition to (6) the condition*

$$\int_{\Gamma} a_{-k-1}(\xi)P_{k-n+1}(\xi)\omega(\xi) = 0 \tag{7}$$

*for every homogeneous polynomial $P_{-k-n+1}(\xi)$ of degree $k - n + 1$.*

**Remark.** This assertion is easily proven. In fact note that Condition 4 can be written in the form

$$(p^k a_{-k-1}(\xi), \psi) \equiv \int_\Gamma \left[\int_{-\infty}^{\infty} \psi(\xi, p) p^k \, dp\right] a_{-k-1}(\xi)\omega(\xi) = 0,$$

where $a_{-k-1}(\xi)$ is arbitrary except that it fulfills Eqs. (6) and (7). In view of the duality of the test function space and the space conjugate to it, the space of all unessential functions is generated by functions of the form given by (5). #

Henceforth in calculating the Radon transform of a generalized function we must always bear in mind that it is defined only up to an unessential function.

## 2.2. Radon Transform of Generalized Functions Concentrated on Points and Line Segments

Let us calculate the Radon transform of $\delta(x) = \delta(x_1, ..., x_n)$. We may think of $\delta(x)$ as the distribution function for a unit mass concentrated at $x = 0$. As we saw in Section 1.1, in order to calculate the Radon transform we must calculate the mass $M(\xi, p)$ in the region $(\xi, x) < p$. The Radon transform is then given by $\check{F}(\xi, p) = \partial M/\partial p$. For our particular case, obviously, $M(\xi, p) = 1$ for $p > 0$, and $M(\xi, p) = 0$ for $p < 0$. For this function we have $\partial M(\xi, p)/\partial p = \delta(p)$, so that the Radon transform of $\delta(x)$ is $\delta(p)$.

Let us now calculate the Radon transform of a generalized function concentrated on a line. Let some finite mass be distributed with density $a(x_1)$ on the $x_1$ axis so that $\int_{-\infty}^{\infty} | a(x_1)| \, dx < \infty$. Then the distribution function may be written

$$F(x) = a(x_1) \, \delta(x_2, ..., x_n).$$

We wish now to find the mass $M(\xi, p)$ in the region $(\xi, x) < p$. Now the $(\xi, x) = p$ hyperplane crosses the $x_1$ axis at $x_1 = p/\xi_1$, so that

$$M(\xi, p) = \int_{-\infty}^{p/\xi_1} a(x) \, dx \qquad \text{for} \quad \xi_1 > 0,$$

$$M(\xi, p) = \int_{p/\xi_1}^{\infty} a(x) \, dx \qquad \text{for} \quad \xi_1 < 0.$$

Taking the derivative, we have $\partial M(\xi, p)/\partial p = | \xi_1 |^{-1} a(p/\xi_1)$. Thus the Radon transform of $a(x_1) \, \delta(x_2, ..., x_n)$, with $\int_{-\infty}^{\infty} | a(x)| \, dx < \infty$, is

$|\xi_1|^{-1}a(p/\xi_1)$. This expression has a rather simple meaning, for $a(p/\xi_1)$ is the mass density at the point where the hyperplane crosses the $x_1$ axis.

We find in particular that the Radon transform of the characteristic function of an interval on the $x_1$ axis [that is, of $a(x_1)\,\delta(x_2, \ldots, x_n)$, where $a(x_1) = 1$ or 0 depending on whether or not $x_1$ is or not on the interval] is $|\xi_1|^{-1}$ if the $(\xi, x) = p$ hyperplane intersects the interval, and zero otherwise. This result makes it possible to write the integral of some $f(x)$ over an interval $[\alpha, \beta]$ of the $x_1$ axis in terms of its integrals $\check{f}(\xi, p)$ over hyperplanes. Specifically, from Plancherel's theorem we have

$$\int_{-\infty}^{\infty} a(x_1) f(x_1, 0, \ldots, 0)\, dx_1$$

$$= \frac{(-1)^{\frac{1}{2}(n-1)}}{2(2\pi)^{n-1}} \int_\Gamma \left[\int_{-\infty}^{\infty} |\xi_1|^{-1} a\left(\frac{p}{\xi_1}\right) \check{f}_p^{(n-1)}(\xi, p)\, dp\right] \omega(\xi). \tag{1}$$

Now let $a(x_1)$ be the characteristic function of the interval $[\alpha, \beta]$, and choose $\Gamma$ to be the two hyperplanes $\xi_1 = \pm 1$. Then (1) becomes

$$\int_\alpha^\beta f(x_1, 0, \ldots, 0)\, dx_1 = \frac{(-1)^{\frac{1}{2}(n-1)}}{(2\pi)^{n-1}} \int \left[\int_\alpha^\beta \check{f}_p^{(n-1)}(\xi, p)\, dp\right] d\xi_2 \cdots d\xi_n,$$

where $\xi = (1, \xi_2, \ldots, \xi_n)$.

### 2.3. Radon Transform of $(x_1)_+^\lambda\, \delta(x_2, \ldots, x_n)$

We have found the Radon transform when $\int_{-\infty}^{\infty} |a(x_1)|\, dx_1 < \infty$. Let us now turn to the new case $a(x_1) = (x_1)_+^\lambda$ (defined for $\lambda \neq -1, -2, \ldots$), so that we are dealing with a generalized function concentrated on the half-line, or ray.[22] Note that the integral $\int_0^\infty x^\lambda\, dx$ fails to converge in the ordinary sense for any value of $\lambda$, so that the previous result will not apply. The desired result can be obtained by a method we shall use often in the future, *the partitioning method* for generalized functions, which is useful in several different contexts in generalized function theory.

[22] Recall that the generalized function $t_+^\lambda$ is defined for $-\infty < t < \infty$ by

$$(t_+^\lambda, f) = \int_0^\infty t^\lambda f(t)\, dt,$$

where the integral converges for $\operatorname{Re}\lambda > -1$ and is defined by analytic continuation in $\lambda$ for $\operatorname{Re}\lambda < -1$. This generalized function is analytic in $\lambda$ everywhere except at $\lambda = -1, -2, \ldots$, where it has simple poles (see Volume 1, Chapter I, Section 3.2).

This method is the following. Write the generalized function $(x_1)_+^\lambda$ as the sum of two terms[23] in the form

$$(x_1)_+^\lambda = (x_1)_+^\lambda\, \theta(1 - x_1) + (x_1)_+^\lambda\, \theta(x_1 - 1),$$

where $\theta$ has its usual meaning. The integrals

$$\int_{-\infty}^{\infty} (x_1)_+^\lambda\, \theta(1 - x_1)\, dx_1 \equiv \int_0^1 x_1^\lambda\, dx_1$$

and

$$\int_{-\infty}^{\infty} (x_1)_+^\lambda\, \theta(x_1 - 1)\, dx_1 \equiv \int_1^{\infty} x_1^\lambda\, dx_1$$

converge, each in a certain region of the $\lambda$ plane. The first converges for $\operatorname{Re}\lambda > -1$, and the second for $\operatorname{Re}\lambda < -1$. Therefore the Radon transforms of $(x_1)_+^\lambda\theta(1 - x_1)\,\delta(x_2, \ldots, x_n)$ and $(x_1)_+^\lambda\theta(x_1 - 1)\,\delta(x_2, \ldots, x_n)$ can each be calculated from the results of Section 2.2. In itself this may seem on first view not to be very helpful, since the two integrals converge for different values of $\lambda$, but the desired result can then be obtained by analytic continuation. In particular, the Radon transform of $(x_1)_+^\lambda\theta(1 - x_1)\,\delta(x_2, \ldots, x_n)$ is

$$|\xi_1|^{-1}\left(\frac{p}{\xi_1}\right)_+^\lambda \theta\left(1 - \frac{p}{\xi_1}\right) \equiv [p_+^\lambda(\xi_1)_+^{-1-\lambda} + p_-^\lambda(\xi_1)_-^{-1-\lambda}]\,\theta\left(1 - \frac{p}{\xi_1}\right)$$

and that of $(x_1)_+^\lambda\theta(x_1 - 1)\,\delta(x_2, \ldots, x_n)$ is

$$|\xi_1|^{-1}\left(\frac{p}{\xi_1}\right)_+^\lambda \theta\left(\frac{p}{\xi_1} - 1\right) \equiv [p_+^\lambda(\xi_1)_+^{-1-\lambda} + p_-^\lambda(\xi_1)_-^{-1-\lambda}]\,\theta\left(\frac{p}{\xi_1} - 1\right).$$

These equations remain meaningful for any noninteger $\lambda$. We may thus add them to obtain the Radon transform of $(x_1)_+^\lambda\,\delta(x_2, \ldots, x_n)$, where $\lambda$ is not an integer, namely

$$|\xi_1|^{-1}(p/\xi_1)_+^\lambda \equiv p_+^\lambda(\xi_1)_+^{-1-\lambda} + p_-^\lambda(\xi_1)_-^{-1-\lambda}. \tag{1}$$

### 2.3a. Radon Transform of $(x_1)_+^k\,\delta(x_2, \ldots, x_n)$ for Nonnegative Integer $k$

The result we have just obtained applies for noninteger $\lambda$. The limit of integer $\lambda$ illustrates an instructive point, and we will therefore go into it in some detail. The generalized function $p_+^\lambda(\xi_1)_+^{-1-\lambda} + p_-^\lambda(\xi_1)_-^{-1-\lambda}$

[23] We used this method of partition in Volume 1, Chapter I, Section 3.8 to calculate $\int_0^\infty x^\lambda\, dx$.

has poles at $\lambda = 0, 1, \ldots$, where $(\xi_1)_+^{-1-\lambda}$ and $(\xi_1)_-^{-1-\lambda}$ also have poles. At first this might seem contradictory, since $(x_1)_+^\lambda \, \delta(x_2, \ldots, x_n)$ is not itself singular at these values of $\lambda$. It will be seen that there is, however, no contradiction when it is recalled that the Radon transform of a generalized function is defined only to within an *unessential* function. Although $p_+^\lambda(\xi_1)_+^{-1-\lambda} + p_-^\lambda(\xi_1)_-^{-1-\lambda}$ has poles at $\lambda = k$ for nonnegative integers $k$, the leading term of its Laurent series about such a point is an unessential function, which we may therefore drop.

Let us therefore calculate the Radon transform of $(x_1)_+^k \, \delta(x_2, \ldots, x_n)$ for nonnegative integer $k$. For this purpose we expand $p_+^\lambda(\xi_1)_+^{-1-\lambda} + p_-^\lambda(\xi_1)_-^{-1-\lambda}$ in a Laurent series in powers of $\lambda - k$ about $\lambda = k$. We recall the Laurent series

$$\begin{aligned}
p_+^\lambda &= p_+^k + (\lambda - k) p_+^k \ln |p| + \cdots \\
p_-^\lambda &= p_-^k + (\lambda - k) p_-^k \ln |p| + \cdots \\
(\xi_1)_+^{-1-\lambda} &= -(-1)^k \frac{\delta^{(k)}(\xi_1)}{k!} \frac{1}{\lambda - k} + (\xi_1)_+^{-1-k} + \cdots \\
(\xi_1)_-^{-1-\lambda} &= -\frac{\delta^{(k)}(\xi_1)}{k!} \frac{1}{\lambda - k} + (\xi_1)_-^{-1-k} + \cdots,
\end{aligned} \tag{1}$$

where $(\xi_1)_+^{-1-k}$ and $(\xi_1)_-^{-1-k}$ are associated homogeneous generalized functions.[24] This yields

$$\begin{aligned}
p_+^\lambda(\xi_1)_+^{-1-\lambda} + p_-^\lambda(\xi_1)_-^{-1-\lambda} = &-\frac{(-1)^k}{k!} p^k \, \delta^{(k)}(\xi_1) \frac{1}{\lambda - k} \\
&+ p_+^k(\xi_1)_+^{-1-k} + p_-^k(\xi_1)_-^{-1-k} \\
&- \frac{(-1)^k}{k!} \delta^{(k)}(\xi_1) p^k \ln |p| + \cdots,
\end{aligned} \tag{2}$$

where we have omitted terms involving positive powers of $\lambda - k$. The first term in this expansion is an unessential function [25] and we shall

[24] These Laurent expansions and the definitions of the associated generalized functions will be found in Volume 1, pp. 338–339.

[25] It is easily verified by direct calculation that $p^k \delta^{(k)}(\xi_1)$ is unessential, i.e., that the functional corresponding to this generalized function vanishes. Indeed, from the definition we have

$$(p^k \delta^{(k)}(\xi_1), \psi) \propto \int_\Gamma \delta^{(k)}(\xi_1) \left( \int_{-\infty}^{\infty} \psi(\xi, p) \, p^k \, dp \right) \omega(\xi).$$

Now recall that our test functions have the property that $\int_{-\infty}^{\infty} \psi(\xi, p) p^k \, dp$ is a polynomial of degree $k - n + 1$ in $\xi$. It then follows immediately that $(p^k \delta^{(k)}(\xi_1), \psi) = 0$.

therefore drop it. Then by going to the limit as $\lambda \to k$, we find that the Radon transform of the generalized function $(x_1)_+^k\,\delta(x_2, ..., x_n)$ with $k = 0, 1, ...$ is

$$p_+^k(\xi_1)_+^{-1-k} + p_-^k(\xi_1)_-^{-1-k} - \frac{(-1)^k}{k!}\,\delta^{(k)}(\xi_1)p^k \ln |p|. \tag{3}$$

Summarizing, we have thus established that the Radon transform of the generalized function $(x_1)_+^\lambda\,\delta(x_2, ..., x_n)$, with $\lambda \neq -1, -2, ...$, which is concentrated on the ray $x_1 > 0$, $x_2 = \cdots = x_n = 0$, is

$$p_+^\lambda(\xi_1)_+^{-1-\lambda} + p_-^\lambda(\xi_1)_-^{-1-\lambda} \quad \text{for} \quad \lambda \neq 0, 1, ..., \tag{4}$$

and

$$p_+^k(\xi_1)_+^{-1-k} + p_-^k(\xi_1)_-^{-1-k} - \frac{(-1)^k}{k!}\,\delta^{(k)}(\xi_1)p^k \ln |p| \quad \text{for} \quad \lambda = k = 0, 1, .... \tag{4'}$$

We find in particular that the Radon transform of the characteristic function of the ray, namely of

$$\theta(x_1)\,\delta(x_2, ..., x_n),$$

is

$$\theta(p)(\xi_1)_+^{-1} + \theta(-p)(\xi_1)_-^{-1} - \delta(\xi_1) \ln |p|.$$

It is interesting to compare the Radon transforms of the characteristic functions of a ray and of an interval. On the set of hyperplanes not parallel to the ray, the Radon transform is of the same form for both the ray and the interval. That is, it is $|\xi_1|^{-1}$ if the $(\xi, x) = p$ hyperplane intersects the ray or line segment, and is zero otherwise. But the Radon transform of the characteristic function of the ray contains the added term $-\delta(\xi_1) \ln |p|$, which is concentrated on the set of hyperplanes parallel to the ray.

**Remark.** The results we have obtained are easily generalized to the case of any ray passing through the origin. Consider the ray with direction vector $x_0$, and let $a_{x_0}(x; \lambda)$ be a generalized function concentrated on it and defined by

$$(a_{x_0}(x; \lambda), f(x)) = \int_0^\infty t^\lambda f(tx_0)\,dt. \tag{5}$$

(This integral converges for $\operatorname{Re}\lambda > -1$, and is obtained by analytic continuation in $\lambda$ for $\operatorname{Re}\lambda < -1$.) For $x_0 = (1, 0, ..., 0)$ this generalized function coincides with $(x_1)_+^\lambda\,\delta(x_2, ..., x_n)$. It is easily shown that the

Radon transform of $a_{x_0}(x; \lambda)$ is then obtained by replacing $\xi_1$ in our previous formulas by the inner product $(\xi, x_0)$. Thus the Radon transform of the generalized function defined by (5) is

$$p_+^\lambda(\xi, x_0)_+^{-1-\lambda} + p_-^\lambda(\xi, x_0)_-^{-1-\lambda} \quad \text{for} \quad \lambda \neq 0, 1, \ldots, \tag{6}$$

and

$$p_+^k(\xi, x_0)_+^{-1-k} + p_-^k(\xi, x_0)_-^{-1-k} - \frac{(-1)^k}{k!}\delta^{(k)}[(\xi, x_0)]p^k \ln|p|$$
$$\text{for} \quad \lambda = k = 0, 1, \ldots . \quad \# \tag{6'}$$

In conclusion, we give the expressions for the Radon transforms of the generalized functions

$$|x_1|^\lambda \delta(x_2, \ldots, x_n) \equiv (x_1)_+^\lambda \delta(x_2, \ldots, x_n) + (x_1)_-^\lambda \delta(x_2, \ldots, x_n)$$

and

$$|x_1|^\lambda \operatorname{sgn} x_1 \delta(x_2, \ldots, x_n) = (x_1)_+^\lambda \delta(x_2, \ldots, x_n) - (x_1)_-^\lambda \delta(x_2, \ldots, x_n),$$

both concentrated on the entire $x_1$ axis. The Radon transform of $|x_1|^\lambda \delta(x_2, \ldots, x_n)$ is

$$|p|^\lambda |\xi_1|^{-1-\lambda}, \quad \text{for} \quad \lambda \neq 2k \quad (k = 0, 1, \ldots), \tag{7}$$

and

$$p^{2k}|\xi_1|^{-1-2k} - \frac{2}{(2k)!}\delta^{(2k)}(\xi_1)p^{2k}\ln|p|, \quad \text{for} \quad \lambda = 2k \quad (k = 0, 1, \ldots) \tag{7'}$$

(here $|\xi_1|^{-1-2k}$ is an associated homogeneous generalized function).

It may be noted that $p^{2k}|\xi_1|^{-1-2k}$ is unessential if $2k < n - 1$, and may therefore be dropped from Eq. (7′) for these $k$.

The Radon transform of the generalized function

$$|x_1|^\lambda \operatorname{sgn} x_1 \delta(x_2, \ldots, x_n)$$

is

$$|p|^\lambda \operatorname{sgn} p\, |\xi_1|^{-1-\lambda} \operatorname{sgn} \xi_1, \quad \text{for} \quad \lambda \neq 2k - 1$$
$$(k = 1, 2, \ldots), \tag{8}$$

and

$$p^{2k-1}\xi_1^{-2k} \operatorname{sgn} \xi_1 + \frac{2}{(2k-1)!}\delta^{(2k-1)}(\xi_1)p^{2k-1}\ln|p|, \quad \text{for} \quad \lambda = 2k - 1$$
$$(k = 1, \ldots) \tag{8'}$$

($\xi_1^{-2k} \operatorname{sgn} \xi_1$ is an associated homogeneous generalized function). For $2k < n$, the first term in (8′) can be dropped, as it is then unessential. [We find in particular that the Radon transform of $\delta(x_2, \ldots, x_n)$ is $-2\delta(\xi_1) \ln | p |$.]

Remark. The Radon transform of the $(n - 1)$-dimensional $\delta$ function, and therefore also the expression for the integral of an arbitrary $f(x)$ along the $x_1$ axis in terms of its integrals along hyperplanes, can also be obtained somewhat more directly by a reduction method. The procedure is the following. Consider all lines parallel to the $x_1$ axis and the integrals of $f(x)$ along these lines. These integrals define a function $f_1(x)$ in the $(n - 1)$-dimensional space $R_{n-1}$. Then the integrals of $f(x)$ over hyperplanes in $R_n$ parallel to the $x_1$ axis are the same as the integrals of $f_1(x)$ over all possible hyperplanes in $R_{n-1}$. This reduces the problem of calculating a function along the $x_1$ axis in $R_n$ to solving the following problem: To determine $f_1(x)$ in $R_{n-1}$ when its integrals over all possible hyperplanes in this space are known. This problem is solved by the Radon transform inversion formula (Section 1.4). This same method can also be iterated to obtain an expression for the integral of a function over a $k$-dimensional subspace in terms of integrals over hyperplanes. We leave this as an exercise for the reader. #

## 2.4. Integral of a Function over a Given Region in Terms of Integrals over Hyperplanes

Consider the following problem. Let $\varphi(\xi, p)$ be the integrals of some $f(x)$ over all possible hyperplanes. We wish to calculate the integral of $f(x)$ over a given region $V$.

To solve this problem we introduce the characteristic function $\chi(x)$ of $V$. We may then write

$$\int_V f(x)\, dx = \int \chi(x) f(x)\, dx.$$

Let $S(\xi, p)$ be the Radon transform of $\chi(x)$. Then from the definition of the Radon transform of a generalized function we have

$$\int \chi(x) f(x)\, dx = \frac{(-1)^{\frac{1}{2}(n-1)}}{2(2\pi)^{n-1}} \int_\Gamma \Big[\int_{-\infty}^{\infty} S(\xi, p) \varphi_p^{(n-1)}(\xi, p)\, dp\Big]\, \omega(\xi). \qquad (1)$$

This equation is the solution of our problem. In other words, in order to express the integral of $f(x)$ over $V$ in terms of integrals over hyperplanes, we need only calculate the Radon transform of the characteristic function of $V$.

Now if $V$ is bounded and our space is Euclidian, as we have seen, $S(\xi, p)$ is the area of the intersection of $V$ with the hyperplane whose normalized equation is $(\xi, x) = p$ (that is, for $|\xi| = 1$). In an affine space, however, the concept of area does not in general exist, although that of volume does. In such an affine space $S(\xi, p)\, dp$ is the *volume* of that part of $V$ which lies between the hyperplanes whose equations are[26] $(\xi, x) = p$ and $(\xi, x) = p + dp$. For simplicity let us agree by convention to call $S(\xi, p)$ the area of the intersection with the plane for a bounded region $V$ in affine spaces also.

Thus the integral of $f(x)$ over a bounded region $V$ may be written in terms of its integrals $\varphi(\xi, p)$ over hyperplanes in the form

$$\int_V f(x)\, dx = \frac{(-1)^{\frac{1}{2}(n-1)}}{2(2\pi)^{n-1}} \int_\Gamma \left[\int_{-\infty}^{\infty} S(\xi, p) \varphi_p^{(n-1)}(\xi, p)\, dp\right] \omega(\xi).$$

Since this formula for the integral does not depend on whether $V$ is bounded or unbounded, it is natural to call $S(\xi, p)$ the generalized area of the intersection also if $V$ is unbounded, and we shall therefore sometimes speak of the area of the intersection of a plane and an unbounded region. This definition for an unbounded region is obviously additive; that is, let $V$ be divided into subregions $V_i$ (with common points only on their boundaries). Then up to an unessential term, the area of the intersection $S(\xi, p)$ of $V$ with the $(\xi, x) = p$ hyperplane is the sum of the areas $S_i(\xi, p)$ of the $V_i$.

The problem is now to calculate these generalized areas for unbounded regions. To do this, we consider an unbounded region $V$ as the limit of a sequence of bounded regions $V_h$, writing $\lim_{h\to\infty} V_h = V$. Then the areas $S_h(\xi, p)$ for the $V_h$ can be calculated from geometrical considerations. Then as $h \to \infty$, the function $S_h(\xi, p)$ converges in the sense of functionals to the Radon transform $S(\xi, p)$ of the characteristic function of $V$. Let us study this convergence.

Assume that for $S_h(\xi, p)$ there exists an asymptotic series in $h$ as $h \to \infty$, of the form

$$S_h(\xi, p) \sim \sum_{i=1}^{\infty} a_i(h) f_i(\xi, p), \tag{2}$$

[26] In other words,

$$S(\xi, p) = \frac{\partial V(\xi, p)}{\partial p},$$

where $V(\xi, p)$ is the volume of that part of $V$ contained in the half-space $(\xi, x) < p$.

where $\lim_{h\to\infty}[a_i(h)/a_{i+1}(h)] = \infty$ and one of the $a_i(h)$ is equal to one.[27] Assume for instance, that $a_k(h) = 1$. Then $\lim_{h\to\infty} a_i(h) = \infty$ for $i < k$, and $\lim_{h\to\infty} a_i(h) = 0$ for $i > k$. Now we know that as $h \to \infty$, the function $S_h(\xi, p)$ converges in the sense of functionals of the Radon transform of the characteristic function of $V$. This means that all of the divergent terms in (2) are unessential, and therefore that the Radon transform of the characteristic function of $V$ is

$$S(\xi, p) = f_k(\xi, p).$$

Note that this process does not define $S(\xi, p)$ uniquely. Specifically, it may depend upon the particular sequence of bounded regions converging to $V$. Nevertheless all the $S(\xi, p)$ obtained in this way give the same functional, i.e., differ only by an unessential function of $\xi$ and $p$. Because of this lack of uniqueness, on the other hand, the area of a given infinite section of an unbounded region is not a well-defined concept. In fact this "area" can be thought of only as a function of the intersecting plane.

As an example, let us calculate the "area" $S(\xi, p)$ of the plane sections through the upper sheet of the two-sheeted hyperboloid[28]

$$\frac{x_1^2}{a_1^2} - \frac{x_2^2}{a_2^2} - \frac{x_3^2}{a_3^2} > 1, \qquad x_1 > 0, \tag{3}$$

in three dimensions. The plane sections of a hyperboloid are ellipses whose areas are easily calculated, hyperbolas, and parabolas (a limiting case). What we shall do is calculate the area of a hyperbola.

Consider the region bounded by the hyperboloid and the plane whose equation is $x_1 = h$, and let $S_h(\xi, p)$ be the area of the intersection of this region with the $(\xi, x) = p$ plane. Then we expand $S_h(\xi, p)$ in an asymptotic series in $h$, and proceed as above to obtain $S(\xi, p)$. Unfortunately this geometrical approach involves steps which, although simple in principle, tend to become quite cumbersome. We shall omit them and present only the final result.

In order to write out this result in affine invariant form, we proceed as follows. Consider the quadratic curve whose equation is

$$a_{11}x_1^2 + 2a_{12}x_1x_2 + a_{22}x_2^2 + 2a_{13}x_1 + 2a_{23}x_2 + a_{33} = 0. \tag{4}$$

[27] The question of when such an asymptotic series exists is itself of interest. We shall not, however, go into it here.

[28] Henceforth in speaking of hyperboloids and cones we shall usually mean not the surfaces themselves but the regions they enclose. Similarly, in speaking of ellipses and hyperbolas we shall mean not the curves themselves but the two-dimensional regions of which they are the boundary.

As is well known, the determinants

$$\delta = \begin{vmatrix} a_{11} & a_{12} \\ a_{21} & a_{22} \end{vmatrix} \quad \text{and} \quad \Delta = \begin{vmatrix} a_{11} & a_{12} & a_{13} \\ a_{21} & a_{22} & a_{23} \\ a_{31} & a_{32} & a_{33} \end{vmatrix}$$

(where we write $a_{ij} = a_{ji}$ for all $i, j$) are invariant under affine transformations (i.e., under parallel displacements and centroaffine transformations with determinant 1). It is not true, however, that $\delta$ and $\Delta$ are invariants of the curve, since when the $a_{ij}$ are replaced by $\lambda a_{ij}$ (with $\lambda \neq 0$), the curve remains invariant although $\delta$ and $\Delta$ are multiplied by $\lambda^2$ and $\lambda^3$, respectively. But it is clear that $\Delta/|\delta|^{\frac{3}{2}}$ will be an affine invariant of the curve.

Now assume that the $(\xi, x) = p$ plane intersects the hyperboloid given by (3) in a hyperbola. Let us choose some coordinate system, for instance $x_1$ and $x_2$, on the plane. Then the $a_{ij}$ of (4) for the hyperbola on this plane will be functions of $\xi$ and $p$, and therefore so will $\delta$ and $\Delta$. It turns out that the area of this hyperbola may be written in terms of $\delta$ and $\Delta$ in the form

$$S(\xi, p) = \tfrac{1}{2} \,|\, \xi_3^{-1} \Delta\, \delta^{-\frac{3}{2}} \,| \cdot \ln |\, \xi_3^{-1} \Delta\, \delta^{-\frac{3}{2}} \,|. \tag{5}$$

The factor $\xi_3^{-1}$ appears here in the following way. Recall that the element of area $\omega$ on the $(\xi, x) = p$ plane is defined by $d(\xi, x)\omega = dx_1\, dx_2\, dx_3$, so that if $x_1$ and $x_2$ are chosen as the coordinates on the plane, $\omega = \xi_3^{-1}\, dx_1 dx_2$. It may be noted that if we choose $y_1 = \alpha_{11}x_1 + \alpha_{12}x_2 + \alpha_{13}x_3$ and $y_2 = \alpha_{21}x_1 + \alpha_{22}x_2 + \alpha_{23}x_3$, as the coordinates on the plane, we arrive at

$$\omega = a(\xi)\, dy_1\, dy_2 \equiv \begin{vmatrix} \xi_1 & \xi_2 & \xi_3 \\ \alpha_{11} & \alpha_{12} & \alpha_{13} \\ \alpha_{21} & \alpha_{22} & \alpha_{23} \end{vmatrix}^{-1} dy_1\, dy_2,$$

and then (5) may be rewritten

$$S(\xi, p) = \tfrac{1}{2} \,|\, a(\xi)\, \Delta \delta^{-\frac{3}{2}} \,| \cdot \ln |\, a(\xi)\, \Delta \delta^{-\frac{3}{2}} \,|.$$

It is interesting to compare (5) with the equation

$$S(\xi, p) = \pi \,|\, \xi_3^{-1} \Delta \delta^{-\frac{3}{2}} \,|$$

tor the area of an ellipse.

The formula we have obtained for the area of a hyperbola is interesting for the following reason. A given hyperbola may be obtained from sections through different hyperboloids. Its area obtained in this way will always be a function of $\xi$ and $p$. It is remarkable, however, that this area can be expressed directly in terms of the coefficients appearing in the equation for the hyperbola and that this expression as given by (5) is independent of the hyperboloid through which the hyperbola is obtained. This property is true also for the area of a bounded figure (for instance, of an ellipse), and this is why it is reasonable to speak of the area of a hyperbola.

A detailed discussion of the Radon transform of the characteristic function of the upper sheet of a hyperboloid will be given in Section 2.6, where we shall obtain a final expression for this transform.

## 2.5. Radon Transform of the Characteristic Function of One Sheet of a Cone

In this section we shall find the Radon transform of the characteristic function of the upper sheet of the cone

$$\frac{x_1^2}{a_1^2} - \frac{x_2^2}{a_2^2} - \frac{x_3^2}{a_3^2} > 0, \qquad x_1 > 0.$$

It will be shown that this Radon transform is

$$\varphi(\xi, p) = \pi a_1 a_2 a_3 [p_+^2 \theta(\xi_1) + p_-^2 \theta(-\xi_1)] Q_+^{-\frac{3}{2}} + a_1 a_2 a_3 p_2 \ln |p| Q_-^{-\frac{3}{2}}, \tag{1}$$

where

$$Q = Q(\xi) \equiv a_1^2 \xi_1^2 - a_2^2 \xi_2^2 - a_3^2 \xi_3^2$$

and we have written, for instance,

$$Q_+^\lambda = \begin{cases} Q^\lambda & \text{for } Q > 0, \\ 0 & \text{for } Q < 0. \end{cases}$$

In order to give some geometric insight into Eq. (1) we make the following remarks. The presence of two terms in the square brackets reflects the fact that the same plane may be expressed either by $(\xi, x) = p$ or $(-\xi, x) = -p$. We need therefore explain only the first term

$$\pi a_1 a_2 a_3 p_+^2 \theta(\xi_1)(a_1^2 \xi_1^2 - a_2^2 \xi_2^2 - a_3^2 \xi_3^2)_+^{-\frac{3}{2}}.$$

This is the area of an ellipse, since for $p > 0$, $\xi_1 > 0$, and $a_1^2 \xi_1^2 - a_2^2 \xi_2^2 - a_3^2 \xi_3^2 > 0$, the the plane intersects with the cone in an ellipse. The second term in the square brackets, as mentioned, has a similar origin. The remaining term

$$a_1a_2a_3p^2 \ln |p| (a_1^2\xi_1^2 - a_2^2\xi_2^2 - a_3^2\xi_3^2)_-^{-\frac{3}{2}}$$

is the area of a hyperbola, since for

$$a_1^2\xi_1^2 - a_2^2\xi_2^2 - a_3^2\xi_3^2 < 0,$$

the intersection is a hyperbola.

It is, of course, possible to derive Eq. (1) by the limiting process described in the previous section. We shall use a different approach here, however, in which we first write down the Radon transform of the generalized function

$$\theta(x_1)\left(\frac{x_1^2}{a_1^2} - \frac{x_2^2}{a_2^2} - \frac{x_3^2}{a_3^2}\right)_+^\lambda$$

for complex $\lambda$, and then obtain the limit as $\lambda \to 0$. For simplicity we shall treat the circular cone

$$x_1^2 - x_2^2 - x_3^2 > 0, \qquad x_1 > 0;$$

the result can then be generalized by affine transformation. We proceed from the following fact. The Radon transform of the generalized function

$$\theta(x_1)P_+^\lambda(x) \equiv \theta(x_1)(x_1^2 - x_2^2 - x_3^2)_+^\lambda$$

is

$$\varphi(\xi, p) = \frac{\pi}{\lambda + 1}[p_+^{2+2\lambda}\theta(\xi_1) + p_-^{2+2\lambda}\theta(-\xi_1)]Q_+^{-\lambda-\frac{3}{2}}(\xi)$$

$$+ \frac{1}{2}\frac{\pi}{(\lambda + 1)\sin \pi\lambda}|p|^{2+2\lambda}Q_-^{-\lambda-\frac{3}{2}}(\xi), \tag{2}$$

where $Q(\xi) = \xi_1^2 - \xi_2^2 - \xi_3^2$. (A nonrigorous derivation of this result will be given in the appendix to this section.) Now by going to the limit as $\lambda \to 0$, we obtain the Radon transform of the characteristic function of the cone. But note that we cannot simply go to the limit in Eq. (2), since the coefficient of the second term diverges at $\lambda = 0$. We know, however, that as a generalized function $\varphi(\xi, p)$ will have no singularity at $\lambda = 0$ [since $\theta(x_1)P_+^\lambda(x)$ has no singularity at this value of $\lambda$]. Therefore the residue of $\varphi(\xi, p)$ at $\lambda = 0$, namely

$$p^2(\xi_1^2 - \xi_2^2 - \xi_3^2)_-^{-\frac{3}{2}},$$

must be an unessential function.

Thus in order to obtain the Radon transform of $\theta(x_1)P_+^0(x)$ from Eq. (2), we must expand $\varphi(\xi, p)$ in a Laurent series about $\lambda = 0$, drop the leading term as unessential, and then find the limit as $\lambda \to 0$. We will then find that the Radon transform of the characteristic function of our circular cone is

$$\varphi(\xi, p) = \pi[p_+^2\theta(\xi_1) + p_-^2\theta(-\xi_1)]Q_+^{-\frac{3}{2}}(\xi)$$
$$+ p^2 \ln|p| Q_-^{-\frac{3}{2}}(\xi) - \tfrac{1}{2}p^2 Q_-^{-\frac{3}{2}}(\xi) \ln Q_-(\xi). \tag{3}$$

The last term in (3) is an unessential function, and we shall therefore drop it. To see this, recall that a function of the form $p^2 a(\xi)$ is unessential if $\int_\Gamma a(\xi)\omega(\xi) = 0$, where the integral is taken over any surface $\Gamma$ enclosing the origin in $\xi$ space. This means that we must show that

$$\int_\Gamma Q_-^{-\frac{3}{2}}(\xi) \ln Q_-(\xi)\omega(\xi) = 0.$$

Now since

$$Q_-^{-\frac{3}{2}}(\xi) \ln Q_-(\xi) = \left.\frac{\partial Q_-^\lambda(\xi)}{\partial \lambda}\right|_{\lambda=-\frac{3}{2}},$$

it is sufficient to show that

$$\int_\Gamma Q_-^\lambda(\xi)\omega(\xi) \equiv \int_\Gamma (\xi_1^2 - \xi_2^2 - \xi_3^2)_-^\lambda \omega(\xi) = 0 \tag{4}$$

for all $\lambda$. To show this we choose $\Gamma$ to be the two planes given by $\xi_1 = \pm 1$. Then the integral in (4) becomes $2\int (1 - \xi_2^2 - \xi_3^2)_-^\lambda \, d\xi_2 \, d\xi_3$. When we transform to polar coordinates, this becomes

$$2\int (1 - \xi_2^2 - \xi_3^2)_-^\lambda \, d\xi_2 \, d\xi_3 = 4\pi \int_1^\infty (r^2 - 1)^\lambda r \, dr = 2\pi \int_0^\infty t^\lambda \, dt = 0.$$

(It was shown in Volume 1, p. 70, that $\int_0^\infty t^\lambda \, dt = 0$.) Thus it is shown that the last term in (3) is an unessential function. When it is dropped, we obtain Eq. (1) with $a_1 = a_2 = a_3 = 1$.

In conclusion, we present without proof the Radon transform of the generalized function $\theta(x_1)(x_1^2 - x_2^2 - \cdots - x_n^2)_+^\lambda$ in $n$ dimensions (for odd $n$):

$$[\theta(x_1)(x_1^2 - x_2^2 - \cdots - x_n^2)_+^\lambda]^\vee$$
$$= \frac{\pi^{\frac{1}{2}(n-1)}}{(\lambda + 1) \cdots [\lambda + \frac{1}{2}(n-1)]} \Big\{ [p_+^{2\lambda+n-1}\theta(\xi_1) + p_-^{2\lambda+n-1}\theta(-\xi_1)]$$
$$\times (\xi_1^2 - \xi_2^2 - \cdots - \xi_n^2)_+^{-\lambda-\frac{1}{2}n}$$
$$- \frac{(-1)^{\frac{1}{2}(n-1)}}{2 \sin \pi\lambda} |p|^{2\lambda+n-1} (\xi_1^2 - \xi_2^2 - \cdots - \xi_n^2)_-^{-\lambda-\frac{1}{2}n} \Big\}.$$

We suggest that the reader use this formula to find the Radon transform of the characteristic function of the upper sheet of the cone whose equation is

$$x_1^2 - x_2^2 - \cdots - x_n^2 > 0, \qquad x_1 > 0,$$

in a space of odd dimension. It may be noted that the case of even dimension is not particularly difficult either.

## Appendix to Section 2.5

Although all of Section 3 will be devoted to calculating the Radon transforms of functions raised to the power $\lambda$, it is perhaps useful to show a nonrigorous but convenient method for calculating the Radon transform of $\theta(x_1)P_+^\lambda(x) \equiv \theta(x_1)(x_1^2 - x_2^2 - x_3^2)_+^\lambda$. It will be seen that this Radon transform can be obtained to within a factor from rather general considerations.

Let us first make use of the homogeneity of $\theta(x_1)P_+^\lambda(x)$. Because it is homogeneous of degree $2\lambda$, its Radon transform $\varphi(\xi, p)$ is homogeneous in $\xi$ of degree $-2\lambda - 3$. Since, further, $\varphi(\xi, p)$ is homogeneous in $\xi, p$ of degree $-1$, it is homogeneous in $p$ of degree $2\lambda + 2$. Now it is known that every generalized function of a single variable $p$, homogeneous of degree $2\lambda + 2$, is a linear combination of $p_+^{2\lambda+2}$ and $p_-^{2\lambda+2}$. Thus the Radon transform of $\theta(x_1)P_+^\lambda(x)$ is of the form

$$\varphi(\xi, p) = p_+^{2\lambda+2}\varphi_+(\xi) + p_-^{2\lambda+2}\varphi_-(\xi),$$

where $\varphi_+(\xi)$ and $\varphi_-(\xi)$ are homogeneous of degree $-2\lambda - 3$ in $\xi$. Since $\varphi(\xi, p)$ is even in $\xi, p$, it follows that $\varphi_-(\xi) = \varphi_+(-\xi)$.

We now use the fact that $\theta(x_1)P_+^\lambda(x)$ is invariant under "hyperbolic rotations," that is, under linear transformations which preserve the quadratic form $x_1^2 - x_2^2 - x_3^2$ and map each sheet of the cone $x_1^2 - x_2^2 - x_3^2 = 0$ to itself. This means that $\varphi_+(\xi)$ and $\varphi_-(\xi)$ must be invariant under such hyperbolic rotations in $\xi$ space, a property which can be used to determine their form. We shall make use of the following result, which we present here without proof.

Every generalized function of $\xi$, homogeneous of degree $\lambda$ (where $\lambda$ is not a negative integer) and invariant under hyperbolic rotations, is a linear combination of the three functions

$$\theta(\xi_1)(\xi_1^2 - \xi_2^2 - \xi_3^2)_+^{\frac{1}{2}\lambda}, \qquad \theta(-\xi_1)(\xi_1^2 - \xi_2^2 - \xi_3^2)_+^{\frac{1}{2}\lambda}, \qquad (\xi_1^2 - \xi_2^2 - \xi_3^2)_-^{\frac{1}{2}\lambda}.$$

(The first two of these are concentrated inside the upper and lower sheets of the cone $\xi_1^2 - \xi_2^2 - \xi_3^2 = 0$, and the last is concentrated outside this cone.) From this we may conclude that

$$\varphi_+(\xi) = \varphi_-(-\xi) = a(\lambda)\theta(\xi_1)(\xi_1^2 - \xi_2^2 - \xi_3^2)_+^{-\lambda-\frac{3}{2}}$$
$$+ b(\lambda)\theta(-\xi_1)(\xi_1^2 - \xi_2^2 - \xi_3^2)_+^{-\lambda-\frac{3}{2}} + c(\lambda)(\xi_1^2 - \xi_2^2 - \xi_3^2)_-^{-\lambda-\frac{3}{2}},$$

where $a(\lambda)$, $b(\lambda)$, and $c(\lambda)$ are analytic functions of $\lambda$ still to be determined. Thus the Radon transform of the generalized function $\theta(x_1)(x_1^2 - x_2^2 - x_3^2)_+^\lambda$ is of the form

$$\varphi(\xi, p) = a(\lambda)[p_+^{2+2\lambda}\theta(\xi_1) + p_-^{2+2\lambda}\theta(-\xi_1)](\xi_1^2 - \xi_2^2 - \xi_3^2)_+^{-\lambda-\frac{3}{2}}$$
$$+ b(\lambda)[p_+^{2+2\lambda}\theta(-\xi_1) + p_-^{2+2\lambda}\theta(\xi_1)](\xi_1^2 - \xi_2^2 - \xi_3^2)_+^{-\lambda-\frac{3}{2}}$$
$$+ c(\lambda)|p|^{2+2\lambda}(\xi_1^2 - \xi_2^2 - \xi_3^2)_-^{-\lambda-\frac{3}{2}}. \qquad (5)$$

We must now calculate $a(\lambda)$, $b(\lambda)$, and $c(\lambda)$.

Note first that the term whose coefficient is $b(\lambda)$ can not occur, so that $b(\lambda) \equiv 0$. This is because if $p$ and $\xi_1$ have opposite signs, the $(\xi, x) = p$ plane does not intersect the $\xi_1^2 - \xi_2^2 - \xi_3^2 > 0$ cone. Now to calculate $a(\lambda)$, consider a plane which intersects the cone in an ellipse, say the $x_1 = 1$ plane; in other words we set $\xi = \xi_0 = (1, 0, 0)$ and $p = 1$. Then Eq. (5) gives

$$\varphi(\xi_0, 1) = a(\lambda).$$

But we can calculate $\varphi(\xi_0, 1)$ directly by integrating $\theta(x_1)P_+^\lambda(x)$ over the plane. This yields

$$\varphi(\xi_0, 1) = \iint\limits_{x_2^2+x_3^2 \leqslant 1} (1 - x_2^2 - x_3^2)^\lambda \, dx_2 \, dx_3 = \frac{\pi}{\lambda + 1}$$

(the integral converging for $\operatorname{Re}\lambda > -1$). Thus

$$a(\lambda) = \frac{\pi}{\lambda + 1}.$$

Finally, we turn to $c(\lambda)$. For this purpose consider a plane intersecting the cone in a hyperbola, say the $x_3 = 1$ plane; that is, we set $\xi = \xi_0 = (0, 0, 1)$ and $p = 1$. Then from (5) we find that

$$\varphi(\xi_0, 1) = c(\lambda).$$

But we can calculate $\varphi(\xi_0, 1)$ directly by integrating $\theta(x_1)P_+^\lambda(x)$ over the plane. This yields

$$c(\lambda) = \varphi(\xi_0, 1) = \iint\limits_{x_1^2 - x_2^2 \geq 1, x_1 > 0} (x_1^2 - x_2^2 - 1)^\lambda \, dx_1 \, dx_2 .$$

Now this integral fails to converge for any $\lambda$. We shall compute it in the following way. We write it as an infinite sum of integrals over intervals of the form $n + 1 > x_1 > n$, where $n$ is a positive integer, and in each of the integrals we write $x_2 = (x_1^2 - 1)^{\frac{1}{2}} t$. Then our integral is represented as the sum of integrals of the form

$$\int_{-1}^{1} (1 - t^2)^\lambda \, dt \int_n^{n+1} (x_1^2 - 1)^{\lambda + \frac{1}{2}} \, dx_1 = \frac{\Gamma(\lambda + 1)\Gamma(\frac{1}{2})}{\Gamma(\lambda + \frac{3}{2})} \int_n^{n+1} (x_1^2 - 1)^{\lambda + \frac{1}{2}} \, dx_1 .$$

Each of these integrals converges for $\operatorname{Re} \lambda > -1$, and each can be continued analytically in $\lambda$ to $\operatorname{Re} \lambda < -1$. By adding all the integrals and factoring out the fraction involving the three gamma functions, we obtain

$$c(\lambda) = \frac{\Gamma(\lambda + 1)\Gamma(\frac{1}{2})}{\Gamma(\lambda + \frac{3}{2})} \int_1^\infty (x_1^2 - 1)^{\lambda + \frac{1}{2}} \, dx_1 .$$

Now

$$\int_1^\infty (x_1^2 - 1)^{\lambda + \frac{1}{2}} \, dx_1 = \frac{1}{2} \frac{\Gamma(-\lambda - 1)\Gamma(\lambda + \frac{3}{2})}{\Gamma(\frac{1}{2})}$$

(the integral converging in the ordinary sense for $-\frac{3}{2} < \operatorname{Re} \lambda < -1$), so that we finally arrive at

$$c(\lambda) = \tfrac{1}{2}\Gamma(\lambda + 1)\Gamma(-\lambda - 1) = \frac{\pi}{2(\lambda + 1) \sin \pi\lambda} .$$

## 2.6. Radon Transform of the Characteristic Function of One Sheet of a Two-Sheeted Hyperboloid

In this section we shall derive the expression for the Radon transform of the characteristic function of one sheet of the two-sheeted hyperboloid, that is, of the region defined by

$$\frac{x_1^2}{a_1^2} - \frac{x_2^2}{a_2^2} - \frac{x_3^2}{a_3^2} > 1, \qquad x_1 > 0. \tag{1}$$

There are two types of plane sections through this hyperboloid, namely ellipses and hyperbolas, and the Radon transform of its characteristic function should contain two terms, one of which is the area of an ellipse, and the other the area of a hyperbola.[29] Our problem is to write the formulas for these areas as functions of $\xi$, $p$ and of the parameters $a_1$, $a_2$, and $a_3$.

In section 2.4 we have already discussed the expressions for these areas in terms of the coefficients in the equation for the relevant curve. What we must now do is write the equation of the curve in some convenient coordinate system (say $x_1$ and $x_2$) and calculate the determinants $\delta$ and $\Delta$ of the coefficients. Then the area of the ellipse is

$$S_{\mathrm{el}}(\xi, p) = \pi \,|\, \xi_3^{-1} \Delta \delta^{-\frac{3}{2}} \,|, \tag{2}$$

and that of the hyperbola is

$$S_{\mathrm{hyp}}(\xi, p) = \tfrac{1}{2} \,|\, \xi_3^{-1} \Delta \delta^{-\frac{3}{2}} \,| \cdot \ln |\, \xi_3^{-1} \Delta \delta^{-\frac{3}{2}} \,|. \tag{3}$$

Equation (3) can be obtained by a limiting procedure from the Radon transform of the characteristic function of a bounded region (as has been discussed in Section 2.4). The derivation is, however, quite complicated, and we shall omit it. Now we wish to express $\delta$ and $\Delta$ in terms of the coefficients appearing in the equation of the intersecting plane:

$$\xi_1 x_1 + \xi_2 x_2 + \xi_3 x_3 = p. \tag{4}$$

For this purpose we calculate the intersection of the plane with the surface whose equation is

$$\frac{x_1^2}{a_1^2} - \frac{x_2^2}{a_2^2} - \frac{x_3^2}{a_3^2} = 1. \tag{5}$$

In order to do this in terms of $x_1$ and $x_2$, we merely eliminate $x_3$ from (4) and (5), which yields

$$\left(\frac{1}{a_1^2} - \frac{\xi_1^2}{a_3^2 \xi_3^2}\right) x_1^2 - 2 \frac{\xi_1 \xi_2}{a_3^2 \xi_3^2} x_1 x_2 - \left(\frac{1}{a_2^2} + \frac{\xi_2^2}{a_3^2 \xi_3^2}\right) x_2^2$$
$$+ 2 \frac{\xi_1 p}{a_3^2 \xi_3^2} x_1 + 2 \frac{\xi_2 p}{a_3^2 \xi_3^2} x_2 - \frac{p^2}{a_3^2 \xi_3^2} - 1 = 0.$$

[29] It can be shown that the parabolic sections, as in the case of the cone, fail to contribute to the Radon transform (although *a priori* one may suppose that certain special cases might give a $\delta$ function contribution; cf. the line in Section 2.3 and the octant in Section 2.8).

From this we find that

$$\delta = (a_1 a_2 a_3)^{-2} \xi_3^{-2} Q, \qquad \Delta = (a_1 a_2 a_3)^{-2} \xi_3^{-2} (p^2 - Q),$$

where

$$Q = Q(\xi) \equiv a_1^2 \xi_1^2 - a_2^2 \xi_2^2 - a_3^2 \xi_3^2 .$$

Thus if the intersection is an ellipse, its area is

$$S_{\text{el}}(\xi, p) = \pi a_1 a_2 a_3 \,|\, p^2 - Q \,|\,|\, Q \,|^{-\frac{3}{2}},$$

and if it is a hyperbola, its area is

$$S_{\text{hyp}}(\xi, p) = \tfrac{1}{2} a_1 a_2 a_3 \,|\, p^2 - Q \,|\,|\, Q \,|^{-\frac{3}{2}} \ln |\, a_1 a_2 a_3 (p^2 - Q) Q^{-\frac{3}{2}} \,|.$$

What remains is to find for which $\xi$ and $p$ the intersection is an ellipse or a hyperbola. We note easily that the ellipse is obtained when $Q(\xi) > 0$ and $p^2 > Q(\xi)$ if $p$ and $\xi_1$ are of the same sign. [If they are of opposite signs, the $(\xi, x) = p$ plane intersects the lower rather than the upper sheet of the hyperboloid.] A hyperbola is obtained when $Q(\xi) < 0$, for arbitrary $p$.

Thus finally we may summarize our result as follows. The Radon transform of the characteristic function of the upper sheet of a two-sheeted hyperboloid, namely of the region defined by Eq. (1), is

$$S(\xi, p) = \pi a_1 a_2 a_3 \theta(p \xi_1)(p^2 - Q)_+ Q_+^{-\frac{3}{2}}$$
$$+ \tfrac{1}{2} a_1 a_2 a_3 (p^2 - Q)_- Q_-^{-\frac{3}{2}} \ln |\, a_1 a_2 a_3 (p^2 - Q) Q^{-\frac{3}{2}} \,|, \tag{6}$$

where

$$Q = Q(\xi) \equiv a_1^2 \xi_1^2 - a_2^2 \xi_2^2 - a_3^2 \xi_3^2 .$$

Note that this result can be obtained in two ways. The first is by going to the limit of the Radon transform of the characteristic function of a bounded region. The second is to find the Radon transform of

$$\theta(x_1) \left( \frac{x_1^2}{a_1^2} - \frac{x_2^2}{a_2^2} - \frac{x_3^2}{a_3^2} - 1 \right)_+^{\lambda}$$

for complex $\lambda$, and then to go to the limit as $\lambda \to 0$. (This method was illustrated in Section 2.5 for a cone rather than a hyperboloid.)

It is easily shown that

$$|\, \xi_3^{-1} \Delta \delta^{-\frac{3}{2}} \,|\, \ln(a_1^2 a_2^2 a_3^2 \xi_3^2 \,|\, \delta \,|) \equiv a_1 a_2 a_3 (p^2 - Q) Q_-^{-\frac{3}{2}} \ln |\, Q \,|$$

is an unessential function (see the discussion following Eq. (3) of Section 2.5), and that therefore the $Q^{-\frac{3}{2}}$ in the logarithm may be dropped in Eq. (6).

Remark. This function is unessential for a hyperbola but not for an ellipse. For an ellipse it must be replaced by zero when the intersection is an imaginary ellipse (i.e., when $p^2 < Q$), and it is therefore not a polynomial in $p$. #

Problem. Use the expression for the area of a hyperbola to calculate the Radon transform of the characteristic function of the hyperbolic paraboloid $x_3 - x_1x_2 > 0$.

## 2.7. Radon Transform of Homogeneous Functions

Let $P(x)$ be an arbitrary nonnegative homogeneous function of degree 1. We define the homogeneous generalized function $P^\lambda(x)$ by

$$(P^\lambda, f) = \int P^\lambda(x) f(x)\, dx. \tag{1}$$

This integral is known to converge for Re $\lambda > 0$. We shall understand it in terms of its analytic continuation in $\lambda$ for Re $\lambda < 0$. We wish to find the Radon transform of $P^\lambda(x)$.

Let us first express $P^\lambda(x)$ in terms of the generalized function $a_{x'}(x; \lambda)$ which we defined in Section 2.3a by

$$(a_{x'}(x;\lambda), f) = \int_0^\infty f(tx')t^\lambda\, dt. \tag{2}$$

To do this we transform in Eq. (1) to polar coordinates,[30] obtaining

$$\begin{aligned}(P^\lambda, f) &= \int_\Gamma \left[\int_0^\infty t^{\lambda+n-1} f(tx')\, dt\right] P^\lambda(x')\omega(x') \\ &= \int_\Gamma (a_{x'}(x; \lambda+n-1), f) P^\lambda(x')\omega(x').\end{aligned} \tag{3}$$

Here the integral is taken over any surface enclosing the origin (for instance, a sphere), and

$$\omega(x') = \sum_{k=1}^{n} (-1)^{k-1} x_k'\, dx_1' \cdots dx_{k-1}'\, dx_{k+1}' \cdots dx_n' .$$

[30] See Section 2.5.

Thus $P^\lambda(x)$ may be written in terms of $a_{x'}(x; \lambda)$ in the form

$$P^\lambda(x) = \int_\Gamma a_{x'}(x; \lambda + n - 1) P^\lambda(x') \omega(x'). \tag{4}$$

This can be simplified somewhat if the hypersurface $P(x) = 1$ is chosen for $\Gamma$. In this case the expression becomes

$$P^\lambda(x) = \int_{P(x')=1} a_{x'}(x; \lambda + n - 1) \omega(x'). \tag{5}$$

With this formula it is easy to find the Radon transform of $P^\lambda(x)$. As we saw in Section 2.3, the Radon transform of $a_{x'}(x; \lambda + n - 1)$ is

$$p_+^{\lambda+n-1}(\xi, x')_+^{-\lambda-n} + p_-^{\lambda+n-1}(\xi, x')_-^{-\lambda-n}.$$

(For simplicity we shall deal only with the case of nonintegral $\lambda$.) Thus by using Eq. (5) we find that the Radon transform of the generalized function $P^\lambda$ is

$$p_+^{\lambda+n-1} \int_{P(x)=1} (\xi, x)_+^{-\lambda-n} \omega(x) + p_-^{\lambda+n-1} \int_{P(x)=1} (\xi, x)_-^{-\lambda-n} \omega(x).$$

If the $P(x) = 1$ hypersurface is bounded, both integrals converge for $\operatorname{Re} \lambda < -n$. For $\operatorname{Re} \lambda > n$, they are defined by analytic continuation in $\lambda$.

## 2.8. Radon Transform of the Characteristic Function of an Octant

Let us consider yet another interesting example of the Radon transform; namely, let us find the Radon transform of the characteristic function of the octant

$$x_1 > 0, \quad x_2 > 0, \quad x_3 > 0.$$

The intersection of a plane with an octant is either a triangle, an "exterior triangle,"[31] or an angle. Therefore the formula for the Radon transform of the characteristic function of an octant will contain terms corresponding to the generalized areas of these figures. Let us first find these areas.

Let us first consider the case in which the $(\xi, x) = p$ plane intersects the octant in a triangle. This will occur when the $\xi_i$ and $p$ all have the

[31] An exterior triangle is a figure consisting of two half-lines and a line segment (the regions labeled $\sigma_i$ in Fig. 2).

same sign. The plane intersects the coordinate axes at the points whose coordinates are $p/\xi_1$, $p/\xi_2$, and $p/\xi_3$. Thus the volume of the tetrahedron bounded by our plane and the three coordinate planes is $\frac{1}{6}p^3/\xi_1\xi_2\xi_3$. Therefore the area of the intersection is[32]

$$\frac{\partial}{\partial p}\left(\frac{1}{6}\frac{p^3}{\xi_1\xi_2\xi_3}\right) = \frac{1}{2}\frac{p^2}{\xi_1\xi_2\xi_3}.$$

Thus the Radon transform of the characteristic function of the octant contains the term

$$\tfrac{1}{2}p_+^2(\xi_1)_+^{-1}(\xi_2)_+^{-1}(\xi_3)_+^{-1} + \tfrac{1}{2}p_-^2(\xi_1)_-^{-1}(\xi_2)_-^{-1}(\xi_3)_-^{-1},$$

corresponding to the area of the triangle.

We now calculate the generalized areas of the exterior triangles and of the angles, not for one octant, but simultaneously for all the octants intersected by the plane. If the plane neither passes through the origin nor is parallel to one of the coordinate axes, it intersects seven octants in one triangle, three external triangles, and three angles. Let $s$ be the area of the triangle, $s_i$ be the generalized areas of the angles, and $\sigma_i$ be the generalized areas of the external triangles ($i = 1, 2, 3$; see Fig. 2). The problem is to express the $\sigma_i$ and $s_i$ in terms of $s$.

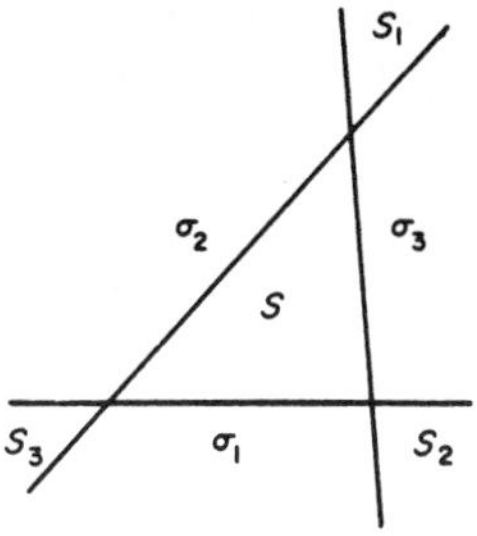


FIG. 2.

First we shall show that the generalized area of the half-plane vanishes. The half-plane is the intersection of a plane with the half-space, and we may thus find its generalized area by calculating the Radon transform of the characteristic function of the half-space. Now to know this Radon transform is equivalent to being able to calculate the integral of any function $f(x)$ over the half-space in terms of integrals over planes, and

[32] If $|\xi| = 1$, this expression is the ordinary one for the area of a triangle in Euclidean space.

it is clear that in order to calculate the integrals over the half-space we need only know the integrals over planes parallel to the boundary, which means that the Radon transform of the characteristic function of the half-space is concentrated on the set of planes of fixed normal (that is, with normal perpendicular to the bounding plane of the half-space). This in turn implies that the generalized area of the half-plane vanishes.

We now make use of additivity, and write the equations that state that the areas of all the half-planes in Fig. 2 vanish. These equations are six in number, namely

$$\begin{aligned} s_2 + s_3 + \sigma_1 &= 0, & \sigma_2 + \sigma_3 + s_1 &= -s,\\ s_3 + s_1 + \sigma_2 &= 0, & \sigma_3 + \sigma_1 + s_2 &= -s,\\ s_1 + s_2 + \sigma_3 &= 0, & \sigma_1 + \sigma_2 + s_3 &= -s. \end{aligned}$$

Of these six only four are independent, and they lead to

$$\begin{aligned} \sigma_1 &= (-\tfrac{2}{3} + \alpha_1)s, & \sigma_2 &= (-\tfrac{2}{3} + \alpha_2)s, & \sigma_3 &= (-\tfrac{2}{3} + \alpha_3)s,\\ s_1 &= (\tfrac{1}{3} + \alpha_1)s, & s_2 &= (\tfrac{1}{3} + \alpha_2)s, & s_3 &= (\tfrac{1}{3} + \alpha_3)s, \end{aligned} \qquad (1)$$

where the $\alpha_i$ are related by the single equation

$$\alpha_1 + \alpha_2 + \alpha_3 = 0.$$

Thus there still remains some arbitrariness in the generalized areas of the external triangles and angles. It will be seen later that this arbitrariness associated with the $\alpha_i$ cannot be removed and that their variation introduces only unessential functions.

These $\alpha_i$ are constants, or rather are independent of the intersecting plane. To see this, let $\varphi_1(\xi, p)$ be the Radon transform of the characteristic function of one of the octants. Now let us perform a dilation in one of the coordinate directions, say the $x_1$ direction:

$$x_1' = \lambda x_1 (\lambda > 0), \qquad x_2' = x_2, \quad x_3' = x_3.$$

This changes $\varphi_1(\xi_1, \xi_2, \xi_3; p)$ into $\lambda^{-1}\varphi_1(\lambda^{-1}\xi_1, \xi_2, \xi_3; p)$ [see Eq. (1) of Section 1.3]. Since, however, the octant is invariant under this transformation, the generalized areas of the intersections must remain invariant. Thus

$$\lambda^{-1}\varphi_1(\lambda^{-1}\xi_1, \xi_2, \xi_3; p) = \varphi_1(\xi_1, \xi_2, \xi_3; p).$$

But then it follows from (1) that the $\alpha_i$ are invariant under replacement of $\xi_1$ by $\lambda^{-1}\xi_1$ and are therefore independent of $\xi_1$. Similarly, they are independent of $\xi_2$ and $\xi_3$.

Having thus calculated the generalized areas of the exterior triangles and the angles, let us find their contributions to the Radon transform of the characteristic function of the octant.

The $(\xi, x) = p$ plane intersects our octant in an exterior triangle when only one of the $\xi_i$ has a sign different from $p$. If, however, two of the $\xi_i$ have signs different from $p$, the intersection is an angle. Then from Eq. (1) we may conclude that the Radon transform of the characteristic function of the octant contains the terms (we now write $\alpha_i$ instead of $\frac{1}{2}\alpha_i$)

$$p_+^2\{(-\tfrac{1}{3} + \alpha_1)(\xi_1)_-^{-1}(\xi_2)_+^{-1}(\xi_3)_+^{-1}\} + p_-^2\{(-\tfrac{1}{3} + \alpha_1)(\xi_1)_+^{-1}(\xi_2)_-^{-1}(\xi_3)_-^{-1}\},$$

corresponding to the areas of the exterior triangles. Here we use the braces to denote the sum of three terms obtained from each other by cyclic permutation of the indices; that is,

$$\{a_1 b_2 c_3\} = a_1 b_2 c_3 + a_2 b_3 c_1 + a_3 b_1 c_2 .$$

Similarly, we find that the angles contribute the terms

$$p_+^2\{(\tfrac{1}{6} + \alpha_1)(\xi_1)_+^{-1}(\xi_2)_-^{-1}(\xi_3)_-^{-1}\} + p_-^2\{(\tfrac{1}{6} + \alpha_1)(\xi_1)_-^{-1}(\xi_2)_+^{-1}(\xi_3)_+^{-1}\}.$$

Thus the total contribution from intersections not parallel to any of the coordinate axes is given by

$$\begin{aligned}\varphi_1(\xi, p) = {} & \tfrac{1}{2}[p_+^2(\xi_1)_+^{-1}(\xi_2)_+^{-1}(\xi_3)_+^{-1} + p_-^2(\xi_1)_-^{-1}(\xi_2)_-^{-1}(\xi_3)_-^{-1}] \\ & + p_+^2\{(-\tfrac{1}{3} + \alpha_1)(\xi_1)_-^{-1}(\xi_2)_+^{-1}(\xi_3)_+^{-1} + (\tfrac{1}{6} + \alpha_1)(\xi_1)_+^{-1}(\xi_2)_-^{-1}(\xi_3)_-^{-1}\} \\ & + p_-^2\{(-\tfrac{1}{3} + \alpha_1)(\xi_1)_+^{-1}(\xi_2)_-^{-1}(\xi_3)_-^{-1} + (\tfrac{1}{6} + \alpha_1)(\xi_1)_-^{-1}(\xi_2)_+^{-1}(\xi_3)_+^{-1}\}, \end{aligned} \tag{2}$$

where, as we have mentioned,

$$\alpha_1 + \alpha_2 + \alpha_3 = 0.$$

This equation thus has a degree of arbitrariness related to the $\alpha_i$. It turns out, further, that this is unavoidable and that the sum of terms containing the $\alpha_i$ is an unessential function.

This may be proved as follows. The sum of such terms is

$$p^2\{\alpha_1(\xi_1)_-^{-1}(\xi_2)_+^{-1}(\xi_3)_+^{-1} + \alpha_1(\xi_1)_+^{-1}(\xi_2)_-^{-1}(\xi_3)_-^{-1}\}. \tag{3}$$

We must show that the residue of the homogeneous function

$$a(\xi) = \{\alpha_1(\xi_1)_-^{-1}(\xi_2)_+^{-1}(\xi_3)_+^{-1} + \alpha_1(\xi_1)_+^{-1}(\xi_2)_-^{-1}(\xi_3)_-^{-1}\},$$

that is, the integral

$$\int_\Gamma a(\xi)\omega(\xi),$$

vanishes. Here $\omega(\xi) = \xi_1\, d\xi_2\, d\xi_3 + \xi_2\, d\xi_3\, d\xi_1 + \xi_3\, d\xi_1\, d\xi_2$, and the integration is over any surface enclosing the origin. Having proven this, we will have shown that (3) is an unessential function (see Section 2.1). The proof is simple: Note that the integral

$$J = \int_I [(\xi_1)_-^{-1}(\xi_2)_+^{-1}(\xi_3)_+^{-1} + (\xi_1)_+^{-1}(\xi_2)_-^{-1}(\xi_3)_-^{-1}]\omega(\xi)$$

is invariant under permutation of the indices. Therefore

$$\int_\Gamma a(\xi)\omega(\xi) = (\alpha_1 + \alpha_2 + \alpha_3)J = 0,$$

where we have used the fact that $\alpha_1 + \alpha_2 + \alpha_3 = 0$.

The geometric considerations we have gone through above are not sufficient to determine the Radon transform of the characteristic function of the octant. This is because the desired Radon transform also contains terms concentrated on the set of planes parallel to the coordinate axes and to obtain these terms from geometrical considerations would be somewhat more difficult. Thus we have actually been able to guess only the "principal part" of the desired Radon transform.

The entire Radon transform can be calculated by considering the generalized function

$$(x_1)_+^\lambda (x_2)_+^\lambda (x_3)_+^\lambda,$$

where $\lambda$ is a complex number. For $\lambda = 0$ this becomes the characteristic function of the octant, and thus by calculating its Radon transform and going to the limit as $\lambda \to 0$, we will find the desired Radon transform. The Radon transform of $(x_1)_+^\lambda (x_2)_+^\lambda (x_3)_+^\lambda$ will be found in the next section. Here we present without proof the result for the octant.

The Radon transform of the characteristic function of the octant

$$x_1 > 0, \quad x_2 > 0, \quad x_3 > 0$$

is

$$\begin{aligned}\varphi(\xi, p) = {} & \tfrac{1}{2}p_+^2(\xi_1)_+^{-1}(\xi_2)_+^{-1}(\xi_3)_+^{-1} - \tfrac{1}{3}p_+^2\{(\xi_1)_-^{-1}(\xi_2)_+^{-1}(\xi_3)_+^{-1}\} + \tfrac{1}{6}p_+^2\{(\xi_1)_+^{-1}(\xi_2)_-^{-1}(\xi_3)_-^{-1}\} \\ & + \tfrac{1}{12}\pi^2 p_+^2\{[-(\xi_1)_+^{-1} + 2(\xi_1)_-^{-1}]\,\delta(\xi_2)\,\delta(\xi_3)\} \\ & -\tfrac{1}{2}p_+^2 \ln p_+\{\delta(\xi_1)\xi_2^{-1}\xi_3^{-1}\} + \cdots. \end{aligned} \tag{4}$$

We have here written out only the terms containing $p_+^2$. The terms containing $p_-^2$ are obtained by replacing $(\xi_i)_+$ everywhere by $(\xi_i)_-$ and vice versa. Here the $(\xi_i)_+^{-1}$ and $(\xi_i)_-^{-1}$ are associated homogeneous generalized functions.[33]

Let us analyze this equation. It contains terms of three different types. Those of the first type are of the forms

$$\tfrac{1}{2}p_+^2(\xi_1)_+^{-1}(\xi_2)_+^{-1}(\xi_3)_+^{-1} \quad \text{and} \quad -\tfrac{1}{3}p_+^2(\xi_1)_-^{-1}(\xi_2)_+^{-1}(\xi_3)_+^{-1},$$

etc. We already know that they represent the generalized areas of intersections with planes not parallel to the edges of the octant.

Now consider the set of terms including

$$\tfrac{1}{12}\pi^2 p_+^2[-(\xi_1)_+^{-1} + 2(\xi_1)_-^{-1}]\,\delta(\xi_2)\,\delta(\xi_3) + \tfrac{1}{12}\pi^2 p_-^2[-(\xi_1)_-^{-1} + 2(\xi_1)_+^{-1}]\,\delta(\xi_2)\,\delta(\xi_3) \tag{5}$$

and the two others obtained from this by cyclic permutation of the indices. These terms are concentrated on the set of planes whose equations are of the form $\xi_1 x_1 = p$, planes parallel to one of the coordinate planes. Thus they form the contribution from one of the boundary planes of the octant to the transform of its characteristic function.

Finally, consider the set of terms including

$$-\tfrac{1}{2}p_+^2 \ln p_+\ \delta(\xi_1)\xi_2^{-1}\xi_3^{-1} - \tfrac{1}{2}p_-^2 \ln p_-\ \delta(\xi_1)\xi_2^{-1}\xi_3^{-1}$$

$$\equiv -\tfrac{1}{2}p^2 \ln |p|\ \delta(\xi_1)\xi_2^{-1}\xi_3^{-1} \tag{6}$$

and two other terms obtained from this by cyclic permutation of the indices. These terms are concentrated on the set of planes whose equations are of the form $\xi_2 x_2 + \xi_3 x_3 = p$, planes parallel to one of the coordinate axes. Thus they form the contribution from one of the edges of the octant to the Radon transform of its characteristic function.

Thus we see how, in Eq. (4), each boundary plane and edge contributes to the Radon transform of the characteristic function of the octant.

Equation (4) is easily generalized to the case of any corner bounded by three planes. Consider such a corner whose vertex is at the origin and

[33] Recall that the associated generalized function $t_+^{-1}$ on the line is defined by (Volume 1, p. 86)

$$(t_+^{-1}, f(t)) = \int_0^1 t^{-1}[f(t) - f(0)]\,dt + \int_1^\infty f(t)\,t^{-1}\,dt.$$

If $f(t)$ is defined on an interval $[a, b]$ rather than on the entire line, $t_+^{-1}$ is defined in the same way except that $f(t)$ is set equal to zero outside of $[a, b]$. The generalized function $t_-^{-1}$ is defined similarly.

whose edges are given by three vectors $e^{(1)}$, $e^{(2)}$, and $e^{(3)}$. We shall assume that the parallelopiped of which the $e^{(i)}$ are the edges has volume one. The Radon transform of the characteristic function of such a corner is obtained by replacing the $\xi_i$ in Eq. (4) by the $(e^{(i)}, \xi)$.

We leave to the reader the problem of calculating the Radon transform of the characteristic function of half of a circular cone (see Fig. 3), though

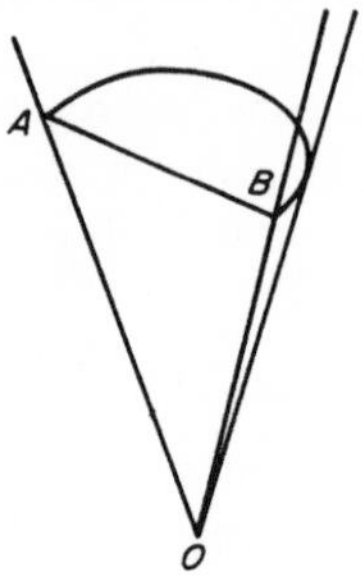


FIG. 3.

we shall describe some considerations that may prove useful. One may first find the contribution to the Radon transform from intersections with planes parallel to neither $OA$ nor $OB$. If the intersection is bounded, its area is easily calculated directly. Thus one must analyze the case in which the intersection is unbounded, which occurs when the plane intersects the entire circular cone in a hyperbola.[34] Now if the $AOB$ surface does not intersect this hyperbola or if it cuts off only a finite part, the generalized area is easily calculated. This is because we already know the area of the whole hyperbola, and what is left may be calculated by using additivity. One must thus find the contribution from intersections with planes whose normals point in certain special directions. Now these special directions are related to the boundary plane $AOB$ and the two edges $OA$ and $OB$. The contribution from $AOB$ can be obtained by replacing the half-cone by a plane-bounded corner as discussed above, one of whose boundary planes is $AOB$. This contribution will be the same for the half-cone and for the plane-bounded corner. One may thus take the desired term from Eq. (4). Next, consider the contribution from $OA$. For this purpose, construct a plane-bounded corner as above with vertex at $O$, and let its boundary planes be the following. First, $AOB$; second, the plane passing through $OA$ tangent to the cone; third, an

[34] One may expect, as in the case of the entire circular cone, that the parabolic intersections do not contribute to the Radon transform.

arbitrarily chosen plane to complete the corner. The contribution from $OA$ is then the same for the plane-bounded corner so constructed and the half-cone, and its contribution can again be obtained from Eq. (4).

## 2.9. The Generalized Hypergeometric Function

In this section we define the generalized hypergeometric function and make some observations concerning it.

Consider the $k$ linear forms $(\xi^{(1)}, x), \ldots, (\xi^{(k)}, x)$ on an $n$-dimensional space and the generalized function

$$(\xi^{(1)}, x)_+^{\lambda_1-1} \cdots (\xi^{(k)}, x)_+^{\lambda_k-1}, \tag{1}$$

where the $\lambda_i$ are complex numbers. *We shall define the generalized hypergeometric function*

$$F(\lambda_1, \ldots, \lambda_k \mid \xi^{(1)}, \ldots, \xi^{(k)} \mid \xi, p)$$

*as the Radon transform of* (1); *in other words,*

$$F(\lambda_1, \ldots, \lambda_k \mid \xi^{(1)}, \ldots, \xi^{(k)} \mid \xi, p) = [(\xi^{(1)}, x)_+^{\lambda_1-1} \cdots (\xi^{(k)}, x)_+^{\lambda_k-1}]^{\vee}. \tag{2}$$

A more general definition of the hypergeometric function is $F = [\Delta_{1+}^{\lambda_1-1} \cdots \Delta_{k+}^{\lambda_k-1}]^{\vee}$, were the $\Delta_i$ are determinants whose elements are the $(\xi^{(\alpha)}, x)$. It seems that all the special functions arising in representation theory can be expressed in terms of such hypergeometric functions.

We can always assume that the number of linearly independent forms among the $(\xi^{(\alpha)}, x)$ of Eq. (2) is no less than the dimension $n$, for if it were less, the Radon transform of $(\xi^{(1)}, x)_+^{\lambda_1-1} \cdots (\xi^{(k)}, x)_+^{\lambda_k-1}$ would reduce to a Radon transform in a space of lower dimension.

In the special case in which the $(\xi^{(\alpha)}, x)$ are all linearly independent, we shall call the Radon transform of $(\xi^{(1)}, x)_+^{\lambda_1-1} \cdots (\xi^{(k)}, x)_+^{\lambda_k-1}$ the *generalized beta function.* The ordinary beta function is obtained in the two-dimensional case. Specifically,

$$B(\lambda_1, \lambda_2) = F(\lambda_1, \lambda_2 \mid \xi_0^{(1)}, \xi_0^{(2)} \mid \xi_0, 1),$$

where $\xi_0^{(1)} = (1, 0)$, $\xi_0^{(2)} = (0, 1)$, $\xi_0 = (1, 1)$.

It may be shown further that the ordinary hypergeometric function is obtained from ours as the special case of two dimensions with three linear forms, two of which are independent. The proof is as follows.

The ordinary hypergeometric function may be given by

$$F(\alpha, \beta, \gamma; t) = \frac{1}{B(\beta, \gamma - \beta)} \int_0^1 x^{\beta-1}(1 - x)^{\gamma-\beta-1}(1 - tx)^{-\alpha}\, dx.$$

Obviously the integral can be thought of as the integral of the function (on a two-dimensional space)

$$(x_1)_+^{\beta-1}(x_2)_+^{\gamma-\beta-1}[(1-t)x_1+x_2]_+^{-}$$

over the plane whose equation is $x_1 + x_2 = 1$. Then

$$F(\alpha, \beta, \gamma; t) = \frac{1}{\mathrm{B}(\beta, \gamma-\beta)} F(\beta, \gamma-\beta, 1-\alpha \mid \xi_0^{(1)}, \xi_0^{(2)}, \xi_0^{(3)} \mid \xi_0, 1), \quad (3)$$

where

$$\xi_0^{(1)} = (1, 0), \quad \xi_0^{(2)} = (0, 1), \quad \xi_0^{(3)} = (1-t, 1), \quad \xi_0 = (1, 1).$$

A somewhat more natural definition of the generalized hypergeometric function is the following. It is the Radon transform of

$$\chi_1((\xi^{(1)}, x)) \cdots \chi_k((\xi^{(k)}, x)),$$

where the $\chi_i$ are characters of the multiplicative group of real numbers. Recall that a character is a function $\chi$ defined on a group such that $\chi(\alpha\beta) = \chi(\alpha)\chi(\beta)$ for any $\alpha$ and $\beta$ in the group (or in our case for numbers $\alpha \neq 0$ and $\beta \neq 0$). It is known that every such function is either $\chi(\alpha) = |\alpha|^\lambda$, or $\chi(\alpha) = |\alpha|^\lambda \operatorname{sgn} \alpha$. This definition of the generalized hypergeometric function can be extended to the complex domain (see Chapter II, Section 3.7).

Consider the following example. Let three independent linear forms $x_1$, $x_2$, and $x_3$ be given in a three-dimensional space. We wish to calculate the generalized beta function

$$\varphi(\xi, p) \equiv [(x_1)_+^{\lambda_1-1}(x_2)_+^{\lambda_2-1}(x\ )_+^{\lambda_3-1}]^{\vee}, \quad (4)$$

where $\lambda_i \neq 0, -1, -2, \ldots$. Then $\varphi(\xi, p)$ will be given by

$$\varphi(\xi, p) = \frac{\Gamma(\lambda_1)\Gamma(\lambda_2)\Gamma(\lambda_3)}{2i\Gamma(\lambda)\sin \pi\lambda}$$
$$\times [e^{i\pi\lambda}(p-i0)^{-1+\lambda}(\xi_1+i0)^{-\lambda_1}(\xi_2+i0)^{-\lambda_2}(\xi_3+i0)^{-\lambda_3}$$
$$-e^{-i\pi\lambda}(p+i0)^{-1+\lambda}(\xi_1-i0)^{-\lambda_1}(\xi_2-i0)^{-\lambda_2}(\xi_3-i0)^{-\lambda_3}], \quad (5)$$

where $\lambda = \lambda_1 + \lambda_2 + \lambda_3$.

If $\xi_3 \neq 0$, we may write

$$\varphi(\xi, p) = |\xi_3|^{-1} \int_{(\xi, x)=p} (x_1)_+^{\lambda_1-1}(x_2)_+^{\lambda_2-1}(x_3)_+^{\lambda_3-1}\, dx_1\, dx_2\,. \quad (6)$$

Since (4) fails to vanish only in the octant $x_i > 0$, the integral is in fact over only that part of the $(\xi, x) = p$ plane that lies in this octant. This part of the plane is either a triangle, an exterior triangle, or an angle. We shall deal with each of these cases separately.

Consider first the case of the triangle. This occurs when the three $p/\xi_i$ are positive. Let us write $x_i\xi_i/p = \alpha_i$ (no sum is implied) in the integrand. This yields

$$\varphi(\xi, p) = \frac{1}{|\xi_3|}\left(\frac{p}{\xi_1}\right)^{\lambda_1}\left(\frac{p}{\xi_2}\right)^{\lambda_2}\left(\frac{p}{\xi_3}\right)^{\lambda_3-1}\int_{\Sigma\alpha_i=1,\alpha_i>0}\alpha_1^{\lambda_1-1}\alpha_2^{\lambda_2-1}\alpha_3^{\lambda_3-1}\,d\alpha_1\,d\alpha_2\,.$$

To calculate this integral, we make the change of variables $\alpha_1 = (1-\alpha_2)t$, so that

$$\int_{\Sigma\alpha_i=1,\alpha_i>0}\alpha_1^{\lambda_1-1}\alpha_2^{\lambda_2-1}\alpha_3^{\lambda_3-1}\,d\alpha_1\,d\alpha_2 = \int_0^1\alpha_2^{\lambda_2-1}(1-\alpha_2)^{\lambda_1+\lambda_3-1}\,d\alpha_2\int_0^1 t^{\lambda_1-1}(1-t)^{\lambda_3-1}\,dt$$
$$= \frac{\Gamma(\lambda_1)\Gamma(\lambda_2)\Gamma(\lambda_3)}{\Gamma(\lambda_1+\lambda_2+\lambda_3)}\,.$$

Thus if the $p/\xi_i$ are all positive,

$$\varphi(\xi, p) = \frac{\Gamma(\lambda_1)\Gamma(\lambda_2)\Gamma(\lambda_3)}{\Gamma(\lambda_1+\lambda_2+\lambda_3)}|p|^{\lambda_1+\lambda_2+\lambda_3-1}|\xi_1|^{-\lambda_1}|\xi_2|^{-\lambda_2}|\xi_3|^{-\lambda_3}\,.$$

Now consider the case of exterior triangle. This occurs if one of the $p/\xi_i$, say $p/\xi_3$, is negative. This time we write

$$x_1\xi_1/p = \alpha_1\,,\qquad x_2\xi_2/p = \alpha_2\,,\qquad -x_3\xi_3/p = \alpha_3$$

in the integrand of (6), which yields

$$\varphi(\xi, p) = \frac{1}{|\xi_3|}\left(\frac{p}{\xi_1}\right)^{\lambda_1}\left(\frac{p}{\xi_2}\right)^{\lambda_2}\left|\frac{p}{\xi_3}\right|^{\lambda_3-1}\int_{\Sigma\alpha_i=1,\alpha_i>0}\alpha_1^{\lambda_1-1}\alpha_2^{\lambda_2-1}\alpha_3^{\lambda_3-1}\,d\alpha_1\,d\alpha_2\,.$$

To calculate this integral we write $\alpha_1 + \alpha_2 = u$ and $\alpha_1 - \alpha_2 = v$. We then have

$$\int_{\Sigma\alpha_i=1,\alpha_i>0}\alpha_1^{\lambda_1-1}\alpha_2^{\lambda_2-1}\alpha_3^{\lambda_3-1}\,d\alpha_1\,d\alpha_2$$
$$= \frac{1}{2}\int_1^\infty\int_{-u}^{u}\left(\frac{u+v}{2}\right)^{\lambda_1-1}\left(\frac{u-v}{2}\right)^{\lambda_2-1}(u-1)^{\lambda_3-1}\,dv\,du$$
$$= \frac{1}{2}\int_1^\infty u^{\lambda_1+\lambda_2-1}(u-1)^{\lambda_3-1}\,du\int_{-1}^{1}\left(\frac{1+t}{2}\right)^{\lambda_1-1}\left(\frac{1-t}{2}\right)^{\lambda_2-1}dt$$

(where we have made the additional change $v = ut$). Both integrals in the last line converge for

$$\lambda_1, \lambda_2, \lambda_3 > 0, \qquad \lambda_1 + \lambda_2 + \lambda_3 < 1$$

and they can be written in terms of the beta function. The final expression obtained in this way is

$$\int_{\Sigma\alpha_i=1,\alpha_i>0} \alpha_1^{\lambda_1-1}\alpha_2^{\lambda_2-1}\alpha_3^{\lambda_3-1}\, d\alpha_1\, d\alpha_2 = \frac{\sin \pi(\lambda_1+\lambda_2)}{\sin \pi(\lambda_1+\lambda_2+\lambda_3)} \frac{\Gamma(\lambda_1)\Gamma(\lambda_2)\Gamma(\lambda_3)}{\Gamma(\lambda_1+\lambda_2+\lambda_3)}.$$

Thus if $p/\xi_3 < 0$, while the other two $p/\xi_i$ are positive, we arrive at

$$\varphi(\xi, p) = \frac{\sin \pi(\lambda_1+\lambda_2)}{\sin \pi(\lambda_1+\lambda_2+\lambda_3)} \frac{\Gamma(\lambda_1)\Gamma(\lambda_2)\Gamma(\lambda_3)}{\Gamma(\lambda_1+\lambda_2+\lambda_3)} \times |p|^{\lambda_1+\lambda_2+\lambda_3-1} |\xi_1|^{-\lambda_1} |\xi_2|^{-\lambda_2} |\xi_3|^{-\lambda_3}.$$

Now consider the final case of an angle, which occurs when two of the $p/\xi_i$ are negative and one is positive. By similar considerations we arrive at

$$\varphi(\xi, p) = \frac{\sin \pi\lambda_3}{\sin \pi(\lambda_1+\lambda_2+\lambda_3)} \frac{\Gamma(\lambda_1)\Gamma(\lambda_2)\Gamma(\lambda_3)}{\Gamma(\lambda_1+\lambda_2+\lambda_3)} \times |p|^{\lambda_1+\lambda_2+\lambda_3-1} |\xi_1|^{-\lambda_1} |\xi_2|^{-\lambda_2} |\xi_3|^{-\lambda_3}$$

(we have assumed $p/\xi_3$ to be positive). Combining all the above we arrive at the following result. The Radon transform of the generalized function

$$(x_1)_+^{\lambda_1-1}(x_2)_+^{\lambda_2-1}(x_3)_+^{\lambda_3-1}$$

is

$$\varphi(\xi, p) = \frac{\Gamma(\lambda_1)\Gamma(\lambda_2)\Gamma(\lambda_3)}{\Gamma(\lambda_1+\lambda_2+\lambda_3)} p_+^{\lambda_1+\lambda_2+\lambda_3-1} \times \Big[(\xi_1)_+^{-\lambda_1}(\xi_2)_+^{-\lambda_2}(\xi_3)_+^{-\lambda_3} + \Big\{\frac{\sin \pi(\lambda_1+\lambda_2)}{\sin \pi(\lambda_1+\lambda_2+\lambda_3)} (\xi_1)_+^{-\lambda_1}(\xi_2)_+^{-\lambda_2}(\xi_3)_-^{-\lambda_3}\Big\} + \Big\{\frac{\sin \pi\lambda_3}{\sin \pi(\lambda_1+\lambda_2+\lambda_3)} (\xi_1)_-^{-\lambda_1}(\xi_2)_-^{-\lambda_2}(\xi_3)_+^{-\lambda_3}\Big\}\Big] + \cdots. \tag{7}$$

The braces here, as before, denote the sum of three terms obtained by cyclic permutation of the indices. For conciseness we have again written out only the terms involving $p_+$. Those involving $p_-$ are obtained by replacing $(\xi_i)_+$ by $(\xi_i)_-$, and vice versa. Of course $p_\pm^\lambda$, $(\xi_i)_\pm^\lambda$ are understood in the sense of generalized functions.

It is found that Eq. (7) becomes much simpler if it is written not in terms of the $p_\pm^\lambda$, $(\xi_i)_\pm^\lambda$ but in terms of the $(p \pm i0)^\lambda$, $(\xi \pm i0)^\lambda$. Recall that these generalized functions are defined on the line by

$$(t + i0)^\lambda = t_+^\lambda + e^{i\pi\lambda} t_-^\lambda ,$$

$$(t - i0)^\lambda = t_+^\lambda + e^{-i\pi\lambda} t_-^\lambda .$$

It is then found that

$$\varphi(\xi, p) = \frac{\Gamma(\lambda_1)\Gamma(\lambda_2)\Gamma(\lambda_3)}{2i\Gamma(\lambda) \sin \pi\lambda}$$

$$\times [e^{i\pi\lambda}(p - i0)^{-1+\lambda}(\xi_1 + i0)^{-\lambda_1}(\xi_2 + i0)^{-\lambda_2}(\xi_3 + i0)^{-\lambda_3}$$

$$-e^{-i\pi\lambda}(p + i0)^{-1+\lambda}(\xi_1 - i0)^{-\lambda_1}(\xi_2 - i0)^{-\lambda_2}(\xi_3 - i0)^{-\lambda_3}], \tag{8}$$

where $\lambda = \lambda_1 + \lambda_2 + \lambda_3$. (Verification of this expression is left to the reader.)

Equation (8) is no longer meaningful when $\lambda$ is a positive integer, since in this case the denominator vanishes. This again is related to the lack of uniqueness in the definition of the Radon transform. In order to calculate $\varphi(\xi, p)$ for positive integer $\lambda$ from (7) one must first eliminate an unessential function from the expression as it stands and then go to the limit. In this way one can use (7) to obtain an expression for

$$[(x_1)_+^0 (x_2)_+^0 (x_3)_+^0]^\vee ,$$

i.e., an expression for the Radon transform of the characteristic function of the octant $x_i > 0$. This expression was given without proof in Section 2.8.

## 3. Radon Transforms of Some Particular Generalized Functions

We shall now turn to the Radon transforms of generalized functions associated with quadratic forms. The discussion will be purely analytic. We will find particularly useful here the method of analytic continuation in the coefficients of the quadratic form, a method developed in Volume 1 when the generalized functions themselves were studied. The results we obtain will not be used in the rest of the book, and if the reader so desires he may omit this section entirely. We suggest it as a supplement to the discussion in Volume 1 of $P^\lambda$ and $(P + c)^\lambda$, where $P$ is a quadratic form. It is particularly instructive to compare our present results with the Fourier transforms of these generalized functions. It is found that their Fourier and Radon transforms can be written in terms of the same kind of functions but that, unlike the Fourier transforms, the Radon transforms do not require the introduction of generalized Bessel functions.

We shall discuss only the case of odd dimension; it is suggested that the reader derive the analogous results for even dimension.

### 3.1. Radon Transforms of the Generalized Functions $(P + i0)^\lambda$, $(P - i0)^\lambda$, and $P_+^\lambda$, for Nondegenerate Quadratic Forms $P$

In Volume 1 we defined the generalized functions $(P + i0)^\lambda$, $(P - i0)^\lambda$, $P_+^\lambda$, and $P_-^\lambda$, where $P = P(x)$ is a nondegenerate quadratic form and $\lambda$ is a complex number. We shall start by restating these definitions. Consider the generalized function $(P + iP_1)^\lambda$, where $P_1$ is a positive definite quadratic form, defined by

$$((P + iP_1)^\lambda, f) = \int (P + iP_1)^\lambda f(x)\, dx. \tag{1}$$

The integral in this equation converges for $\operatorname{Re}\lambda > 0$ and is an analytic function of $\lambda$. It is understood in terms of analytic continuation in $\lambda$ for $\operatorname{Re}\lambda < 0$. Then $(P + i0)^\lambda$ is defined as the limit of $(P + iP_1)^\lambda$ as all the coefficients of $P_1$ converge to zero. The generalized function $(P - i0)^\lambda$ is defined analogously. Finally, $P_+$ and $P_-$ are given in terms of the generalized functions just defined by

$$(P + i0)^\lambda = P_+^\lambda + e^{\pi i\lambda}P_-^\lambda, \qquad (P - i0)^\lambda = P_+^\lambda + e^{-\pi i\lambda}P_-^\lambda. \tag{2}$$

Let us calculate the Radon transforms of these functions on the assumption that $\lambda$ is neither an integer nor a half-integer.

We shall start with $P^\lambda \equiv (P \pm i0)^\lambda$, where $P$ is a positive definite quadratic form. Recall the following property of the Radon transform established in Section 1.3. If $\check{F}(\xi, p)$ is the Radon transform of $F(x)$, the Radon transform of $F(A^{-1}x)$, where $A$ is a nonsingular linear transformation, is

$$\check{F}_A(\xi, p) \equiv |\det A|\, \check{F}(A'\xi, p).$$

Now let $\check{F}(\xi, p)$ be the Radon transform of $P^\lambda$. Then if $A$ is any linear transformation that leaves $P$ invariant, it follows from the above that

$$\check{F}(A'\xi, p) = \check{F}(\xi, p),$$

where $A'$ is the transpose of $A$. Hence we may write

$$\check{F}(\xi, p) = \varphi(Q, p), \tag{3}$$

where $Q = Q(\xi)$ is the quadratic form dual to $P(x)$, i.e., such that its matrix is the inverse of that of $P$. On the other hand, it also follows from the above property of the Radon transform that for $\alpha > 0$ the Radon transform of $\alpha^{2\lambda}P^\lambda(x) = P^\lambda(\alpha x)$ is $\alpha^{2\lambda}\varphi(Q, p) = \alpha^{-n}\varphi(\alpha^{-2}Q, p)$. Therefore $\varphi(Q, p)$ is a homogeneous function of $Q$ of degree $-\lambda - \frac{1}{2}n$ so that it may be written in the form

$$\varphi(Q, p) = \varphi(p)Q^{-\lambda-\frac{1}{2}n}. \tag{4}$$

Finally, in order to find $\varphi(p)$ we use the fact that $\check{F}(\xi, p)$ is an even homogeneous function of $\xi, p$ of degree $-1$. It follows from this that $\varphi(p)$ in Eq. (4) is an even homogeneous function of $p$ of degree $2\lambda + n - 1$. Hence $\varphi(p) = c(\lambda) \mid p \mid^{2\lambda+n-1}$ (see Volume 1, Chapter I, Section 3.11).

It thus follows that the Radon transform of $P^\lambda$, where $P = P(x)$ is a positive definite quadratic form, is

$$\check{F}(\xi, p) = c(\lambda)| p |^{2\lambda+n-1}Q^{-\lambda-\frac{1}{2}n}, \tag{5}$$

where $Q = Q(\xi)$ is the quadratic form dual to $P$.

It is relatively simple to show that the $c(\lambda)$ in (5) is

$$c(\lambda) = \frac{\pi^{\frac{1}{2}(n-1)}\Gamma(-\lambda - \frac{1}{2}n + \frac{1}{2})}{\Gamma(-\lambda)\Delta^{\frac{1}{2}}}, \tag{6}$$

where $\Delta$ is the discriminant of $P$. Without loss of generality we may assume for $P$ the canonical form $P = \sum \alpha_i^2 x_i^2$. Then $Q = \sum \alpha_i^{-2}\xi_i^2$. Now we choose $\xi = \xi_0 = (0, ..., 0, \alpha_n)$, $p = 1$ in (5). This yields

$$\check{F}(\xi_0, 1) = c(\lambda).$$

On the other hand, by definition we have

$$\begin{aligned}\check{F}(\xi_0, 1) &= \int_{\alpha_n x_n = 1} (\textstyle\sum \alpha_i^2 x_i^2)^\lambda \omega \\ &= \alpha_n^{-1} \int (\alpha_1^2 x_1^2 + \cdots + \alpha_{n-1}^2 x_{n-1}^2 + 1)^\lambda \, dx_1 \cdots dx_n \\ &= (\alpha_1 \cdots \alpha_n)^{-1} \int (t_1^2 + \cdots + t_{n-1}^2 + 1) \, dt_1 \cdots dt_{n-1}\end{aligned}$$

which leads immediately to the desired result.

Let us now use the method of analytic continuation in the coefficients of $P$ (as described in the appendix to this section). On so doing, we arrive at the following result.

Assume the nondegenerate quadratic form $P(x)$ to have $k$ positive and $l$ negative coefficients in the canonical form. Then the Radon transform of $(P + i0)^\lambda$ is

$$\frac{e^{-\frac{1}{2}l\pi i}\pi^{\frac{1}{2}(n-1)}\Gamma(-\lambda - \frac{1}{2}n + \frac{1}{2})}{\Gamma(-\lambda)|\Delta|^{\frac{1}{2}}} |p|^{2\lambda+n-1}(Q - i0)^{-\lambda-\frac{1}{2}n}, \tag{7}$$

where $Q = Q(\xi)$ is the quadratic form dual to $P$, and $\Delta$ is the discriminant of $P$.

Similarly, the Radon transform of $(P - i0)^\lambda$ is

$$\frac{e^{\frac{1}{2}l\pi i}\pi^{\frac{1}{2}(n-1)}\Gamma(-\lambda - \frac{1}{2}n + \frac{1}{2})}{\Gamma(-\lambda)|\Delta|^{\frac{1}{2}}} |p|^{2\lambda+n-1}(Q + i0)^{-\lambda-\frac{1}{2}n}. \tag{7'}$$

Further, by using the relation

$$P_+^\lambda = -\frac{1}{2i \sin \pi\lambda}[e^{-\pi\lambda i}(P + i0)^\lambda - e^{\pi\lambda i}(P - i0)^\lambda]$$

we find the Radon transform of $P_+^\lambda$ is

$$-\frac{\pi^{\frac{1}{2}(n-1)}\Gamma(-\lambda - \frac{1}{2}n + \frac{1}{2})}{2i \sin \pi\lambda\Gamma(-\lambda)|\Delta|^{\frac{1}{2}}} |p|^{2\lambda+n-1}$$
$$\times [e^{-\frac{1}{2}\pi(l+2\lambda)i}(Q - i0)^{-\lambda-\frac{1}{2}n} - e^{\frac{1}{2}\pi(l+2\lambda)i}(Q + i0)^{-\lambda-\frac{1}{2}n}]. \tag{8}$$

This expression can be restated somewhat by writing it in terms of $Q_+^{-\lambda-\frac{1}{2}n}$ and $Q_-^{-\lambda-\frac{1}{2}n}$. Elementary manipulations lead to the following expression for the Radon transform of $P_+^\lambda$:

$$\frac{(-1)^{\frac{1}{2}(n-1)}\pi^{\frac{1}{2}(n-1)}}{(\lambda + 1) \cdots (\lambda + \frac{1}{2}n - \frac{1}{2})|\Delta|^{\frac{1}{2}} \sin \pi\lambda} |p|^{2\lambda+n-1}$$
$$\times [\sin \tfrac{1}{2}\pi(l + 2\lambda)\, Q_+^{-\lambda-\frac{1}{2}n} - \sin \tfrac{1}{2}\pi k\, Q_-^{-\lambda-\frac{1}{2}n}], \tag{9}$$

where $Q = Q(\xi)$ is the quadratic form dual to $P$, and $k$, $l$ are the number of positive and negative terms, respectively, in the canonical form of $P$. The Radon transform of $P_-^\lambda$ is obtained from this expression by interchanging $k$ with $l$ and $Q_+$ with $Q_-$.

An interesting special case is the one in which $k$ is even and $l$ is odd. Then the Radon transform of $P_+^\lambda$ becomes

$$\frac{(-1)^{\frac{1}{2}k}\pi^{\frac{1}{2}(n-1)}\cot \pi\lambda}{(\lambda+1)\cdots(\lambda+\frac{1}{2}n-\frac{1}{2})}|p|^{2\lambda+n-1}Q_+^{-\lambda-\frac{1}{2}n}.$$

This implies that if $P$ has an even number of positive terms and an odd number of negative terms in its canonical form, all one needs in order to calculate the integral

$$\int P_+^\lambda(x)f(x)\,dx$$

are the integrals of $f(x)$ only over those $(\xi, x) = p$ hyperplanes for which $Q(\xi) > 0$ (recall that by assumption $\lambda$ is neither an integer nor a half-integer).

## Appendix to Section 3.1

We wish to describe in some detail the method of analytic continuation in the coefficients of $P$.[35] We wish to find the Radon transform of $P^\lambda$ where $P = P_1 + iP_2$ is a quadratic form with complex coefficients such that $\operatorname{Im} P = P_2$ is positive definite. The generalized function $P^\lambda$ is analytic not only in $\lambda$ but also in the coefficients of $P$. Therefore the Radon transform of $P^\lambda$ is also an analytic function of these coefficients in the "upper half-plane" of quadratic forms, i.e., for $\operatorname{Im} P > 0$. This means that the Radon transform of $P^\lambda$ is determined uniquely by its values on the "positive imaginary axis" of quadratic forms, i.e., for $P = iP_2$ with $P_2$ positive definite. Therefore it is sufficient for our purposes to find the Radon transforms of generalized functions of the form $(iP_2)^\lambda$. But we have already actually found the Radon transforms of such generalized functions. In fact from Eq. (5) we may deduce that the Radon transform of $(iP_2)^\lambda$ is

$$\frac{e^{-\frac{1}{4}\pi n i}\pi^{\frac{1}{2}(n-1)}\Gamma(-\lambda-\frac{1}{2}n+\frac{1}{2})}{\Gamma(-\lambda)[(-i)^n\Delta]^{\frac{1}{2}}}|p|^{2\lambda+n-1}Q^{-\lambda-\frac{1}{2}n}, \tag{10}$$

where $Q$ is the quadratic form dual to $P = iP_2$, and $\Delta$ is the discriminant of $P$. This expression must now be continued onto the upper half-plane

[35] See Volume 1, Chapter III, Section 2.2, where this method was used to find the residue of $P_+$ at singular points in the $\lambda$ plane.

of quadratic forms. But the analytic continuation of $Q$ is known, since its coefficients depend analytically on those of $P$. Therefore the problem is solved when we are able to define $[(-i)^n \Delta]^{\frac{1}{2}}$ as a single-value analytic function on the upper half-plane of quadratic forms. This definition was given in Volume 1, Chapter III, Section 2.3. For completeness we shall go through this again here.

We write $P$ in the form

$$P = P_1 + iP_2 ,$$

where $P_1$ and $P_2$ are quadratic forms with real coefficients and $P_2$ is positive definite. There exists a nonsingular linear transformation

$$x_k = \sum_{l=1}^{n} b_{kl} y_l$$

with real coefficients which transforms $P_1$ and $P_2$ to the forms

$$P_1 = \lambda_1 y_1^2 + \cdots + \lambda_n y_n^2 ,$$

$$P_2 = y_1^2 + \cdots + y_n^2 .$$

The $\lambda_i$ are real coefficients independent of the particular transformation, and are thus invariants of $P$ itself. Then $[(-i)^n \Delta]^{\frac{1}{2}}$ can be defined by the expression

$$[(-i)^n \Delta]^{\frac{1}{2}} = | b |^{-1}(1 - \lambda_1 i)^{\frac{1}{2}} \cdots (1 - \lambda_n i)^{\frac{1}{2}} , \tag{11}$$

where $b$ is the determinant of the matrix $\| b_{kl} \|$, and the square roots are defined by

$$z^{\frac{1}{2}} = | z |^{\frac{1}{2}} \exp(\tfrac{1}{2} i \arg z), \qquad -\pi < \arg z < \pi.$$

Thus the Radon transform of $P^\lambda$, where $P = P(x)$ is any quadratic form with complex coefficients and positive definite imaginary part, is given by Eq. (10) with the square root in the denominator defined by Eq. (11).

Now by definition $(P + i0)^\lambda$ is the limit of $(P + iP_1)^\lambda$ as the positive definite quadratic form $P_1$ approaches zero. Thus by going to the limit in (10), we easily arrive at the desired expression for the Radon transform of $(P + i0)^\lambda$.

## 3.2. Radon Transforms of $(P + c + i0)^\lambda$, $(P + c - i0)^\lambda$, and $(P + c)_+^\lambda$, for Nondegenerate Quadratic Forms

Let us start by calculating the Radon transform of $(x_1^2 + \cdots + x_n^2 + c)^\lambda$, $c > 0$. This is given by the integral, convergent for $\operatorname{Re} \lambda < -\frac{1}{2}(n - 1)$,

$$\begin{aligned} \check{F}(\xi, p) &= \int_{(\xi,x)=p} (x_1^2 + \cdots + x_n^2 + c)^\lambda \omega \\ &= \int_{|\xi| x_n = |p|} (x_1^2 + \cdots + x_n^2 + c)^\lambda \omega \\ &= |\xi|^{-1} \int (x_1^2 + \cdots + x_{n-1}^2 + p^2 |\xi|^{-2} + c)^\lambda \, dx_1 \cdots dx_{n-1}, \end{aligned} \tag{1}$$

where $|\xi|^2 = \sum \xi_i^2$. The last integral is easily calculated in spherical coordinates and leads to the result that the desired Radon transform is

$$\frac{\pi^{\frac{1}{2}(n-1)} \Gamma(-\lambda - \frac{1}{2}n + \frac{1}{2})}{\Gamma(-\lambda)} \frac{1}{|\xi|} \left(\frac{p^2}{|\xi|^2} + c\right)^{\lambda + \frac{1}{2}(n-1)} \tag{2}$$

From this we easily find that the Radon transform of $(P + c)^\lambda$, where $c > 0$ and $P = P(x)$ is any positive definite quadratic form, is

$$\frac{\pi^{\frac{1}{2}(n-1)} \Gamma(-\lambda - \frac{1}{2}n + \frac{1}{2})}{\Gamma(-\lambda) \Delta^{\frac{1}{2}}} Q^{-\frac{1}{2}} \left(\frac{p^2}{Q} + c\right)^{\lambda + \frac{1}{2}(n-1)} \tag{3}$$

Here $Q$ is the quadratic form dual to $P$, and $\Delta$ is the discriminant of $P$. We may now proceed by analytic continuation (see Section 1.2) from Eq. (3) to find the Radon transforms of $(P + c + i0)^\lambda$ and $(P + c - i0)^\lambda$, where $P$ is any nondegenerate quadratic form and $c$ is any real number. We present the final result without actually going through the derivation.

Assume the nondegenerate quadratic form $P = P(x)$ to have $k$ positive and $l$ negative coefficients in its canonical form ($k + l = n$). Then the Radon transform of $(P + c + i0)^\lambda$ is

$$\frac{e^{-\frac{1}{2}\pi l i} \pi^{\frac{1}{2}(n-1)} \Gamma(-\lambda - \frac{1}{2}n + \frac{1}{2})}{\Gamma(-\lambda) |\Delta|^{\frac{1}{2}}} (Q - i0)^{-\frac{1}{2}} \left(\frac{p^2}{Q} + c + i0\right)^{\lambda + \frac{1}{2}(n-1)} \tag{4}$$

Similarly, the Radon transform of $(P + c - i0)^\lambda$ is

$$\frac{e^{\frac{1}{2}\pi l i} \pi^{\frac{1}{2}(n-1)} \Gamma(-\lambda - \frac{1}{2}n + \frac{1}{2})}{\Gamma(-\lambda) |\Delta|^{\frac{1}{2}}} (Q + i0)^{-\frac{1}{2}} \left(\frac{p^2}{Q} + c - i0\right)^{\lambda + \frac{1}{2}(n-1)} \tag{4'}$$

From these we may also derive the Radon transform of

$$(P+c)_+^\lambda = -\frac{1}{2i\sin\pi\lambda}\left[e^{-\pi\lambda i}(P+c+i0)^\lambda - e^{\pi\lambda i}(P+c-i0)^\lambda\right],$$

which is found to be

$$\frac{(-1)^{\frac{1}{2}(n-1)}\pi^{\frac{1}{2}(n-1)}}{(\lambda+1)\cdots(\lambda+\frac{1}{2}n-\frac{1}{2})\sin\pi\lambda\,|\,\Delta\,|^{\frac{1}{2}}}$$

$$\times\left(-\frac{1}{2i}\right)\left[e^{-\frac{1}{2}\pi(l+2\lambda)i}(Q-i0)^{-\frac{1}{2}}\left(\frac{p^2}{Q}+c+i0\right)^{\lambda+\frac{1}{2}(n-1)}\right.$$

$$\left.-e^{\frac{1}{2}\pi(l+2\lambda)i}(Q+i0)^{-\frac{1}{2}}\left(\frac{p^2}{Q}+c-i0\right)^{\lambda+\frac{1}{2}(n-1)}\right]. \tag{5}$$

**Remark.** As before, it is convenient to rewrite this equation in terms of $Q_\pm^{-\frac{1}{2}}$ and $[(p^2/Q)+c]_\pm^{\lambda+\frac{1}{2}(n-1)}$. We then find that the Radon transform of $(P+c)_+^\lambda$ may be written

$$\frac{(-1)^{\frac{1}{2}(n-1)}\pi^{\frac{1}{2}(n-1)}}{(\lambda+1)\cdots(\lambda+\frac{1}{2}n-\frac{1}{2})\sin\pi\lambda\,|\,\Delta\,|^{\frac{1}{2}}}\left[\sin\pi\left(\lambda+\frac{l}{2}\right)Q_+^{-\frac{1}{2}}\left(\frac{p^2}{Q}+c\right)_+^{\lambda+\frac{1}{2}(n-1)}\right.$$

$$-\sin\frac{\pi k}{2}Q_-^{-\frac{1}{2}}\left(\frac{p^2}{Q}+c\right)_-^{\lambda+\frac{1}{2}(n-1)}$$

$$+\sin\pi\left(\lambda+\frac{l-1}{2}\right)Q_-^{-\frac{1}{2}}\left(\frac{p^2}{Q}+c\right)_+^{\lambda+\frac{1}{2}(n-1)}$$

$$\left.-\sin\frac{\pi(k-1)}{2}Q_+^{-\frac{1}{2}}\left(\frac{p^2}{Q}+c\right)_-^{\lambda+\frac{1}{2}(n-1)}\right], \tag{6}$$

or, in another notation,

$$\frac{(-1)^{\frac{1}{2}(n-1)}\pi^{\frac{1}{2}(n-1)}}{(\lambda+1)\cdots(\lambda+\frac{1}{2}n-\frac{1}{2})\sin\pi\lambda\,|\,\Delta\,|^{\frac{1}{2}}}\left[\sin\pi\left(\lambda+\frac{l}{2}\right)Q_+^{-\lambda-\frac{1}{2}n}(p^2+cQ)_+^{\lambda+\frac{1}{2}(n-1)}\right.$$

$$-\sin\frac{\pi k}{2}Q_-^{-\lambda-\frac{1}{2}n}(p^2+cQ)_+^{\lambda+\frac{1}{2}(n-1)}$$

$$+\sin\pi\left(\lambda+\frac{l-1}{2}\right)Q_-^{-\lambda-\frac{1}{2}n}(p^2+cQ)_-^{\lambda+\frac{1}{2}(n-1)}$$

$$\left.-\sin\frac{\pi(k-1)}{2}Q_+^{-\lambda-\frac{1}{2}n}(p^2+cQ)_-^{\lambda+\frac{1}{2}(n-1)}\right]. \tag{6'}$$

Note that for $c > 0$, the fourth term in the square brackets vanishes, while for $c < 0$, the third term vanishes. Similarly, the Radon transform of $(P + c)_-^\lambda$ is

$$\frac{(-1)^{\frac{1}{2}(n-1)}\pi^{\frac{1}{2}(n-1)}}{(\lambda+1)\cdots(\lambda+\frac{1}{2}n-\frac{1}{2})\sin\pi\lambda\,|\varDelta|^{\frac{1}{2}}}\Bigg[\sin\pi\left(\lambda+\frac{k}{2}\right)Q_-^{-\frac{1}{2}}\left(\frac{p^2}{Q}+c\right)_-^{\lambda+\frac{1}{2}(n-1)}$$
$$-\sin\frac{\pi l}{2}Q_+^{-\frac{1}{2}}\left(\frac{p^2}{Q}+c\right)_+^{\lambda+\frac{1}{2}(n-1)}$$
$$+\sin\pi\left(\lambda+\frac{k-1}{2}\right)Q_+^{-\frac{1}{2}}\left(\frac{p^2}{Q}+c\right)_-^{\lambda+\frac{1}{2}(n-1)}$$
$$-\sin\frac{\pi(l-1)}{2}Q_-^{-\frac{1}{2}}\left(\frac{p^2}{Q}+c\right)_+^{\lambda+\frac{1}{2}(n-1)}\Bigg], \qquad (7)$$

or, in another notation,

$$\frac{(-1)^{\frac{1}{2}(n-1)}\pi^{\frac{1}{2}(n-1)}}{(\lambda+1)\cdots(\lambda+\frac{1}{2}n-\frac{1}{2})\sin\pi\lambda\,|\varDelta|^{\frac{1}{2}}}\Bigg[\sin\pi\left(\lambda+\frac{k}{2}\right)Q_-^{-\lambda-\frac{1}{2}n}(p^2+cQ)_+^{\lambda+\frac{1}{2}(n-1)}$$
$$-\sin\frac{\pi l}{2}Q_+^{-\lambda-\frac{1}{2}n}(p^2+cQ)_+^{\lambda+\frac{1}{2}(n-1)}$$
$$+\sin\pi\left(\lambda+\frac{k-1}{2}\right)Q_+^{-\lambda-\frac{1}{2}n}(p^2+cQ)_-^{\lambda+\frac{1}{2}(n-1)}$$
$$-\sin\frac{\pi(l-1)}{2}Q_-^{-\lambda-\frac{1}{2}n}(p^2+cQ)_+^{\lambda+\frac{1}{2}(n-1)}\Bigg]. \quad \# \qquad (7')$$

## 3.3. Radon Transforms of the Characteristic Functions of Hyperboloids and Cones

We shall now make use of the results of Sections 3.1 and 3.2 to calculate the Radon transforms of the characteristic functions of hyperboloids and cones in odd-dimensional spaces. In this way we will define the generalized areas of sections through these regions.

Let us start with $x_1^2 - x_2^2 - \cdots - x_n^2 + 1 > 0$, the interior of a single-sheeted hyperboloid. The characteristic function is

$$(x_1^2 - x_2^2 - \cdots - x_n^2 + 1)_+^0 .$$

According to Section 3.2, its Radon transform is

$$\frac{\pi^{\frac{1}{2}(n-1)}}{\Gamma(\frac{1}{2}n+\frac{1}{2})}Q_+^{-\frac{1}{2}n}(p^2+Q)_+^{\frac{1}{2}(n-1)} - \frac{(-1)^{\frac{1}{2}(n-1)}\pi^{\frac{1}{2}(n-1)}}{\Gamma(\frac{1}{2}n+\frac{1}{2})}\lim_{\lambda\to 0} U, \qquad (1)$$

where

$$U = [Q_-^{-\lambda-\frac{1}{2}n}(p^2+Q)_+^{\lambda+\frac{1}{2}(n-1)} + (-1)^{\frac{1}{2}(n-1)}\cos\pi\lambda Q_-^{\lambda-\frac{1}{2}n}(p^2+Q)_-^{\lambda+\frac{1}{2}(n-1)}] \times [\sin\pi\lambda]^{-1}$$

and $Q = \xi_1^2 - \xi_2^2 - \cdots - \xi_n^2$. On going to the limit, we obtain the result

$$\frac{\pi^{\frac{1}{2}(n-1)}}{\Gamma(\frac{1}{2}n+\frac{1}{2})} Q_+^{-\frac{1}{2}n}(p^2+Q)^{\frac{1}{2}(n-1)} - \frac{(-1)^{\frac{1}{2}(n-1)}\pi^{\frac{1}{2}(n-3)}}{\Gamma(\frac{1}{2}n+\frac{1}{2})} Q_-^{-\frac{1}{2}n}(p^2+Q)^{\frac{1}{2}(n-1)} \ln\left|\frac{p^2+Q}{Q}\right| \tag{2}$$

for the interior of the single-sheeted hyperboloid.

For the other regions of interest the results are the following.

The Radon transform of $(x_1^2 - x_2^2 - \cdots - x_n^2 + 1)_-^0$, the characteristic function of the exterior of a single-sheeted hyperboloid, is

$$\frac{(-1)^{\frac{1}{2}(n-1)}\pi^{\frac{1}{2}(n-3)}}{\Gamma(\frac{1}{2}n+\frac{1}{2})} Q_-^{-\frac{1}{2}n}(p^2+Q)^{\frac{1}{2}(n-1)} \ln\left|\frac{p^2+Q}{Q}\right|, \tag{3}$$

This expression shown that if one wants to calculate the integral of some $f(x)$ over the region outside a single-sheeted hyperboloid, one need not know the integrals over those $(\xi, x) = p$ hyperplanes for which $Q > 0$, that is, those hyperplanes whose intersection with the interior of the hyperboloid has finite area.

The Radon transform of $(x_1^2 - x_2^2 - \cdots - x_n^2 - 1)_+^0$, the characteristic function of the interior of a two-sheeted hyperboloid, is

$$\frac{\pi^{\frac{1}{2}(n-1)}}{\Gamma(\frac{1}{2}n+\frac{1}{2})} Q_+^{-\frac{1}{2}n}(p^2-Q)_+^{\frac{1}{2}(n-1)} - \frac{(-1)^{\frac{1}{2}(n-1)}\pi^{\frac{1}{2}(n-3)}}{\Gamma(\frac{1}{2}n+\frac{1}{2})} Q_-^{-\frac{1}{2}n}(p^2-Q)^{\frac{1}{2}(n-1)} \ln\left|\frac{p^2-Q}{Q}\right|. \tag{4}$$

This function vanishes for those hyperplanes for which $Q > 0$ and $p^2 - Q < 0$, a result which should cause no surprise since these are the hyperplanes which fail to intersect the region.

The Radon transform of $(x_1^2 - x_2^2 - \cdots - x_n^2 - 1)_-^0$, the characteristic function of the exterior of a two-sheeted hyperboloid, is

$$\frac{(-1)^{\frac{1}{2}(n-1)}\pi^{\frac{1}{2}(n-1)}}{\Gamma(\frac{1}{2}n+\frac{1}{2})} Q_+^{-\frac{1}{2}n}(p^2-Q)_-^{\frac{1}{2}(n-1)} + \frac{(-1)^{\frac{1}{2}(n-1)}\pi^{\frac{1}{2}(n-3)}}{\Gamma(\frac{1}{2}n+\frac{1}{2})} Q_-^{-\frac{1}{2}n}(p^2-Q)^{\frac{1}{2}(n-1)} \ln\left|\frac{p^2-Q}{Q}\right|. \tag{5}$$

This function vanishes for those hyperplanes for which $Q > 0$ and $p^2 - Q > 0$, those which intersect the hyperboloid with finite area. This means that if one wants to calculate the integral of some $f(x)$ over the region outside a two-sheeted hyperboloid, one need not know its integral over hyperplanes whose intersection with the hyperboloid has finite area.

Let us now turn to the cone given by $x_1^2 - x_2^2 - \cdots - x_n^2 > 0$, that is, let us calculate the Radon transform of $P_+^0$, where $P = x_1^2 - x_2^2 - \cdots - x_n^2$. According to Eq. (9) of Section 3.1, the Radon transform of $P_+^\lambda = (x_1^2 - x_2^2 - \cdots - x_n^2)_+^\lambda$

$$\frac{\pi^{\frac{1}{2}(n-1)}}{(\lambda+1)\cdots(\lambda+\frac{1}{2}[n-1])} \mid p \mid^{2\lambda+n-1} \left[Q_+^{-\lambda-\frac{1}{2}n} - \frac{(-1)^{\frac{1}{2}(n-1)}}{\sin \pi\lambda} Q_-^{-\lambda-\frac{1}{2}n}\right],$$

where $Q = \xi_1^2 - \xi_2^2 - \cdots - \xi_n^2$. We now go to the limit as $\lambda \to 0$ and find that the Radon transform of $P_+^0$ is

$$\frac{\pi^{\frac{1}{2}(n-1)}}{\Gamma(\frac{1}{2}n+\frac{1}{2})} p^{n-1} Q_+^{-\frac{1}{2}n} - \frac{(-1)^{\frac{1}{2}(n-1)}\pi^{\frac{1}{2}(n-1)}}{\Gamma(\frac{1}{2}n+\frac{1}{2})} \lim_{\lambda\to 0} \frac{\mid p \mid^{2\lambda+n-1} Q_-^{-\lambda-\frac{1}{2}n}}{\sin \pi\lambda}. \tag{6}$$

**Remark.** The expression whose limit is here being taken has a denominator which vanishes in the limit. Therefore the generalized function in the numerator must also vanish in the limit, or

$$\lim_{\lambda\to 0} \mid p \mid^{2\lambda+n-1} Q_-^{-\lambda-\frac{1}{2}n} = p^{n-1} Q_-^{-\frac{1}{2}n}$$

must be an unessential function, as is easily verified. In fact for this purpose it is sufficient to show that

$$\int_\Gamma Q_-^{-\frac{1}{2}n}(\xi)\omega(\xi) = 0, \tag{7}$$

where $\Gamma$ is any surface enclosing the origin in $\xi$ space. The integral here must be understood in terms of its regularization, namely,

$$\int_\Gamma Q_-^{-\frac{1}{2}n}(\xi)\omega(\xi) = \frac{(-1)^{\frac{1}{2}(n+1)}}{2i} \left[\int_\Gamma (Q+i0)^{-\frac{1}{2}n}\omega(\xi) - \int_\Gamma (Q-i0)^{-\frac{1}{2}n}\omega(\xi)\right].$$

The integrals on the right-hand side converge, and it is easily shown that

$$\int (Q+i0)^{-\frac{1}{2}n}\omega(\xi) = \int (Q-i0)^{-\frac{1}{2}n}\omega(\xi),$$

which completes the proof. #

We thus find that the Radon transform of the characteristic function of the cone is

$$\frac{\pi^{\frac{1}{2}(n-1)}}{\Gamma(\frac{1}{2}n + \frac{1}{2})} p^{n-1} Q_+^{-\frac{1}{2}n} - \frac{(-1)^{\frac{1}{2}(n-1)} \pi^{\frac{1}{2}(n-3)}}{\Gamma(\frac{1}{2}n + \frac{1}{2})} p^{n-1} Q_-^{-\frac{1}{2}n} \ln \frac{p^2}{Q_-} . \tag{8}$$

The results of this section can be used to express the integral of a function of $x$ over any of the regions discussed in terms of its Radon transform. For instance, let $\varphi(\xi, p)$ be the Radon transform of $f(x)$. Then

$$\int_{x_1^2 - x_2^2 - \cdots - x_n^2 > 0} f(x)\, dx = \frac{(-1)^{\frac{1}{2}(n-1)}}{2(2\pi)^{n-1}} \int_\Gamma \left[ \int_{-\infty}^{\infty} S(\xi, p) \frac{\partial^{n-1} \varphi(\xi, p)}{\partial p^{n-1}}\, dp \right] \omega(\xi), \tag{9}$$

where $S(\xi, p)$ is given by (8).

### 3.4. Radon Transform of a Delta Function Concentrated on a Quadratic Surface

Consider the following problem. We wish to express the integral of some $f(x)$ over a quadratic surface in terms of its integrals over hyperplanes. As before, we shall assume the space to have odd dimension. By using Plancherel's theorem, we can reduce this problem to calculating the Radon transform of the $\delta$ function concentrated on this surface.

Let us first find the Radon transform of

$$\delta(x_1^2 + \cdots + x_n^2 - 1),$$

which is concentrated on the surface of the unit sphere. Now it has been established that

$$\delta(x_1^2 + \cdots + x_n^2 - 1) = \operatorname*{res}_{\lambda=-1} (x_1^2 + \cdots + x_n^2 - 1)_+^\lambda ,$$

so that we may use the results of Section 3.2, from which it is found that the Radon transform is

$$\frac{(-1)^{\frac{1}{2}(n-1)} \pi^{\frac{1}{2}(n-1)}}{(\frac{1}{2}n - \frac{3}{2})!} (\xi_1^2 + \cdots + \xi_n^2)^{-\frac{1}{2}n+1} (p^2 - \xi_1^2 - \cdots - \xi_n^2)_+^{\frac{1}{2}(n-3)}. \tag{1}$$

We could have also attempted to calculate this Radon transform by using the fact that

$$\delta(x_1^2 + \cdots + x_n^2 - 1) = \operatorname*{res}_{\lambda=-1} (x_1^2 + \cdots + x_n^2 - 1)_-^\lambda .$$

We would then have obtained

$$\frac{\pi^{\frac{1}{2}(n-1)}}{(\frac{1}{2}n-\frac{3}{2})!}(\xi_1^2+\cdots+\xi_n^2)^{-\frac{1}{2}n+1}(p^2-\xi_1^2-\cdots-\xi_n^2)_-^{\frac{1}{2}(n-3)}. \qquad (1')$$

Appearances to the contrary, these two results are not contradictory. Indeed, their difference is

$$\frac{\pi^{\frac{1}{2}(n-1)}}{(\frac{1}{2}n-\frac{3}{2})!}(\xi_1^2+\cdots+\xi_n^2)^{-\frac{1}{2}n+1}(p^2-\xi_1^2-\cdots-\xi_n^2)^{\frac{1}{2}(n-3)},$$

a polynomial of degree $n-3$ in $p$, which is therefore an unessential function.

Thus if one wishes to calculate the integral of some $f(x)$ over a sphere, it is sufficient to know its integrals either only over hyperplanes that intersect the sphere or only over hyperplanes that do not intersect the sphere.

In a similar way, the results of Section 3.2 can be used to find that the Radon transform of $\delta(x_1^2-x_2^2-\cdots-x_n^2+1)$, which is concentrated on the surface of a one-sheeted hyperboloid, is

$$\frac{(-1)^{\frac{1}{2}(n-1)}\pi^{\frac{1}{2}(n-3)}}{(\frac{1}{2}n-\frac{3}{2})!}Q_-^{-\frac{1}{2}n+1}(p^2+Q)^{\frac{1}{2}(n-3)}\ln\left|\frac{p^2+Q}{Q}\right|, \qquad (2)$$

where $Q=\xi_1^2-\xi_2^2-\cdots-\xi_n^2$.

Thus if one wishes to express the integral of some $f(x)$ over the surface of a single-sheeted hyperboloid in terms of its integrals over hyperplanes, one need not know its integrals over those hyperplanes whose intersection with the hyperboloid has finite area.

The Radon transform of $\delta(x_1^2-x_2^2-\cdots-x_n^2-1)$, which is concentrated on the surface of a two-sheeted hyperboloid, is given by either of the two equivalent expressions

$$\frac{\pi^{\frac{1}{2}(n-1)}}{(\frac{1}{2}n-\frac{3}{2})!}Q_+^{-\frac{1}{2}n+1}(p^2-Q)_+^{\frac{1}{2}(n-3)} + \frac{(-1)^{\frac{1}{2}(n-1)}\pi^{\frac{1}{2}(n-3)}}{(\frac{1}{2}n-\frac{3}{2})!}Q_-^{-\frac{1}{2}n+1}(p^2-Q)^{\frac{1}{2}(n-3)}\ln\left|\frac{p^2-Q}{Q}\right| \qquad (3)$$

or

$$\frac{(-1)^{\frac{1}{2}(n-1)}\pi^{\frac{1}{2}(n-1)}}{(\frac{1}{2}n-\frac{3}{2})!}Q_+^{-\frac{1}{2}n+1}(p^2-Q)_-^{\frac{1}{2}(n-3)} + \frac{(-1)^{\frac{1}{2}(n-1)}\pi^{\frac{1}{2}(n-3)}}{(\frac{1}{2}n-\frac{3}{2})!}Q^{-\frac{1}{2}n+1}(p^2-Q)^{\frac{1}{2}(n-3)}\ln\left|\frac{p^2+Q}{Q}\right|. \qquad (3')$$

Thus if one wishes to calculate the integral of some $f(x)$ over the surface of a two-sheeted hyperboloid, one need not know its integrals either over hyperplanes that do not intersect the hyperboloid or over hyperplanes whose intersection has finite area.

We now turn to the Radon transform of

$$\delta(x_1^2 - x_2^2 - \cdots - x_n^2),$$

which is concentrated on the surface of a cone. It satisfies the two relations[36]

$$\delta(x_1^2 - x_2^2 - \cdots - x_n^2) = \operatorname*{res}_{\lambda=-1} (x_1^2 - x_2^2 - \cdots - x_n^2)_+^\lambda$$

$$= \operatorname*{res}_{\lambda=-1} (x_1^2 - x_2^2 - \cdots - x^2)_-^\lambda .$$

According to Section 3.1, the Radon transform of $(x_1^2 - x_2^2 - \cdots - x_n^2)_-^\lambda$ is

$$\frac{(-1)^{\frac{1}{2}(n-1)}\pi^{\frac{1}{2}(n-1)}}{(\lambda+1)\cdots(\lambda+\frac{1}{2}[n-1])} \cot \pi\lambda \mid p \mid^{2\lambda+n-1}(\xi_1^2 - \xi_2^2 - \cdots - \xi_n^2)_-^{-\lambda-\frac{1}{2}n}.$$

From this it follows that the Radon transform of $\delta(x_1^2 - x_2^2 - \cdots - x_n^2)$ is[37]

$$\frac{(-1)^{\frac{1}{2}(n-1)}2\pi^{\frac{1}{2}(n-3)}}{(\frac{1}{2}n - \frac{3}{2})!} p^{n-3} \ln \mid p \mid Q^{-\frac{1}{2}n+1}. \tag{4}$$

Equation (4) can then be used to write the integral of $f(x)$ over the surface of a hypercone in terms of its Radon transform $\check{f}(\xi, p)$:

$$\int_{x_1^2-x_2^2-\cdots-x_n^2=0} f(x)\,dx = \frac{2^{-n+1}\pi^{-\frac{1}{2}(n+1)}}{(\frac{1}{2}n-\frac{3}{2})!}\int_\Gamma\int_{-\infty}^{\infty} p^{n-3}\ln\mid p\mid(\xi_1^2-\xi_2^2-\cdots-\xi_n^2)_-^{-\frac{1}{2}n+1}$$

$$\times \check{f}_p^{(n-1)}(\xi, p)\,dp\,\omega(\xi).$$

We see that this does not require knowing the integrals of $f(x)$ over hyperplanes tangent to the cone or over hyperplanes parallel to these.

[36] See Volume 1, p. 278.

[37] We have dropped the unessential term

$$cp^{n-3}Q_-^{-\frac{1}{2}n+1}\ln Q_- .$$

## 4. Summary of Radon Transform Formulas

I. NOTATION

$f(x)$ is a function in $n$-dimensional affine space, where we write $x = (x_1, ..., x_n)$, $dx \equiv dx_1 \cdots dx_n$. $\check{f}(\xi, p) \equiv [f(x)]^{\vee}$ is the Radon transform of $f(x)$.

$$\xi = (\xi_1, ..., \xi_n), \qquad (\xi, x) \equiv \xi_1 x_1 + \cdots + \xi_n x_n,$$

$$\omega(\xi) = \sum (-1)^{k-1} \xi_k \, d\xi_1 \cdots d\xi_{k-1} \, d\xi_{k+1} \cdots d\xi_n,$$

$$\theta(t) = \begin{cases} 1, & \text{for } t > 0, \\ 0, & \text{for } t < 0, \end{cases}$$

$P(x)$ is a nondegenerate quadratic form, and $Q(x)$ is its dual, i.e. the quadratic from whose matrix is the inverse of that of $P(x)$.

II. DEFINITION OF RADON TRANSFORM

$$\check{f}(\xi, p) = \int f(x) \, \delta(p - (\xi, x)) \, dx.$$

III. FUNDAMENTAL PROPERTIES OF RADON TRANSFORMS

1. $\check{f}(\alpha\xi, \alpha p) = |\alpha|^{-1} \check{f}(\xi, p)$, for any $\alpha \neq 0$;
2. $\{f(x + a)\}^{\vee} = \check{f}(\xi, p + (\xi, a))$;
3. $\{f(A^{-1}x)\}^{\vee} = |\det A| \, \check{f}(A'\xi, p)$,

where $A$ is a nonsingular linear transformation and $A'$ is its transpose.

4. $$\left\{ \left(a, \frac{\partial}{\partial x}\right) f(x) \right\}^{\vee} = (a, \xi) \frac{\partial \check{f}(\xi, p)}{\partial p}.$$

5. $$\frac{\partial}{\partial p} \{(a, x) f(x)\}^{\vee} = -\left(a, \frac{\partial}{\partial \xi}\right) \check{f}(\xi, p).$$

6. $$\{f_1 * f_2\}^{\vee} = \int_{-\infty}^{\infty} \check{f}_1(\xi, t) \check{f}_2(\xi, p - t) \, dt.$$

IV. INVERSE RADON TRANSFORM FORMULA

For dimension $n$ odd we have

$$f(x) = \frac{(-1)^{\frac{1}{2}(n-1)}}{2(2\pi)^{n-1}} \int_{\Gamma} \check{f}_p^{(n-1)}(\xi, (\xi, x)) \omega(\xi).$$

For dimension $n$ even we have

$$f(x) = \frac{(-1)^{\frac{1}{2}n}(n-1)!}{(2\pi)^n} \int_\Gamma \left\{ \int_{-\infty}^{\infty} \check{f}(\xi, p)[p - (\xi, x)]^{-n} \, dp \right\} \omega(\xi).$$

Here $\Gamma$ is any surface enclosing the origin.

V. ANALOG OF PLANCHEREL'S THEOREM

For dimension $n$ odd we have

$$\int f(x)\bar{g}(x) \, dx = \tfrac{1}{2}(2\pi)^{1-n} \int_\Gamma \left\{ \int_{-\infty}^{\infty} \check{f}_p^{(m)}(\xi, p)\bar{\check{g}}_p^{(m)}(\xi, p) \, dp \right\} \omega(\xi),$$

where $m = \frac{1}{2}(n-1)$. For dimension $n$ even we have

$$\int f(x)\bar{g}(x) \, dx = \frac{(-1)^{\frac{1}{2}n}(n-1)!}{(2\pi)^{-n}} \int_\Gamma \left\{ \int\int \check{f}(\xi, p_1)\bar{\check{g}}(\xi, p_2)(p_1 - p_2)^{-n} \, dp_1 \, dp_2 \right\} \omega(\xi).$$

## VI. Table of Radon Transforms of Some Generalized Functions for Dimension $n$ Odd

The Radon transform of a generalized function is defined in Section 2.1 of this chapter. The asterisk denotes results not derived in the book.

| Entry no. | Function | Radon transform |
|---|---|---|
| 1 | 1* | $p^{n-1}a(\xi)$, where $a(\xi)$ is any even function, homogeneous of degree $-n$, such that $\int_\Gamma a(\xi)\,\omega(\xi) = (-1)^{\frac{1}{2}(n-1)}2^n\pi^{n-1}\Gamma^{-1}(n)$ |
| 2 | $\theta(x_1)$* | $\frac{1}{2}\pi^{n-\frac{3}{2}}\Gamma(1-\frac{1}{2}n)\,\Gamma^{-1}(\frac{1}{2}n+\frac{1}{2}) \times [p_+^{n-1}(\xi_1)_+^{-1} + p_-^{n-1}(\xi_1)_-^{-1}]\,\delta(\xi_2, \ldots, \xi_n)$ |
| 3 | $\delta(x_1, \ldots, x_n)$ | $\delta(p)$ |
| 4 | $\delta(x_1, \ldots, x_k)$*, $k$ odd $(k < n)$ | $\pi^{n-k-\frac{1}{2}}\Gamma\left(\frac{k-n+1}{2}\right)\Gamma^{-1}\left(\frac{n-k}{2}\right)\lvert p\rvert^{n-k-1} \times \delta(\xi_{k+1}, \ldots, \xi_n)$ |
| 5 | $\delta(x_1, \ldots, x_k)$*, $k$ even | $(-1)^{\frac{1}{2}(n-k+1)}2\pi^{n-k-\frac{1}{2}}\Gamma^{-1}\left(\frac{n-k+1}{2}\right) \times \Gamma^{-1}\left(\frac{n-k}{2}\right)p^{n-k-1}\ln\lvert p\rvert\,\delta(\xi_{k+1}, \ldots, \xi_n)$ |
| 6 | $a(x_1)\,\delta(x_2, \ldots, x_n)$, where $\int_{-\infty}^{\infty}\lvert a(x_1)\rvert\,dx_1 < \infty$ | $\lvert\xi_1\rvert^{-1}a(p/\xi_1)$ |
| 7 | $(x_1)_+^\lambda\,\delta(x_2, \ldots, x_n)$, $\lambda$ noninteger | $p_+^\lambda\,(\xi_1)_+^{-1-\lambda} + p_-^\lambda\,(\xi_1)_-^{-1-\lambda}$ |
| 8 | $(x_1)_+^k\,\delta(x_2, \ldots, x_n)$, $k = 0, 1, \ldots$ | $p_+^k(\xi_1)_+^{-1-k} + p_-^k(\xi_1)_-^{-1-k} - \frac{(-1)^k}{k!}p^k\ln\lvert p\rvert\,\delta^{(k)}(\xi_1)$ |
| 9 | $\lvert x_1\rvert^\lambda\delta(x_2, \ldots, x_n)$, $\lambda \neq -1, -2, \ldots$, $\lambda \neq 2k$ $(k = 0, 1, \ldots)$ | $\lvert p\rvert^\lambda\,\lvert\xi_1\rvert^{-1-\lambda}$ |

*Continued*

| Entry no. | Function | Radon transform |
|---|---|---|
| 10 | $x_1^{2k}\delta(x_2, \ldots, x_n)$<br>$k = 0, 1, \ldots$ | $p^{2k} \| \xi_1 \|^{-1-2k} - \frac{2}{(2k)!} p^{2k} \ln \| p \| \delta^{(2k)}(\xi_1)$ |
| 11 | $\| x_1 \|^{\lambda} \operatorname{sgn} x_1 \delta(x_2, \ldots, x_n)$<br>$\lambda \neq -1, -2, \ldots$<br>$\lambda \neq 2k - 1 \quad (k = 1, 2, \ldots)$ | $p \|^{\lambda} \operatorname{sgn} p \cdot \| \xi_1 \|^{-1-\lambda} \operatorname{sgn} \xi_1$ |
| 12 | $x_1^{2k-1}\delta(x_2, \ldots, x_n)$<br>$k = 1, 2, \ldots$ | $p^{2k-1} \| \xi_1 \|^{-2k} \operatorname{sgn} \xi_1$<br>$+ \frac{2}{(2k-1)!} p^{2k-1} \ln \| p \| \delta^{(2k-1)}(\xi_1)$ |
| 13 | Characteristic function of the positive cone in three dimensions.<br>$x_1^2 - x_2^2 - x_3^2 > 0, \quad x_1 > 0$ | $\pi[p_+^2 \theta(\xi_1) + p_-^2 \theta(-\xi_1)](\xi_1^2 - \xi_2^2 - \xi_3^2)_+^{-\frac{3}{2}}$<br>$+ p^2 \ln \| p \| (\xi_1^2 - \xi_2^2 - \xi_3^2)_-^{-\frac{3}{2}}$ |
| 14 | Characteristic function of the upper sheet of a hyperboloid in three dimensions.<br>$x_1^2 - x_2^2 - x_3^2 > 1, \quad x_1 > 0$ | $\pi\theta(\xi, p)(\xi_1^2 - \xi_2^2 - \xi_3^2)_+^{-\frac{3}{2}}(p^2 - \xi_1^2 + \xi_2^2 + \xi_3^2)_+$<br>$+ \frac{1}{2}(\xi_1^2 - \xi_2^2 - \xi_3^2)_-^{-\frac{3}{2}}(p^2 - \xi_1^2 + \xi_2^2 + \xi_3^2)$<br>$\times \ln \| p^2 - \xi_1^2 + \xi_2^2 + \xi_3^2 \|$ |
| 15 | $P_+^{\lambda}(x)$ | $\frac{(-1)^{\frac{1}{2}(n-1)} \pi^{\frac{1}{2}(n-1)} \| p \|^{2\lambda+n-1}}{(\lambda + 1) \cdots (\lambda + \frac{1}{2}n - \frac{1}{2}) \| \Delta \|^{\frac{1}{2}} \sin \pi\lambda}$<br>$\times [\sin \pi(\frac{1}{2}l + \lambda) Q_+^{-\lambda-\frac{1}{2}n}(\xi)$<br>$-\sin \frac{1}{2}\pi k Q_-^{-\lambda-n\frac{1}{2}}(\xi)],$<br>where $k$ and $l$ are, respectively, the number of positive and negative terms in the canonical form of $P(x)$, and $\Delta$ is its discriminant. |

*Continued*

| Entry no. | Function | Radon transform |
|---|---|---|
| 16 | $\theta(x_1)(x_1^2 - x_2^2 - \cdots - x_n^2)_+^{\lambda *}$ | $\dfrac{\pi^{\frac{1}{2}(n-1)}[p_+^{2\lambda+n-1}\theta(\xi_1) + p_-^{2\lambda+n-1}\theta(-\xi_1)]}{(\lambda+1)\cdots(\lambda+\frac{1}{2}n-\frac{1}{2})}$<br>$\times\,(\xi_1^2 - \xi_2^2 - \cdots - \xi_n^2)_+^{-\lambda-\frac{1}{2}n}$<br>$-\dfrac{(-1)^{\frac{1}{2}(n-1)}\pi^{\frac{1}{2}(n-1)}\lvert p\rvert^{2\lambda+n-1}}{2(\lambda+1)\cdots(\lambda+\frac{1}{2}n-\frac{1}{2})\sin\pi\lambda}$<br>$\times\,(\xi_1^2 - \xi_2^2 - \cdots - \xi_n^2)_-^{-\lambda-\frac{1}{2}n}$ |
| 17 | $[P(x) + c]_+^{\lambda}$ | $\dfrac{(-1)^{\frac{1}{2}(n-1)}\pi^{\frac{1}{2}(n-1)}}{(\lambda+1)\cdots(\lambda+\frac{1}{2}n-\frac{1}{2})\lvert\Delta\rvert^{\frac{1}{2}}\sin\pi\lambda}$<br>$\times\,[\sin\pi(\tfrac{1}{2}l+\lambda)Q_+^{-\lambda-\frac{1}{2}n}(p^2+cQ)_+^{\lambda+\frac{1}{2}(n-1)}$<br>$-\sin\tfrac{1}{2}\pi k\,Q_-^{-\lambda-\frac{1}{2}n}(p^2+cQ)_+^{\lambda+\frac{1}{2}(n-1)}$<br>$+\sin\pi(\tfrac{1}{2}l-\tfrac{1}{2}+\lambda)Q_-^{-\lambda-\frac{1}{2}n}(p^2+cQ)_-^{\lambda+\frac{1}{2}(n-1)}$<br>$-\sin\tfrac{1}{2}\pi(k-1)Q_+^{-\lambda-\frac{1}{2}n}(p^2+cQ)_-^{\lambda+\frac{1}{2}(n-1)}$ |
| 18 | $\theta(x_1)(x_1^2 - x_2^2 - x_3^2 - 1)_+^{\lambda *}$ | $\dfrac{\pi}{\lambda+1}\theta(p\xi_1)(\xi_1^2 - \xi_2^2 - \xi_3^2)_+^{-\lambda-\frac{3}{2}}$<br>$\times\,(p^2 - \xi_1^2 + \xi_2^2 + \xi_3^2)_+^{\lambda+1} + \dfrac{\pi}{2(\lambda+1)\sin\pi\lambda}$<br>$\times\,(\xi_1^2 - \xi_2^2 - \xi_3^2)_-^{-\lambda-\frac{3}{2}}(p^2 - \xi_1^2 + \xi_2^2 + \xi_3^2)_+^{\lambda+1}$ |
| 19 | $\delta(x_1^2 - x_2^2 - \cdots - x_n^2)$ | $(-1)^{\frac{1}{2}(n-1)}2\pi^{\frac{1}{2}(n-3)}\Gamma^{-1}(\tfrac{1}{2}n-\tfrac{1}{2})p^{n-3}\ln\lvert p\rvert$<br>$\times\,(\xi_1^2 - \xi_2^2 - \cdots - \xi_n^2)_-^{\frac{1}{2}n+1}$ |
| 20 | $\delta(x_1^2 + \cdots + x_n^2 - 1)$ | $(-1)^{\frac{1}{2}(n-1)}\pi^{\frac{1}{2}(n-1)}\Gamma^{-1}(\tfrac{1}{2}n-\tfrac{1}{2})$<br>$\times\,(\xi_1^2 + \cdots + \xi_n^2)^{-\frac{1}{2}n+1}(p^2 - \xi_1^2 - \cdots - \xi_n^2)_+^{\frac{1}{2}(n-3)}$ |

*Continued*

| Entry no. | Function | Radon transform |
|---|---|---|
| 21 | $\delta(x_1^2 - x_2^2 - \cdots - x_n^2 - 1)$ | $\pi^{\frac{1}{2}(n-1)}\Gamma^{-1}(\frac{1}{2}n - \frac{1}{2})(\xi_1^2 - \xi_2^2 - \cdots - \xi_n^2)_+^{-\frac{1}{2}n+1}$ $\times (p^2 - \xi_1^2 + \xi_2^2 + \cdots + \xi_n^2)_+^{\frac{1}{2}(n-3)}$ $+(-1)^{\frac{1}{2}(n-1)}\pi^{\frac{1}{2}(n-3)}\Gamma^{-1}(\frac{1}{2}n - \frac{1}{2})$ $\times (\xi_1^2 - \xi_2^2 - \cdots - \xi_n^2)_-^{-\frac{1}{2}n+1}$ $\times (p^2 - \xi_1^2 + \xi_2^2 + \cdots + \xi_n^2)^{\frac{1}{2}(n-3)}$ $\times \ln\left\lvert\dfrac{p^2 - \xi_1^2 + \xi_2^2 + \cdots + \xi_n^2}{\xi_1^2 - \xi_2^2 - \cdots - \xi_n^2}\right\rvert$ |

CHAPTER II

# INTEGRAL TRANSFORMS IN THE COMPLEX DOMAIN

This chapter will be concerned with an analysis of some problems of integral geometry in the complex domain.

We first state the general problem of integral geometry. Consider some space $X$ in which are given certain manifolds, assumed to be analytic and to depend analytically on certain parameters $\lambda_1, \ldots, \lambda_k$:

$$\mathfrak{M} = \mathfrak{M}(\lambda) = \mathfrak{M}(\lambda_1, \ldots, \lambda_k).$$

With a function $f(x)$ defined on $X$ we associate its integrals over these manifolds:

$$\varphi(\lambda) = \int_{\mathfrak{M}(\lambda)} f(x)\, d\sigma. \tag{1}$$

This gives a new function $\varphi(\lambda)$ defined on the set of manifolds. Our problem shall be the following: Given $\varphi$, find the original function $f(x)$. More exactly, we shall be interested in the following questions.

1. When is the mapping $f \to \varphi$ defined by (1) for "arbitrary" $f(x) \in X$ a one-to-one mapping? When this mapping is one-to-one, what is the inversion formula that gives $f(x)$ in terms of $\varphi(\lambda)$?

2. What functions of $\lambda$ can be represented by an integral such as (1) with $f(x)$ an "arbitrary" function on $X$?

Although speaking of "arbitrary" functions, we shall in fact deal with classes of functions defined either by some smoothness conditions or by asymptotic properties, but in any case classes containing locally all infinitely differentiable functions.[1] Such a class may be, for instance, the

[1] That is, every point has a neighborhood $U$ such that for every infinitely differentiable function $\varphi$ on $U$ the class contains a function of $\psi$ that coincides with $\varphi$ on $U$.

class of all infinitely differentiable functions of bounded support or the class of rapidly decreasing infinitely differentiable functions.

One would naturally expect that to specify uniquely a function of $n$ variables requires a function of the same number of variables. In other words, in a reasonably stated problem of integral geometry, the manifolds $\mathfrak{M}(\lambda)$ over which we integrate in Eq. (1) should depend on $n$ parameters, where $n$ is the dimension of $X$. It is just these problems which form the basic content of integral geometry as we understand it in this work. Previously integral geometry was understood in a different sense, being involved primarily with calculating invariant measures in one or another homogeneous space (see, for instance, Blyashke [2]).

Problems in integral geometry arise in a natural way in dealing with the representation theory of Lie groups and essentially form the basis of this theory. The reader will understand this connection when he has read Chapters IV and VI of this volume.

Every problem of integral geometry has a local analog, in which the manifolds are replaced by tangent planes. The local problem of integral geometry has been solved in its entirety only for the simplest case of a family of $(n-1)$-dimensional hyperplanes in an $n$-dimensional affine space. It would be very interesting if some one were to come forward with a solution of this local problem in the general case of a family of $k$-dimensional hyperplanes in an $n$-dimensional affine space. One should start, of course, with a family of lines. It would be most reasonable to start with the problem in a complex space, since the complex case is actually simpler than the real one. So far only the first step has been taken in this direction: the problem of integral geometry has been solved for several families of lines in three complex dimensions. The solution will be given in Section 1 of this chapter.

Section 2 of this chapter gives the solution of the problem of integral geometry for a family of lines lying on a quadric in four complex dimensions. This problem has not been chosen randomly. Later, in Chapter V, the reader will see that it lies at the basis of the theory of the representations of the Lorentz group.

In Section 3 we study the Radon transform in the complex domain. That section could have been treated as an appendix to Chapter I, which deals with the Radon transform in the real domain. In the real case most of the formulas involving the Radon transform differ for spaces of even and odd dimension. It will be seen that the complex case is more closely related to odd dimension than to the more refined case of even dimension.

## 1. Line Complexes in a Space of Three Complex Dimensions and Related Integral Transforms

In this section we shall consider the following problem of integral geometry. Given a certain family of lines in an affine space of three complex dimensions, we are asked to construct a function $f(z) = f(z_1, z_2, z_3)$ from a knowledge of its integrals over these lines. Since $f$ depends on three complex variables, we shall deal with three-parameter families of lines, which we shall call line complexes. The solution will be given here for a certain special class of line complexes for which the solution is simpler and such that the formula giving $f$ in terms of its integrals is a local one (see Section 1.4).

### 1.1. Plücker Coordinates of a Line

We shall describe lines by means of their *Plücker coordinates*, which are defined in the following way. Let

$$(z_1, z_2, z_3) \quad \text{and} \quad (z_1', z_2', z_3')$$

be any two points on a line. The Plücker coordinates of the line are the six numbers

$$\begin{aligned} \alpha_1 &= z_1' - z_1, & p^1 &= z_2' z_3 - z_3' z_2, \\ \alpha_2 &= z_2' - z_2, & p^2 &= z_3' z_1 - z_1' z_3, \\ \alpha_3 &= z_3' - z_3, & p^3 &= z_1' z_2 - z_2' z_1. \end{aligned} \tag{1}$$

The vectors $\alpha = (\alpha_1, \alpha_2, \alpha_3)$ and $p = (p^1, p^2, p^3)$ have the following geometrical meanings: $\alpha$ is a direction-vector of the line, and $p = [\alpha, z]$ is the *vector product* of $\alpha$ with the "radius vector" $z$ of any point on the line. Obviously if $\alpha$, $p$ are the Plücker coordinates of a line, any point $z$ on the line satisfies the equation

$$p = [\alpha, z]. \tag{2}$$

This implies that the line is uniquely determined by its Plücker coordinates. Conversely, the Plücker coordinates are determined uniquely by the line *to within a constant factor*. In other words, the Plücker coordinates are homogeneous coordinates of the line. Indeed, let $p = [\alpha, z]$ and $p' = [\alpha', z']$ be two equations for the same line. Since $\alpha$ and $\alpha'$ are both direction vectors of the line, we have

$$\alpha' = \lambda\alpha, \qquad \lambda \neq 0.$$

But this means that $p' = \lambda[\alpha, z] = \lambda p$. Thus if $\alpha, p$ and $\alpha', p'$ are both the Plücker coordinates of a line, they are proportional.

The coordinates $\alpha, p$ are related to each other according to

$$(\alpha, p) \equiv \alpha_1 p^1 + \alpha_2 p^2 + \alpha_3 p^3 = 0, \tag{3}$$

as follows immediately from the fact that $p = [\alpha, z]$ is orthogonal to $\alpha$. It is easily shown, conversely, that given any six numbers $\alpha_i$ and $p^j$ satisfying (3), they can be considered the Plücker coordinates of some line. The equation of this line is then given by (2).[2]

Under affine transformations, the Plücker coordinates are transformed as follows:

I. Under the translation

$$z' = z + z_0$$

the Plücker coordinates transform according to

$$\alpha' = \alpha, \qquad p' = p + [\alpha, z_0].$$

II. Under the centroaffine (homogeneous) transformation

$$z'_i = a^j_i z_j$$

the Plücker coordinates transform according to

$$\alpha'_i = a^j_i \alpha_i\,, \qquad p'^i = b^j_i p^j,$$

where $b^i_j$ is the cofactor of $a^i_j$.

These formulas follow simply from the definitions of the Plücker coordinates.

### 1.2. Line Complexes

The set of all lines in a three-dimensional complex space can be considered a manifold. Specifically, we shall treat the set of numbers $(\alpha_1, \alpha_2, \alpha_3, p^1, p^2, p^3)$ as homogeneous coordinates of points in a five-dimensional complex projective space. These will be the Plücker coordinates of a line if and only if

$$(\alpha, p) = 0. \tag{1}$$

[2] Strictly speaking we shall always be assuming also that $\alpha \neq 0$. This assumption need not, however, be made, for we may speak of lines in a projective, rather than in an affine space. Then $\alpha = 0$ denotes the lines at infinity.

Thus the set of all lines in a three-dimensional complex space generates the quadratic surface given by (1) in five-dimensional complex projective space. From this surface we omit the points corresponding to lines at infinity, i.e., points for which $\alpha = 0$.

In what follows we shall consider *line complexes*, i.e., families of lines depending on three complex parameters. Every line complex is given by a single equation of the form

$$F(\alpha, p) = 0. \tag{2}$$

relating its Plücker coordinates, which must not be a consequence of (1) ($F$ is a homogeneous function). The only complexes we shall deal with will be algebraic; in other words, we shall assume that $F$ is a (homogeneous) polynomial in irreducible form. The degree of $F$ will be called the order of the complex.

The lines tangent to an arbitrary (algebraic) surface are an example of a complex. Another example is the complex of lines intersecting a given algebraic curve (located, perhaps, at infinity). A third is formed by the lines on a one-parameter family of planes. Complexes of the second and third types can be treated as degenerate cases of those of the first type in which the surface "contracts" to a line or when it is a developable surface.

We list some properties of line complexes.

The set of all lines of a complex passing through a given point $z_0$ form a one-parameter set. This is because if a line $\alpha$, $p$ passes through $z_0$, then

$$p = [\alpha, z_0], \tag{3}$$

which imposes two additional conditions on the Plücker coordinates.

Consider the equation of the cone generated by the lines of a complex passing through some point $z_0$. Let Eq. (2) be the equation of the complex. If $z$ is a point of the cone, the Plücker coordinates of the generator passing through $z$ will be

$$\alpha = z - z_0, \qquad p = [\alpha, z_0] = [z, z_0].$$

We now insert these into (2) and find that the cone generated by the lines of the $F(\alpha, p) = 0$ complex that pass through $z_0$ is given by

$$F(z - z_0, [z, z_0]) = 0. \tag{4}$$

Obviously this conical surface is in general of the same order as the original complex.

It is easily shown also that the lines of a complex with a given direction vector $\alpha$ form a one-parameter set, unless the equation of the complex does not depend explicitly on $p$. In this exceptional case the complex consists of all lines parallel to the generators of a certain cone. Therefore either it contains no lines with a given direction vector, or else all lines with that direction vector belong to the complex.

The equation of the cylindrical surface formed by all lines of a complex with a given direction $\alpha$ is obtained by replacing $p$ by $p = [\alpha, z]$ in Eq. (2). This yields

$$F(\alpha, [\alpha, z]) = 0. \tag{5}$$

Obviously the order of this surface is no greater than the degree of $F$ with respect to the Plücker coordinates $p^1, p^2, p^3$.

### 1.3. A Special Class of Complexes

Let us now subject our complexes to an additional requirement, which we first formulate geometrically.

Consider any two points $z^{(1)}$ and $z^{(2)}$ on one of the lines of a complex, together with all those lines of the complex which pass through $z^{(1)}$ or $z^{(2)}$ or both. This will yield two cones with vertices at the two chosen points (except for the special case in which every line passing through one of these points lies in the complex). The original line, passing through both these points, is a common generator of the cones. We shall require that *any two such cones constructed of lines of the complex and having a common generator must also have a common tangent plane along this generator*. (We shall consider this requirement fulfilled formally in the special case in which at least one of the pairs of cones degenerates to the set of all lines passing through the given point.)

Shortly we shall formulate this condition algebraically, but first note that if the vertex of a cone constructed of lines of the complex is allowed to move off to infinity, in the limit the cone is transformed to a cylinder. Thus for the special class of complexes we are now considering, a cylinder and cone, both composed of lines of the complex and having a common generator, will also be tangent along this common generator.

Section 1.2 contained several examples of complexes. All these examples satisfy this requirement.[3] Let us verify that this is the case for the complex of lines tangent to a certain surface $S$. Consider two cones tangent to $S$ and having a common generator $l$. Obviously the

[3] The converse can also be shown to be true; namely, it can be shown that the three examples given exhaust all irreducible complexes satisfying our additional requirement.

plane tangent to $S$ at the point where $l$ is tangent to $S$ is simultaneously tangent to both cones of which $l$ is the common generator.

It is easily shown also that the complex of lines intersecting a given curve also satisfies our requirement. Note, however, that a special case occurs when one of the points chosen is on the curve; in this case every line passing through the point belongs to the complex.

We now turn to the algebraic statement of this condition. Let $F(\alpha, p) = 0$ be the equation of the complex, where $F$ is an irreducible polynomial. *In order that two arbitrary cones constructed of lines of the complex and having a common generator be tangent to each other, it is necessary and sufficient that the Plücker coordinates of the complex satisfy the relation*

$$(F_\alpha, F_p) \equiv \sum_{i=1}^{3} F_{\alpha_i}(\alpha, p) F_{p^i}(\alpha, p) = 0. \tag{1}$$

In other words it is necessary and sufficient that Eq. (1) be a consequence of $F(\alpha, p) = 0$ and of the orthogonality condition $(\alpha, p) = 0$.

For the proof, consider two cones generated by lines of the complex, one with vertex at $z^0$ and the other with vertex at $z'$. Their equations are

$$F(z - z^0, [z, z^0]) = 0 \tag{2}$$

and

$$F(z - z', [z, z']) = 0. \tag{2'}$$

Now let the line passing through $z^0$ and $z'$, namely, the line whose Plücker coordinates are

$$\alpha = z' - z^0, \qquad p = [z', z^0],$$

belong to the complex, so that $F(z' - z^0, [z', z^0]) = 0$, which implies that it is a common generator. The direction or orientation of the plane tangent along $\alpha, p$ to the cone whose equation is (2) is given by the partial derivatives of $\Phi(z) \equiv F(z - z^0, [z, z^0])$, or by the vector[4]

$$F_\alpha(\alpha, p) + [z^0, F_p(\alpha, p)].$$

If (2) and (2′) are to be tangent along their common generator $\alpha, p$, it is necessary and sufficient that $F_\alpha + [z^0, F_p]$ and $F_\alpha + [z', F_p]$ be vectors that are collinear on this generator,[5] that is, that

$$\lambda(F_\alpha + [z^0, F_p]) = \mu(F_\alpha + [z', F_p]). \tag{3}$$

[4] This is a covariant vector. By analogy with the case of a real Euclidean space, it may be called the normal vector.

[5] If $\alpha, p$ is a general line in the complex and $z^0$, $z'$ are general points on this line, these vectors will not vanish. This follows from the irreducibility of $F$.

Now it can be shown that this implies that $(F_\alpha, F_p) = 0$. Indeed, if $\lambda \neq \mu$, we take the scalar product of both sides of (3) with $F_p$, and the result follows immediately. Now let $\lambda = \mu$. Then it follows from (3) that $[\alpha, F_p] = [z' - z^0, F_p] = 0$, which means that $F_p$ is collinear with $\alpha$. On the other hand, by applying Euler's theorem on homogeneous functions to the polynomial $F(\alpha) = F(\alpha, [\alpha, z^0])$, we obtain

$$(\alpha, F_\alpha + [z^0, F_p]) = kF(\alpha) = 0 \tag{4}$$

[where we taken $F(\alpha)$ to be homogeneous of degree $k$ in $\alpha$]. Now we may replace $\alpha$ by the collinear vector $F_p$ in this equation, which again yields $(F_\alpha, F_p) = 0$.

This proves that the condition that cones with a common generator be tangent implies Eq. (1). Conversely, Eq. (1) implies that cones with a common generator are tangent, or that the vectors

$$F_\alpha + [z^0, F_p] \qquad \text{and} \qquad F_\alpha + [z', F_p]$$

are collinear. Indeed, it follows from (1) that

$$(F_p, F_\alpha + [z^0, F_p]) = 0.$$

From this and Eq. (4) (obtained from Euler's theorem on homogeneous functions) we may conclude that $F_\alpha + [z^0. F_p]$ is collinear with $[\alpha, F_p]$. For the same reason $F_\alpha + [z', F_p]$ is also collinear with $[\alpha, F_p]$. If, therefore, $[\alpha, F_p] \neq 0$, these vectors are collinear with each other. In the special case $[\alpha, F_p] \equiv [z', F_p] - [z^0, F_p] = 0$ these vectors simply coincide.

### 1.4. The Problem of Integral Geometry for a Line Complex

Consider a line complex

$$F(\alpha, p) = 0 \tag{1}$$

(where $F$ is an irreducible polynomial) such that the Plücker coordinates of each line in the complex satisfy the relation

$$(F_\alpha, F_p) = \sum_{i=1}^{3} F_{\alpha_i} F_{p^i} = 0. \tag{2}$$

(The geometrical meaning of this requirement has been discussed in Section 1.3.) We shall now solve the following problem of integral geometry for such a complex.

Let $f(z) = f(z_1, z_2, z_3)$ be an infinitely differentiable rapidly decreasing function (recall that all of its derivatives are also rapidly decreasing). We take its integral over each of the lines of the complex, writing

$$\varphi(\alpha, p) = \tfrac{1}{2}i \int f(\alpha t + \beta)\, dt\, d\bar{t}, \tag{3}$$

where $\alpha$, $p$ are the Plücker coordinates of the line, and $\beta$ is any point on the line, that is, $p = [\alpha, \beta]$.[6] The problem is the following. *We wish to be able to determine $f(z)$ if we know $\varphi(\alpha, p)$ for all the lines of the complex.* It will be seen that this problem has a unique solution and that this solution is a local one. By this we mean that in order to find the value of $f(z)$ at some point $z^0$ it is sufficient to know only its integrals over those lines of the complex that pass through $z^0$ and over lines (also in the complex) infinitesimally close to these. Moreover, we shall see that in order to find $f(z)$ at some $z^0$ we need consider only those lines of the complex which are parallel to the ones passing through $z^0$. It will be shown in Section 1.5 that the complexes we are dealing with are in a certain sense the most general complexes for which the problem of integral geometry has a unique and local solution.[7]

We now proceed to solve the problem. Let us attempt to find the value of $f(z)$ at some point $z^0$, proceeding in the following way. We first consider all possible lines of the complex that pass through $z^0$, i.e., lines whose Plücker coordinates $\alpha$ and $p = [\alpha, z^0]$ satisfy the equation $F(\alpha, p) = 0$. As we have seen, these lines form a cone whose equation is

$$F(z - z^0, [z, z^0]) = 0. \tag{4}$$

We now subject each of these lines to an infinitesimal parallel displacement. This will not change $\alpha$, and $p$ will increase by $dp = (dp^1, dp^2, dp^3)$.

We now show that $dp$ is collinear with the vector

$$a = F_\alpha + [z^0, F_p], \tag{5}$$

which gives the direction of the tangent plane of the cone. (An exception will be the case in which the direction of $dp$ is not uniquely defined, but in this case it can always be chosen collinear with $a$ anyway.)

[6] Note that the integral over the line as defined by Eq. (3) depends not only on the line itself, but also on the choice of its Plücker coordinates $\alpha$, $p$, since for any $\lambda \neq 0$ we clearly have

$$\varphi(\lambda\alpha, \lambda p) = \lambda^{-1}\bar{\lambda}^{-1}\varphi(\alpha, p).$$

[7] The simplest complex for which the problem does not have a unique solution is the general complex of order 1 (see Gel'fand and Graev (15)).

To proceed to the proof we remark that $p$ satisfies two relations. The first of these is the identity relating the Plücker coordinates, namely

$$(\alpha, p) = 0, \tag{6}$$

and the second is the requirement that $\alpha, p$ belong to the complex, namely

$$F(\alpha, p) = 0. \tag{7}$$

Differentiation of (6) and (7) leads to the following two relations for $dp$

$$(\alpha, dp) = 0, \tag{8}$$

$$(F_p, dp) = 0, \tag{9}$$

which uniquely determine its direction. An exception occurs when $F_p$ is collinear with $\alpha$. We must now verify that $a$, the vector given by (5), satisfies the same relations as $dp$. It will then follow that $dp$ and $a$ are collinear. But that $(\alpha, a) = 0$ follows immediately from Euler's theorem on homogeneous functions as applied to the polynomial $F(\alpha, [\alpha, z^0])$ [see Eq. (4) of Section 1.3]. Next,

$$(F_p, a) \equiv (F_p, F_\alpha) + (F_p, [z^0, F_p]) = 0$$

is a direct consequence of Eq. (2).

Thus we have shown that *parallel displacement of a line of the complex passing through $z^0$ gives rise to a vector $dp$ collinear with the vector $a$ of Eq.* (5).

We now introduce the operators which we shall call the "infinitesimal parallel displacement operators for lines of the complex passing through $z^0$," namely

$$L = \left(a, \frac{\partial}{\partial p}\right) \equiv a_1 \frac{\partial}{\partial p^1} + a_2 \frac{\partial}{\partial p^2} + a_3 \frac{\partial}{\partial p^3} \tag{10}$$

and

$$\bar{L} = \left(\bar{a}, \frac{\partial}{\partial \bar{p}}\right) \equiv \bar{a}_1 \frac{\partial}{\partial \bar{p}^1} + \bar{a}_2 \frac{\partial}{\partial \bar{p}^2} + \bar{a}_3 \frac{\partial}{\partial \bar{p}^3}, \tag{10'}$$

where $a$ is given by (5). We shall show that the average of $L\bar{L}\varphi(\alpha, p)$ over the set of lines in the complex passing through $z^0$ is, to within a factor, the desired value of $f$ at $z^0$. More explicitly, we formulate this statement in the following theorem.

**Theorem.** *Assume that all the lines of the complex $F(\alpha, p) = 0$, where $F$ is an irreducible polynomial, satisfy the condition*

$$(F_\alpha, F_p) = 0$$

*and that the cone generated by the lines of the complex passing through $z^0$ has no singular points other than $z^0$. Then the value of $f(z)$ at $z^0$ is given in terms of the integrals $\varphi(\alpha, p)$ of $f(z)$ over lines of the complex by the inversion formula*

$$f(z^0) = c_{z^0} \frac{i}{2} \int_\Gamma L\bar{L}\varphi(\alpha, [\alpha, z^0])\omega_{z^0}(\alpha)\bar{\omega}_{z^0}(\alpha), \tag{11}$$

*where $L$ and $\bar{L}$ are the infinitesimal parallel displacement operators defined by* (10) *and* (10′). *The integral is taken along any contour $\Gamma$ on the cone $F(\alpha, [\alpha, z^0]) = 0$ in $\alpha$ space which intersects each generator of this cone exactly once. The differential form $\omega_{z^0}(\alpha)$ is given by*[8]

$$\omega_{z^0}(\alpha) = \frac{\alpha_1\, d\alpha_2 - \alpha_2\, d\alpha_1}{a_3} = \frac{\alpha_2\, d\alpha_3 - \alpha_3\, d\alpha_2}{a_1} = \frac{\alpha_3\, d\alpha_1 - \alpha_1\, d\alpha_3}{a_2}, \tag{12}$$

*where $a \equiv F_\alpha + [z^0, F_p]$.* It will be shown later also that $c_{z^0}$ is given by

$$c_{z^0}^{-1} = -\pi\Delta \frac{i}{2} \int_\Gamma \frac{B(a, a)}{A^2(\alpha, \alpha)}\, \omega_{z^0}(\alpha)\bar{\omega}_{z^0}(\alpha), \tag{13}$$

where $A(\alpha, \alpha)$ is a Hermitian positive definite quadratic form, and $B(a, a)$ is its dual (that is, the matrix of $B$ is the inverse of the matrix of $A$), and $\Delta$ is the discriminant of $A$.

Strictly speaking, there exists no contour $\Gamma$ intersecting each of the generators of the cone exactly once, and the contour integral "along $\Gamma$" is understood in the following way. The space of the generators is divided into sufficiently small regions in each of which we choose a contour $\Gamma_i$ crossing all the generators of this region exactly once. Then the integral of (11) is defined as the sum of the integrals over the $\Gamma_i$. This definition is valid because it does not depend on the choice of the $\Gamma_i$ or on the way in which the space of generators is divided up. Indeed, it is easily verified that the integrand of (11) is homogeneous of degree (0, 0) on the cone. It is therefore invariant under deformation of the contour (see also the similar discussion in the definition of the residue for the homogeneous function, Volume 1, p. 395).

[8] The differential form $\omega_{z^0}(\alpha)$ on the cone is defined by

$$dF(\alpha)\omega_{z^0}(\alpha) = \alpha_1\, d\alpha_2\, d\alpha_3 + \alpha_2\, d\alpha_3\, d\alpha_1 + \alpha_3\, d\alpha_1\, d\alpha_2,$$

where $F(\alpha) = 0$ is the equation of the cone. It is easily verified directly that each of the expressions given in (12) satisfies this relation.

### 1.5. The Inversion Formula. Proof of the Theorem of Section 1.4

The inversion formula need be calculated only for some fixed point, say for $z^0 = 0$. In other words, it is sufficient to show that

$$f(0) = c_0 \frac{i}{2} \int_\Gamma \left(F_\alpha, \frac{\partial}{\partial p}\right)\left(\bar{F}_\alpha, \frac{\partial}{\partial \bar{p}}\right) \varphi(\alpha, 0)\omega(\alpha)\bar{\omega}(\alpha), \tag{1}$$

where $\omega(\alpha)$ is given by

$$\omega(\alpha) = \frac{\alpha_1\, d\alpha_2 - \alpha_2\, d\alpha_1}{F_{\alpha_3}} = \frac{\alpha_2\, d\alpha_3 - \alpha_3\, d\alpha_2}{F_{\alpha_1}} = \frac{\alpha_3\, d\alpha_1 - \alpha_1\, d\alpha_3}{F_{\alpha_2}},$$

and $\Gamma$ is a contour on the surface of the $F(\alpha, 0) = 0$ cone. But since $\varphi(\alpha, p)$ is obtained by integrating $f(z)$ over the lines of the complex, we may think of the integral

$$(\Phi, f) = \frac{i}{2} \int_\Gamma \left(F_\alpha, \frac{\partial}{\partial p}\right)\left(\bar{F}_\alpha, \frac{\partial}{\partial \bar{p}}\right) \varphi(\alpha, 0)\omega(\alpha)\bar{\omega}(\alpha) \tag{2}$$

as a linear functional on the space of the $f(z)$, and this functional is obviously concentrated on the surface of the $F(z, 0) = 0$ cone. What we must show, then, is that

$$\Phi = c\, \delta(z).$$

Note first that $\Phi$ is a homogeneous generalized function of degree $(-3, -3)$. To see this, replace $f(z)$ by $f(\lambda z)$ with $\lambda \neq 0$. Then $\varphi(\alpha, p)$ is transformed to

$$\varphi_1(\alpha, p) = \varphi(\lambda\alpha, \lambda^2 p) = \lambda^{-1}\bar{\lambda}^{-1}\varphi(\alpha, \lambda p),$$

and then it is clear that

$$\frac{\partial^2 \varphi_1(\alpha, 0)}{\partial p^i\, \partial \bar{p}^j} = \frac{\partial^2 \varphi(\alpha, 0)}{\partial p^i\, \partial \bar{p}^j}.$$

Therefore the integrand of (2) does not change when $\varphi(\alpha, p)$ is replaced by $\varphi_1(\alpha, p)$. This proves that $(\Phi, f(\lambda z)) = (\Phi, f(z))$, or that $\Phi$ is homogeneous of degree $(-3, -3)$ (recall that the test functions are functions of three variables). Now it is known that there exists only one kind of such homogeneous function concentrated at the origin, namely $c\,\delta(z)$ (see Volume 1, p. 392). In order, therefore, to prove that $\Phi = c\,\delta(z)$, it is sufficient to show that $\Phi$ is concentrated at the vertex of the cone. In other words, we must prove the following assertion.

**Lemma.** *If $f(z) = 0$ in a neighborhood of $z = 0$, then $(\Phi, f) = 0$.*

Proof. It is obviously sufficient to prove the lemma for some $f(z)$ concentrated in an arbitrarily small neighborhood of some point $z^0 \neq 0$. Let, therefore, $z^0 \neq 0$ be any point on the $F(z, 0) = 0$ cone. By assumption, the three first partial derivatives of $F$ with respect to the $z_i$ are not all zero at $z^0$. It is therefore always possible to find a coordinate $z_i$ and a derivative $F_{z_j}$ with $i \neq j$, such that both are nonzero at $z^0$.[9] Assume, for example, that $F_{z_2} \neq 0$ and $z_3 \neq 0$ at $z^0$. Now pick an arbitrarily small neighborhood of $z^0$ such that neither $F_{z_2}$ nor $z_3$ vanishes anywhere in this neighborhood, and let $f(z)$ be any function with support in this neighborhood. What we must show is that $(\Phi, f) = 0$. This is easily done if we write $(\Phi, f)$ explicitly in terms of $f$.

The integrals over the lines may be written in the form

$$\varphi(\alpha, p) = \frac{i}{2} \int f(\alpha_1 t + v_1, \alpha_2 t + v_2, \alpha_3 t) \, dt \, d\bar{t},$$

where $(v_1, v_2, 0)$ is the point of intersection of the particular line with the $z_3 = 0$ plane. The $v_i$ can be written directly in terms of the Plücker coordinates of the line: in fact

$$v_1 = p^2/\alpha_3, \qquad v_2 = -p^1/\alpha_3.$$

Hence

$$\varphi(\alpha, p) = \frac{i}{2} \int f\left(\alpha_1 t + \frac{p^2}{\alpha_3}, \alpha_2 t - \frac{p^1}{\alpha_3}, \alpha_3 t\right) dt \, d\bar{t}. \tag{3}$$

Observe that in this notation $\varphi$ is independent of $p^3$, a result obtainable because the orthogonality relation

$$\alpha_1 p^1 + \alpha_2 p^2 + \alpha_3 p^3 = 0$$

makes it possible always to eliminate one of the Plücker coordinates. Now we insert this expression for $\varphi(\alpha, p)$ into Eq. (2) for $(\Phi, f)$. This yields the desired expression in terms of $f$, namely

$$(\Phi, f) = \left(\frac{i}{2}\right)^2 \int_\Gamma \left[\int \frac{1}{|\alpha_3|^2} \left(-F_{\alpha_1} \frac{\partial}{\partial z_2} + F_{\alpha_2} \frac{\partial}{\partial z_1}\right) \right.$$
$$\left. \times \left(-\bar{F}_{\alpha_1} \frac{\partial}{\partial \bar{z}_2} + \bar{F}_{\alpha_2} \frac{\partial}{\partial \bar{z}_1}\right) f(\alpha t) \, dt \, d\bar{t}\right] \omega(\alpha)\bar{\omega}(\alpha), \tag{4}$$

[9] This is obvious if none or one of the coordinates vanishes at $z_0$. If, on the other hand, for instance $z_1 = z_2 = 0$ and $z_3 \neq 0$ at $z_0$, then $z_1 F_{z_1} + z_2 F_{z_2} + z_3 F_{z_3} = 0$ implies that $F_{z_3} = 0$. Therefore either $F_{z_1} \neq 0$ or $F_{z_2} \neq 0$.

where the outer integral is taken along a contour on the $F(\alpha, 0) = 0$ cone and $\omega(\alpha)$ is the differential form defined below Eq. (1). We note that actually the integration in (4) is over the entire surface of the $F(z, 0) = 0$ cone. Let us change the variables in the integrand, choosing the new variables to be the coordinates $z_i = \alpha_i t$ of the points on the cone. By assumption $F_{z_2} \neq 0$ on the support of $f(z)$, so that we may choose the independent coordinates of the cone to be $z_1$ and $z_3$. We then have

$$dz_1\, dz_3 = -t\, dt(\alpha_3\, d\alpha_1 - \alpha_1\, d\alpha_3).$$

In the new variables Eq. (4) becomes

$$(\Phi, f) = \left(\frac{i}{2}\right)^2 \int \frac{1}{|z_3|^2}\left(\frac{\partial}{\partial z_1} - \frac{F_{z_1}}{F_{z_2}}\frac{\partial}{\partial z_2}\right)\left(\frac{\partial}{\partial \bar{z}_1} - \frac{\bar{F}_{z_1}}{\bar{F}_{z_2}}\frac{\partial}{\partial \bar{z}_2}\right)$$

$$\times f(z)\, dz_1\, d\bar{z}_1\, dz_3\, d\bar{z}_3$$

$$= \left(\frac{i}{2}\right)^2 \int \frac{1}{|z_3|^2} f''_{z_1\bar{z}_1}\, dz_1\, d\bar{z}_1\, dz_3\, d\bar{z}_3,$$

where $f''_{z_1\bar{z}_2}$ is the total derivative of $f$ with respect to $z_1$ and $\bar{z}_1$ on the surface of the $F(z, 0) = 0$ cone.

This integral vanishes by Stokes' theorem, which completes the proof of the lemma. We have thus shown that the generalized function $\Phi$ is concentrated on the vertex of the cone, which establishes Eq. (1) for the inversion formula.

What remains is to calculate the constant $c_0$ in Eq. (1). To do this we apply our inversion formula to the test function

$$f(z) = e^{-A(z,z)},$$

where

$$A(z, z) = \sum g_{ij} z_i \bar{z}_j$$

is an arbitrary positive definite Hermitian form. We start by calculating the integrals of $f(z)$ along straight lines. These are

$$\varphi(\alpha, [\alpha, \beta]) = \frac{i}{2} \int f(\alpha t + \beta)\, dt\, d\bar{t}$$

$$= \frac{i}{2} \int \exp[-t\bar{t}A(\alpha, \alpha) - tA(\alpha, \beta) - \bar{t}A(\beta, \alpha) - A(\beta, \beta)]\, dt\, d\bar{t}$$

$$= \frac{\pi}{A(\alpha, \alpha)} \exp\left[-\frac{A(\beta, \beta)A(\alpha, \alpha) - |A(\alpha, \beta)|^2}{A(\alpha, \alpha)}\right].$$

It is easily verified that

$$A(\beta, \beta)A(\alpha, \alpha) - | A(\alpha, \beta)|^2 = \Delta B([\alpha, \beta], [\alpha, \beta]),$$

where

$$B(z, z) = \sum g^{ij} z_i \bar{z}_j$$

is the Hermitian form whose matrix is the inverse of that of $A$, and $\Delta$ is the discriminant of $A$. We thus arrive at

$$\varphi(\alpha, p) = \frac{\pi}{A(\alpha, \alpha)} \exp\left[-\frac{B(p, p)}{A(\alpha, \alpha)}\right].$$

When we apply the infinitesimal parallel translation operators $L$ and $\bar{L}$ [see Section 1.4, Eqs. (10) and (10′)] to this function, we obtain

$$L\bar{L}\varphi(\alpha, 0) = -\frac{\pi \Delta B(F_\alpha, F_\alpha)}{A^2(\alpha, \alpha)}.$$

On inserting this into the formula for $f(0)$, we arrive at

$$c_0^{-1} = -\pi\Delta \frac{i}{2} \int_\Gamma \frac{B(F_\alpha(\alpha, 0), F_\alpha(\alpha, 0))}{A^2(\alpha, \alpha)} \omega(\alpha)\bar{\omega}(\alpha). \tag{5}$$

By a similar procedure for an arbitrary point $z^0$ we would have obtained

$$c_{z^0}^{-1} = -\pi\Delta \frac{i}{2} \int_\Gamma \frac{B(a, a)}{A^2(\alpha, \alpha)} \omega_{z^0}(\alpha)\bar{\omega}_{z^0}(\alpha), \tag{6}$$

where $a = F_\alpha + [z^0, F_p]$.

## 1.6. Examples of Complexes

In this section we shall consider two examples of line complexes.

**Example 1.** Consider the complex of lines consisting of the generators of the quadratic cone whose equation is

$$z_1^2 + z_2 z_3 = 0$$

and of all lines parallel to them. Obviously the equation of this complex is

$$F(\alpha, p) \equiv \alpha_1^2 + \alpha_2\alpha_3 = 0.$$

The inversion formula for this case becomes

$$f(z) = c\frac{i}{2}\int_{\Gamma} L\bar{L}\varphi(\alpha, [\alpha, z])\frac{d\alpha_3\, d\bar{\alpha}_3}{4\,|\,\alpha_1^2\,|},$$

where

$$L = 2\alpha_1\frac{\partial}{\partial p^1} + \alpha_3\frac{\partial}{\partial p^2} + \alpha_2\frac{\partial}{\partial p^3},$$

and the integral is along the contour $\alpha_1^2 + \alpha_2\alpha_3 = 0$, $\alpha_2 = 1$. Let us calculate $c$ by Eq.(5) of Section 1.5. We set $A(z, z) = 2z_1\bar{z}_1 + z_2\bar{z}_2 + z_3\bar{z}_3$ in that equation, arriving at

$$\begin{aligned} c^{-1} &= -\pi\frac{i}{2}\int_{\Gamma}\frac{2d\alpha_3\, d\bar{\alpha}_3}{4\,|\,\alpha_1^2\,|(2\,|\,\alpha_1\,|^2 + |\,\alpha_2\,|^2 + |\,\alpha_3\,|^2)} \\ &= -\pi\frac{i}{2}\int\frac{d\alpha\, d\bar{\alpha}}{|\,\alpha\,|(|\,\alpha\,| + 1)^2} \\ &= -2\pi^2. \end{aligned}$$

Thus since $\alpha_1^2 = -\alpha_3$ on $\Gamma$, we find that

$$f(z) = -\frac{1}{8\pi^2}\frac{i}{2}\int_{\Gamma} L\bar{L}\varphi(\alpha, [\alpha, z])\frac{d\alpha_3\, d\bar{\alpha}_3}{|\,\alpha_3\,|}.$$

Example 2. Now consider the complex of all lines intersecting the quadratic curve

$$z_1 z_2 = 1, \qquad z_3 = 0. \tag{1}$$

Each of the lines of the complex is determined by its direction vector $\alpha$ and the point $(\lambda, \lambda^{-1}, 0)$ at which it intersects the curve. It can therefore be specified by the parametric equations

$$z_1 = \alpha_1 t + \lambda, \qquad z_2 = \alpha_2 t + \lambda^{-1}, \qquad z_3 = \alpha_3 t. \tag{2}$$

We find from this equation that the Plücker coordinates $p^i$ of the line specified by $\alpha$ and $\lambda$ are

$$p^1 = -\alpha_3\lambda^{-1}, \qquad p^2 = \alpha_3\lambda, \qquad p^3 = \alpha_1\lambda^{-1} - \alpha_2\lambda. \tag{3}$$

The equation of the complex is obtained by eliminating $\lambda$ from any two of Eqs. (3). By eliminating it from the first two we obtain

$$F(\alpha, p) \equiv \alpha_3^2 + p^1 p^2 = 0. \tag{4}$$

It is thus seen that the complex is of order two.

Consider a point $z^0$, and let us find all lines passing through $z^0$. This can be done by eliminating $t$ from Eqs. (2), which yields

$$\lambda = \frac{\alpha_3 z_1^0 - \alpha_1 z_3^0}{\alpha_3}, \qquad \lambda^{-1} = \frac{\alpha_3 z_2^0 - \alpha_2 z_3^0}{\alpha_3}. \tag{5}$$

Now $\lambda$ is easily eliminated by multiplying these two equations, and we find that the direction vectors $\alpha$ of all lines passing through $z_0$ are then given by

$$\Phi(\alpha) \equiv \alpha_3^2 + (\alpha_3 z_1^0 - \alpha_1 z_3^0)(\alpha_2 z_3^0 - \alpha_3 z_2^0) = 0. \tag{6}$$

Thus those lines of the complex that pass through $z^0$ are all lines satisfying Eq. (2) whose direction vectors satisfy (6) and for which $\lambda$ is defined by (5).

The integrals of $f(z)$ along lines of the complex are given by

$$\varphi(\alpha; \lambda) = \frac{i}{2} \int f(\alpha_1 t + \lambda, \alpha_2 t + \lambda^{-1}, \alpha_3 t)\, dt\, dt. \tag{7}$$

Now let us use the inversion formula of Section 1.5 to express $f(z)$ in terms of $\varphi(\alpha; \lambda)$. This requires knowing the form of the infinitesimal parallel translation operator $L$ for our complex. As we know, this operator is

$$L = a_1 \frac{\partial}{\partial p^1} + a_2 \frac{\partial}{\partial p^2} + a_3 \frac{\partial}{\partial p^3},$$

where the $a_i$ are the coordinates of the vector $a = F_\alpha + [z, F_p]$. For our complex we have

$$a_1 = -p^1 z_3, \qquad a_2 = p^2 z_3, \qquad a_3 = 2\alpha_3 + p^1 z_1 - p^2 z_2.$$

We now use Eqs. (2) and (3) to write the $a_i$ in terms of $\alpha$, $z_3$, and $\lambda$, namely

$$a_1 = \alpha_3 z_3 \lambda^{-1}, \qquad a_2 = \alpha_3 z_3 \lambda, \qquad a_3 = -(\alpha_1 \lambda^{-1} + \alpha_2 \lambda) z_3.$$

Therefore

$$L = \lambda z_3 \left[\alpha_3 \lambda^{-2} \frac{\partial}{\partial p^1} + \alpha_3 \frac{\partial}{\partial p^2} - (\alpha_1 \lambda^{-2} + \alpha_2) \frac{\partial}{\partial p^3}\right].$$

Now note that

$$p^1 = -\alpha_3 \lambda^{-1}, \qquad p^2 = \alpha_3 \lambda, \qquad p^3 = \alpha_1 \lambda^{-1} - \alpha_2 \lambda,$$

so that the expression in brackets is just the derivative with respect to $\lambda$. Hence we may write

$$L = z_3\lambda \frac{\partial}{\partial\lambda}.$$

Then with the inversion formula of Section 1.5 we arrive finally at

$$f(z) = c_z \frac{i}{2} \int_\Gamma | z_3 |^2 | \lambda |^2 \varphi_{\lambda\bar{\lambda}} \left(\alpha; \frac{\alpha_3 z_1 - \alpha_1 z_3}{\alpha_3}\right) \omega(\alpha)\bar{\omega}(\alpha),$$

where $\omega(\alpha) = (\alpha_3\, d\alpha_1 - \alpha_1\, d\alpha_3)/\alpha_2 z_3 \lambda$ and the integral is along a contour $\Gamma$ on the surface of the cone whose equation is

$$\Phi(\alpha) \equiv \alpha_3^2 + (\alpha_3 z_1 - \alpha_1 z_3)(\alpha_2 z_3 - \alpha_3 z_2) = 0.$$

Let us choose in particular $\Gamma$ to be the intersection of the cone with the $\alpha_3 = 1$ plane. Then the inversion formula becomes

$$f(z) = c_z \frac{i}{2} \int_{(z_1-\alpha_1 z_3)(z_2-\alpha_2 z_3)=1} \varphi_{\lambda\bar{\lambda}}(\alpha_1, \alpha_2, 1; z_1 - \alpha_1 z_3)\, d\alpha_1\, d\bar{\alpha}_1 .$$

It is lert to the reader to calculate $c_z$.

### 1.7. Note on Translation Operators

The inversion formula we have derived can be generalized somewhat. Instead of making a parallel translation of the generators of the cone with vertex at $z^0$, let us displace them in the complex in an arbitrary manner. Then instead of the infinitesimal parallel translation operators, the more general operators

$$L\varphi(\alpha, [\alpha, z^0]) = \sum u_i \frac{\partial\varphi}{\partial\alpha_i} + v_i \frac{\partial\varphi}{\partial p^i}, \tag{1}$$

$$\bar{L}\varphi(\alpha, [\alpha, z^0]) = \sum \bar{u}_i \frac{\partial\varphi}{\partial\bar{\alpha}_i} + \bar{v}_i \frac{\partial\varphi}{\partial\bar{p}^i} \tag{1'}$$

will appear in the inversion formula, where the $u_i$ and $v_i$ are homogeneous analytic functions of $\alpha$ and define the displacement. Now form the function

$$\psi(\alpha, z^0) = [L + w(\alpha)][\bar{L} + \bar{w}(\alpha)]\varphi(\alpha, [\alpha, z^0]), \tag{2}$$

where $w(\alpha)$ is some homogeneous function, and consider its average over a set of generators of the cone whose vertex is at $z^0$. The $u_i$, $v_i$, and $w$ are functions of $\alpha$ that must be chosen so that the average is $cf(z^0)$, that is, so that the inversion formula is

$$\int_{\Gamma} (L + w)(\bar{L} + \bar{w})\varphi(\alpha, [\alpha, z^0])\omega_{z^0}(\alpha)\bar{\omega}_{z^0}(\alpha) = cf(z^0), \tag{3}$$

where $c \neq 0$. Necessary and sufficient conditions on the $u_i$, $v_i$, and $w$ can be established for Eq. (3) to hold. We shall not, however, state these conditions, but shall restrict ourselves to the following remark: an inversion formula of the form of (3) can be obtained only for the special complexes to which we have restricted ourselves in Section 1.3 and 1.4. Let us sketch a proof of this assertion. For simplicity, set $z^0 = 0$, and assume that the complex

$$F(\alpha, p) = 0$$

yields the inversion formula

$$cf(0) = \int_{\Gamma} (L + w)(\bar{L} + \bar{w})\varphi(\alpha, 0)\omega(\alpha)\bar{\omega}(\alpha), \qquad c \neq 0, \tag{4}$$

where $L$ and $\bar{L}$ are given by Eqs. (1) and (1′) and the integral is along a contour $\Gamma$ on the surface of the $F(\alpha, 0) = 0$ cone. It is easily shown that Eq. (4) is possible only when the infinitesimal displacement $d\alpha$, $dp$ of the generators of the cone lies in the tangent surface to the cone (that is, when $\sum u_i F_{\alpha_i} = 0$ and $\sum v_i F_{\alpha_i} = 0$). In addition, $c$ will be different from zero only if $dp \neq 0$. This imposes certain requirements on the complex. That $d\alpha$, $dp$ lie in the tangent plane may be written in the form

$$\begin{gathered} F_{\alpha_i}\, d\alpha_1 + F_{\alpha_i}\, d\alpha_2 + F_{\alpha_i}\, d\alpha_3 = 0, \\ dp^1/F_{\alpha_1} = dp^2/F_{\alpha_2} = dp^3/F_{\alpha_3}. \end{gathered} \tag{5}$$

That the displacement lies within the complex, on the other hand, may be written in the form

$$\sum F_{\alpha_i}\, d\alpha_i + \sum F_{p^i}\, dp^i = 0. \tag{6}$$

Now since by assumption $dp \neq 0$, Eqs. (5) and (6) yield

$$\sum F_{\alpha_i} F_{p^i} = 0.$$

But this is just the condition that we imposed on the complex in Sections 1.3 and 1.4.

## 2. Integral Geometry on a Quadratic Surface in a Space of Four Complex Dimensions

### 2.1. Statement of the Problem

Consider the quadratic surface

$$z_1^2 + z_2^2 + z_3^2 + z_4^2 = 1 \tag{1}$$

in a space of four complex variables. This is a ruled surface, which means that at every point of the surface there exist complex lines lying entirely within the surface (linear generators). In this respect this surface resembles not so much a sphere in a real space as a single-sheeted hyperboloid. Since, however, in a complex space all nondegenerate quadratic forms are equivalent (that is, can be transformed into each other by linear transformations), we may with equal validity call the surface of Eq. (1) either a sphere or a hyperboloid.

Let us calculate the number of parameters on which the set of all lines on this sphere depends. A line is given by a direction vector $\alpha$, determined up to a factor, and the position vector $\beta$ of a point on the line, determined up to an additive multiple of $\alpha$. Thus the family of all lines in four dimensions depends on $8 - 2 = 6$ complex parameters. What we must find, then, is the number of additional conditions placed on $\alpha$ and $\beta$ by the requirement that the lines lie on the sphere given by (1). For this purpose we write the equation of a line in the parametric form

$$z = \alpha t + \beta$$

(where $t$ is a complex parameter) and insert this expression for $z$ into the equation of the surface. The result yields three additional conditions on $\alpha$ and $\beta$, namely,

$$\alpha_1^2 + \alpha_2^2 + \alpha_3^2 + \alpha_4^2 = 0,$$

$$\alpha_1\beta_1 + \alpha_2\beta_2 + \alpha_3\beta_3 + \alpha_4\beta_4 = 0,$$

$$\beta_1^2 + \beta_2^2 + \beta_3^2 + \beta_4^2 = 1.$$

Thus the family of lines lying on the surface of the sphere depends on $6 - 3 = 3$ complex parameters, and is therefore of the same dimension as the surface itself.

We now consider the following problem of integral geometry. *With every function $f(z)$ defined on the surface of the sphere we associate its integrals along the line generators of the sphere (these integrals will be* *defined more accurately in Section* 2.3). *We wish to be able to find* $f(z)$ *when only these integrals are known.* This problem, whose solution will be stated in Section 2.4, is similar to the one we have been considering of line complexes in three dimensions, and we shall use the same methods to solve it. The only difference is that instead of the affine space of three complex dimensions, we are now dealing with a three-dimensional complex manifold.

We shall see later, in Chapter IV, that this problem lies at the basis of the whole theory of the representations of the Lorentz group (that is, the group of unimodular complex two-by-two matrices, which define a hyperboloid in four dimensions).

## 2.2. Line Generators of Quadratic Surfaces

Let us first choose a canonical form for the equation of a quadratic surface. Since all nondegenerate quadratic forms in a complex space are equivalent, that is, are obtainable from each other by linear transformations, the equation of a "sphere" can be expressed in terms of any nondegenerate quadratic form. We choose, therefore, to write the equation of a quadratic surface in the form

$$z_1 z_4 - z_2 z_3 = 1. \tag{1}$$

It will be seen later that this equation is in many ways more convenient than Eq. (1) of Section 2.1.

Proceeding, we wish to establish a convenient way to specify a particular line on the surface whose equation is (1). First consider the intersection of this surface with a two-dimensional hyperplane. This intersection is a quadratic curve which may sometimes degenerate to a pair of straight lines and sometimes to just a single straight line (when the second line of the pair lies at infinity).

Let us find the conditions under which this intersection is a single line. For simplicity we shall temporarily write the equations of the hyperplane in a form in which they are solved for $z_1$ and $z_4$, namely,

$$\begin{aligned} z_1 &= \alpha_1 z_2 + \beta_1 z_3 + \gamma_1 , \\ z_4 &= \alpha_2 z_2 + \beta_2 z_3 + \gamma_2 . \end{aligned} \tag{2}$$

Now we insert these expressions into (1), obtaining the equation

$$(\alpha_1 z_2 + \beta_1 z_3 + \gamma_1)(\alpha_2 z_2 + \beta_2 z_3 + \gamma_2) - z_2 z_3 = 1 \tag{3}$$

for the intersection. This intersection will be a straight line when Eq. (3) degenerates to a linear equation, i.e., when $\alpha_1\alpha_2 = \beta_1\beta_2 = 0$ and $\alpha_1\beta_2 + \alpha_2\beta_1 = 1$. Thus either $\alpha_1 = \beta_2 = 0$ and $\alpha_2\beta_1 = 1$, or $\alpha_2 = \beta_1 = 0$ and $\alpha_1\beta_2 = 1$. In the first of these cases the equations of the hyperplane are

$$z_1 = \beta_1 z_3 + \gamma_1 , \qquad z_4 = z_2/\beta_1 + \gamma_2 , \tag{4}$$

and in the second they are

$$z_1 = \alpha_1 z_2 + \gamma_1 , \qquad z_4 = z_3/\alpha_1 + \gamma_2 . \tag{4'}$$

Here $\gamma_1 = \gamma_2 = 0$ is not allowed, since if this were so the hyperplane would simply not intersect the quadratic surface.

We thus obtain two families of hyperplanes which intersect the surface in a straight line. Henceforth, we shall write the equations of these families in homogeneous form. The equations of the first family are

$$\begin{aligned} uz_1 + vz_3 &= u', \\ uz_2 + vz_4 &= v'; \end{aligned} \tag{5}$$

and those of the second are

$$\begin{aligned} u'z_4 - v'z_3 &= u, \\ -u'z_2 + v'z_1 &= v. \end{aligned} \tag{5'}$$

(We are excluding the case in which either $u = v = 0$ or $u' = v' = 0$.)

We arrived at these equations on the assumption that it is possible to solve the equations of the hyperplane for $z_1$ and $z_4$. It is not difficult to show, however, that this involves no loss of generality and that every hyperplane that intersects our surface in a straight line is of the form (5) or (5'). It is also easily shown that *all* lines on the surface are obtained by its intersection with hyperplanes of the form of (5) or (5') (for every line on the surface intersects two lines at infinity which also lie on the surface). In fact, moreover, all these lines can be obtained from intersections with the hyperplanes of just *one* of these two families. This is because a hyperplane whose equation is (5) and another whose equation is (5') intersect the surface in the same line. This may be seen as follows. Multiply the first of Eqs. (5) by $z_4$, the second by $-z_3$, and add, obtaining

$$u'z_4 - v'z_3 = u(z_1z_4 - z_2z_3) = u.$$

Similarly,

$$-u'z_2 + v'z_1 = v(-z_2z_3 + z_1z_4) = v.$$

Consequently Eqs. (5′) are implied by (5). The converse is also true. Thus on the $z_1z_4 - z_2z_3 = 1$ surface these sets of equations are equivalent and therefore intersect the surface in the same line.

It is thus established that a line on the surface of Eq. (1) may be given by specifying a hyperplane by Eq. (5) [with $(u, v) \neq (0, 0)$ and $(u', v') \neq (0, 0)$], the line being the intersection of the surface and the hyperplane. It is easily seen further that two different hyperplanes in this family will intersect the surface in two different lines. Thus the four numbers $(u, v; u', v')$ in Eqs. (5) may be thought of as homogeneous coordinates for the lines lying on the $z_1z_4 - z_2z_3 = 1$ surface.

Note that $(u, v; u', v')$ are not arbitrary, for we have excluded the case $u = v = 0$ and the case $u' = v' = 0$ [as in each of these cases the hyperplane whose equation is (5) fails to intersect the surface]. *Thus the manifold of lines on a quadratic surface in a space of four complex dimensions forms a three-dimensional (complex) projective space from which are removed two nonintersecting one-dimensional hypersurfaces, namely* $u = v = 0$ *and* $u' = v' = 0$, *where* $(u, v; u', v')$ *are homogeneous coordinates in the space.*

As an example, let us find the homogeneous coordinates of the lines passing through a given point. Every line on the surface passing through some point $z^0$ with coordinates $z_i^0$ is given by the set of equations

$$uz_1 + vz_3 = uz_1^0 + vz_3^0,$$

$$uz_2 + vz_4 = uz_2^0 + vz_4^0.$$

The coordinates of this line are therefore

$$(u, v; uz_1^0 + vz_3^0, uz_2^0 + vz_4^0).$$

Let us now find the conditions under which two lines on our quadratic surface are parallel. If the homogeneous coordinates of a certain line are $(u, v; u', v')$, the equation of the line is (5) or, equivalently, (5′). Its direction vector $\alpha = (\alpha_1, \alpha_2, \alpha_3, \alpha_4)$ will obviously satisfy the relations

$$u\alpha_1 + v\alpha_3 = 0, \qquad u'\alpha_4 - v'\alpha_3 = 0,$$

$$u\alpha_2 + v\alpha_4 = 0, \qquad -u'\alpha_2 + v'\alpha_1 = 0,$$

from which we obtain

$$\frac{\alpha_1}{\alpha_3} = \frac{\alpha_2}{\alpha_4} = -\frac{u}{v}; \qquad \frac{\alpha_3}{\alpha_4} = \frac{\alpha_1}{\alpha_2} = \frac{u'}{v'}.$$

Thus two lines on the $z_1z_4 - z_2z_3 = 1$ surface are parallel if and only if their homogeneous coordinates $(u_1, v_1; u_1', v_1')$ and $(u_2, v_2; u_2', v_2')$ are related according to

$$u_1/v_1 = u_2/v_2 \quad \text{and} \quad u_1'/v_1' = u_2'/v_2' .$$

Remark. It is interesting to write down the homogeneous coordinates of all the lines in the intersection of the $z_1z_4 - z_2z_3 = 1$ surface with an arbitrary hyperplane. We shall write $(z, z) = z_1z_4 - z_2z_3$, so that the equation of our surface becomes $(z, z) = 1$; by $(u, v)$ we shall denote the bilinear form corresponding to the quadratic form $(z, z)$. Let the equation of the hyperplane be

$$(z, z^0) = \cosh \tau,$$

where $z^0$ is a point on the surface and $\tau$ is a complex number. For $\tau = 0$ this hyperplane intersects the surface in a cone whose vertex is at $z^0$. We have already seen that the lines generating a cone have homogeneous coordinates

$$(u, v; uz_1^0 + vz_3^0, uz_2^0 + vz_3).$$

If $\tau \neq 0$, the intersection is a nondegenerate two-dimensional quadratic hypersurface. This hypersurface has two families of line generators. It is a simple matter to show that the homogeneous coordinates of these two families are, respectively,

$$(u, v; e^{\tau}[uz_1^0 + vz_3^0], e^{\tau}[uz_2^0 + vz_4^0]),$$

and

$$(u, v; e^{-\tau}[uz_1^0 + vz_3^0], e^{-\tau}[uz_2^0 + vz_4^0]).$$

This is left as a problem for the reader. #

## 2.3. Integrals of $f(z)$ over Quadratic Surfaces and along Complex Lines

Consider the quadratic surface

$$z_1z_4 - z_2z_3 = 1 \tag{1}$$

and some function $f(z)$ defined on it. We wish to define the integral of $f(z)$ over this surface. For this purpose we associate with the surface the differential form $\sigma(z)$ defined by

$$dz_1\, dz_2\, dz_3\, dz_4 = d(z_1z_4 - z_2z_3) \cdot \sigma(z).$$

It is easily seen that in suitable coordinate systems on our surface this differential form can be written in any of the following forms:

$$\begin{aligned}\sigma(z) &= -\frac{dz_1\,dz_2\,dz_3}{z_1} = \frac{dz_2\,dz_3\,dz_4}{z_4} \\ &= \frac{dz_1\,dz_3\,dz_4}{z_3} = -\frac{dz_1\,dz_2\,dz_4}{z_2}.\end{aligned} \tag{2}$$

*Then the integral $I$ of $f(z)$ over the surface of Eq.* (1) *is defined by*

$$I = \left(\frac{i}{2}\right)^3 \int_{z_1 z_4 - z_2 z_3 = 1} f(z)\sigma(z)\bar{\sigma}(z). \tag{3}$$

We shall now associate a differential form with each line generator of the surface, thereby defining the integral of $f(z)$ along such a line. Let $(u, v; u', v')$ be the homogeneous coordinates of this line, so that it is given by the equation of the surface and either

$$\begin{aligned} uz_1 + vz_3 &= u', \\ uz_2 + vz_4 &= v', \end{aligned} \tag{4}$$

or (equivalently)

$$\begin{aligned} u'z_4 - v'z_3 &= u, \\ -u'z_2 + v'z_1 &= v. \end{aligned} \tag{4'}$$

These equations imply a relation between the $dz_i$ on the line, namely

$$\begin{aligned} u\,dz_1 + v\,dz_3 &= 0, \\ u\,dz_2 + v\,dz_4 &= 0, \\ u'\,dz_4 - v'\,dz_3 &= 0, \\ u'\,dz_2 - v'\,dz_1 &= 0. \end{aligned}$$

On the line with coordinates $(u, v; u', v')$ we therefore have

$$\frac{dz_1}{vu'} = \frac{dz_2}{vv'} = -\frac{dz_3}{uu'} = -\frac{dz_4}{uv'} = \omega. \tag{5}$$

It is this differential form $\omega$ that we shall associate with the line in question.

*Then the integral of $f(z)$ along the generator $(u, v; u', v')$ of the $z_1z_4 - z_2z_3 = 1$ surface, that is, along the line given by the equation of the surface and Eqs.* (4), *is defined by*

$$\varphi(u, v; u', v') = \frac{i}{2} \int f(z)\omega\bar{\omega}, \tag{6}$$

*where the differential form $\omega$ on the line is defined by* (5). With this definition $\varphi(u, v; u', v')$ is a homogeneous function of $u, v; u', v'$ of degree $(-2, -2)$, so that

$$\varphi(\alpha u, \alpha v; \alpha u', \alpha v') = \alpha^{-2}\bar{\alpha}^{-2}\varphi(u, v; u', v')$$

for any $\alpha \neq 0$.

**Remark.** It is helpful to note that $\omega$ satisfies the relations

$$\sigma(z) = d(uz_1 + vz_3)\, d(uz_2 + vz_4) \cdot \omega \tag{7}$$

and

$$\sigma(z) = d(u'z_4 - v'z_3)\, d(-u'z_2 + v'z_1) \cdot \omega, \tag{7'}$$

where $\sigma(z)$ is the differential form on our surface, defined by (2). The validity of these relations is easily established simply by inserting into them the expression for $\omega$ and $\sigma(z)$. Needless to say, (7) or (7′) could have been used as a definition of $\omega$. #

## 2.4. Expression for $f(z)$ on a Quadratic Surface in Terms of Its Integrals along Line Generators

Let us state again the problem of integral geometry which we shall be solving here. Consider the quadratic surface given by

$$z_1z_4 - z_2z_3 = 1 \tag{1}$$

in a space of four complex dimensions. Each line generator of this surface is specified by

$$\begin{aligned} uz_1 + vz_3 &= u', \\ uz_2 + vz_4 &= v', \end{aligned} \tag{2}$$

or by the equivalent equations

$$\begin{aligned} u'z_4 - v'z_3 &= u, \\ -u'z_2 + v'z_1 &= v. \end{aligned} \tag{2'}$$

Either (2) or (2′) defines the homogeneous coordinates $(u, v; u', v')$ of the generator. Let $f(z)$ be some infinitely differentiable function of bounded support on our surface. Then we define the integral of $f(z)$ over the line given by (2) as

$$\varphi(u, v; u', v') = \frac{i}{2}\int f(z)\omega\bar{\omega}, \tag{3}$$

where

$$\omega = \frac{dz_1}{vu'} = \frac{dz_2}{vv'} = -\frac{dz_3}{uu'} = -\frac{dz_4}{uv'}.$$

Our problem is to find $f(z)$ if $\varphi(u, v; u', v')$ is known.

Remark. This problem has a simple local analog. Consider a point $z^0$ on the surface of Eq. (1) and construct the tangent hyperplane at this point. This tangent plane intersects the surface in a quadratic cone with vertex at $z^0$. Let us now "blow up" the surface holding $z^0$ fixed, that is, we perform the similarity transformation $z \to \lambda z + (1 - \lambda)z^0$; and let $|\lambda| \to \infty$. In the limit the surface is transformed into the tangent plane, and its generators to the lines of the tangent plane that are parallel to the generators of the cone. We then have the following local analog of the problem formulated above: To find the value of $f(z)$ at the vertex of a quadratic cone in three dimensions, knowing its integrals along the generators of the cone and along lines parallel to them. For instance, we may choose $z^0 = (1, 0, 0, 1)$ so that the tangent plane intersects the surface along the cone whose equation is

$$t_1^2 + t_2 t_3 = 0, \tag{4}$$

where $t_1 = 1 + z_1$, $t_2 = z_2$, $t_3 = z_3$. Then the local problem is to find the value of our function at the origin in three dimensions when we know its integrals along the generators and lines parallel to the generators of the cone whose equation is (4). This problem was already solved in Section 1.6, where we obtained

$$f(0) = -\frac{1}{8\pi^2}\frac{i}{2}\int_\Gamma L\bar{L}\varphi(\alpha, 0)\frac{d\alpha_3\, d\bar{\alpha}_3}{|\alpha_3|}, \tag{5}$$

where $\varphi(\alpha, p)$ is the integral of $f(z)$ along the line whose Plücker coordinates are $\alpha$, $p$, and

$$L = 2\alpha_1\frac{\partial}{\partial p_1} + \alpha_3\frac{\partial}{\partial p_2} + \alpha_2\frac{\partial}{\partial p_3};$$

the integral is taken along the contour whose equation is $\alpha_1^2 + \alpha_2\alpha_3 = 0$, $\alpha_2 = 1$. #

We now return to the nonlocal problem. We shall solve it first in the same way as we did the earlier problem of a line complex in three dimensions. Later, in Section 2.6, we will solve it also in a different way.

First we introduce the infinitesimal parallel translation operators on the set of generators of the surface. Recall that the condition that two lines be parallel is that their homogeneous coordinates be related by

$$u_1/v_1 = u_2/v_2\,, \qquad u_1'/v_1' = u_2'/v_2'$$

(see Section 2.2). Let the coordinates of a line be $(u, v; u', v')$. Then any parallel line has coordinates $(\lambda u, \lambda v; \mu u', \mu v')$, and the homogeneity of the coordinates can be used to choose $\lambda$ and $\mu$ so that $\lambda\mu = 1$, so that we may write $\lambda = e^{-\tau}$ and $\mu = e^{\tau}$. Thus every line parallel to $(u, v; u', v')$ has coordinates of the form

$$(e^{-\tau}u, e^{-\tau}v; e^{\tau}u', e^{\tau}v'). \tag{6}$$

Under infinitesimal parallel translation the coordinates $(u, v; u', v')$ of a line will then change by

$$du = -u\,d\tau, \qquad dv = -v\,d\tau, \qquad du' = u'\,d\tau, \qquad dv' = v'\,d\tau$$

and we may accordingly define the *infinitesimal parallel translation operators on the set of line generators of our surface as*

$$\begin{aligned} L &= -u\frac{\partial}{\partial u} - v\frac{\partial}{\partial v} + u'\frac{\partial}{\partial u'} + v'\frac{\partial}{\partial v'}\,, \\ \bar{L} &= -\bar{u}\frac{\partial}{\partial \bar{u}} - \bar{v}\frac{\partial}{\partial \bar{v}} + \bar{u}'\frac{\partial}{\partial \bar{u}'} + \bar{v}'\frac{\partial}{\partial \bar{v}'} \end{aligned} \tag{7}$$

Let us write

$$\psi(u, v; u', v') = L\bar{L}\varphi(u, v; u', v'). \tag{8}$$

We shall see that the average of $\psi$ over those generators that pass through some given point $z$ will be, up to a constant factor, the desired $f(z)$.

We have as yet, however, to define the average of $\psi$ over the set of generators passing through $z$. Recall that on the surface every line passing through a given point $z$ has coordinates

$$(u, v; uz_1 + vz_3\,, uz_2 + vz_4)$$

(see Section 2.2). Thus what we want is the average of

$$\psi(u, v; uz_1 + vz_3, uz_2 + vz_4)$$

over $u$ and $v$. Now the function we have here is homogeneous of degree $(-2, -2)$ in $u$, $v$, and therefore it is natural to define its average as its *residue* with respect to $u$, $v$, that is, as the integral

$$\frac{i}{2}\int_\Gamma \psi(u, v; uz_1 + vz_3, uz_2 + vz_4)(u\,dv - v\,du)(\bar{u}\,d\bar{v} - \bar{v}\,d\bar{u}).$$

[The integral is over any contour $\Gamma$ in the $(u, v)$ plane that crosses each complex line of the form $\alpha u + \beta v = 0$ at one point.[10]] Thus in analogy with the problem for a line complex in three dimensions, we may expect to obtain the following result.

*The function $f(z)$ defined on the $z_1z_4 - z_2z_3 = 1$ surface is given in terms of its integrals $\varphi(u, v; u', v')$ along the lines on the surface by the inversion formula*

$$f(z) = -\frac{1}{8\pi^2}\frac{i}{2}\int_\Gamma L\bar{L}\varphi(u, v; uz_1 + vz_3, uz_2 + vz_4)$$
$$\times (u\,dv - v\,du)(\bar{u}\,d\bar{v} - \bar{v}\,d\bar{u}), \qquad (9)$$

*where $L$ is given by* (7) *and the integral is along any contour $\Gamma$ on the $(u, v)$ plane that crosses each line of the form $\alpha u + \beta v = 0$ at one point.*

In the following section we derive this result, and then in Section 2.6 we derive it again in a different way.

## 2.5. Derivation of the Inversion Formula

It is sufficient to derive the inversion formula for any fixed point on the surface, and we shall choose this point to be $z^0 = (1, 0, 0, 1)$. Thus we wish to show that

$$\frac{i}{2}\int L\bar{L}\varphi(u, v; u, v)(u\,dv - v\,du)(\bar{u}\,d\bar{v} - \bar{v}\,d\bar{u}) = c^{-1}f(1, 0, 0, 1), \qquad (1)$$

[10] See Volume 1, p. 395 or the last paragraph of Section 1.4 of the present chapter.

where

$$L = -u\frac{\partial}{\partial u} - v\frac{\partial}{\partial v} + u'\frac{\partial}{\partial u'} + v'\frac{\partial}{\partial v'}.$$

The integral of Eq. (1) is a continuous functional defined on the infinitely differentiable functions $f(z)$ of bounded support on the $z_1z_4 - z_2z_3 = 1$ surface. We shall thus write

$$(F,f) = \frac{i}{2}\int_\Gamma L\bar{L}\varphi(u, v; u, v)(u\,dv - v\,du)(\bar{u}\,d\bar{v} - \bar{v}\,d\bar{u}). \tag{2}$$

Obviously $F$ is a functional concentrated on the surface of the cone generated by those line generators of the surface that pass through $z^0$. It can be shown that this functional is in fact concentrated on the vertex of this cone. More formally we make the following assertion.

**Lemma.** *If $f(z) = 0$ in a neighborhood of $(1, 0, 0, 1)$, then $(F, f) = 0$.*

**Proof.** Assume the support of $f$ to lie within a sufficiently small neighborhood that does not contain $z^0 = (1, 0, 0, 1)$. A coordinate system may be introduced in this neighborhood in the following way. Note first that $z_2$ and $z_3$ may not vanish simultaneously at $z \neq z^0$ on the surface of the cone whose vertex is at $z^0$. This is because this cone is the intersection of the surface $z_1z_4 - z_2z_3 = 1$ and the hyperplane $z_1 + z_4 = 2$. If, therefore, someplace on this cone $z_2 = z_3 = 0$, then $z_1 = z_4 = 1$, and the point in question is the vertex. Now assume that, for instance, $z_3$ fails to vanish on the support of $f$. Then by using the equation of the surface we can obtain $z_2$ as a continuous function of $z_1$, $z_3$, and $z_4$ in this neighborhood, and therefore these three coordinates can be used as local coordinates on the support of $f$. Thus we may write $f = f_1(z_1, z_3, z_4)$. Further, $z_1$ and $z_4$ can be eliminated by using the equations

$$\begin{aligned} uz_1 + vz_3 &= u', & u'z_4 - v'z_3 &= u, \\ uz_2 + vz_4 &= v', & -u'z_2 + v'z_1 &= v, \end{aligned} \tag{3}$$

and then the integral along the line $(u, v; u', v')$ can be written in terms of $z_3$ alone. We have

$$z_1 = -\frac{v}{u}z_3 + \frac{u'}{u}, \qquad z_4 = \frac{v'}{u'}z_3 + \frac{u}{u'},$$

and therefore[11]

$$\varphi(u, v; u', v')$$
$$= \frac{1}{|u|^2 |u'|^2} \frac{i}{2} \int f_1\left(-\frac{v}{u} z_3 + \frac{u'}{u}, z_3, \frac{v'}{u'} z_3 + \frac{u}{u'}\right) dz_3\, d\bar{z}_3 \qquad (4)$$
$$= \frac{i}{2} \int f_1\left(-vu'z + \frac{u'}{u}, uu'z, uv'z + \frac{u}{u'}\right) dz\, d\bar{z}.$$

We now insert this expression for $\varphi$ into Eq. (2). A simple direct calculation yields

$$L\varphi(u, v; u, v)$$
$$\equiv \left(-u\frac{\partial}{\partial u} - v\frac{\partial}{\partial v} + u'\frac{\partial}{\partial u'} + v'\frac{\partial}{\partial v'}\right) \varphi(u, v; u', v')\Big|_{u'=u,\, v'=v}$$
$$= 2\frac{i}{2} \int \left(\frac{\partial}{\partial z_1} - \frac{\partial}{\partial z_4}\right) f_1(-uv\, z + 1, u^2 z, uv\, z + 1)\, dz\, d\bar{z}.$$

Thus if we let $u = 1$ describe the contour of integration $\Gamma$, Eq. (2) becomes

$$(F, f) = 4\left(\frac{i}{2}\right)^2 \int \left(\frac{\partial}{\partial z_1} - \frac{\partial}{\partial z_4}\right)\left(\frac{\partial}{\partial \bar{z}_1} - \frac{\partial}{\partial \bar{z}_4}\right)$$
$$\times f_1(-vz + 1, z, vz + 1)\, dz\, d\bar{z}\, dv\, d\bar{v}$$
$$= 4\left(\frac{i}{2}\right)^2 \int \left(\frac{\partial}{\partial z_1} - \frac{\partial}{\partial z_4}\right)\left(\frac{\partial}{\partial \bar{z}_1} - \frac{\partial}{\partial \bar{z}_4}\right) \qquad (5)$$
$$\times f_1(1 - t, z, 1 + t)\, dt\, d\bar{t} \frac{dz\, d\bar{z}}{|z|^2}$$

It is now easy to show that this integral vanishes. Indeed, the integrand contains

$$\left(\frac{\partial}{\partial z_1} - \frac{\partial}{\partial z_4}\right)\left(\frac{\partial}{\partial \bar{z}_1} - \frac{\partial}{\partial \bar{z}_4}\right) f_1(1 - t, z, 1 + t),$$

which is the total derivative of $f_1(1 - t, z, 1 + t)$ with respect to $t$ and $\bar{t}$. Hence $(F, f)$ vanishes by Stokes' theorem, which proves the lemma.

[11] Note that we may assume $u$ and $u'$ to be nonzero. Indeed, let $u = 0$. Since we are dealing only with lines passing through (1, 0, 0, 1), it follows that $u' = u = 0$. Then Eq. (3) implies that on this line $z_3 = 0$, and the line will therefore not pass through the support of $f$.

We have thus shown that the functional $F$ defined by Eq. (2) is concentrated at $z^0 = (1, 0, 0, 1)$, and therefore that $F$ is a linear combination of the $\delta$ function and its derivatives. We shall now show further that $(F, f) = kf(1, 0, 0, 1)$, for which we need only show that $F$ is homogeneous of degree $(-3, -3)$. This may be seen as follows. Assume that $f$ has support in an arbitrarily small neighborhood of $z^0$. We choose local coordinates in this neighborhood on the $z_1 z_4 - z_2 z_3 = 1$ surface to be $t_1 = z_1 - 1$, $t_2 = z_2$, $t_3 = z_3$ and write $f = f_1(t_1, t_2, t_3)$. In these coordinates the integral along $(u, v; u', v')$ may be written

$$\begin{aligned}\varphi(u, v; u', v') &= \frac{1}{|u'|^2 |v|^2} \\ &\quad \times \frac{i}{2} \int f_1\left(t_1, \frac{v'}{u'}[t_1 + 1] - \frac{v}{u'}, -\frac{u}{v}[t_1 + 1] + \frac{u'}{v}\right) dt_1\, d\bar{t}_1 \\ &= \frac{i}{2} \int f_1\left(vu't, vv't + \frac{v' - v}{u'}, -uu't + \frac{u' - u}{v}\right) dt\, d\bar{t}.\end{aligned}$$

In order to insert this into Eq. (2) for $(F, f)$, we first calculate

$$\begin{aligned}&L\varphi(u, v; u, v) \\ &\equiv \left(-u \frac{\partial}{\partial u} - v \frac{\partial}{\partial v} + u' \frac{\partial}{\partial u'} + v' \frac{\partial}{\partial v'}\right) \varphi(u, v; u', v')\Big|_{u'=u,\, v'=v} \\ &= 2 \frac{i}{2} \int \left(\frac{v}{u} \frac{\partial}{dt_2} + \frac{u}{v} \frac{\partial}{\partial t_3}\right) f_1(uvt, v^2 t, -u^2 t)\, dt\, d\bar{t}.\end{aligned}$$

Again if $u = 1$ is chosen as the contour of integration $\Gamma$, Eq. (2) becomes

$$\begin{aligned}(F, f) = 4 \left(\frac{i}{2}\right)^2 \int &\left(v \frac{\partial}{\partial t_2} + \frac{1}{v} \frac{\partial}{\partial t_3}\right)\left(\bar{v} \frac{\partial}{\partial \bar{t}_2} + \frac{1}{\bar{v}} \frac{\partial}{\partial \bar{t}_3}\right) \\ &\times f_1(vt, v^2 t, -t)\, dt\, d\bar{t}\, dv\, d\bar{v}.\end{aligned} \tag{6}$$

It is then immediately seen that $F$ is homogeneous of degree $(-3, -3)$, for the integral in (6) is invariant under the replacement of $f_1(t_1, t_2, t_3)$ by $f_1(\alpha t_1, \alpha t_2, \alpha t_3)$. In other words,

$$(F, f_1(\alpha t_1, \alpha t_2, \alpha t_3)) = (F, f_1(t_1, t_2, t_3)).$$

which proves that $(F, f) = kf_1(0, 0, 0)$ and thereby proves Eq. (1). The value of $c$ may be calculated by noting that it must be the same as in the inversion formula for the corresponding local problem [see Eq. (5) of Section 2.4]. Consequently $c = -1/(8\pi^2)$.[12]

## 2.6. Another Derivation of the Inversion Formula[13]

In this section we give another derivation of the inversion formula, based on the properties of the generalized function $|z_1^2 + z_2^2 + z_3^2|^\lambda$. Recall that our problem is to write $f(z)$, an infinitely differentiable function with bounded support defined on the $(z, z) = 1$ surface, in terms of its integrals along the linear generators of the surface. We shall do this by reducing it to another problem of integral geometry which we shall then proceed to solve. Let us start by stating this other problem.

Let $z^0$ be a point on the quadratic surface $(z, z) = 1$ and consider the intersection of this surface with the three-dimensional hyperplanes (which, as before, we shall call simply planes)

$$(z, z^0) = \cosh \tau.$$

When $\cosh \tau = 1$, this intersection is a cone with vertex at $z^0$, for $(z, z^0) = 1$ intersects the surface at $z^0$. When $\cosh \tau = -1$, this intersection is a cone with vertex at $-z^0$. When $\cosh \tau \neq \pm 1$, this intersection is a nondegenerate two-dimensional quadratic surface.

Let the integral $\psi(\tau, z^0)$ of $f(z)$ over the intersection between the surface and the plane be defined by

$$\psi(\tau, z^0) = \left(\frac{i}{2}\right)^2 \int f(z)\omega_{z^0}\bar{\omega}_{z^0}, \tag{1}$$

where $\omega_{z^0}$ is a differential form defined by

$$dz_1\, dz_2\, dz_3\, dz_4 = d(z, z)\, d(z, z^0)\omega_{z^0}. \tag{2}$$

Our problem is to find the value of $f(z)$ at $z^0$ if we know $\psi(\tau, z^0)$.[14] We

[12] This constant could also have been calculated less indirectly by applying the inversion formula to a conveniently chosen function, for instance to

$$f(z) = \exp\{-|z_1|^2 - |z_1|^2 - |z_3|^2 - |z_4|^2\}.$$

[13] In order to follow the present development the reader should be familiar with the properties of the generalized functions $P^\lambda(z)\bar{P}^\mu(z)$ of Volume 1, Appendix B, Section 2.7.

[14] A local analog of this problem is to find the value of $f(z)$ at the origin in a three-dimensional complex affine space when the integrals of $f(z)$ over the $z_1^2 + z_2^2 + z_3^2 = \tau^2$ surfaces are known.

shall see that the solution is

$$f(z^0) = -\frac{1}{8\pi^2}\frac{\partial^2\psi(\tau, z^0)}{\partial\tau\,\partial\bar{\tau}}\bigg|_{\tau=0} \tag{3}$$

We shall first explain how this will lead to a solution of the original problem. Note that the intersection of the $(z, z) = 1$ surface and the $(z, z^0) = \cosh\tau$ plane ($\cosh\tau \neq \pm 1$) can be stratified into nonintersecting straight lines. If, therefore, we know the integrals $\varphi(u, v; u', v')$ of $f(z)$ along straight lines, we can calculate the integral of $f(z)$ over the intersection of the surface and the plane. In fact it can be shown (we omit the proof) that

$$\psi(\tau, z^0) = 4\left(\frac{i}{2}\right)\int_\Gamma \varphi(e^{-\frac{1}{2}\tau}u, e^{-\frac{1}{2}\tau}v, e^{\frac{1}{2}\tau}[uz_1^0 + vz_3^0], e^{\frac{1}{2}\tau}[uz_2^0 + vz_4^0])$$
$$\times (u\,dv - v\,du)(\bar{u}\,d\bar{v} - \bar{v}\,d\bar{u}). \tag{4}$$

The integral here is along a contour $\Gamma$ on the $(u, v)$ plane which intersects every line of the form $\alpha u + \beta v = 0$ at a single point.[15] Now if we insert this expression into Eq. (3), we obtain

$$f(z^0) = -\frac{1}{8\pi^2}\frac{i}{2}\int_\Gamma L\bar{L}\varphi(u, v; uz_1^0 + vz_3^0, uz_2^0 + vz_4^0)$$
$$\times (u\,dv - v\,du)(\bar{u}\,d\bar{v} - \bar{v}\,d\bar{u}).$$

But this is exactly the inversion formula stated in Section 2.4 and derived in Section 2.5.

Thus what we wish to do is to establish Eq. (3). To start, consider the integral

$$F(\lambda) = \left(\frac{i}{2}\right)^3 \int_{(z,z)=1} f(z)|(z, z^0)^2 - 1|^{2\lambda}\sigma(z)\bar{\sigma}(z), \tag{5}$$

where $\sigma(z)$ is a differential form on the $(z, z) = 1$ surface defined by

$$dz_1\,dz_2\,dz_3\,dz_4 = d(z, z)\sigma(z).$$

Now the factor multiplying $f(z)$ in the integrand is constant on the $(z, z^0) = \cosh\tau$ plane, so that this integral can be expressed in terms of

[15] See the last paragraph of Section 1.4.

the integrals $\psi(\tau, z^0)$ over the intersections of these planes with the $(z, z) = 1$ surface. In fact we have

$$F(\lambda) = \frac{1}{2}\left(\frac{i}{2}\right) \int_{|\mathrm{Im}\tau|<\pi} \psi(\tau, z^0) | \cosh^2 \tau - 1 |^{2\lambda} d(\cosh \tau)\, d(\overline{\cosh \tau})$$
$$= \frac{1}{2}\left(\frac{i}{2}\right) \int_{|\mathrm{Im}\tau|<\pi} \psi(\tau, z^0) | \sinh \tau |^{4\lambda+2} d\tau\, d\bar{\tau} \tag{6}$$

This integral converges for $\mathrm{Re}\,\lambda > -1$, for which it is analytic in $\lambda$. For $\mathrm{Re}\,\lambda < -1$ we define $F(\lambda)$ by analytic continuation in $\lambda$. It will be shown that $F(\lambda)$ has a simple pole at $\lambda = -\frac{3}{2}$; we shall calculate this residue first by Eq. (6) and then by Eq. (5). Comparison of these two expressions for the residue will yield the desired expression for $f(z^0)$.

Assume temporarily that $f(z)$ is nonzero only for sufficiently small values of $(z, z^0) - 1$ (it will be seen later that this implies no loss of generality), and let us proceed to calculate the residue using Eq. (6). For this purpose we rewrite (6) in the form

$$F(\lambda) = \frac{1}{2} \cdot \frac{i}{2} \int_{|\mathrm{Im}\tau|<\pi} \psi(\tau, z^0) \left| \frac{\sinh \tau}{\tau} \right|^{4\lambda+2} | \tau |^{4\lambda+2} d\tau\, d\bar{\tau}.$$

By the assumption concerning $f(z)$,

$$\psi(\tau, z^0) \left| \frac{\sinh \tau}{\tau} \right|^{4\lambda+2}$$

is continuous and infinitely differentiable in $\tau$ for $\lambda = -\frac{3}{2}$ in the strip $-\pi \leqslant \mathrm{Im}\,\tau \leqslant \pi$.[16] On the other hand, at this value of $\lambda$ the generalized function $| \tau |^{4\lambda+2}$ has a simple pole whose residue is

$$\tfrac{1}{2}\delta^{(1,1)}(\tau, \bar{\tau})$$

(see Volume 1, Appendix B, Section 1.3). Therefore

$$\underset{\lambda=-\frac{3}{2}}{\mathrm{res}}\, F(\lambda) = \frac{\pi}{4} \frac{\partial^2}{\partial \tau\, \partial \bar{\tau}} \left[ \psi(\tau, z^0) \left| \frac{\sinh \tau}{\tau} \right|^{-4} \right]_{\tau=0}$$
$$= \frac{\pi}{4} \frac{\partial^2 \psi(\tau, z^0)}{\partial \tau\, \partial \bar{\tau}} \bigg|_{\tau=0}. \tag{7}$$

[16] The infinite differentiability of $\psi(\tau, z^0)$ follows directly from Eq. (4). As for $|(\sinh \tau)/\tau|^{4\lambda+2}$, it has singularities at $\tau = \pm i\pi$; but at these values of $\tau$, the other factor $\psi(\tau, z^0)$ vanishes.

This result has been obtained by using Eq. (6). Let us now find the same residue by using Eq. (5). Note first that this residue depends only on the values of $f$ in an arbitrarily small neighborhood of $z^0$. To prove this we make use of the fact that $|(z, z^0)^2 - 1|^{2\lambda}$ is of the form $G^\lambda(z)\bar{G}^\lambda(z)$, a type of generalized function whose properties have been studied (see Volume 1, Appendix B, Section 2.2). It is known that if the $G = 0$ surface has no singular points, $G^\lambda \bar{G}^\lambda$ is a generalized function whose singularities in $\lambda$ occur at $\lambda = -1, -2, \ldots$ . This means that its residue at $\lambda = -\frac{3}{2}$ is a generalized function concentrated on the set of singular points of the $G = 0$ surface. We now apply this result to our case, in which $G = (z, z^0)^2 - 1$, so that the $G = 0$ surface is composed of the intersection of the $(z, z) = 1$ surface with the two planes $(z, z^0) = 1$ and $(z, z^0) = -1$. These intersections are cones with vertices at $z^0$ and $-z^0$, respectively, and therefore the only singular points of the $G = 0$ surface are $z^0$ and $-z^0$. This shows that the generalized function on the hyperboloid

$$\operatorname*{res}_{\lambda=-\frac{3}{2}} |(z, z^0)^2 - 1|^{2\lambda}$$

is concentrated at the two points $z^0$ and $-z^0$. But we have assumed that $f(z)$ vanishes in a neighborhood of $-z^0$, so that

$$\operatorname*{res}_{\lambda=-\frac{3}{2}} F(\lambda) = \operatorname*{res}_{\lambda=-\frac{3}{2}} \left(\frac{i}{2}\right)^3 \int f(z)|(z, z^0)^2 - 1|^{2\lambda}\sigma(z)\bar{\sigma}(z)$$

is uniquely determined by the values of $f$ in an arbitrarily small neighborhood of $z^0$. Thus in calculating the residue of $F(\lambda)$ at $\lambda = -\frac{3}{2}$, we may assume $f(z)$ to have support in an arbitrarily small neighborhood of $z^0$.

Let us now perform a linear transformation in the integrand of (5) to change the equation $(z, z) = 1$ of the surface to the canonical form $z_1^2 + z_2^2 + z_3^2 + z_4^2 = 1$ so that $z^0$ takes on the coordinates $(0, 0, 0, 1)$. Then we have

$$\begin{aligned} F(\lambda) &= 16\left(\frac{i}{2}\right)^3 \int f_1(z)|z_4^2 - 1|^{2\lambda}\sigma_1(z)\bar{\sigma}_1(z) \\ &= 16\left(\frac{i}{2}\right)^3 \int f_1(z)|z_1^2 + z_2^2 + z_3^2|^{2\lambda}\sigma_1(z)\bar{\sigma}_1(z), \end{aligned} \tag{8}$$

where $f_1(z)$ is the function $f(z)$ expressed in the new variables, and $\sigma_1(z)$ is defined by $dz_1\, dz_2\, dz_3\, dz_4 = d(z_1^2 + z_2^2 + z_3^2 + z_4^2)\sigma_1(z)$. We choose $z_1$, $z_2$, and $z_3$ to be the local coordinates of the surface in the neighborhood of $z^0$. In these coordinates

$$\sigma_1(z)\bar{\sigma}_1(z) = \tfrac{1}{4}\, dz_1\, dz_2\, dz_3\, d\bar{z}_1\, d\bar{z}_2\, d\bar{z}_3\, |1 - z_1^2 - z_2^2 - z_3^2|^{-1}.$$

At $\lambda = -\frac{3}{2}$ the generalized function $| z_1^2 + z_2^2 + z_3^2 |^{2\lambda}$ has a pole whose residue is

$$-\tfrac{1}{2}\pi^3 \, \delta(z)$$

(see Volume 1, Appendix B, Section 2.7). Then from Eq. (8) we have

$$\operatorname*{res}_{\lambda=-\frac{3}{2}} F(\lambda) = -2\pi^3 f_1(0, 0, 0, 1) = -2\pi^3 f(z^0).$$

We now compare this result with Eq. (7). Equation (3), namely

$$f(z^0) = -\frac{1}{8\pi^2} \frac{\partial^2 \psi(\tau, z^0)}{\partial \tau \, \partial \bar{\tau}} \bigg|_{\tau=0},$$

follows immediately. Thus we have expressed $f(z^0)$ in terms of integrals of $f(z)$ over the plane intersections of the $(z, z) = 1$ surface with the $(z, z^0) = \cosh \tau$ planes.

Recall that in deriving this expression we required that $f(z)$ be nonzero only for sufficiently small values of $(z, z^0) - 1$. From the inversion formula we have obtained, it is seen that this requirement represents no restriction, since the final result involves only integrals of $f(z)$ over those intersections on which $(z, z^0)$ is arbitrarily close to 1.

## 2.7. Rapidly Decreasing Functions on Quadratic Surfaces. The Paley-Wiener Theorem

We shall start by introducing the concept of a *rapidly decreasing function* (with rapidly decreasing derivatives) on the $(z, z) = z_1 z_4 - z_2 z_3 = 1$ surface. This will define the analog for this surface of the test function space we have elsewhere called $S$.

We shall say that $f(z)$ *decreases rapidly on the* $(z, z) = 1$ *surface if for every polynomial* $P(z)$ *in the variables* $z_k$ *and* $\bar{z}_k$ *the function* $P(z)f(z)$ *is bounded on the* $(z, z) = 1$ *surface.*

We shall use the term *rapidly decreasing*, however, for a somewhat narrower class of functions, which may be said to be rapidly decreasing together with all their derivatives. This class may be defined as follows. With each infinitely differentiable function $f(z)$ on the $(z, z) = 1$ surface we associate the generalized function

$$F(z) = f(z) \, \delta((z, z) - 1), \tag{1}$$

which is concentrated on this surface. Let

$$P(z, \partial/\partial z)$$

be an operator which is a polynomial in the $z_k$, $\bar{z}_k$, and the differential operators $\partial/\partial z_k$, $\partial/\partial \bar{z}_k$. We shall call $P(z, \partial/\partial z)$ *an internal operator with respect to the* $(z, z) = 1$ *surface* if, when applied to any generalized function of the form of (1) it transforms it to another one of the same form, i.e., if

$$P(z, \partial/\partial z)[f(z)\, \delta((z, z) - 1)] = g(z)\, \delta((z, z) - 1),$$

where $g(z)$ is a function defined on the same surface. In this way an internal operator $P$ associates with every $f(z)$ on the surface another function $g(z)$ on the same surface, and we may therefore consider it an operator on the functions defined on the $(z, z) = 1$ surface.

*Then we shall call $f(z)$ defined on the $(z, z) = 1$ surface rapidly decreasing (together with all its derivatives) if for every internal operator $P(z, \partial/\partial z)$ the function*

$$P(z, \partial/\partial z) f(z)$$

*is bounded on the $(z, z) = 1$ surface.*

The (infinitely differentiable) rapidly decreasing functions on the $(z, z) = 1$ surface form a linear space which we shall denote by $S$. A topology may be introduced on this space by the set of norms

$$\| f \|_P = \sup_z | P(z, \partial/\partial z) f(z) |, \tag{2}$$

where $P$ runs through the set of internal operators on the $(z, z) = 1$ surface.

**Remark.** It would be interesting to define for the $(z, z) = 1$ surface the analogs of the spaces we have called $S_\alpha^\beta$ in Volume 2. These can undoubtedly be defined similarly to $S$ in terms of sets of norms related to certain elements of the ring of internal operators. #

Let us now turn to the following problem. In Section 2.7 we associated with each $f(z)$ on the $(z, z) = 1$ surface its integrals $\varphi(u, v; u', v')$ along the lines on this surface. We may now ask for the set of $\varphi(u, v; u', v')$ functions obtained from infinitely differentiable rapidly decreasing functions.

Recall first that each $\varphi$ obtained in this way is always homogeneous:

$$\varphi(\alpha u, \alpha v; \alpha u', \alpha v') = \alpha^{-2} \bar{\alpha}^{-2} \varphi(u, v; u', v')$$

for any $\alpha \neq 0$. This follows immediately from the definition. In addition, differentiability and rapid decrease of $f$ imply certain smoothness and asymptotic properties of $\varphi$. We state these properties here without proof (proof will be given in Chapter IV).

*Differentiability condition.* $\varphi(u, v; u', v')$ is infinitely differentiable with respect to $u, v, u', v'$ and their complex conjugates everywhere except where $u = v = 0$ or $u' = v' = 0$.

*Asymptotic condition.* For every real number $k$ the function

$$\left(\frac{|u'|^2 + |v'|^2}{|u|^2 + |v|^2}\right)^k (|u|^2 + |v|^2)(|u'|^2 + |v'|^2)\varphi(u, v; u', v')$$

is bounded in $u, v, u', v'$. The same is true of the $\varphi_P$, which correspond to $Pf(z)$ where $P$ is any internal operator.

It is found that these conditions on the $\varphi(u, v; u', v')$ functions are not sufficient. If $\varphi$ is to be obtainable by integration of some infinitely differentiable rapidly decreasing function $f(z)$ along a straight line, it must satisfy some subsidiary conditions. These conditions and the above two then form a set of necessary and sufficient conditions on $\varphi$. They will be obtained from group-theoretical considerations in Section 5 of Chapter IV. We shall nevertheless state them here without derivation.

**Remark.** It is interesting to recall that the same situation arises also in connection with the Radon transform in an $n$-dimensional affine space (Chapter I, Section 1.6). Specifically, the Radon transform of an infinitely differentiable rapidly decreasing function in $n$ dimensions satisfies, in addition to the natural smoothness and asymptotic conditions, additional subsidiary conditions. #

The subsidiary conditions on $\varphi(u, v; u', v')$ are more easily stated in terms of its Mellin transform rather than $\varphi$ itself. Introducing the inhomogeneous coordinate system

$$z_1 = u/v, \qquad z_2 = u'/v', \qquad \lambda = v'/v$$

on the set of lines, we write

$$\varphi_1(z_1, z_2, \lambda) = \lambda\bar{\lambda}\varphi(z_1, 1; \lambda z_2, \lambda). \tag{3}$$

The coordinates $z_1$ and $z_2$ determine the direction of the line. For $z_1$ and $z_2$ fixed, therefore, $\varphi_1$ is a function on the set of all lines parallel to a certain line. We now integrate $\varphi_1$ with the weight factor $\lambda^{n_1-1}\bar{\lambda}^{n_2-1}$ over such a set of parallel lines, writing

$$K(z_1, z_2; n_1, n_2) = \frac{i}{2}\int \varphi_1(z_1, z_2, \lambda)\lambda^{n_1-1}\bar{\lambda}^{n_2-1}\, d\lambda\, d\bar{\lambda}, \tag{4}$$

which is naturally called the *Mellin transform* of $\varphi_1(z_1, z_2, \lambda)$. Here $(n_1, n_2)$ is any pair of complex numbers such that $n_1 - n_2$ is an integer. (This is a necessary requirement for $\lambda^{n_1-1}\bar{\lambda}^{n_2-1}$ to be a single-valued function of $\lambda$.)

*Subsidiary conditions.* The subsidiary conditions on $\varphi(u, v; u', v')$ may be stated in terms of its Mellin transform in the following way.

*Condition 1.* If $n_1, n_2$ are not integers of the same sign, then

$$\frac{i}{2}\int K(z, z_2; n_1, n_2)(z - z_1)^{-n_1-1}(\bar{z} - \bar{z}_1)^{-n_1-1}\, dz\, d\bar{z}$$
$$= \frac{i}{2}\int K(z_1, z; -n_1, -n_2)(z_2 - z)^{-n_1-1}(\bar{z}_2 - \bar{z})^{-n_1-1}\, dz\, d\bar{z}. \qquad (5)$$

Both integrals here must be understood in terms of their regularizations.

*Condition 2.* If $n_1 = 1, 2, \ldots$, then

$$\frac{\partial^{n_1}}{\partial z_1^{n_1}} K(z_1, z_2; n_1, n_2) = (-1)^{n_1} \frac{\partial^{n_1}}{\partial z_2^{n_1}} K(z_1, z_2; -n_1, n_2). \qquad (6)$$

*Condition 3.* If $n_2 = 1, 2, \ldots$, then

$$\frac{\partial^{n_2}}{\partial \bar{z}_1^{n_2}} K(z_1, z_2; n_1, n_2) = (-1)^{n_2} \frac{\partial^{n_2}}{\partial \bar{z}_2^{n_2}} K(z_1, z_2; n_1, -n_2). \qquad (6')$$

The group-theoretical content of these conditions will be discussed in Chapter IV. We shall sketch here, however, how these subsidiary conditions could have been obtained from geometrical considerations.

Condition 1 is related to the following geometrical fact. Consider the intersection of the $(z, z) = 1$ surface with an arbitrary three-dimensional hyperplane. This intersection is a two-dimensional quadratic surface with two families of linear generators. Therefore whether we integrate $\varphi$ over the first or second family of generators, we obtain the same result, namely the integral of $f(z)$ over the two-dimensional hypersurface. This implies a certain condition on $\varphi$. Other conditions can be obtained by treating $Pf$ instead of $f$, where $P$ is an internal operator. It can be shown that the conditions obtained in this way are equivalent to Eq. (5).

Let us now turn to Conditions 2 and 3. We have seen that $f(z)$ may be expressed in terms of $\varphi$ by the inversion formula [Eq. (9) of Section 2.4]

$$f(z) = -\frac{1}{8\pi^2}\frac{i}{2}\int_{\Gamma} \psi(u, v; uz_1 + vz_3, uz_2 + vz_4)$$
$$\times (u\, dv - v\, du)(\bar{u}\, d\bar{v} - \bar{v}\, d\bar{u}), \qquad (7)$$

where

$$\psi(u, v; u', v') = L\bar{L}\varphi(u, v; u', v'),$$

$$L = -u\frac{\partial}{\partial u} - v\frac{\partial}{\partial v} + u'\frac{\partial}{\partial u'} + v'\frac{\partial}{\partial v'}.$$

Now the integral in (7) is taken over the set of all lines on the $(z, z) = 1$ surface that pass through the given $z$. As $z$ is allowed to move off to infinity, this set of lines goes over into a set of parallel lines on the surface. Thus the condition that $f(z) \to 0$ as $|z| \to \infty$ implies that the integral of $\psi$ over a set of parallel lines on the $(z, z) = 1$ surface must vanish. It can be shown that each of the conditions $|z|^k f(z) \to 0$ as $|z| \to \infty$ (for positive integer $k$) implies that if a certain differential operator is applied to $\psi$ and the result is integrated over a set of parallel lines, the integral must vanish. It can be shown that the resulting conditions on $\psi$ are equivalent to Eqs. (6) and (6′).

## 3. The Radon Transform in the Complex Domain

### 3.1. Definition of the Radon Transform

We wish now to study the relation between functions $f(z)$ on a complex affine space and the integrals of these functions over all possible hyperplanes. In a sense this section is an appendix to Chapter I, where we discussed similar questions in a real space.

Let us start by defining the integral over a hyperplane in a complex affine space. Consider an $n$-dimensional complex affine space consisting of points $z = (z_1, \ldots, z_n)$. The hyperplanes are given by equations of the form

$$(\zeta, z) = s,$$

where we now write $(\zeta, z) = \zeta_1 z_1 + \cdots + \zeta_n z_n$. With each such hyperplane we associate a differential form $\omega$ defined by

$$dz_1 \cdots dz_n = d(\zeta, z)\omega. \tag{1}$$

An expression for this differential form is easily obtained in any coordinate system on the hyperplane. For instance, if we choose the coordinates on the $(\zeta, z) = s$ hyperplane to be all the $z_j$ except $z_k$, it takes the form

$$\omega = (-1)^{k-1}\zeta_k^{-1}\, dz_1 \cdots dz_{k-1}\, dz_{k+1} \cdots dz_n . \tag{2}$$

Let $f(z)$ be an infinitely differentiable rapidly decreasing (together with all of its derivatives) function. *Then the integral of $f(z)$ over the $(\zeta, z) = s$ hyperplane shall be defined by*[17]

$$\check{f}(\zeta, s) = \left(\frac{i}{2}\right)^{n-1} \int_{(\zeta,z)=s} f(z)\omega\bar{\omega}, \tag{3}$$

[17] The power of $i/2$ is related to the differential form in the integrand and is here chosen so as to make the differential form real (see Volume 1, Appendix B, Section 2.1).

*where* $\omega$ *is the differential form defined by Eq.* (1) *and* $\bar{\omega}$ *is its complex conjugate*

$$\bar{\omega} = (-1)^{k-1}\bar{\zeta}_k^{-1}\, d\bar{z}_1 \cdots d\bar{z}_{k-1}\, d\bar{z}_{k+1} \cdots d\bar{z}_n\,.$$

We shall call $\check{f}(\zeta, s)$, defined on the set of hyperplanes, the Radon transform of $f(z)$. It is also convenient to express $\check{f}(\zeta, s)$ as an integral over all space by using the $\delta$ function. We then have

$$\check{f}(\zeta, s) = \left(\frac{i}{2}\right)^n \int f(z)\, \delta(s - (\zeta, z))\, dz\, d\bar{z}, \tag{4}$$

where $dz = dz_1 \cdots dz_n$. It is seen from this expression, among other things, that $\check{f}(\zeta, s)$ is homogeneous in $\zeta, s$ of degree $(-1, -1)$. This means that for any complex number $\alpha \neq 0$ we have

$$\check{f}(\alpha\zeta, \alpha s) = \alpha^{-1}\bar{\alpha}^{-1}\check{f}(\zeta, s). \tag{5}$$

Thus $\check{f}(\zeta, s)$ actually depends upon the same number of complex variables as does $f(z)$.

Remark. The hyperplanes we are dealing with in the complex $n$-dimensional space have $(2n - 2)$ real dimensions. Thus the Radon transform in $n$ complex dimensions, when stated in terms of $2n$ real dimensions, involves integrating $f$ over certain $(2n - 2)$-dimensional hyperplanes forming a $2n$-parameter family. #

We state without proof some of the elementary properties of the Radon transform. The assertions can be verified essentially exactly as in the real case (see Chapter I, Section 1.3), and we shall therefore omit the proofs.

(a) *Let* $A$ *be a nonsingular linear transformation of the* $z_j$. *Then the Radon transform of*

$$f_A(z) = f(A^{-1}z)$$

*is*

$$\check{f}_A(\zeta, s) = |\det A|^2 \check{f}(A'\zeta, s), \tag{6}$$

*where* $\check{f}(\zeta, s)$ *is the Radon transform of* $f(z)$, *and* $A'$ *is the transpose of* $A$ [*i.e.*, $(\zeta, Az) = (A'\zeta, z)$].

(b) *The Radon transform of* $f(z + a)$ *is*

$$\check{f}(\zeta, s + (\zeta, a)),$$

(c) *The Radon transform of*

$$\left[\left(a, \frac{\partial}{\partial z}\right) + \left(b, \frac{\partial}{\partial \bar{z}}\right)\right] f(z) = \sum_{k=1}^{n} \left(a_k \frac{\partial}{\partial z_k} + b_k \frac{\partial}{\partial \bar{z}_k}\right) f(z)$$

*is*

$$(a, \zeta) \frac{\partial \check{f}(\zeta, s)}{\partial s} + (b, \bar{\zeta}) \frac{\partial \check{f}(\zeta, s)}{\partial \bar{s}} . \tag{7}$$

A consequence of (c) is that if $P(\xi, \eta)$ is a polynomial homogeneous of degree $k$ in $\xi$ and homogeneous of degree $l$ in $\eta$, the Radon transform of

$$P\left(\frac{\partial}{\partial z}, \frac{\partial}{\partial \bar{z}}\right) f(z)$$

is

$$P(\zeta, \bar{\zeta}) \frac{\partial^{k+l} \check{f}(\zeta, s)}{\partial s^k \, \partial \bar{s}^l} .$$

(d) *The Radon transform of the convolution*

$$f(z) = \left(\frac{i}{2}\right)^n \int f_1(z - z') f_2(z') \, dz' \, d\bar{z}'$$

*is given by*

$$\check{f}(\zeta, s) = \frac{i}{2} \int \check{f}_1(\zeta, s - t) \check{f}_2(\zeta, t) \, dt \, d\bar{t}, \tag{8}$$

*where* $\check{f}_1$ *and* $\check{f}_2$ *are the Radon transforms of* $f_1$ *and* $f_2$ .

## 3.2. Representation of $f(z)$ in Terms of Its Radon Transform

Let $f(z)$ be a function in an $n$-dimensional complex space, and let $\check{f}(\zeta, s)$ be its Radon transform

$$\check{f}(\zeta, s) = \left(\frac{i}{2}\right)^n \int f(z) \, \delta(s - (\zeta, z)) \, dz \, d\bar{z}. \tag{1}$$

We wish to obtain the inverse Radon transform formula, that is, to invert Eq. (1).

**Remark.** A similar problem was solved in Chapter I for the real case. It was found there that the inversion formula is different for even and odd dimensions. We shall see that in the complex case no such difference will arise, and that the inversion formula is analogous to the odd-dimensional real case. #

In order to obtain the inversion formula we shall repeat essentially the considerations of Chapter I for the odd-dimensional real case. We first differentiate $\check{f}(\zeta, s)$, $n-1$ times with respect to $s$ and $n-1$ times with respect to $\bar{s}$.[18] The function so obtained will be denoted by

$$\psi(\zeta, s) = \check{f}_s^{(n-1,n-1)}(\zeta, s). \tag{2}$$

Next we average $\psi(\zeta, s)$ over the set of hyperplanes passing through a given point $z$. It will be shown that up to a constant factor this gives the desired value of $f$ at $z$.

The average of $\psi(\zeta, s)$ over the hyperplanes passing through $z$ means the following. Note first that every hyperplane passing through some point $z^0$ is given by an equation of the form $(\zeta, z) = (\zeta, z^0)$. Thus what we are dealing with is the average of $\psi(\zeta, (\zeta, z))$ over $\zeta$. But this function is homogeneous of degree $(-n, -n)$ in $\zeta$ [since $\check{f}$, whose derivative was taken in order to obtain $\psi$, is homogeneous of degree $(-1, -1)$]. Because of this homogeneity, it is natural to define the average of $\psi(\zeta, (\zeta, z))$ as its residue, i.e., as the integral

$$\left(\frac{i}{2}\right)^{n-1} \int_\Gamma \psi(\zeta, (\zeta, z))\omega(\zeta)\bar{\omega}(\zeta) \tag{3}$$

over any surface $\Gamma$ in $\zeta$ space which crosses each complex line passing through the origin once.[19] Here the differential form $\omega(\zeta)$ is defined by

$$\omega(\zeta) = \sum_{k=1}^{n} (-1)^{k-1}\zeta_k \, d\zeta_1 \cdots d\zeta_{k-1} \, d\zeta_{k+1} \cdots d\zeta_n \tag{4}$$

(see Volume 1, Appendix B, Section 2.5). Note that since the integrand is homogeneous of degree $(0, 0)$ this integral may be treated as an integral in the projective space of complex lines passing through $\zeta = 0$. What we wish to show, therefore, is the following.

*Let $\check{f}(\zeta, s)$ be the Radon transform of $f(z)$. Then $f(z)$ can be given in terms of its Radon transform by*

$$\left(\frac{i}{2}\right)^{n-1} \int_\Gamma \check{f}_s^{(n-1,n-1)}(\zeta, (\zeta, z))\omega(\zeta)\bar{\omega}(\zeta) = cf(z), \tag{5}$$

*where*

$$\omega(\zeta) = \sum_{k=1}^{n} (-1)^{k-1}\zeta_k \, d\zeta_1 \cdots d\zeta_{k-1} \, d\zeta_{k+1} \cdots d\zeta_n \,,$$

[18] It is natural to call $\partial/\partial s$ and $\partial/\partial \bar{s}$ the infinitesimal parallel translation operators for the hyperplane, since as $s$ varies the $(\zeta, s) = s$ hyperplane moves parallel to itself.

[19] The meaning of this integral, requiring construction of $\Gamma$ out of pieces of analytic manifolds, is discussed, for instance, at the end of Section 1.4, or in Volume 1, p. 395.

*and the integral is over any surface $\Gamma$ in $\zeta$ space which intersects once every complex line passing through the origin.* The constant $c$ will be determined later.

We prove Eq. (5) first for $z = 0$; that is, we prove that

$$\left(\frac{i}{2}\right)^{n-1} \int_\Gamma \check{f}_s^{(n-1,n-1)}(\zeta, 0)\omega(\zeta)\bar{\omega}(\zeta) = cf(0). \tag{6}$$

Equation (6) defines a continuous functional $F$ on the space of infinitely differentiable rapidly decreasing functions:

$$(F, f) = \left(\frac{i}{2}\right)^{n-1} \int_\Gamma \check{f}_s^{(n-1,n-1)}(\zeta, 0)\omega(\zeta)\bar{\omega}(\zeta). \tag{7}$$

What we must show is that $F = c\,\delta(z)$. Now it is easily established that $F$ satisfies the condition

$$(F, f(Az)) = (F, f(z)), \tag{8}$$

where $A$ is any nonsingular linear transformation.[20] The left side of (8) can be averaged over the set of transformations $A$ which leave invariant the quadratic form

$$|z_1|^2 + \cdots + |z_n|^2$$

(that is, over the set of unitary transformations), which yields

$$(F, f_1(z)) = (F, f(z)), \tag{9}$$

where $f_1(z)$ is the average of $f(z)$ over spheres of the form

$$|z_1|^2 + \cdots + |z_n|^2 = r^2.$$

This means that in calculating $(F, f)$ we may replace $f$ by its average over such spheres. In other words $F$ may be considered a functional in the space of functions on the half-line $0 \leqslant r < \infty$. But on this half-line $F$ is homogeneous of degree $-1$, for $(F, f(\alpha z)) = (F, f(z))$ for $\alpha \neq 0$. Now it is easily shown that up to a factor any homogeneous generalized function of degree $-1$ on the half-line is the $\delta$ function (cf. Volume 1, Chapter I, Section 3.11). This proves that $F = c\,\delta(z)$, or Eq. (6).

The expression for $f(z)$ at any other point $z^0$ in terms of its Radon transform can now be obtained from Eq. (6). All we need do is apply

[20] See the similar discussion for the real case in Chapter I, Section 1.4.

Eq. (6) to $f_1(z) = f(z + z^0)$. Since the Radon transform of $f_1(z)$ is $\check{f}(\zeta, s + (\zeta, z_0))$ (see Section 3.1), we arrive at

$$\left(\frac{i}{2}\right)^{n-1} \int_{\Gamma} \check{f}_s^{(n-1,n-1)}(\zeta, (\zeta, z_0))\omega(\zeta)\bar{\omega}(\zeta) = cf(z_0). \tag{10}$$

We have yet to calculate the value of $c$ in this inversion formula. To do this, we apply it to the function

$$f(z) = \exp(-| z_1 |^2 - \cdots -| z_n |^2).$$

The Radon transform of this $f(z)$ is

$$\check{f}(\zeta, s) = \left(\frac{i}{2}\right)^n \int \exp(-| z_1 |^2 - \cdots -| z_n |^2)\, \delta(s - \zeta_1 z_1 - \cdots - \zeta_n z_n) \times dz_1\, d\bar{z}_1 \cdots dz_n\, d\bar{z}_n .$$

By applying a unitary transformation to the $z_j$, this integral can be written in the form

$$\check{f}(\zeta, s) = \left(\frac{i}{2}\right)^n \int \exp(-| z_1 |^2 - \cdots -| z_n |^2)\, \delta(s - | \zeta | z_1) \times dz_1\, d\bar{z}_1 \cdots dz_n\, d\bar{z}_n ,$$

where $| \zeta |^2 = | \zeta_1 |^2 + \cdots + | \zeta_n |^2$. Integrating, we obtain

$$\check{f}(\zeta, s) = \frac{\pi^{n-1}}{| \zeta |^2} \exp\left(-\frac{s\bar{s}}{| \zeta |^2}\right).$$

We may take a series expansion of the exponential to arrive at

$$\check{f}_s^{(n-1,n-1)}(\zeta, 0) = (-1)^{n-1}(n-1)!\pi^{n-1} | \zeta |^{-2n}.$$

Now we put $f(0)$ and $\check{f}_s^{(n-1,n-1)}(\zeta, 0)$ into Eq. (6), from which we find that

$$c = (-1)^{n-1}(n-1)!\pi^{n-1} \left(\frac{i}{2}\right)^{n-1} \int_{\Gamma} \frac{\omega(\zeta)\bar{\omega}(\zeta)}{| \zeta |^{2n}} .$$

The integral here is easily calculated. For instance, choose $\Gamma$ to be $\zeta_1 = 1$. Then

$$\left(\frac{i}{2}\right)^{n-1} \int \frac{\omega(\zeta)\bar{\omega}(\zeta)}{| \zeta |^{2n}} = \left(\frac{i}{2}\right)^{n-1} \int \frac{d\zeta_2\, d\bar{\zeta}_2 \cdots d\zeta_n\, d\bar{\zeta}_n}{(1 + | \zeta_2 |^2 + \cdots + | \zeta_n |^2)^n}$$

$$= \Omega_{2n-2} \int_0^{\infty} \frac{r^{2n-3}\, dr}{(1 + r^2)^n} = \frac{\pi^{n-1}}{(n-1)!}$$

(where $\Omega_{2n-2}$ is the area of the unit sphere in a space of $2n - 2$ real dimensions). Consequently

$$c = (-1)^{n-1}\pi^{2n-2}.$$

We thus arrive at the following final result. *Let $f(z)$ be an infinitely differentiable rapidly decreasing function of $n$ complex variables, and let $\check{f}(\zeta, s)$ be its Radon transform. Then*

$$f(z) = \frac{(-1)^{n-1}}{\pi^{2n-2}} \left(\frac{i}{2}\right)^{n-1} \int_\Gamma \check{f}_s^{(n-1,n-1)}(\zeta, (\zeta, z))\omega(\zeta)\bar{\omega}(\zeta), \tag{11}$$

*where*

$$\omega(\zeta) = \sum_{k=1}^{n} (-1)^{k-1}\zeta_k \, d\zeta_1 \cdots d\zeta_{k-1} \, d\zeta_{k+1} \cdots d\zeta_n \,,$$

*and the integral is taken over any surface $\Gamma$ in $\zeta$ space which intersects each complex line passing through the origin just once.*

### 3.3. Analog of Plancherel's Theorem for the Radon Transform

Let $f(z)$ and $g(z)$ be two infinitely differentiable rapidly decreasing functions in an $n$-dimensional complex affine space, and let $\check{f}(\zeta, s)$ and $\check{g}(\zeta, s)$ be their Radon transforms. We shall prove that *these functions and their Radon transforms satisfy the analog of Plancherel's theorem*

$$\left(\frac{i}{2}\right)^n \int f(z)\bar{g}(z) \, dz \, d\bar{z}$$

$$= \frac{(-1)^{n-1}}{\pi^{2n-2}} \left(\frac{i}{2}\right)^n \int_\Gamma \left[\int \check{f}(\zeta, s)\bar{\check{g}}_s^{(n-1,n-1)}(\zeta, s) \, ds \, d\bar{s}\right] \omega(\zeta)\bar{\omega}(\bar{\xi}), \tag{1}$$

*or its equivalent*

$$\left(\frac{i}{2}\right)^n \int f(z)\bar{g}(z) \, dz \, d\bar{z}$$

$$= \frac{1}{\pi^{2n-2}} \left(\frac{i}{2}\right)^n \int_\Gamma \left[\int \frac{\partial^{n-1}\check{f}(\zeta, s)}{\partial s^{n-1}} \frac{\partial^{n-1}\bar{\check{g}}(\zeta, s)}{\partial \bar{s}^{n-1}} \, ds \, d\bar{s}\right] \omega(\zeta)\bar{\omega}(\zeta), \tag{2}$$

*where $\omega(\zeta)$ is the differential form previously defined and $\Gamma$, as usual, is any surface in $\zeta$ space which intersects every complex line passing through* *the origin just once.*[21] [Equation (2) can be obtained from (1) by integration by parts.]

**Proof.** Let us apply the inversion formula for the Radon transform of Section 3.2 to the convolution

$$F(z) = \left(\frac{i}{2}\right)^n \int f(u)\bar{g}(u - z)\, du\, d\bar{u} \tag{3}$$

of $f(z)$ and $g^*(z) = \bar{g}(-z)$. It is seen from the elementary properties of the Radon transform listed at the end of Section 3.1 that the Radon transform of $F(z)$ is

$$\check{F}(\zeta, s) = \frac{i}{2} \int \check{f}(\zeta, t)\bar{\check{g}}(\zeta, t - s)\, dt\, d\bar{t}.$$

Then from the inversion formula we obtain

$$F(0) = \frac{(-1)^{n-1}}{\pi^{2n-1}} \left(\frac{i}{2}\right)^{n-1} \int_\Gamma \check{F}_s^{(n-1,n-1)}(\zeta, 0)\omega(\zeta)\bar{\omega}(\zeta)$$

$$= \frac{(-1)^{n-1}}{\pi^{2n-2}} \left(\frac{i}{2}\right)^{n} \int_\Gamma \left[\int \check{f}(\zeta, t)\bar{\check{g}}_s^{(n-1,n-1)}(\zeta, t)\, dt\, d\bar{t}\right] \omega(\zeta)\bar{\omega}(\zeta).$$

On the other hand,

$$F(0) = (\tfrac{1}{2}i)^n \int f(u)\bar{g}(u)\, du\, d\bar{u},$$

which leads immediately to the stated result.

We have obtained this result for rapidly decreasing $f(z)$ and $g(z)$. By going to the limit in the usual way, however, we may extend it to any $f(z)$ and $g(z)$ with integrable square modulus.

**Remark.** Plancherel's theorem implies that the correspondence

$$f(z) \to \check{f}_s^{(n-1,n-1)}(\zeta, s)$$

is an isometric mapping of the square integrable functions of $z$, i.e., those that satisfy

$$\|f\|^2 = \left(\frac{i}{2}\right)^n \int |f(z)|^2\, dz\, d\bar{z} < \infty, \tag{4}$$

[21] Recall that in a real space the form of Plancherel's theorem differs depending on whether the dimension of the space is odd or even. The simpler, odd dimensional case is the analog of the result for the complex space.

into the space of homogeneous functions $\psi(\zeta, s)$ of degree $(-n, -n)$ that satisfy

$$\| \psi \|^2 = \frac{1}{\pi^{2n-2}} \left(\frac{i}{2}\right)^n \int_\Gamma \left[\int | \psi(\zeta, s)|^2 \, ds \, d\bar{s}\right] \omega(\zeta)\bar{\omega}(\zeta) < \infty. \tag{5}$$

It can be shown that this mapping is *onto* the latter space, but we shall not go into the proof here. #

### 3.4. Analog of the Paley-Wiener Theorem for the Radon Transform

If $\check{f}(\zeta, s)$ is to be the Radon transform of some infinitely differentiable rapidly decreasing function $f(z)$ of $n$ complex variables, it must satisfy the following necessary and sufficient conditions.

*Condition 1.* $\check{f}(\zeta, s)$ is homogeneous in $\zeta$, $s$ of degree $(-1, -1)$; i.e.,

$$\check{f}(\alpha\zeta, \alpha s) = \alpha^{-1}\bar{\alpha}^{-1}\check{f}(\zeta, s)$$

for any $\alpha \neq 0$.

*Condition 2.* $\check{f}(\zeta, s)$ is infinitely differentiable with respect to $\zeta$, $s$ and $\bar{\zeta}$, $\bar{s}$ for $\zeta \neq 0$.

*Condition 3.* Asymptotically as $|s| \to \infty$ and for any $k > 0$, we have

$$|\check{f}(\zeta, s)| = o(|s|^{-k})$$

uniformly in $\zeta$ for $\zeta$ in any bounded closed region not containing the origin. This equation holds also for each derivative of $\check{f}(\zeta, s)$ to any order with respect to $\zeta$, $s$ or their complex conjugates.

*Condition 4.* For nonnegative integers $k$, $l$, the integral

$$\frac{i}{2} \int \check{f}(\zeta, s) s^k \bar{s}^l \, ds \, d\bar{s}$$

is a polynomial in $\zeta$, $\bar{\zeta}$, homogeneous of degree $(k, l)$.

We saw in Chapter I, Section 1.6 that similar conditions hold in the real case.

We shall not prove these necessary and sufficient conditions, since the proof is exactly the same as in the real case.

### 3.5. Radon Transform of Generalized Functions

The Radon transform of a generalized function will be defined so as to coincide with the usual definition for test functions. Let us rewrite Plancherel's theorem

$$\left(\frac{i}{2}\right)^n \int F(z)f(z)\,dz\,d\bar{z}$$
$$= \frac{(-1)^{n-1}}{\pi^{2n-2}}\left(\frac{i}{2}\right)^n \int_\Gamma \left[\int \check{F}(\zeta, s)\check{f}_s^{(n-1,n-1)}(\zeta, s)\,ds\,d\bar{s}\right]\omega(\zeta)\bar{\omega}(\zeta), \qquad (1)$$

discussed in Section 3.3 (written here, unlike the form in Section 3.3, without taking the complex conjugate of the second factor in the integrand) in the form

$$(F, f) = \frac{(-1)^{n-1}}{\pi^{2n-2}}(\check{F}, \psi), \qquad (2)$$

where

$$\psi(\zeta, s) = \check{f}_s^{(n-1,n-1)}(\zeta, s). \qquad (3)$$

This can now be used to define the Radon transform of a function $F$ as a functional on the space of the $\psi(\zeta, s)$ functions, and this definition is easily extended to generalized functions $F$. Thus let $F$ be a generalized function on the space of test functions $f(z)$. We shall call the Radon transform of $F$ the functional $\check{F}$ on the $\psi(\zeta, s)$ of Eq. (3), where $\check{f}$ is the Radon transform of a test function $f$. The functional $\check{F}$ is defined by an equation formally the same as (2), namely

$$(F, f) = \frac{(-1)^{n-1}}{\pi^{2n-2}}(\check{F}, \psi). \qquad (4)$$

Let our test function space be $S$ (the infinitely differentiable rapidly decreasing functions). In Section 3.4 we learned which $\check{f}(\zeta, s)$ can be Radon transforms of functions in $S$, and we may thus calculate the properties of their derivatives $\psi(\zeta, s)$. Specifically, these derivatives may be characterized in the following way.

*Condition 1.* $\psi(\zeta, s)$ is homogeneous in $\zeta, s$ of degree $(-n, -n)$; i.e.,

$$\psi(\alpha\zeta, \alpha s) = \alpha^{-n}\bar{\alpha}^{-n}\psi(\zeta, s)$$

for all $\alpha \neq 0$.

*Condition 2.* $\psi(\zeta, s)$ is infinitely differentiable with respect to $\zeta, s$ and their complex conjugates for $\zeta \neq 0$.

*Condition 3.* Asymptotically as $|s| \to \infty$ and for any $k > 0$, we have

$$|\psi(\zeta, s)| = o(|s|^{-k})$$

uniformly in $\zeta$ for $\zeta$ in any bounded closed region not containing the origin. This equation holds also for each derivative of $\psi$ to any order with respect to $\zeta$, $s$ or their complex conjugates.

*Condition 4.* For nonnegative integers $k$, $l$ the integral

$$\int \psi(\zeta, s) s^k \bar{s}^l \, ds \, d\bar{s}$$

is a polynomial in $\zeta$, $\bar{\zeta}$, homogeneous of degree $(k - n + 1, l - n + 1)$. (If either $k$ or $l$ is less than $n - 1$, the integral vanishes.)

Thus the Radon transform of a generalized function $F$ is a functional on the space of functions $\psi(\zeta, s)$ satisfying Conditions 1–4. By the theorem on extending functionals, $\check{F}$ can be extended in several different ways to the space of all infinitely differentiable homogeneous functions of degree $(-n, -n)$ satisfying only Condition 3 (rapid decrease in $s$); a natural topology can be introduced in this function space. In this way the Radon transform of a generalized function becomes a generalized function in the usual sense, except that it is not uniquely defined. It is natural then to ask which functions are unessential, that is, which generalized functions correspond to functionals which vanish on the subspace satisfying Condition 4. As is easily shown, *the subspace of unessential functions is generated by functions of the form*

$$s^k \bar{s}^l a_{-k-1,-l-1}(\zeta),$$

*for nonnegative integers $k$, $l$, where $a_{-k-1,-l-1}(\zeta)$ is a homogeneous function of degree $(-k-1, -l-1)$. For $k, l \geqslant n - 1$, this function must satisfy the additional condition*

$$\int a_{-k-1,-l-1}(\zeta) P_{k-n+1,l-n+1}(\zeta) \omega(\zeta) \bar{\omega}(\zeta) = 0 \tag{5}$$

*for any homogeneous polynomial $P_{k-n+1,l-n+1}(\zeta)$ of degree $(k - n + 1, l - n + 1)$.*

The proof of this assertion is exactly the same as in the real case (see Chapter I, Section 2.1), and we shall therefore omit it.

## 3.6. Examples

We now give some examples of the Radon transforms of generalized functions.

**Example 1.** The Radon transform of $F(z) \equiv 1$ is

$$\check{F}(\zeta, s) = \frac{(-1)^{n-1}\pi^{2n-2}}{\Gamma^2(n)} s^{n-1}\bar{s}^{n-1}a(\zeta),$$

where $a(\zeta)$ is any homogeneous function of degree $(-n, -n)$ with residue 1, that is, such that

$$\left(\frac{i}{2}\right)^{n-1} \int_\Gamma a(\zeta)\omega(\zeta)\bar{\omega}(\zeta) = 1.$$

**Example 2.** The Radon transform of $F(z) \equiv \delta(z)$ is $\check{F}(\zeta, s) = \delta(s)$. This is easily shown simply by testing it in the definition

$$(F, f) = \frac{(-1)^{n-1}}{\pi^{2n-2}} (\check{F}, \check{f}_s^{(n-1,n-1)})$$

of the Radon transform of a generalized function.

**Example 3.** The Radon transform of $F(z) \equiv P^\lambda$, where $P = P(z)$ is a positive definite Hermitian form, is

$$\check{F}(\zeta, s) = c(\lambda)\Delta^{-1}s^{\lambda+n-1}\bar{s}^{\lambda+n-1}Q^{-\lambda-n}(\zeta), \tag{6}$$

where $Q(\zeta)$ is the dual Hermitian form [i.e., whose matrix is the inverse of that $P(z)$], $\Delta$ is the discriminant of $P$, and

$$c(\lambda) = \frac{\pi^{n-1}\Gamma(-\lambda - n + 1)}{\Gamma(-\lambda)}.$$

To prove this, we use the fact if $\check{F}(\zeta, s)$ is the Radon transform of $F(z)$, then the Radon transform of $F_A(z) \equiv F(A^{-1}z)$ is $\check{F}_A(\zeta, s) = |\det A|^2 \check{F}(A'\zeta, s)$, for $A$ a nonsingular linear transformation. This means that if $A$ has determinant 1 and leaves $P$ invariant, then

$$\check{F}(A'\zeta, s) = \check{F}(\zeta, s).$$

Therefore $\check{F}$ is of the form

$$\check{F}(\zeta, s) = \varphi(Q, s),$$

where $Q = Q(\zeta)$ is the dual Hermitian form. On the other hand, if $P^\lambda(z)$ is replaced by $|\alpha|^{2\lambda}P^\lambda(z) \equiv P^\lambda(\alpha z)$, its Radon transform $\varphi(Q, s)$ goes over into $|\alpha|^{2\lambda}\varphi(Q, s) \equiv |\alpha|^{-2n}\varphi(|\alpha|^{-2}Q, s)$. Therefore, $\varphi(Q, s)$ is homogeneous in $Q$ of degree $-\lambda-n$, which means that we may write

$$\varphi(Q, s) = \varphi(s)Q^{-\lambda-n}.$$

Finally, when we remember that this must be homogeneous in $\zeta$, $s$ of degree $(-1, -1)$ we arrive at the desired equation (6).

We have now to calculate $c(\lambda)$. Without loss of generality we may assume that $P = \sum \alpha_k^2 z_k \bar{z}_k$ so that $Q = \sum \alpha_k^{-2} \zeta_k \bar{\zeta}_k$. Now let us set $\zeta = \zeta_0 = (0, \ldots, 0, \alpha_n)$, $s = 1$ in Eq. (6). It then reads

$$\check{F}(\zeta_0, 1) = c(\lambda) \Delta^{-1}.$$

By definition, however,

$$\begin{aligned}\check{F}(\zeta_0, 1) &= \left(\frac{i}{2}\right)^{n-1} \int_{\alpha_n z_n = 1} \left(\sum \alpha_k^2 z_k \bar{z}_k\right)^\lambda \omega\bar{\omega} \\ &= \left(\frac{i}{2}\right)^{n-1} \frac{1}{\alpha_n^2} \int (\alpha_1^2 z_1 \bar{z}_1 + \cdots + \alpha_{n-1}^2 z_{n-1} \bar{z}_{n-1} + 1)^\lambda \\ &\quad \times dz_1\, d\bar{z}_1 \cdots dz_{n-1}\, d\bar{z}_{n-1} \\ &= \frac{1}{\alpha_1^2 \cdots \alpha_n^2} \int (t_1^2 + \cdots + t_{2n-2}^2 + 1)^\lambda\, dt_1 \cdots dt_{2n-2},\end{aligned}$$

or

$$\check{F}(\zeta_0, 1) = \frac{\pi^{n-1} \Gamma(-\lambda - n + 1)}{\Gamma(-\lambda)} \frac{1}{\Delta}.$$

Our expression for $c(\lambda)$ follows immediately.

It is suggested that the reader calculate also the Radon transforms of $(P + i0)^\lambda$, $(P - i0)^\lambda$, $P_+^\lambda$, and $P_-^\lambda$, for $P = P(z)$ any nondegenerate Hermitian form. [Hint: make use of Eq. (6) and the method of analytic continuation in the coefficients of the quadratic form.]

Example 4. The Radon transform of $F(z) \equiv P^\lambda \bar{P}^\mu$, where $P = P(z)$ is a nondegenerate quadratic form and $\lambda - \mu$ is an integer, is

$$\check{F}(\zeta, s) = c(\lambda, \mu) |\, \Delta\, |^{-1} s^{2\lambda+n-1} \bar{s}^{2\mu+n-1} Q^{-\lambda-\frac{1}{2}n} \bar{Q}^{-\mu-\frac{1}{2}n}, \tag{7}$$

where $Q = Q(\zeta)$ is the dual quadratic form, $\Delta$ is the discriminant of $P$, and

$$c(\lambda, \mu) = (-1)^{-n-1} 2^{-n+1} \pi^{n-2} \frac{\sin[\pi(\lambda + \frac{1}{2}n)] \Gamma(\lambda + 1) \Gamma(\mu + 1)}{\cos \pi\lambda\, \Gamma(\lambda + \frac{1}{2}n + \frac{1}{2}) \Gamma(\mu + \frac{1}{2}n + \frac{1}{2})}. \tag{8}$$

Equation (7) is obtained by the same sort of considerations as were used in Example 3. To calculate $c(\lambda, \mu)$ is, however, somewhat less trivial. We may assume without loss of generality that $P = \sum_{k=1}^{n} z_k^2$ and therefore $Q = \sum_{n}^{k=1} \zeta_k^2$. Let us apply Plancherel's theorem to the pair of functions $P^\lambda \bar{P}^\lambda$ and $\exp(-|z|^2)$, where $|z|^2 = \sum |z_k|^2$. It is easily shown that the Radon transform of $\exp(-|z|^2)$ is

$$\pi^{n-1} |\zeta|^{-2} \exp(-|s/\zeta|^2).$$

Therefore Plancherel's theorem states that

$$\begin{aligned}\left(\frac{i}{2}\right)^n \int |P|^{2\lambda} \exp(-|z|^2)\, dz\, d\bar{z} \\ = c(\lambda, \lambda) \frac{(-1)^{n-1}}{\pi^{n-1}} \frac{\Gamma^2(2\lambda + n)}{\Gamma^2(2\lambda + 1)} \left(\frac{i}{2}\right)^n \int_\Gamma \Big[\int s^{2\lambda} \bar{s}^{2\lambda} \frac{|Q(\zeta)|^{-2\lambda - n}}{|\zeta|^2} \\ \times \exp\left(-\frac{|s|^2}{|\zeta|^2}\right) ds\, d\bar{s}\Big]\, \omega(\zeta)\bar{\omega}(\zeta).\end{aligned} \tag{9}$$

The integral on the left-hand side has been calculated in Volume 1, pp. 401–402. There we found that

$$\left(\frac{i}{2}\right)^n \int |P|^{2\lambda} \exp(-|z|^2)\, dz\, d\bar{z} = \frac{\pi^n 2^{2\lambda} \Gamma(\lambda + 1)\Gamma(\lambda + \frac{1}{2}n)}{\Gamma(\frac{1}{2}n)}. \tag{10}$$

The integration over $s$ and $\bar{s}$ on the right-hand side of (9) is easily carried through. Then (9) becomes

$$\begin{aligned}\frac{\pi^n 2^{2\lambda} \Gamma(\lambda + 1)\Gamma(\lambda + \frac{1}{2}n)}{\Gamma(\frac{1}{2}n)} = c(\lambda, \lambda) \frac{(-1)^{n-1} \Gamma^2(2\lambda + n)}{\pi^{n-2} \Gamma(2\lambda + 1)} \\ \times \left(\frac{i}{2}\right)^{n-1} \int_\Gamma |Q(\zeta)|^{-2\lambda - n} |\zeta|^{2\lambda} \omega(\zeta)\bar{\omega}(\zeta).\end{aligned} \tag{11}$$

Now to calculate the integral on the right-hand side here, we shall proceed in the following way. In general,

$$\begin{aligned}\left(\frac{i}{2}\right)^n \int |Q(z)|^{-2\lambda - n} \exp(-|z|^2)\, dz\, d\bar{z} \\ = \left(\frac{i}{2}\right)^n \int_\Gamma \Big[\int |Q(t\zeta)|^{-2\lambda - n} \exp(-|t\zeta|^2) t^{n-1} \bar{t}^{n-1}\, dt\, d\bar{t}\Big]\, \omega(\zeta)\bar{\omega}(\zeta)\end{aligned} \tag{12}$$

(a transition to generalized polar coordinates). The integral on the left-hand side can be calculated by (10), while on the right-hand side the integration over $t$, $\bar{t}$ is relatively simple.

As a result we find that

$$\left(\frac{i}{2}\right)^{n-1}\int_{I}|Q(\zeta)|^{-2\lambda-n}\,|\,\zeta\,|^{4\lambda}\omega(\zeta)\bar{\omega}(\zeta)$$
$$=\frac{\pi^{n-1}2^{-2\lambda-n}\Gamma(-\lambda-\frac{1}{2}n+1)\Gamma(-\lambda)}{\Gamma(\frac{1}{2}n)\Gamma(-2\lambda)}.$$

When this is inserted into (11), we arrive at

$$\begin{aligned}c(\lambda,\lambda)&=\frac{(-1)^{n-1}\pi^{n-1}2^{4\lambda+n}\Gamma(\lambda+1)\Gamma(\lambda+\frac{1}{2}n)\Gamma(2\lambda+1)\Gamma(-2\lambda)}{\Gamma^2(2\lambda+n)\Gamma(-\lambda-\frac{1}{2}n+1)\Gamma(-\lambda)}\\&=(-1)^{n-1}2^{-n+1}\pi^{n-2}\frac{\sin[\pi(\lambda+\frac{1}{2}n)]\Gamma^2(\lambda+1)}{\cos\pi\lambda\;\Gamma^2(\lambda+\frac{1}{2}n+\frac{1}{2})}.\end{aligned}\tag{13}$$

We can now calculate $c(\lambda, \mu)$ by writing[22] $\mu = \lambda + k$ and using the recursion relation

$$(\textstyle\sum \partial^2/\partial \bar{z}_k^2)P^{\lambda}\bar{P}^{\mu+1} = 4(\mu+1)(\mu+\frac{1}{2}n)P^{\lambda}\bar{P}^{\mu}.$$

On taking the Radon transform of both sides and canceling some terms, we obtain a recursion relation for $c(\lambda, \mu)$, namely

$$c(\lambda,\mu+1)=\frac{\mu+1}{\mu+\frac{1}{2}n+\frac{1}{2}}\,c(\lambda,\mu).$$

Thus

$$c(\lambda,\mu)\equiv c(\lambda,\lambda+k)=\frac{\Gamma(\lambda+k+1)}{\Gamma(\lambda+1)}\,\frac{\Gamma(\lambda+\frac{1}{2}n+\frac{1}{2})}{\Gamma(\lambda+k+\frac{1}{2}n+\frac{1}{2})}\cdot c(\lambda,\lambda).$$

We now use (13), arriving immediately at Eq. (8).

Example 5. Let $f(z) = f(z_1, \ldots, z_n)$ be an entire analytic function. Such a function can be treated as a generalized function on the space of all infinitely differentiable functions *with bounded support* (in $n$ complex variables).[23] Then the Radon transform $\check{f}(\zeta, s)$ of $f(z)$ is defined. We may now ask for the properties of such a Radon transform.

It is found that *the Radon transform of an entire analytic function is necessarily of the form*

$$\check{f}(\zeta,s)=s^{n-1}\bar{s}^{n-1}\varphi(\zeta,s),\tag{14}$$

[22] Recall that $\mu - \lambda$ must be an integer.

[23] This example differs from the previous ones in that the test function space is no longer $S$.

*where $\varphi(\zeta, s)$ is an entire analytic function of the complex variable $s$ and satisfies the homogeneity condition*

$$\varphi(\alpha\zeta, \alpha s) = \alpha^{-n}\bar{\alpha}^{-n}\varphi(\zeta, s)$$

*for any $\alpha \neq 0$.*

We shall give a simple but nonrigorous proof of this assertion. The analyticity of $f(z)$ may be stated in the form

$$\frac{\partial f}{\partial \bar{z}_i} = 0, \qquad i = 1, \ldots, n.$$

Now let us take the Radon transforms of these equations. We find that $\bar{\zeta}_i[\partial \check{f}(\zeta, s)/\partial \bar{s}]$ is equal to zero in the sense of generalized functions $(i = 1, \ldots, n)$. This means that

$$\bar{\zeta}_i \frac{\partial \check{f}(\zeta, s)}{\partial \bar{s}} = \sum_{k,l=0}^{\infty} s^k \bar{s}^l a_{ikl}(\zeta), \tag{15}$$

where the $a_{ikl}(\zeta)$ are homogeneous functions of degree $(-k-1, -l-1)$ such that for $k, l \geqslant n-1$

$$\int_\Gamma a_{ikl}(\zeta) P_{k-n+1, l-n+1}(\zeta)\omega(\zeta)\bar{\omega}(\zeta) = 0 \tag{16}$$

for any polynomial $P_{k-n+1, l-n+1}(\zeta)$ homogeneous of degree $(k-n+1, l-n+1)$. Integrating (15) with respect to $\bar{s}$, we arrive at

$$\check{f}(\zeta, s) = \sum_{k,l=0}^{\infty} s^k \bar{s}^l a_{kl}(\zeta), \tag{17}$$

where

$$l a_{kl}(\zeta) = \bar{\zeta}_i^{-1} a_{ik,l-1}(\zeta). \tag{18}$$

(We may assume that $a_{kl}(\zeta)$ is independent of the index $i$.)

Now let us drop those terms in (17) which vanish in the sense of generalized functions. First, this will include all terms for which $k < n-1$ or $l < n-1$. If, further, $k \geqslant n-1$ and $l > n-1$, Eq. (16) implies that

$$\int_\Gamma a_{kl}(\zeta) P_{k-n+1, l-n+1}(\zeta)\omega(\zeta)\bar{\omega}(\zeta) = 0.$$

This means that all terms with $k \geqslant n-1$ and $l > n-1$ may also be dropped. We are left, therefore, only with those terms for which $k \geqslant n-1$ and $l = n-1$. Thus

$$\check{f}(\zeta, s) = s^{n-1}\bar{s}^{n-1}\varphi(\zeta, s),$$

where $\varphi(\zeta, s)$ is an entire function of the complex variable $s$.

## 3.7. The Generalized Hypergeometric Function in the Complex Domain

In Chapter I, Section 2.9 we defined the generalized hypergeometric function in the real domain. We now wish to extend it to the complex domain.

For this purpose we introduce the concept of a *character of the multiplicative group of complex numbers* $z \neq 0$. We shall call a *character* any continuous function $\chi(z)$ on the group of complex numbers $z \neq 0$ such that

$$\chi(z_1 z_2) = \chi(z_1)\chi(z_2) \tag{1}$$

for $z_1$ and $z_2$ in the group. Let us find all the characters of the group of complex numbers. In polar coordinates we write $z_k = r_k \exp(i\varphi_k)$. Then Eq. (1) implies that

$$\chi(r_1 r_2) = \chi(r_1)\chi(r_2)$$

and

$$\chi(\exp[i(\varphi_1 + \varphi_2)]) = \chi(\exp[i\varphi_1])\chi(\exp[i\varphi_2]).$$

This gives $\chi(r) = r^\alpha$ and $\chi(e^{i\varphi}) = e^{in\varphi}$, where $\alpha$ is any complex number and $n$ is any integer. Therefore every character of the multiplicative group of complex numbers is of the form

$$\chi(z) = | z |^\alpha \exp(in \arg z), \tag{2}$$

which may more conveniently be written

$$\chi(z) = z^\lambda \bar{z}^\mu, \tag{3}$$

where $\lambda$ and $\mu$ are complex numbers whose difference is an integer (in fact, $\lambda + \mu = \alpha$, $\lambda - \mu = n$). Recall that in Volume 1, Appendix B, Section 1.3 it was established that this function may be considered a homogeneous generalized function of the variable $z$. With this we may proceed to the definition of the generalized hypergeometric function.

Consider the $p$ linear forms

$$(\zeta^{(1)}, z), \ldots, (\zeta^{(p)}, z), \tag{4}$$

in $n$ complex variables, where we write $z = (z_1, \ldots, z_n)$, $(\zeta^{(k)}, z) = \zeta_1^{(k)} z_1 + \cdots + \zeta_n^{(k)} z_n$.

We shall define *the generalized hypergeometric function*

$$F(\chi_1, \ldots, \chi_p \mid \zeta^{(1)}, \ldots, \zeta^{(p)} \mid \zeta, s)$$

*as the Radon transform of the generalized function*

$$\chi_1[(\zeta^{(1)}, z)] \cdots \chi_p[(\zeta^{(p)}, z)],$$

*where the $\chi_k$ are characters; in other words,*

$$F(\chi_1, \ldots, \chi_p \mid \zeta^{(1)}, \ldots, \zeta^{(p)} \mid \zeta, s) = \{\chi_1[(\zeta^{(1)}, z)] \cdots \chi_p[(\zeta^{(p)}, z)]\}^{\vee}. \tag{5}$$

We can always assume as we did in the real case that the number of linearly independent linear forms is at least $n$, where $n$ is the dimension of the space. When all the linear forms are linearly independent, it is natural to call $F$ the *generalized beta function.*

Let us turn in particular to the two-dimensional case. Let $\zeta^{(1)} = (\zeta_1^{(1)}, \zeta_2^{(1)})$, $\zeta^{(2)} = (\zeta_1^{(2)}, \zeta_2^{(2)})$ be two linearly independent vectors. For any other vector we shall write

$$\zeta = \zeta_1 \zeta^{(1)} + \zeta_2 \zeta^{(2)}.$$

Let $\chi_1$ and $\chi_2$ be characters, which we write in the form

$$\chi_1(z) = z^{m_1 - 1} \bar{z}^{m_2 - 1}, \qquad \chi_2(z) = z^{n_1 - 1} \bar{z}^{n_2 - 1}.$$

We then have

$$F(\chi_1, \chi_2 \mid \zeta^{(1)}, \zeta^{(2)} \mid \zeta, s)$$

$$= \frac{\sin \pi m_1 \sin \pi n_1}{\sin \pi(m_1 + n_1)} B(m_1, n_1) B(m_2, n_2) |\Delta|^{-2} \left| \frac{s}{\zeta_1 \zeta_2} \right|^2 \chi_1\left(\frac{s}{\zeta_1}\right) \chi_2\left(\frac{s}{\zeta_2}\right), \tag{6}$$

where

$$\Delta = \begin{vmatrix} \zeta_1^{(1)} & \zeta_2^{(1)} \\ \zeta_1^{(2)} & \zeta_2^{(2)} \end{vmatrix}$$

and $B$ is the ordinary beta function. The product of the last three factors in (6) must, of course, be understood in the sense of generalized functions.

The derivation of (6) is left as a problem for the reader.

CHAPTER III

# REPRESENTATIONS OF THE GROUP OF COMPLEX UNIMODULAR MATRICES IN TWO DIMENSIONS

In the next two chapters we shall develop the representation theory and harmonic analysis (the analog of the Fourier integral) for the group $G$ of two-dimensional complex matrices

$$\begin{Vmatrix} \alpha & \beta \\ \gamma & \delta \end{Vmatrix}$$

with determinant $\alpha\delta - \beta\gamma = 1$. This group is interesting for several reasons. First, it is (locally) isomorphic to important groups such as the Lobachevskian motions and the Lorentz transformations (as we shall establish in the first section). Second, it is the simplest of the so-called simple Lie groups, which exhibit particularly clearly the difference between harmonic analysis (the Fourier integral) on the group and harmonic analysis in Euclidean space (the ordinary Fourier integral). At first sight it may seem that rather than this group we might study the even simpler group of two-dimensional real *unimodular* matrices (i.e., with determinant one), which is also of some interest (it is locally isomorphic to the group of Lorentz transformations of the plane and to the group of Lobachevskian motions of the plane). This is not so, however, for complex groups are simpler than real ones (real groups exhibit specific additional pecularities). In any case, in Chapter VII we shall consider the representations of the group of two-dimensional real unimodular matrices.

Chapters III and IV will thus form an indivisible unit in which our basic purpose will be to construct the Fourier integral on the group of two-dimensional complex unimodular matrices. If we were undertaking this task for the usual Fourier integral, however, we would be in a better position, for the functions used in its construction ($e^{i\lambda x}$, the trigonometric functions) are quite familiar. Now, however, we must start from the very beginning to develop the analog of the $e^{i\lambda x}$. Chapter III is devoted to this preliminary problem, and Chapter IV to harmonic analysis itself. It may be remarked that generalized functions are particularly helpful in developing the exponential function for the group, so that Chapter III is of interest not only as it pertains to Chapter IV; it has inherent interest as an illustration of the power of generalized functions.

The analogs of exponentials on a group are called representations. These will be defined more accurately in the second section; we shall not, however, attempt to develop the entire theory of representations in this chapter, to say nothing of the many possible applications of the theory.

An introduction to representation theory, including also applications, will be found in I. M. Gel'fand, R. A. Minlos, and Z. Ya. Shapiro [16], "Representations of the Rotation and Lorentz Groups," whose intersection with the present book is quite small. Together these two books contain many of the facts concerning the representations of the group of two-dimensional complex unimodular matrices. Much useful material is contained also in M. A. Naimark [36], "Linear Representations of the Lorentz Group."

Understanding of the present chapter requires a knowledge of generalized functions of a single complex variable (Volume 1, Appendix B, Section 1).

## 1. The Group of Complex Unimodular Matrices in Two Dimensions and Some of Its Realizations

### 1.1. Connection with the Proper Lorentz Group

Consider the set of Hermitian matrices

$$a = \begin{Vmatrix} x_0 - x_3 & x_2 - ix_1 \\ x_2 + ix_1 & x_0 + x_3 \end{Vmatrix}.$$

With each such matrix $a$ we associate the vector $x = (x_0, x_1, x_2, x_3)$ in a four-dimensional real space $R_4$. This association is one-to-one and linear. Now with each complex unimodular matrix

$$g = \begin{Vmatrix} \alpha & \beta \\ \gamma & \delta \end{Vmatrix}, \qquad \alpha\delta - \beta\gamma = 1,$$

we associate the transformation

$$a' = g^* a g$$

on the Hermitian matrices, where the asterisk denotes the Hermitian conjugate. In this way with each $g$ we associate a linear transformation in $R_4$, and we shall call this transformation $A_g$. It is obvious that this association has the *group property* $A_{g_1g_2} = A_{g_1}A_{g_2}$; in other words, the product of $g_1$ with $g_2$ corresponds to the product of the corresponding linear transformations. Let us now find which linear transformation $A_g$ corresponds to $g$.

Since $\det g = 1$, the transformation from $a$ to $a' = g^*ag$ on the Hermitian matrices leaves invariant the determinant

$$\det a = x_0^2 - x_1^2 - x_2^2 - x_3^2 .$$

In other words, *the quadratic form*

$$x_0^2 - x_1^2 - x_2^2 - x_3^2$$

*is invariant under the* $A_g$. It can be shown, moreover, that $\det A_g = 1$. Indeed, since $A_g$ does not alter the quadratic form, its determinant is either $+1$ or $-1$. If transformations $A_g$ existed with both of these determinants, the $A_g$ would form a disconnected set. But this is impossible, since the set of $g$ matrices is connected. Finally, *the set of* $A_g$ *maps what we shall call the positive cone,* (in physics usage, the *forward* or *future* cone)

$$x_0^2 - x_1^2 - x_2^2 - x_3^2 = 0, \qquad x_0 > 0, \tag{1}$$

*into itself*. Indeed, the inequalities $x_0 > 0$, $x_0^2 - x_1^2 - x_2^2 - x_3^2 > 0$ are necessary and sufficient conditions for $a$ to be positive definite. But it is known that the transformation $a' = g^*ag$ transforms positive definite matrices into positive definite matrices.

We have thus shown that to each two-dimensional complex unimodular matrix $g$ corresponds a linear transformation $A_g$ on $R_4$ with the following properties.

1. $A_g$ leaves invariant the quadratic form

$$x_0^2 - x_1^2 - x_2^2 - x_3^2 .$$

2. $\det A_g = 1$.
3. $A_g$ leaves invariant the positive cone

$$x_0^2 - x_1^2 - x_2^2 - x_3^2 = 0, \qquad x_0 > 0.$$

Linear transformations in $R_4$ satisfying these conditions are called *proper Lorentz transformations*, and the group of all such transformations is called *the proper Lorentz group*. It can be shown that *every* proper Lorentz transformation corresponds to some two-dimensional complex unimodular matrix.

Let us now find all $g$ corresponding to the unit matrix of the Lorentz group. If $g$ is such a matrix, then obviously

$$a = g^*ag \tag{2}$$

for any Hermitian matrix $a$. Let us choose $a = e$ to be the two-dimensional unit matrix. Then $g^*g = e$, or $g^* = g^{-1}$. Then we may rewrite Eq. (2) in the form

$$ga = ag,$$

so that $g$ commutes with every Hermitian matrix $a$. It then follows that $g$ is a multiple of the unit matrix. The requirement that $\det g = 1$ then implies that this multiple must be $\pm 1$. Thus the *identity transformation corresponds to the two matrices*

$$g = \pm \begin{Vmatrix} 1 & 0 \\ 0 & 1 \end{Vmatrix}.$$

This implies immediately that two matrices $g_1$ and $g_2$ correspond to a givent Lorentz transformation if an only if $g_1 = \pm g_2$.

*We have thus established that there exists a correspondence between the two-dimensional complex unimodular matrices and the proper Lorentz transformations. This correspondence is such that to each two-dimensional complex unimodular matrix g corresponds one Lorentz transformation, and to each Lorentz transformation correspond two matrices differing only in sign. This correspondence preserves multiplication.* Because of this correspondence we will henceforth call the group of complex unimodular two-dimensional matrices the Lorentz group.

It is a simple matter to write out the Lorentz transformation corresponding to a given $g$. Let

$$a = \begin{Vmatrix} x_0 - x_3 & x_2 - ix_1 \\ x_2 + ix_1 & x_0 + x_3 \end{Vmatrix}$$

and $a' = g^*ag$, where

$$g = \begin{Vmatrix} \alpha & \beta \\ \gamma & \delta \end{Vmatrix}.$$

Then the elements of $a'$ are

$$a'_{11} = (|\alpha|^2 + |\gamma|^2)x_0 + i(\alpha\bar{\gamma} - \bar{\alpha}\gamma)x_1 + \alpha\bar{\gamma} + \bar{\alpha}\gamma)x_2 + (|\gamma|^2 - |\alpha|^2)x_3;$$

$$a'_{12} = (\bar{\alpha}\beta + \bar{\gamma}\delta)x_0 + i(\bar{\gamma}\beta - \bar{\alpha}\delta)x_1 + (\bar{\gamma}\beta + \bar{\alpha}\delta)x_2 \\ + (\bar{\gamma}\delta - \bar{\alpha}\beta)x_3;$$

$$a'_{21} = \bar{a}'_{12};$$

$$a'_{22} = (|\beta|^2 + |\delta|^2)x_0 + i(\beta\bar{\delta} - \bar{\beta}\delta)x_1 + (\beta\bar{\delta} + \bar{\beta}\delta)x_2 \\ + (|\delta|^2 - |\beta|^2)x_3 .$$

To this matrix corresponds the point $x' = (x'_0, x'_1, x'_2, x'_3)$ in $R_4$ with coordinates

$$x'_0 = \frac{a'_{11} + a'_{22}}{2}, \qquad x'_1 = \operatorname{Im} a'_{21},$$

$$x'_2 = \operatorname{Re} a'_{21}, \qquad x'_3 = \frac{a'_{22} - a'_{11}}{2}.$$

By inserting the above expressions for the matrix elements of $a'$, we obtain the linear transformation in $R_4$ from $x = (x_0, x_1, x_2, x_3)$, the point associated with $a$, to $x'$, namely

$$x'_0 = \tfrac{1}{2}(|\alpha|^2 + |\beta|^2 + |\gamma|^2 + |\delta|^2)x_0 + \operatorname{Im}(\bar{\alpha}\gamma + \bar{\beta}\delta)x_1 \\ + \operatorname{Re}(\bar{\alpha}\gamma + \bar{\beta}\delta)x_2 + \tfrac{1}{2}(|\gamma|^2 + |\delta|^2 - |\alpha|^2 - |\beta|^2)x_3; \tag{$3_0$}$$

$$x'_1 = \operatorname{Im}(\alpha\bar{\beta} + \gamma\bar{\delta})x_0 + \operatorname{Re}(\alpha\bar{\delta} - \bar{\gamma}\beta)x_1 \\ - \operatorname{Im}(\bar{\gamma}\beta + \bar{\alpha}\delta)x_2 + \operatorname{Im}(\alpha\bar{\beta} - \gamma\bar{\delta})x_3; \tag{$3_1$}$$

$$x'_2 = \operatorname{Re}(\bar{\alpha}\beta + \bar{\gamma}\delta)x_0 + \operatorname{Im}(\gamma\bar{\beta} - \alpha\bar{\delta})x_1 \\ + \operatorname{Re}(\bar{\gamma}\beta + \bar{\alpha}\delta)x_2 + \operatorname{Re}(\bar{\gamma}\delta - \bar{\alpha}\beta)x_3; \tag{$3_2$}$$

$$x_3 = \tfrac{1}{2}(|\beta|^2 + |\delta|^2 - |\alpha|^2 - |\gamma|^2)x_0 + \operatorname{Im}(\bar{\beta}\delta - \bar{\alpha}\gamma)x_1 \\ + \operatorname{Re}(\beta\bar{\delta} - \bar{\alpha}\gamma)x_2 + \tfrac{1}{2}(|\alpha|^2 + |\delta|^2 - |\beta|^2 - |\gamma|^2)x_3 . \tag{$3_4$}$$

## 1.2. Connection with Lobachevskian and Other Motions

We have seen that each complex unimodular two-dimensional matrix $g$ induces a Lorentz transformation on $R_4$ according to $a' = g^*ag$, where

$$a = \begin{Vmatrix} x_0 - x_3 & x_2 - ix_1 \\ x_2 + ix_1 & x_0 + x_3 \end{Vmatrix}. \tag{1}$$

These Lorentz transformations map the surfaces

$$x_0^2 - x_1^2 - x_2^2 - x_3^2 = c$$

into themselves, for they preserve the corresponding quadratic form. There are three types of such surfaces. These are either sheet of a two-sheeted hyperboloid (when $c > 0$), a single-sheeted hyperboloid (when $c < 0$), and either the positive or negative cone (when $c = 0$).

Remark. If instead of points $x$ in $R_4$ we were dealing with Hermitian matrices $a$, the surfaces named above would be the following three types of manifolds in the space of Hermitian matrices: all positive definite (or negative definite) Hermitian matrices with fixed determinant $c > 0$, all Hermitian matrices with fixed determinant $c < 0$, and all Hermitian matrices $a \geqslant 0$ (or $a \leqslant 0$)[1] with determinant zero. #

The Lorentz transformations induce transformations that we shall call *motions* of these surfaces. In this way to each complex unimodular two-dimensional matrix $g$ there corresponds a motion on each of the surfaces of the following three types.

(a) Either sheet of the two-sheeted hyperboloid $x_0^2 - x_1^2 - x_2^2 - x_3^2 = c > 0$.

(b) The single-sheeted hyperboloid $x_0^2 - x_1^2 - x_2^2 - x_3^2 = c < 0$.

(c) Either sheet of the cone $x_0^2 - x_1^2 - x_2^2 - x_3^2 = 0$.

It is easily shown that a given motion corresponds to two matrices $g_1$ and $g_2$ if and only if $g_1 = \pm g_2$.

The upper sheet of a two-sheeted hyperboloid together with the motions defined in this way is one model of *Lobachevskian space*. (More detailed discussion of various models of Lobachevskian space and of the relation between them will be found at the beginning of Chapter V.) This means that *the group of complex two-dimensional unimodular matrices is locally isomorphic to the group of Lobachevskian motions*. In addition to Lobachevskian space there exist two related spaces with groups of motions locally isomorphic to the same group of matrices. Models of these spaces are the single-sheeted hyperboloid and the positive cone.

The group of motions on each of these surfaces is *transitive*, i.e., every point of the space can be transformed by some motion to any other point. We shall prove this assertion for the upper sheet of the two-sheeted hyperboloid

$$x_0^2 - x_1^2 - x_2^2 - x_3^2 = 1.$$

[1] That is, matrices $a$ whose corresponding Hermitian form takes on nonnegative (or nonpositive) values.

(The proof for the other surfaces is similar.) As before, we use Eq. (1) to associate the Hermitian matrix $a$ with the point $x = (x_0, x_1, x_2, x_3)$ in $R_4$. Then the points on our surface correspond to positive definite *unimodular* Hermitian matrices. Now it is well known that every such matrix can be written in the form

$$a = g^* g = g^* \begin{Vmatrix} 1 & 0 \\ 0 & 1 \end{Vmatrix} g,$$

where $g$ is a complex unimodular matrix. This proves that there exists a motion transforming the fixed unit matrix into $a$.

## 2. Representations of the Lorentz Group Acting on Homogeneous Functions of Two Complex Variables

### 2.1. Representations of Groups

We shall start by giving a slightly inexact definition of a representation of a group.

A representation of a topological group is a continuous function $T(g)$ on the group such that

$$T(g_1 g_2) = T(g_1) T(g_2) \tag{1}$$

for any two elements $g_1$ and $g_2$ of the group. For the group of real numbers under addition, there exist scalar solutions of Eq. (1), namely the exponential functions. If a group is noncommutative, however, the scalar solutions of Eq. (1) may be very few indeed. It is then natural to require that rather than a scalar function $T(g)$ be a linear operator acting on some linear topological space. Such operator functions can then be thought of as generalizations to arbitrary groups of the idea of an exponential function.

With these preliminaries we may give an accurate definition of a representation of a group. Let $G$ be a topological group and $E$ be some linear topological space. We shall call *a representation of $G$ on $E$ an association such that to each element $g \in G$ there corresponds a linear operator $T(g)$ acting on $E$ and depending continuously on $g$ and satisfying the functional equation*

$$T(g_1 g_2) = T(g_1) T(g_2)$$

*and such that the unit operator corresponds to the identity element of the group*. If $E$ is a finite-dimensional space, the representation is called finite dimensional.

The analogs of the exponentials $e^{i\lambda x}$ with *real* $\lambda$ are the so-called unitary representations, which may be defined as follows. *Let a scalar product* $(\xi, \eta)$ *be introduced on* $E$ (*i.e., a continuous positive Hermitian functional*). *A representation* $T(g)$ *on* $E$ *is called unitary* (*with respect to this scalar product*) *if*

$$(T(g)\xi, T(g)\eta) = (\xi, \eta)$$

*for every* $g \in G$ *and for every pair* $\xi, \eta \in E$.

Our goal in this section is to find the analogs of the exponentials for the group $G$ of two-dimensional complex unimodular matrices.

Consider first the simple example of the finite-dimensional representation of $G$ in which each $g$ is taken to represent itself. Another example is the following. Consider the homogeneous polynomials of degree $n_1 - 1$ of two complex variables $z_1, z_2$, that is, functions of the form

$$P(z_1, z_2) = a_0 z_1^{n_1-1} + a_1 z_1^{n_1-2} z_2 + \cdots + a_{n_1-1} z_2^{n_1-1}.$$

These form a linear space of dimension $n_1$. With every matrix $g \in G$ we associate a linear transformation $T(g)$ on the space of these polynomials, obtaining a matrix of dimension $n_1$. Let $T(g)$ be given by the formula

$$T(g)P(z_1, z_2) = P(\alpha z_1 + \gamma z_2, \beta z_1 + \delta z_2).$$

It is easily verified that $T(g)$ satisfies the functional equation (1) and thereby forms a representation of $G$. Finally, a more general finite-dimensional representation of $G$ is obtained by considering polynomials not only of $z_1$ and $z_2$, but also of their complex conjugates. Let these polynomials be of degree $n_1 - 1$ in $z_1$ and $z_2$, and of degree $n_2 - 1$ in $\bar{z}_1$ and $\bar{z}_2$; they then form a space of dimension $n_1 n_2$. If, for instance, $n_1 = n_2 = 2$, they are of the form

$$P(z_1, z_2; \bar{z}_1, \bar{z}_2) = a_{11} z_1 \bar{z}_1 + a_{12} z_1 \bar{z}_2 + a_{21} z_2 \bar{z}_1 + a_{22} z_2 \bar{z}_2,$$

that is, they are quadratic forms.

**Remark.** The representations on the spaces of polynomials of $z_1, z_2$ and their complex conjugates exhaust all of the so-called *irreducible* finite-dimensional representations of $G$. #

In what follows we shall deal mostly with infinite-dimensional representations. Of course $G$ has very many representations and of these many differ in an essential way from each other. But there are those whose difference is illusory, a consequence only of the fact that we have not given a very informative definition of which we mean by essentially different representations. For instance, the same formulas for the representation can be obtained on a space of infinitely differentiable functions as on a space of functions differentiable only five times.

We shall construct here those representations of $G$ which play a basic role. These will be the representations on the spaces called $D_\chi$, the natural generalizations of the spaces of homogeneous polynomials. It can be shown that every infinite-dimensional *irreducible* representation of the Lorentz group, with some natural additional assumptions, contains as a "nucleus" (that is, as a representation on an everywhere dense subspace with a stronger topology) a representation on $D_\chi$. This is in fact the basic role of the representations on the $D_\chi$.

## 2.2. The $D_\chi$ Spaces of Homogeneous Functions

We wish to define the spaces $D_\chi$ on which we will thereafter construct the representations of the two-dimensional complex unimodular group. The $D_\chi$ are spaces of homogeneous functions $f(z_1, z_2; \bar{z}_1, \bar{z}_2)$ of two complex variables. For notational convenience we shall henceforth write these functions simply $f(z_1, z_2)$. Similarly, a function $\varphi(z, \bar{z})$ of a single complex variable will be written simply $\varphi(z)$.

Recall that $f(z_1, z_2)$ is called homogeneous of degree $(\lambda, \mu)$, where $\lambda$ and $\mu$ are complex numbers differing by an integer, if for every complex number $a \neq 0$ we have[2]

$$f(az_1, az_2) = a^\lambda \bar{a}^\mu f(z_1, z_2). \tag{1}$$

Let $n_1$ and $n_2$ be a pair of complex numbers whose difference is an integer. For simplicity we shall denote this pair of numbers by the single symbol $\chi$, so that

$$\chi = (n_1, n_2).$$

Now with each pair $\chi = (n_1, n_2)$ of complex numbers we shall associate the space $D_\chi$ of functions $f(z_1, z_2)$ fulfilling the following requirements.

1. $f(z_1, z_2)$ is homogeneous of degree $(n_1 - 1, n_2 - 1)$.[3]
2. $f(z_1, z_2)$ is infinitely differentiable in $z_1, z_2$ and their complex

[2] We require that $\lambda - \mu$ be an integer since only then will

$$a^\lambda \bar{a}^\mu = |a|^{\lambda+\mu} \exp\{i(\lambda - \mu) \arg a\}$$

be a single-valued function of $a$.

[3] We choose the degree to be $(n_1 - 1, n_2 - 1)$ rather than $(n_1, n_2)$ for reasons that will become evident in the future.

conjugates throughout the entire region of variation of $z_1$ and $z_2$, except at (0, 0).

We now introduce a topology into $D_\chi$. A sequence $\{f_m(z_1, z_2)\}$ of functions in $D_\chi$ will be said to converge to zero if on every closed bounded set not containing (0, 0) these functions converge uniformly to zero together with all their derivatives. We shall show later that $D_\chi$ is complete with respect to this topology.

**Remark.** Other realizations of $D_\chi$ also exist. Consider a "contour" $\Omega$, in the space of two complex dimensions, which crosses each complex line $a_1 z_1 + a_2 z_2 = 0$ once. Because the functions in $D_\chi$ are homogeneous, they are uniquely determined by their values on $\Omega$. In this way $D_\chi$ may be thought of not as the space of homogeneous functions of two complex variables, but as the space of functions on $\Omega$. Note also that if a line $a_1 z_1 + a_2 z_2 = 0$ intersects a contour at more than one point, the values of a homogeneous function at the points of intersection differ by a factor depending only on the degree of homogeneity. #

### 2.3. Two Useful Realizations of the $D_\chi$

Consider the complex line $z_2 = 1$ in a space of two complex dimensions. This line intersects each line passing through the origin (except $z_2 = 0$) at one and only one point. Therefore each $f(z_1 z_2) \in D_\chi$ is determined uniquely by its values on this line. In this way with each $f(z_1, z_2) \in D_\chi$ let us associate the function

$$\varphi(z) = f(z, 1). \tag{1}$$

It is then clear that $f(z_1, z_2)$ is given in terms of $\varphi(z)$ by

$$f(z_1, z_2) = z_2^{n_1-1} \bar{z}_2^{n_2-1} \varphi(z_1/z_2). \tag{2}$$

Now the functions $\varphi(z)$ form a certain function space, and in view of Eq. (2) we may call this space also $D_\chi$. This space has the following intrinsic properties. Because $f(z_1, z_2)$ is infinitely differentiable, $\varphi(z)$ is infinitely differentiable in $z$ and $\bar{z}$. Further, it satisfies an additional condition in the neighborhood of $z = \infty$. This condition may be obtained from (2) by writing

$$f(1, z) = z^{n_1-1} \bar{z}^{n_2-1} \varphi(z^{-1}), \tag{3}$$

so that the function

$$\hat{\varphi}(z) = z^{n_1-1} \bar{z}^{n_2-1} \varphi(-z^{-1}) \tag{4}$$

is infinitely differentiable in $z$ and $\bar{z}$. Thus if $\varphi(z) = f(z, 1)$, for $f(z_1, z_2) \in D_\chi$, then $\varphi(z)$ and $\hat{\varphi}(z)$ are infinitely differentiable in $z$ and $\bar{z}$. Let us call $\hat{\varphi}(z)$ the *inversion* of $\varphi(z)$.

The converse assertion is easily verified: if $\varphi(z)$ and its inversion $\hat{\varphi}(z)$ are infinitely differentiable in $z$ and $\bar{z}$, the function $f(z_1, z_2)$ defined by Eq. (2) is in $D_\chi$. Thus Eq. (2) establishes a one-to-one correspondence between the infinitely differentiable homogeneous functions $f(z_1, z_2)$ of degree $(n_1 - 1, n_2 - 1)$ and the functions $\varphi(z)$ such that $\varphi(z)$ and $\hat{\varphi}(z)$ are infinitely differentiable.

The topology we have introduced for the $f(z_1, z_2)$ functions induces uniquely by a topology for the $\varphi(z)$ functions. A sequence $\{\varphi_m(z)\}$ is said to converge to zero if in every finite region every sequence of derivatives to any given order both of the $\varphi_m(z)$ and of the $\hat{\varphi}_m(z)$ converge uniformly to zero.

Asymptotically as $|z| \to \infty$, a function $\varphi(z) \in D_\chi$ goes as

$$\varphi(z) \sim C z^{n_1-1} \bar{z}^{n_2-1}. \tag{5}$$

This is because according to Eq. (3) $\varphi(z) = z^{n_1-1}\bar{z}^{n_2-1} f(1, z^{-1})$, so that as $|z| \to \infty$ we have $\varphi(z) \sim f(1, 0) z^{n_1-1}\bar{z}^{n_2-1}$.

**Remark.** In fact by using the infinite differentiability of $\varphi(z)$ we can obtain a more exact result: As $|z| \to \infty$ the function $\varphi(z)$ can be expanded in an asymptotic series of the form

$$\varphi(z) \sim z^{n_1-1}\bar{z}^{n_2-1} \sum_{j,k=0}^{\infty} a_{jk} z^{-j} \bar{z}^{-k}. \tag{6}$$

Indeed, since $\hat{\varphi}(z)$ is infinitely differentiable at $z = 0$, we may write the asymptotic Taylor's series

$$\hat{\varphi}(z) \sim \sum_{j,k=0}^{\infty} b_{jk} z^j \bar{z}^k \tag{7}$$

as $z \to 0$. Then (6) follows immediately from (4) and (7) with $a_{jk} = (-1)^{n_1-n_2+j+k} b_{jk}$. In general, asymptotic series cannot be differentiated term by term. The asymptotic Taylor's series of (7) can, however, and the series obtained in this way are asymptotic series for the derivatives of $\hat{\varphi}(z)$. This leads immediately to the result that asymptotic series for the derivatives of $\varphi(z)$ can be obtained through term-by-term differentiation of Eq. (6). The converse is also true. If $\varphi(z)$ is an infinitely differentiable function of $z$ and $\bar{z}$ and has an asymptotic series such as (6) that can be differentiated term by term to yield asymptotic series for its derivatives, then $\varphi(z)$ is in $D_\chi$. We shall omit a proof of this assertion. #

Another useful realization of $D_\chi$ is the following. Consider the manifold $\Omega$ of all points $(z_1, z_2)$ such that $|z_1|^2 + |z_2|^2 = 1$. Every line passing through the origin and intersecting this manifold at $(\omega_1, \omega_2)$ contains also all points of the form $(e^{i\gamma}\omega_1, e^{i\gamma}\omega_2)$, where $\gamma$ is a real number such that $0 \leqslant \gamma < 2\pi$. From homogeneity we have

$$f(e^{i\gamma}\omega_1, e^{i\gamma}\omega_2) = \exp[i(n_1 - n_2)\gamma]f(\omega_1, \omega_2). \tag{8}$$

Consequently with every $f(z_1, z_2) \in D_\chi$ we can associate an infinitely differentiable function $f(\omega_1, \omega_2)$ on the sphere[4] $\Omega$ whose equation is $|\omega_1|^2 + |\omega_2|^2 = 1$; this function satisfies the homogeneity condition of Eq. (8). Obviously $f(z_1, z_2)$ is obtained from $f(\omega_1, \omega_2)$ by the equation

$$f(z_1, z_2) = r^{n_1+n_2-2}f(z_1/r, z_2/r), \tag{9}$$

where $r = (|z_1|^2 + |z_2|^2)^{\frac{1}{2}}$. Conversely, if $f(\omega_1, \omega_2)$ is any infinitely differentiable function on $\Omega$ and satisfies Eq. (8), the function $f(z_1, z_2)$ defined by Eq. (9) belongs to $D_\chi$. The realization of $D_\chi$ in terms of $f(\omega_1, \omega_2)$ is particularly useful because $\Omega$ is a compact manifold.

Remark. Having established this realization, it is not difficult to show that $D_\chi$ is complete. Note that the topology on $D_\chi$ in this realization is defined as follows: a sequence $\{f_m(\omega_1, \omega_2)\}$ converges to zero if and only if together with their derivatives of all orders the functions of this sequence converge uniformly to zero on $\Omega$. The proof of the completeness of $D_\chi$ proceeds in exactly the same way as the proof of the completeness of the infinitely differentiable functions on an interval.[5] #

### 2.4. Representation of $G$ on $D_\chi$

We now go on to determine the representation of $G$, the complex unimodular group in two dimensions, on $D_\chi$. Each $g \in G$, of the form

$$g = \begin{Vmatrix} \alpha & \beta \\ \gamma & \delta \end{Vmatrix}, \qquad \alpha\delta - \beta\gamma = 1,$$

induces the linear transformation

$$z_1' = \alpha z_1 + \gamma z_2, \qquad z_2' = \beta z_1 + \delta z_2$$

[4] If the complex space of the variables $z_1$ and $z_2$ is thought of as a four-dimensional real space, then $\Omega$ is the unit sphere in this space.

[5] See Volume II, Chapter I, Section 3.3.

on the space of the two complex variables $z_1$ and $z_2$. This transformation, in turn, may be associated with the transformation

$$f(z_1, z_2) \to f(\alpha z_1 + \gamma z_2, \beta z_1 + \delta z_2)$$

on $D_\chi$, for it is clear that $f(z_1, z_2) \in D_\chi$ implies $f(\alpha z_1 + \gamma z_2, \beta z_1 + \delta z_2) \in D_\chi$, since the latter function is also infinitely differentiable and homogeneous of degree $(n_1 - 1, n_2 - 1)$.

Let us denote this linear transformation on $D_\chi$ by $T_\chi(g)$, so that

$$T_\chi(g)f(z_1, z_2) = f(\alpha z_1 + \gamma z_2, \beta z_1 + \delta z_2), \tag{1}$$

where $f \in D_\chi$. Obviously $T_\chi(g)$ is continuous in $D_\chi$, meaning that if a sequence $\{f_m(z_1, z_2)\}$ in $D_\chi$ converges to zero, then the sequence $\{T_\chi(g)f_m(z_1, z_2)\}$ also converges to zero. Further, $T_\chi(g)$ depends continuously also on the element $g$ of $G$, meaning that if $\lim_{m\to\infty} g_m = g$, then

$$\lim_{m\to\infty} T_\chi(g_m)f(z_1, z_2) = T_\chi(g)f(z_1, z_2). \tag{2}$$

for any $f \in D_\chi$.

It is easily verified also that

$$T_\chi(g_1 g_2) = T_\chi(g_1)T_\chi(g_2) \tag{3}$$

for any two elements $g_1, g_2 \in G$. Thus the set of transformations $T_\chi(g)$ on $D_\chi$ forms a representation of $G$. We thus arrive at the following assertion. *To every pair of complex numbers $\chi = (n_1, n_2)$ whose difference is an integer there corresponds a representation $T_\chi(g)$ of $G$. This representation is realized on the space $D_\chi$ of infinitely differentiable functions $f(z_1, z_2)$, homogeneous of degree $(n_1 - 1, n_2 - 1)$ in accordance with Eq.* (1). We shall call $\chi$ *the weight of $T(g)$*.

## 2.5. The $T_\chi(g)$ Operators in Other Realizations of $D_\chi$

We saw in Section 2.3 that $D_\chi$ can be realized as the space of all infinitely differentiable functions $\varphi(z)$ of $z$ and $\bar{z}$ with infinitely differentiable inversion $\hat{\varphi}(z) = z^{n_1-1}\bar{z}^{n_2-1}\varphi(-z^{-1})$. Let us find the operators on these functions which correspond to the $T_\chi(g)$. Recall that $\varphi(z)$ corresponds to the function

$$f(z_1, z_2) = z_2^{n_1-1}\bar{z}_2^{n_2-1}\varphi(z_1/z_2) \tag{1}$$

of two variables, and that $T_\chi(g)$ transforms this function into

$$f_g(z_1, z_2) = T_\chi(g)f(z_1, z_2) = f(\alpha z_1 + \gamma z_2, \beta z_1 + \delta z_2). \tag{2}$$

We now use the homogeneity of $f(z_1, z_2)$, obtaining

$$\begin{aligned} f_g(z_1, z_2) &= f(\alpha z_1 + \gamma z_2, \beta z_1 + \delta z_2) \\ &= (\beta z_1 + \delta z_2)^{n_1-1}(\bar{\beta}\bar{z}_1 + \bar{\delta}\bar{z}_2)^{n_2-1} f\left(\frac{\alpha z_1 + \gamma z_2}{\beta z_1 + \delta z_2}, 1\right) \\ &= (\beta z_1 + \delta z_2)^{n_1-1}(\bar{\beta}\bar{z}_1 + \bar{\delta}\bar{z}_2)^{n_2-1} \varphi\left(\frac{\alpha z_1 + \gamma z_2}{\beta z_1 + \delta z_2}\right). \end{aligned}$$

Now $\varphi_g(z) = f_g(z, 1)$ is the function of one variable corresponding to $f_g(z_1, z_2)$, and by the above equation it is given in terms of $\varphi(z)$ by

$$\varphi_g(z) = (\beta z + \delta)^{n_1-1}(\bar{\beta}\bar{z} + \bar{\delta})^{n_2-1} \varphi\left(\frac{\alpha z + \gamma}{\beta z + \delta}\right).$$

Thus when $D_\chi$ is realized as the space of functions $\varphi(z)$ of a single complex variable, $T_\chi(g)$ is given by

$$T_\chi(g)\varphi(z) = (\beta z + \delta)^{n_1-1}(\bar{\beta}\bar{z} + \bar{\delta})^{n_2-1} \varphi\left(\frac{\alpha z + \gamma}{\beta z + \delta}\right). \tag{3}$$

**Remark.** It is also possible to write out $T_\chi(g)$ when $D_\chi$ is the space of functions on some arbitrary "contour" $\Omega$ crossing each line of the form $a_1 z_1 + a_2 z_2 = 0$ once. To every two complex numbers $z_1$ and $z_2$ not simultaneously equal to zero corresponds a point $(\omega_1, \omega_2)$ on $\Omega$ and a number $\rho(z_1, z_2) \neq 0$, such that

$$z_1 = \rho(z_1, z_2)\omega_1, \qquad z_2 = \rho(z_1, z_2)\omega_2.$$

Let $g \in G$, and write

$$\rho_g(\omega_1, \omega_2) = \rho(\alpha\omega_1 + \gamma\omega_2, \beta\omega_1 + \delta\omega_2). \tag{4}$$

Then if $D_\chi$ is realized as the space of functions on $\Omega$, the operator $T_\chi(g)$ in this realization will be given by

$$\begin{aligned} &T_\chi(g)f(\omega_1, \omega_2) \\ &= [\rho_g(\omega_1, \omega_2)]^{n_1-1}[\bar{\rho}_g(\omega_1, \omega_2)]^{n_2-1} f\left(\frac{\alpha\omega_1 + \gamma\omega_2}{\rho_g(\omega_1, \omega_2)}, \frac{\beta\omega_1 + \delta\omega_2}{\rho_g(\omega_1, \omega_2)}\right). \end{aligned} \tag{5}$$

If certain lines given by $a_1 z_1 + a_2 z_2 = 0$ intersect $\Omega$ at several points, $\rho_g(\omega_1, \omega_2)$ is not uniquely defined. But in this case $f(\omega_1 \omega_2)$ is also not arbitrary, but must satisfy certain additional conditions. Specifically, its values at the points where a given line intersects the contour must differ by a definite factor. It is then easily seen that $T_\chi(g)$ is uniquely defined by (5). #

## 2.6. The Dual Representations

Let $D'_\chi$ be the space dual to $D_\chi$, that is, the space of linear functionals on $D_\chi$. Then a bilinear functional

$$(F, \varphi),$$

where $F \in D'_\chi$ and $\varphi \in D_\chi$, is defined in the natural way for these two spaces. Specifically, $(F, \varphi)$ is the value of the functional $F$ on the function $\varphi$.

A representation of $G$ can be defined on $D'_\chi$. An operator $\tilde{T}_\chi(g)$ of such a representation will be defined by

$$(\tilde{T}_\chi(g)F, \varphi) = (F, T_\chi(g^{-1})\varphi). \tag{1}$$

[That the $\tilde{T}_\chi(g)$ form a representation, that is, that $\tilde{T}_\chi(g_1 g_2) = \tilde{T}_\chi(g_1)\tilde{T}_\chi(g_2)$, is trivially verified.] We shall call $\tilde{T}_\chi(g)$ the representation *dual* to $T_\chi(g)$. Note that if we replace $\varphi$ by $T_\chi(g)\varphi$ in Eq. (1), we obtain

$$(\tilde{T}_\chi(g)F, T_\chi(g)\varphi) = (F, \varphi). \tag{2}$$

This means that the bilinear functional $(F, \varphi)$ is invariant under $T_\chi(g)$ and its dual $\tilde{T}_\chi(g)$.

Consider two space $D_{\chi_1}$ and $D_{\chi_2}$ and an invariant bilinear functional $B(\varphi, \psi)$, $\varphi \in D_{\chi_1}$, $\psi \in D_{\chi_2}$, that is, one for which

$$B(T_{\chi_1}(g)\varphi, T_{\chi_2}(g)\psi) = B(\varphi, \psi)$$

for any $g \in G$. Let us assume that this functional is nondegenerate on $D_\chi$, by which we mean that if

$$B(\varphi, \psi) = 0$$

for fixed $\varphi \in D_{\chi_1}$ and for all $\psi \in D_{\chi_2}$, then $\varphi = 0$. We assert that then $D_{\chi_1}$ can be mapped in a one-to-one way into $D'_{\chi_2}$ so that $T_{\chi_1}(g)$ is mapped into $\tilde{T}_{\chi_2}(g)$. In this way $\tilde{T}_{\chi_2}(g)$, the representation dual to $T_{\chi_2}(g)$, may be thought of as an extension of $T_{\chi_1}(g)$.

**Proof.** Each function $\varphi \in D_{\chi_1}$ defines a linear functional $B(\varphi, \psi)$ on $D_{\chi_2}$. In this way $D_{\chi_1}$ is mapped into $D'_{\chi_2}$. Since by assumption

$B(\varphi, \psi)$ is nondegenerate on $D_{\chi_1}$, only $\varphi = 0$ is mapped into zero under this mapping. Thus $D_{\chi_1} \to D'_{\chi_2}$ is a one-to-one mapping.

Let us verify that $\tilde{T}_{\chi_2}(g)$ is an extension of $T_{\chi_1}(g)$, that is, that on $D_{\chi_1}$ it coincides with $T_{\chi_1}(g)$. Consider a function $\varphi$ to be an element of $D'_{\chi_2}$. Then

$$(\tilde{T}_{\chi_2}(g)\varphi, \psi) = (\varphi, T_{\chi_2}(g^{-1})\psi).$$

On the other hand, from the definition of the mapping $D_{\chi_1} \to D'_{\chi_2}$ we have

$$(\varphi, \psi) = B(\varphi, \psi).$$

Therefore,

$$(\varphi, T_{\chi_2}(g^{-1})\psi) = B(\varphi, T_{\chi_2}(g^{-1})\psi) = B(T_{\lambda_1}(g)\varphi, \psi) = (T_{\chi_1}(g)\varphi, \psi).$$

We thus find that $(\tilde{T}_{\chi_2}(g)\varphi, \psi) = (T_{\chi_1}(g)\varphi, \psi)$, or that $T_{\chi_1}(g)\varphi = \tilde{T}_{\chi_2}(g)\varphi$.

In Section 4 we shall show that for every pair of complex numbers $\chi = (n_1, n_2)$ the representations $T_\chi(g)$ and $T_{-\chi}(g)$, where $-\chi = (-n_1, -n_2)$, possess a nondegenerate invariant bilinear functional. This means that the dual representation $\tilde{T}_\chi(g)$ is an extension of $T_{-\chi}(g)$.

## 3. Summary of Basic Results concerning Representations on $D_\chi$

We shall devote the next few pages to stating the basic results concerning representations on $D_\chi$, results which will be proved in Sections 4–6. Their proof is of interest from the point of view of generalized functions, but is not necessary for the development of harmonic analysis on the group. Therefore the reader who so desires may read Section 3 and then skip over directly to Chapter IV.

### 3.1. Irreducibility of Representations on the $D_\chi$ and the Role of Integer Points

There exist several possible definitions of an irreducible representation. The simplest of these, which we shall all *subspace irreducibility*, is the following. *A representation on a space E is called subspace irreducible if E contains no closed invariant subspaces under the representation.* This definition is a natural one for finite-dimensional representations, where no complications having to do with the structure of the representation space can arise (recall that all $n$-dimensional spaces are isomorphic to each other). Of course, in infinite-dimensional spaces, as well, we shall require the absence of invariant subspaces for a representation to be truly irreducible. But this requirement will not be sufficient.

The other kind of irreducibility we wish to introduce is *operator irreducibility*. *A representation on a space E will be called operator irreducible if every closed operator*[6] *on E that commutes with all operators of the representation is a multiple of the unit operator.*

Remark. For finite-dimensional representations, operator irreducibility is equivalent to subspace irreducibility. For infinite-dimensional representations this is no longer true. We will see later that a representation on a $D_\chi$, $\chi = (n_1, n_2)$, is operator irreducible if $n_1$ and $n_2$ are integers of the same sign, although $D_\chi$ has an invariant subspace. Therefore operator irreducibility is not true irreducibility. We shall nevertheless refer to it in this way, in order to avoid having to say every time, "a space in which every closed operator commuting with the representation is a multiple of the unit operator." #

Irreducibility can also be defined differently, in a geometric way which includes both absence of invariant subspaces and operator irreducibility, Let $T(g)$ be a representation on $E$. Consider the direct sum $E \oplus E$. Then $T(g)$ induces a representation on this direct sum. *We shall call* $T(g)$ *irreducible on E if every invariant proper subspace of* $E \oplus E$ *consists of points of the form* $[c_1\xi, c_2\xi]$ *where* $\xi$ *runs through all of E, and* $c_1$ *and* $c_2$ *are fixed numbers.*

We prove first that this definition implies subspace irreducibility, i.e., that $E$ has no invariant subspaces. Let $H$ be an invariant subspace of $E$. Then the set of all pairs of the form $[\xi, \eta]$, where $\xi$ runs through $H$ and $\eta$ through all of $E$, is an invariant subspace of $E \oplus E$. Obviously this is possible only if $H = 0$ or if $H = E$. Next, we prove that this definition implies operator irreducibility. Let $A$ be a closed operator in $E$ that commutes with all the $T(g)$. Consider pairs of the form $[\xi, A\xi]$, where $\xi$ runs through the domain of $A$. By assumption the set of these pairs is a closed subspace in the direct sum $E \oplus E$. Commutativity of $A$ with the $T(g)$ implies first that if $\xi$ is in the domain of $A$, then so is $T(g)\xi$, and second that $AT(g)\xi = T(g)A\xi$. But then $[\xi, A\xi]$ forms an invariant subspace of $E \oplus E$. But by assumption every invariant subspace of $E \oplus E$ consists of points of the form $[c_1\xi, c_2\xi]$, where $\xi$ runs through all of $E$; hence $A$ is a multiple of the unit operator.

Remark. The correct definition of irreducibility should be taken as the following. Let $T(g)$ be a representation on a nuclear space $E$.[7]

[6] An operator $A$ on $E$ is called closed if its graph (i.e., the set of all pairs $[\xi, A\xi]$, where $\xi$ runs through the domain of definition of $A$) is a closed set in the direct sum $E \oplus E$.

[7] For a discussion of nuclear spaces, see Volume 4, Chapter I, Section 3.

Consider the tensor product $E \otimes F$, where $F$ is any arbitrary space. Then a representation of the group is also defined on this space. The representation on $E$ is called irreducible if every invariant subspace of $E \otimes F$ is of the form $E \otimes F_1$, where $F_1$ is a subspace of $F$. If $F$ is two-dimensional, this leads to the previous definition of irreducibility (since in this case $E \otimes F$ coincides with $E \oplus E$). If $F$ is one-dimensional, this is the statement that $E$ have no invariant subspaces. #

In this chapter we shall prove only operator irreducibility of the representations on the $D_\chi$. In fact we shall prove even a weaker result, since we shall be dealing not with all closed operators $A$ but only with continuous ones. Specifically, we shall show in Section 5 that *if $A$ is a continuous operator on $D_\chi$ and if it commutes with the representation $T_\chi(g)$, that is, if*

$$AT_\chi(g) = T_\chi(g)A,$$

*then $A$ is a multiple of the unit operator.*

For the purposes of the present book this result is quite sufficient. A stronger result can, however, be derived. Specifically *a representation on $D_\chi$ for noninteger $\chi$ (see below) is irreducible in every sense.* Subspace irreducible, operator irreducible, and even irreducible as defined in terms of the tensor product. We shall not, however, prove this result here.

A special role will be played by the $T_\chi(g)$ representations in which $\chi$ is an *integer point*. We shall say that $\chi = (n_1, n_2)$ is an integer point if $n_1$ and $n_2$ are integers of the same sign (and both nonzero). We shall soon see that for integer points $D_\chi$ will have invariant subspaces. Consider first the case in which $n_1$ and $n_2$ are positive integers. Recall that $D_\chi$ is the space of infinitely differentiable functions $f(z_1, z_2)$, homogeneous of degree $(n_1 - 1, n_2 - 1)$. Obviously if $n_1$ and $n_2$ are positive integers, $D_\chi$ contains the invariant subspace $E_\chi$ of homogeneous polynomials of degree $(n_1 - 1, n_2 - 1)$. This is because the operators of the representation, acting according to $T_\chi(g)f(z_1, z_2) = f(\alpha z_1 + \gamma z_2, \beta z_1 + \delta z_2)$, transform such polynomials into other such polynomials. If $D_\chi$ is realized in terms of functions of a single variable $z$, the subspace of homogeneous polynomials becomes the subspace of polynomials (in $z$ and $\bar{z}$) of degree no higher than $n_1 - 1$ in $z$ and no higher than $n_2 - 1$ in $\bar{z}$, or the subspace whose basis is $\{z^k \bar{z}^l\}$, where $k = 0, ..., n_1 - 1$, and $l = 0, ..., n_2 - 1$. Obviously the dimension of this polynomial subspace $E_\chi$ is $n_1 n_2$.

Since $E_\chi$ is invariant, $G$ can also be represented on the factor space $D_\chi/E_\chi$, or the space of sets of functions in $D_\chi$ defined up to a polynomial. It can be shown that $D_\chi/E_\chi$ has no invariant subspaces. Thus if $n_1$ and $n_2$ are positive integers, $D_\chi \equiv D_{n_1, n_2}$ has "two levels," that is, it has a single invariant subspace such that the factor space with respect to this subspace has no invariant subspaces. In this sense the representation $T_\chi(g)$ on such a space resembles a Jordan lattice. We may now ask for the form of the representation on the factor space

$$D_{n_1,n_2}/E_{n_1,n_2}.$$

It will be shown in Section 3.2 that

$$D_{n_1,n_2}/E_{n_1,n_2} \cong D_{-n_1,n_2} \cong D_{n_1,-n_2}.$$

Consider now the structure of $D_{-\chi} \equiv D_{-n_1,-n_2}$, where $n_1$ and $n_2$ are still positive integers. Here the situation is the opposite. Specifically, we find an infinite-dimensional invariant subspace

$$F_{-n_1,-n_2} \cong D_{n_1,n_2}/E_{n_1,n_2}$$

and a finite-dimensional factor space

$$D_{-n_1,-n_2}/F_{-n_1,-n_2} \cong E_{n_1,n_2}.$$

Let us describe them. Consider the subspace $F_{-\chi} \equiv F_{-n_1,-n_2}$ of $D_{-\chi}$ consisting of functions $\varphi(z)$ such that

$$b_{jk} = \frac{i}{2} \int \varphi(z) z^j \bar{z}^k \, dz \, d\bar{z} = 0,$$

where $j = 0, 1, ..., n_1 - 1$, and $k = 0, 1, ..., n_2 - 1$. It is a simple matter to show that when $T_{-\chi}(g)$ is applied to any $\varphi(z)$, the $b_{jk}$ are subjected to a linear transformation. Thus $F_{-\chi}$ is invariant.

Since $F_{-\chi}$ is invariant, $G$ can also be represented on the factor space $D_{-\chi}/F_{-\chi}$, that is, on the space of sets of functions in $D_{-\chi}$ defined up to a function in $F_{-\chi}$. Obviously the elements of this factor space are uniquely determined by their moments $b_{jk}$. Thus the factor space $D_{-\chi}/F_{-\chi}$ has the finite dimension $n_1 n_2$.

Remark. Recall that we saw in Section 2 that $D_{-\chi}$ is contained in $D'_\chi$, the dual of $D_\chi$. We could therefore have studied the structure of $D_{-\chi}$ in terms of that of $D_\chi$ simply by going over to the dual space. #

## 3.2. Equivalence of Representations on the $D_\chi$ and the Role of Integer Points

We shall call two representations $T_{\chi_1}(g)$ and $T_{\chi_2}(g)$ *equivalent* if there exists a linear operator $A$ which is a one-to-one bicontinuous mapping of $D_{\chi_1}$ onto $D_{\chi_2}$ such that

$$A T_{\chi_1}(g) = T_{\chi_2}(g) A. \tag{1}$$

In this case we shall call $T_{\chi_2}(g)$ the image of $T_{\chi_1}(g)$ under $A$, and say that $A$ is an *intertwining operator* for the representations $T_{\chi_1}$ and $T_{\chi_2}$. We shall consider essentially different only nonequivalent representations. It is natural, then, to ask when $T_{\chi_1}(g)$ and $T_{\chi_1}(g)$ are equivalent.

It will be shown in Section 5 that $T_{\chi_1}(g)$ and $T_{\chi_2}(g)$ are equivalent if and only if one of the following is true:

*Case* 1. $\chi_1 = \chi_2$ (the trivial case).

*Case* 2. $\chi_1 = (n_1, n_2)$, $\chi_2 = (-n_1, -n_2)$, where $\chi_1$ is not an integer point.

It will also be shown in Section 5 that the intertwining operator $A$ is determined in each of these cases uniquely up to a factor. In the trivial Case 1 this operator is a multiple of the unit operator.[8] The form of $A$ in Case 2 may be described as follows. Let $D_{\chi_1}$ and $D_{\chi_2}$ be realized in terms of functions $\varphi(z)$ of a single complex variable. Then the operator $A$ that maps $D_\chi$ into $D_{-\chi}$ may be defined by

$$A\varphi(z) = \frac{i}{2}\int (z_1 - z)^{-n_1-1}(\bar{z}_1 - \bar{z})^{-n_2-1}\varphi(z_1)\, dz_1\, d\bar{z}_1\,. \tag{2}$$

This integral is to be understood in the sense of its regularization.[9] Similarly, the operator $B$ that maps $D_{-\chi}$ into $D_\chi$ is obtained from Eq. (2) replacing $n_1, n_2$ by $-n_1, -n_2$, so that

$$B\varphi(z) = \frac{i}{2}\int (z_1 - z)^{n_1-1}(\bar{z}_1 - \bar{z})^{n_2-1}\varphi(z_1)\, dz_1\, d\bar{z}_1\,. \tag{2'}$$

Note that $BA$ is an operator on $D_\chi$ which commutes with every operator of the representation. Because the representation is operator irreducible, it follows that there exists a number $\mu$ such that $BA = \mu E$, where $E$ is the unit operator. We shall see in Section 5, in fact, that

$$\mu = (-1)^{n_1-n_2}4\pi^2(n_1 + n_2 + |n_1 - n_2|)^{-1}(-n_1 - n_2 + |n_1 - n_2|)^{-1}.$$

**Remark.** The following theorem can be proven. Let $T(g)$ be an irreducible representation on some space $E$, such that for every continuous rapidly decreasing function $f(g)$ the integral $\int f(g)T(g)\, dg$ converges. Then it is possible to choose a $\chi$ such that there exists a continuous one-to-one mapping of $D_\chi$ into $E$ which intertwines $T(g)$ and

[8] Actually this result is the statement of operator irreducibility of the representation on $D_\chi$, according to which any operator that commutes with the operators of the representation must be a multiple of the unit operator.

[9] This regularization will be defined in more detail in Section 5.

$T_\chi(g)$. We shall call the representation on $E$ equivalent to the corresponding one on $D_\chi$. It turns our that even with this weaker definition of equivalence, there exist no other pairs of equivalent representations on the $D_\chi$. #

Once we have agreed to call essentially different only inequivalent representations, we may speak of the set of all inequivalent representations of $G$. The elements of this set will actually be classes of equivalent representations. Since each representation $T_\chi(g)$ is given by a pair of complex numbers $n_1$, $n_2$ differing by an integer, we may think of this set as a "*Riemann surface.*" The points of this Riemann surface are given by a pair of complex numbers whose difference is an integer. Thus the universal covering space of this representation space is a countable sequence of complex planes. We shall call a branch point of this Riemann surface a point such that each of its neighborhoods contains equivalent representations. The order of the branch point is defined in the natural way. In view of what has been said already, the only branch point on the Riemann surface is $n_1 = 0$, $n_2 = 0$. Its order is 2. As will be seen later, integer $\chi$ play a special role on this surface.

## 3.3. The Problem of Equivalence at Integer Points

Consider now the pair of spaces $D_\chi$ and $D_{-\chi}$ where $\chi = (n_1, n_2)$ and $-\chi = (-n_1, -n_2)$ are integer points (i.e., $n_1$ and $n_2$ are nonzero integers of the same sign). We shall see in Section 5 that the operators $A$ and $B$ such that

$$D_\chi \xrightarrow{A} D_{-\chi} \quad \text{and} \quad D_{-\chi} \xrightarrow{B} D_\chi$$

which have been defined in Eqs. (2) and (2′) of Section 3.2 intertwine the representations also in this special case.

**Remark.** Strictly speaking the definition

$$A\varphi(z) = \frac{i}{2} \int (z_1 - z)^{-n_1-1} (\bar{z}_1 - \bar{z})^{-n_2-1} \varphi(z_1)\, dz_1\, d\bar{z}_1$$

of $A$ has no meaning when $n_1$ and $n_2$ are positive integers, since the right-hand side, considered a function of $n_1$ and $n_2$, has poles at such points. We could, however, have multiplied the integral by the appropriate product of $\Gamma$ functions, thus defining the operator $A$ at these singular points. It will be shown in Section 5 that this new operator is of the form $A = \partial^{n_1+n_2}/\partial z^{n_1} \partial \bar{z}^{n_2}$. #

As we shall see, however, $A$ and $B$ are no longer isomorphic mappings of the spaces on each other. Let us assume, to be specific, that $n_1$ and $n_2$ are positive integers. First consider $B$, which maps $D_{-\chi}$ into $D_\chi$. It is given by

$$B\varphi(z) = \frac{i}{2}\int (z_1 - z)^{n_1-1}(\bar{z}_1 - \bar{z})^{n_2-1}\varphi(z_1)\, dz_1\, d\bar{z}_1 . \tag{1}$$

The right-hand side of this equation is a polynomial of degree $n_1 - 1$ in $z$ and $n_2 - 1$ in $\bar{z}$. Thus $B$ maps all of $D_{-\chi}$ into the subspace $E_\chi \equiv E_{n_1,n_2}$ of $D_\chi$. It is also seen from Eq. (1) that $B$ will map $\varphi(z)$ into zero if and only if the moments

$$b_{jk} = \frac{i}{2}\int z^j \bar{z}^k \varphi(z)\, dz\, d\bar{z}$$

all vanish for $j = 0, 1, ..., n_1 - 1$, and $k = 0, 1, ..., n_2 - 1$. We have already denoted the subspace of these functions (see Section 3.1) by $F_{-\chi}$. Thus $B$ establishes the isomorphism discussed in Section 3.1, namely

$$D_{-n_1,-n_2}/F_{-n_1,-n_2} \cong E_{n_1,n_2} .$$

Now consider the mapping $A$, which maps $D_\chi$ into $D_{-\chi}$. As has already been mentioned, $A = \partial^{n_1+n_2}/\partial z^{n_1}\partial \bar{z}^{n_2}$. Obviously this operator annihilates the subspace $E_\chi$ of polynomials of degree no higher than $n_1 - 1$ in $z$ and $n_2 - 1$ in $\bar{z}$. It can be shown (and will be in Section 5) that $A$ maps $D_\chi$ not onto $D_{-\chi}$, but onto its invariant subspace $F_{-\chi}$. Thus $A$ establishes the isomorphism

$$D_{n_1,n_2}/E_{n_1,n_2} \cong F_{-n_1,-n_2} .$$

Thus we see that the same relations hold at integer $\chi$ as at noninteger $\chi$, relations between the representations in $D_\chi$ and $D_{-\chi}$. It is remarkable, however, that additional relations occur at integer $\chi$. Specifically, the existence at integer points of the following intertwining operators for the representations will be shown in Section 5:

$$D_{n_1,n_2} \xrightarrow{A_1} D_{-n_1,n_2}, \qquad D_{n_1,n_2} \xrightarrow{A_2} D_{n_1,-n_2},$$

and

$$D_{-n_1,n_2} \xrightarrow{B_1} D_{-n_1,-n_2}, \qquad D_{n_1,-n_2} \xrightarrow{B_2} D_{-n_1,-n_2} .$$

(recall that $n_1$ and $n_2$ are *positive* integers by assumption). The operators that give rise to these mappings are

$$A_1 = B_2 = \partial^{n_1}/\partial z^{n_1}, \qquad A_2 = B_1 = \partial^{n_2}/\partial \bar{z}^{n_2}.$$

It will be shown also that $A_1$ and $A_2$ are mappings onto the entire space and annihilate the subspace $E_\chi$. Thus these operators establish the isomorphisms

$$D_{n_1,n_2}/E_{n_1,n_2} \cong D_{-n_1,n_2} \qquad \text{and} \qquad D_{n_1,n_2}/E_{n_1,n_2} \cong D_{n_1,-n_2}.$$

It will be shown also that $B_1$ and $B_2$ are one-to-one mappings of $D_{-\chi}$ onto $F_{-\chi}$, and that they therefore establish the isomorphisms

$$D_{-n_1,n_2} \cong F_{-n_1,-n_2} \qquad \text{and} \qquad D_{n_1,-n_2} \cong F_{-n_1,-n_2}.$$

It should be mentioned that $B_1 A_1 = A$ maps $D_\chi$ into $D_{-\chi}$.

The results can be represented in the single diagram of Fig. 4. In

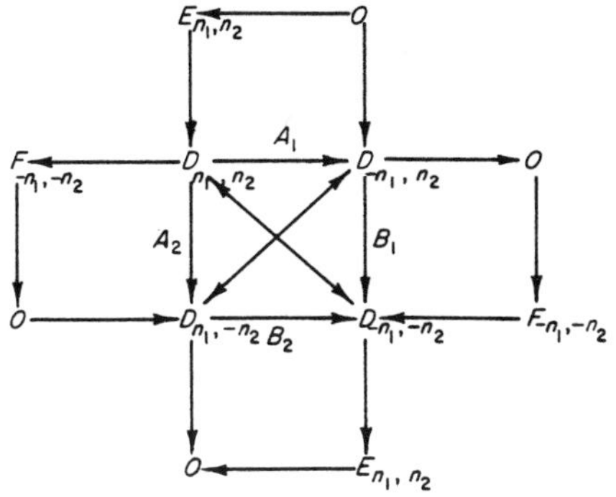


FIG. 4.

this diagram all the (directed) sequences containing two (and only two) of the $D_\chi$ are exact. (Recall again that $n_1$ and $n_2$ are assumed to be positive integers.)

We shall see in Section 5 that there exist no representations connected by such mappings other than those described in Sections 3.2 and 3.3.

Remark. Let us call two irreducible representations $T_1(g)$ and $T_2(g)$ *closest relatives* if they are contained in the same operator-irreducible representation; more accurately, if there exists an operator-irreducible representation on a space $D$, and if $D$ contains an invariant subspace

$D_1$ such that the representation on $D_1$ is equivalent to $T_1(g)$, while the representation on the factor space $D/D_1$ is equivalent to $T_2(g)$. We shall say that two representations are *relatives* or *related* if we can go from one to the other along a finite chain of closest relatives. From the preceding discussion it is seen that the finite-dimensional representations of $G$ on $E_{n_1,n_2}$ are closest relatives to the representations on $D_{-n_1,n_2} \cong D_{n_1,-n_2}$. It can be shown that there exist no other relations between the irreducible representations of $G$ (i.e., except for these, none are even distant relatives of any others). In Chapter VII we shall see that the relationship between the representations of the group of real matrices is somewhat more complicated. A general definition of related representations is given in I. M. Gel'fand, "Some Questions of Analysis and Differential Equations," *Uspekhi matem. nauk*, **14**, No. 3 (1959) (in Russian), in which a topology is introduced on the set of representations. From this topology it is seen that integer $\chi$ are in a certain sense singular points on the "Riemann surface" of the representations. #

## 3.4. Unitary Representations

Under certain conditions there may exist in a $D_\chi$ a Hermitian bilinear functional $(\varphi, \psi)$ which is invariant, i.e., such that

$$(T_\chi(g)\varphi, T_\chi(g)\psi) = (\varphi, \psi) \tag{1}$$

for all $\varphi(z)$ and $\psi(z)$ in $D_\chi$ and for any element $g \in G$. If this functional is positive definite, we will use it for a scalar product, and then the representation itself will be unitary with respect to this scalar product.

We may ask when such a scalar product can be introduced into one of the $D_\chi$. This question will be discussed in Section 6, where it will be shown that it can be done in the following two cases.

*Case 1.* When $n_1 = -\bar{n}_2$. For this case the scalar product is given by

$$(\varphi, \psi) = \frac{i}{2} \int \varphi(z)\bar{\psi}(z)\, dz\, d\bar{z}. \tag{2}$$

We shall call a representation on such a $D_\chi$ *a representation of the principal series.*

*Case 2.* When $n_1 = n_2 = \rho$, where $\rho \neq 0$ is a real number such that $-1 < \rho < 1$. In this case the scalar product is given by

$$(\varphi, \psi) = \left(\frac{i}{2}\right)^2 \int | z_1 - z_2 |^{-2\rho-2} \varphi(z_1)\bar{\psi}(z_2)\, dz_1\, d\bar{z}_1\, dz_2\, d\bar{z}_2. \tag{3}$$

We shall call a representation on such a $D_\chi$ *a representation of the supplementary series.*

## 4. Invariant Bilinear Functionals

### 4.1. Statement of the Problem and the Basic Results

In Section 3 we summarized the results concerning the representations $T_\chi(g)$ on the $D_\chi$. All these results will be obtained by a single method involving the study of invariant bilinear functionals. To find invariant bilinear functionals is in itself an interesting problem of representation theory. We shall engage in no discussion of the equally interesting problem of finding invariant multilinear functionals.

Let us start with a general definition of an invariant bilinear functional for a pair of representations. Consider two representations $T_1(g)$ and $T_2(g)$ of a group $G$, acting on the linear topological spaces $E_1$ and $E_2$, respectively. Let $B(\varphi, \psi)$ be a *continuous bilinear functional* whose arguments $\varphi$ and $\psi$ may be arbitrary elements of $E_1$ and $E_2$. By this we mean that for any $\varphi \in E_1$ and any $\psi \in E_2$, we have

$$B(a\varphi_1 + b\varphi_2, \psi) = aB(\varphi_1, \psi) + bB(\varphi_2, \psi),$$

$$B(\varphi, a\psi_1 + b\psi_2) = aB(\varphi, \psi_1) + bB(\varphi, \psi_2),$$

and that $B(\varphi, \psi)$ is a continuous function of $\varphi$ and $\psi$ in the direct sum of the topological spaces $E_1$ and $E_2$. Then $B(\varphi, \psi)$ is called *invariant under the representations $T_1(g)$ and $T_2(g)$* if for arbitrary $\varphi \in E_1$ and $\psi \in E_2$ and for any $g \in G$, we have

$$B(\varphi, \psi) = B(T_1(g)\varphi, T_2(g)\psi).$$

In Section 2 of this chapter we constructed the representations $T_\chi(g)$ of $G$ on certain linear topological spaces $D_\chi$. All later analysis in this chapter will be based on the study we are about to undertake of bilinear functionals invariant under pairs of representations $T_{\chi_1}(g)$, $T_{\chi_2}(g)$. The basic results we shall obtain can be stated in the form of two theorems.

Consider two representations of our group $G$, namely

$$T_{\chi_1}(g)\varphi(z) = (\beta z + \delta)^{n_1 - 1}(\bar{\beta}\bar{z} + \bar{\delta})^{n_2 - 1}\varphi\left(\frac{\alpha z + \gamma}{\beta z + \delta}\right)$$

and

$$T_{\chi_2}(g)\varphi(z) = (\beta z + \delta)^{m_1 - 1}(\bar{\beta}\bar{z} + \bar{\delta})^{m_2 - 1}\varphi\left(\frac{\alpha z + \gamma}{\beta z + \delta}\right),$$

acting on $D_{\chi_1}$ and $D_{\chi_2}$, respectively. We wish to find a bilinear functional invariant with respect to $T_{\chi_1}(g)$ and $T_{\chi_2}(g)$. The answer will be found to depend on the numbers $s_1 = -\frac{1}{2}(n_1 + m_1)$ and $s_2 = -\frac{1}{2}(n_2 + m_2)$.

For $s_1$ and $s_2$ not negative integers or zero, there exists an invariant bilinear functional if and only if $m_1 = n_1$ and $m_2 = n_2$. When it exists (Section 4.4, Theorem 1) it is of the form

$$B(\varphi, \psi) = \left(\frac{i}{2}\right)^2 \int (z_1 - z_2)^{-n_1-1}(\bar{z}_1 - \bar{z}_2)^{-n_2-1}\varphi(z_1)\psi(z_2)\, dz_1\, d\bar{z}_1\, dz_2\, d\bar{z}_2 .$$

For $s_1$ and $s_2$ negative integers or zero, we obtain the following result (Section 4.5, Theorem 2). An invariant bilinear functional exists if and only if one of the following conditions is fulfilled:

*Case 1.* $m_1 = n_1$, $m_2 = n_2$;

*Case 2.* $m_1 = -n_1$, $m_2 = -n_2$;

*Case 3.* $m_1 = n_1$, $m_2 = -n_2$, where $n_1 = 0, 1, \ldots$;

*Case 4.* $m_1 = -n_1$, $m_2 = n_2$, where $n_2 = 0, 1, \ldots$.

If it exists, it is given by the following formulas:

*Case 1* (in which case $n_1$ and $n_2$ are nonnegative integers).[10]

$$B(\varphi, \psi) = \frac{i}{2}\int \varphi^{(n_1, n_2)}(z)\psi(z)\, dz\, d\bar{z}.$$

*Case 2.*

$$B(\varphi, \psi) = \frac{i}{2}\int \varphi(z)\psi(z)\, dz\, d\bar{z}.$$

*Case 3.*

$$B(\varphi, \psi) = \frac{i}{2}\int \varphi^{(n_1, 0)}(z)\psi(z)\, dz\, d\bar{z}.$$

*Case 4.*

$$B(\varphi, \psi) = \frac{i}{2}\int \varphi^{(0, n_2)}(z)\psi(z)\, dz\, d\bar{z}.$$

It should be emphasized that for $m_1 = n_1$, $m_2 = n_2$, and for $m_1 = -n_1$, $m_2 = -n_2$, invariant bilinear functionals exist for arbitrary $n_1$, $n_2$ while in Cases 3 and 4 they exist only at discrete points.

[10] We have here written

$$\varphi^{(n_1, n_2)}(z) = \frac{\partial^{n_1+n_2}\varphi(z)}{\partial z^{n_1}\, \partial \bar{z}^{n_2}} .$$

The proof of these results is based on the following preliminary remark. With each matrix

$$g = \begin{Vmatrix} \alpha & \beta \\ \gamma & \delta \end{Vmatrix},$$

with $\alpha\delta - \beta\gamma = 1$, one may associate the linear-fractional transformation $w = (\alpha z + \gamma)/(\beta z + \delta)$ of the complex plane. This association has the property that matrix multiplication corresponds to multiplication of these transformations, and both $e$ and $-e$ (where $e$ is the identity matrix) are associated with the identity transformation, while no others are. It is well known that every linear-fractional transformation of the complex plane can be obtained by combining the following three types of transformations:

1. Parallel translation,

$$z \to z + z_0 \qquad \left(g = \begin{Vmatrix} 1 & 0 \\ z_0 & 1 \end{Vmatrix}\right).$$

2. Dilation,

$$z \to \alpha^2 z \qquad \left(g = \begin{Vmatrix} \alpha & 0 \\ 0 & \alpha^{-1} \end{Vmatrix}\right).$$

3. Inversion,

$$z \to -1/z \qquad \left(g = \begin{Vmatrix} 0 & 1 \\ -1 & 0 \end{Vmatrix}\right).$$

Therefore it is sufficient, in trying to establish whether a given bilinear functional is invariant, to study only those operators of the representations which correspond to these three types of transformations.

In the analysis which follows, the conditions that a bilinear functional be invariant under the transformations corresponding to translation and dilation are the same for all cases, so we shall study these separately in Section 4.2.

## 4.2. Necessary Condition for Invariance under Parallel Translation and Dilation

Consider two representations

$$T_{\chi_1}(g)\varphi(z) = (\beta z + \delta)^{n_1 - 1}(\bar{\beta}\bar{z} + \bar{\delta})^{n_2 - 1}\varphi\left(\frac{\alpha z + \gamma}{\beta z + \delta}\right) \tag{1}$$

and

$$T_{\chi_1}(g)\psi(z) = (\beta z + \delta)^{m_1 - 1}(\bar{\beta}\bar{z} + \bar{\delta})^{m_2 - 1}\psi\left(\frac{\alpha z + \gamma}{\beta z + \delta}\right) \tag{1'}$$

of our group $G$, acting on $D_{\chi_1}$ and $D_{\chi_2}$, respectively. Let us find first the form of a bilinear functional if it is to be invariant under parallel translation. We shall start by discussing functionals only on $K$, that is, for $\varphi(z)$ and $\psi(z)$ in the space of infinitely differentiable functions of bounded support.

We assert that *a bilinear functional on K which is invariant under parallel translation is of the form*

$$B(\varphi, \psi) = (B_0, \omega), \tag{2}$$

*where $B_0$ is some generalized function on K, and $\omega(z)$ is given by*

$$\omega(z) = \frac{i}{2} \int \varphi(z_1)\psi(z + z_1)\, dz_1\, d\bar{z}_1 . \tag{3}$$

**Proof.** The operators $T_{\chi_1}(g)$ and $T_{\chi_2}(g)$ corresponding to parallel translation are given by

$$T_{\chi_1}(g)\varphi(z) = \varphi(z + z_0),$$

$$T_{\chi_2}(g)\psi(z) = \psi(z + z_0).$$

The invariance of $B(\varphi, \psi)$ under all $T_{\chi_1}(g)$ and $T_{\chi_2}(g)$ implies that it must be invariant also under these particular operators, which we shall call *displacements*, so that

$$B(\varphi, \psi) = B(\varphi(z + z_0), \psi(z + z_0)). \tag{4}$$

In Volume 4 (Chapter II, Section 3.5) we obtained the following result. *If $B(\varphi, \psi)$ is a Hermitian bilinear functional on K which is invariant under displacements, then it is of the form*

$$B(\varphi, \psi) = \left(B_0, \int \varphi(x)\bar{\psi}(x - y)\, dx\right),$$

*where $B_0$ is some generalized function on K.* Obviously this result can be reformulated to fit the case of functionals which are linear in each argument (i.e., not Hermitian), and we then obtain Eqs. (2) and (3).

Having obtained the general form of a functional invariant under displacements, let us add the requirement that it be invariant also under dilations of the form $z \to \alpha^2 z$, and this will then give us the form of $B_0$. We shall see, in fact, that $B_0$ is a homogeneous generalized function of $z$ and $\bar{z}$ of degree $(s_1 - 1, s_2 - 1)$, where $s_1 = -\frac{1}{2}(n_1 + m_1)$ and $s_2 = -\frac{1}{2}(n_2 + m_2)$.

For dilation $T_{\chi_1}(g)$ and $T_{\chi_2}(g)$ are given by

$$T_{\chi_1}(g)\varphi(z) = \alpha^{-n_1+1}\bar{\alpha}^{-n_2+1}\varphi(\alpha^2 z) \tag{5}$$

and

$$T_{\chi_1}(g)\varphi(z) = \alpha^{-m_1+1}\bar{\alpha}^{-m_2+1}\varphi(\alpha^2 z). \tag{5'}$$

Then the statement that $B(\varphi, \psi)$ is invariant under these operators may be written

$$B(\varphi, \psi) = \alpha^{-n_1-m_1+2}\bar{\alpha}^{-n_2-m_2+2}B(\varphi(\alpha^2 z), \psi(\alpha^2 z)).$$

Now let us use Eqs. (2) and (3), bearing in mind that

$$\frac{i}{2}\int \varphi(\alpha^2 z_1)\psi(\alpha^2[z + z_1])\, dz_1\, d\bar{z}_1$$
$$= |\,\alpha\,|^{-4}\frac{i}{2}\int \varphi(z_1)\psi(\alpha^2 z + z_1)\, dz_1\, d\bar{z}_1 = |\,\alpha\,|^{-4}\omega(\alpha^2 z).$$

We then arrive at

$$(B_0\,, \omega) = \alpha^{-n_1-m_1}\bar{\alpha}^{-n_2-m_2}(B_0\,, \omega(\alpha^2 z)). \tag{6}$$

This means that $B_0$ is homogeneous of degree $(s_1 - 1, s_2 - 1)$ as stated.

Remark. Note that $s_1 - s_2$ must be an integer. This may be seen by writing $\alpha = -1 = e^{\pi i}$ in Eq. (6), which then yields $\exp[\pi(-n_1 - m_1 + n_2 + m_2)i] = 1$. Therefore $-(n_1 + m_1) + (n_2 + m_2) = 2(s_1 - s_2)$ must be an even integer. #

Homogeneous generalized functions were discussed in Appendix B of Volume 1. It was shown there that to each pair of complex numbers $(s_1\,, s_2)$ whose difference is an integer there corresponds (up to a constant factor) one and only one homogeneous generalized function of degree $(s_1 - 1, s_2 - 1)$. This homogeneous function corresponds to the ordinary function $Cz^{s_1-1}\bar{z}^{s_2-1}$ for $z \neq 0$, except if $s_1\,, s_2 = 0, -1, -2, \ldots$ . For these values of the $s_i$ the generalized function becomes $\delta^{(-s_1,-s_2)}(z, \bar{z})$.[11] We shall call this case, i.e.,when $s_1$ and $s_2$ are simultaneously negative integers or zero, a *singular case*.[12] Thus in a nonsingular case

$$(B_0\,, \omega) = \frac{i}{2}\int z^{s_1-1}\bar{z}^{s_2-1}\omega(z)\, dz\, d\bar{z}, \tag{7}$$

[11] We remind the reader that $\delta^{(-s_1,-s_2)}(z, \bar{z})$ denotes the partial derivative

$$\partial^{-s_1-s_2}\,\delta(z, \bar{z})/\partial z^{-s_1}\,\partial\bar{z}^{-s_2}$$

[12] This division into singular and nonsingular cases is made only for convenience, and is not essential.

where for $\operatorname{Re}(s_1 + s_2) < 0$ the integral is to be understood in the sense of its regularization.

Thus if $s_1 = -\frac{1}{2}(n_1 + m_1)$ and $s_2 = -\frac{1}{2}(n_2 + m_2)$ are not simultaneously negative integers or zero, every bilinear functional invariant under parallel translations and dilations is given, for every pair of functions $\varphi, \psi \in K$, by the expression[13]

$$B(\varphi, \psi) = \left(\frac{i}{2}\right)^2 \int z^{s_1-1}\bar{z}^{s_2-1} \int \varphi(z + z_1)\psi(z_1)\, dz\, d\bar{z}\, dz_1\, d\bar{z}_1$$

$$= \left(\frac{i}{2}\right)^2 \int (z_1 - z_2)^{s_1-1}(\bar{z}_1 - \bar{z}_2)^{s_2-1}\varphi(z_1)\psi(z_2)\, dz_1\, d\bar{z}_1\, dz_2\, d\bar{z}_2 . \tag{8}$$

On the other hand, if $s_1$ and $s_2$ are simultaneously negative integers or zero, the invariant bilinear functional is given by the expression

$$B(\varphi, \psi) = \left(\frac{i}{2}\right)^2 \int \delta^{(-s_1, -s_2)}(z) \int \varphi(z + z_1)\psi(z_1)\, dz_1\, d\bar{z}_1\, dz\, d\bar{z}$$

$$= \left(\frac{i}{2}\right)^2 \int \delta^{(-s_1, -s_2)}(z_1 - z_2)\varphi(z_1)\psi(z_2)\, dz_1\, d\bar{z}_1\, dz_2\, d\bar{z}_2$$

$$= \frac{i}{2} \int \varphi^{(-s_1, -s_2)}(z)\psi(z)\, dz\, d\bar{z}. \tag{10}$$

We shall find the conditions under which $B(\varphi, \psi)$ is invariant under inversion in Section 4.3.

Remark. Recall that $D_\chi$ can also be realized as the space of functions $f(z_1, z_2)$ of two complex variables, homogeneous of degree $(n_1 - 1, n_2 - 1)$ and related to the $\varphi(z)$ by

$$f(z_1, z_2) = z_2^{n_1-1}\bar{z}_2^{n_2-1}\varphi(z_1/z_2).$$

Then the expressions for the bilinear functional in terms of these coordinates can be obtained by replacing $z$ by $z_1/z_2$ and $z'$ by $z_1'/z_2'$ in Eq. (8). This yields

$$B(f_1, f_2) = \int_\Gamma \int_{\Gamma'} (z_1 z_2' - z_1' z_2)^{s_1-1}(\bar{z}_1\bar{z}_2' - \bar{z}_1'\bar{z}_2)^{s_2-1} f_1(z_1, z_2) f_2(z_1', z_2')\omega\omega', \tag{11}$$

[13] For $\operatorname{Re}(s_1 + s_2) < 0$ the integral is to be understood in the sense of its regularization. For instance, if $-m < \operatorname{Re}(s_1 + s_2) < -m + 1$, this regularization is defined by

$$B(\varphi, \psi) = \left(\frac{i}{2}\right)^2 \int z^{s_1-1}\bar{z}^{s_2-1} \int \psi(z_1) \left[\varphi(z + z_1) - \sum_{i+j=0}^{m-1} \frac{\varphi^{(i,j)}(z_1) z^i \bar{z}^j}{i!\, j!}\right] dz\, d\bar{z}\, dz_1\, d\bar{z}_1 . \tag{9}$$

where

$$\omega = \frac{i}{2}(z_1\, dz_2 - z_2\, dz_1)(\bar{z}_1\, d\bar{z}_2 - \bar{z}_2\, d\bar{z}_1),$$

$$\omega' = \frac{i}{2}(z_1'\, dz_2' - z_2'\, dz_1')(\bar{z}_1'\, d\bar{z}_2' - \bar{z}_2'\, d\bar{z}_1'),$$

and $\Gamma$ and $\Gamma'$ are contours intersecting once each line passing through the origin. (The exact meaning of this requirement has been explained in Chapter II, as well as in Volume 1, Appendix B, Section 2.5.) Note that $z_1 z_2' - z_1' z_2$ is invariant when the points $(z_1, z_2)$ and $(z_1', z_2')$ are simultaneously subjected to a transformation of the group

$$\begin{aligned} z_1 &\to \alpha z_1 + \gamma z_2, \\ z_2 &\to \beta z_1 + \delta z_2, \end{aligned} \qquad \alpha\delta - \beta\gamma = 1. \tag{12}$$

It would be of some interest to obtain Eqs. (11) directly by starting with the realizations of $D_{\chi_1}$ and $D_{\chi_2}$ as spaces of homogeneous functions. #

## 4.3. Conditions for Invariance under Inversion

We now wish to find what further conditions are needed in order that $B(\varphi, \psi)$ be invariant also under the inversion $z \to -1/z$. In the present section we shall deal with the case in which $s_1$ and $s_2$ are not both negative integers or zero. The singular cases will be treated in Section 4.5.

The first question we ask is, for what values of $\chi_1 = (n_1, n_2)$ and $\chi_2 = (m_1, m_2)$ will such an invariant bilinear functional exist? Let us then assume that it exists. As was shown in Section 2, it must then be given by

$$B(\varphi, \psi) = \left(\frac{i}{2}\right)^2 \int (z_1 - z_2)^{s_1 - 1}(\bar{z}_1 - \bar{z}_2)^{s_2 - 1}\varphi(z_1)\psi(z_2)\, dz_1\, d\bar{z}_1\, dz_2\, d\bar{z}_2 \tag{1}$$

for $\varphi(z)$ and $\psi(z)$ in $K$. The requirement that $B$ be invariant under operators $T_{\chi_1}(g_0)$ and $T_{\chi_2}(g_0)$, where $g_0$ is an inversion, places certain further requirements on $s_1$ and $s_2$, and these are what we wish to find. First note that our operators are given by

$$T_{\chi_1}(g_0)\varphi(z) = z^{n_1 - 1}\bar{z}^{n_2 - 1}\varphi(-z^{-1}), \tag{2}$$

$$T_{\chi_2}(g_0)\psi(z) = z^{m_1 - 1}\bar{z}^{m_2 - 1}\psi(-z^{-1}). \tag{2'}$$

Since in general these equations transform functions in $K$ into functions not in $K$, we must narrow the class of functions we are dealing with.

Let us require, therefore, that $\varphi$ and $\psi$ vanish in a neighborhood of $z = 0$. It then follows that the functions appearing in (2) and (2′) are also in $K$. We shall assume further that the supports of $\varphi(z)$ and $\psi(z)$ have no points in common. This means that there exist nonintersecting compact sets $A_1$ and $A_2$ such that $\varphi(z)$ vanishes outside $A_1$ and $\psi(z)$ vanishes outside $A_2$.

The invariance of $B(\varphi, \psi)$ under the transformations of (2) and (2′) may now be written

$$\left(\frac{i}{2}\right)^2 \int (z_1 - z_2)^{s_1-1}(\bar{z}_1 - \bar{z}_2)^{s_2-1}\varphi(z_1)\psi(z_2)\, dz_1\, d\bar{z}_1\, dz_2\, d\bar{z}_2$$
$$= \left(\frac{i}{2}\right)^2 \int (z_1 - z_2)^{s_1-1}(\bar{z}_1 - \bar{z}_2)^{s_2-1}$$
$$\times\, z_1^{n_1-1}\bar{z}_1^{n_2-1}\varphi(-z_1^{-1})z_2^{m_1-1}\bar{z}_2^{m_2-1}\psi(-z_2^{-1})\, dz_1\, d\bar{z}_1\, dz_2\, d\bar{z}_2 . \tag{3}$$

Since by assumption $\varphi$ and $\psi$ have nonintersecting supports, each integral converges in the ordinary sense. Let us now write $z_k = -1/w_k$, $k = 1, 2$, on the right-hand side of Eq. (3), which then becomes

$$\left(\frac{i}{2}\right)^2 \int (z_1 - z_2)^{s_1-1}(\bar{z}_1 - \bar{z}_2)^{s_2-1}\varphi(z_1)\psi(z_2)\, dz_1\, d\bar{z}_1\, dz_2\, d\bar{z}_2$$
$$= \left(\frac{i}{2}\right)^2 \int (w_1 - w_2)^{s_1-1}(\bar{w}_1 - \bar{w}_2)^{s_2-1}w_1^{-s_1-n_1}\bar{w}_1^{-s_2-n_2}\varphi(w_1)$$
$$\times\, w_2^{-s_1-m_1}\bar{w}_2^{-s_2-m_2}\psi(w_2)\, dw_1\, dw_1\, dw_2\, d\bar{w}_2 , \tag{3′}$$

where we have used the fact that $n_1 - n_2 + m_1 - m_2$ is even, as established in Section 4.2. This equation is possible only if

$$-s_1 - n_1 = 0, \qquad -s_2 - n_2 = 0,$$
$$-s_1 - m_1 = 0, \qquad -s_2 - m_2 = 0.$$

Now inserting $s_k = -\frac{1}{2}(n_k + m_k)$ we find that $n_k = m_k = -s_k$, $k = 1, 2$. Recall moreover that by assumption the $s_k$ are not both negative integers or zero. It then follows that the $n_k$ may not be nonnegative integers.

Summarizing, we have the following result.

*Let $T_{\chi_1}(g)$, $\chi_1 = (n_1, n_2)$, and $T_{\chi_2}(g)$, $\chi_2 = (m_1, m_2)$, be two representations of $G$, and assume that $s_k = -\frac{1}{2}(n_k + m_k)$, $k = 1, 2$, are not both negative integers or zero. Then a bilinear functional $B(\varphi, \psi)$ invariant under $T_{\chi_1}(g)$ and $T_{\chi_2}(g)$ can exist only if $\chi_1 = \chi_2$, that is, if $n_k = m_k$. If it exists it is given for $\varphi(z)$ and $\psi(z)$ in $K$ by the expression*

$$B(\varphi, \psi) = \left(\frac{i}{2}\right)^2 \int (z_1 - z_2)^{-n_1-1}(\bar{z}_1 - \bar{z}_2)^{-n_2-1}\varphi(z_1)\psi(z_2)\, dz_1\, d\bar{z}_1\, dz_2\, d\bar{z}_2 . \tag{4}$$

[*For* $\operatorname{Re}(n_1 + n_2) > 0$ *the integral is to be understood in the sense of its regularization.*][14]

## 4.4. Sufficiency of Conditions for the Existence of Invariant Bilinear Functionals (Nonsingular Case)

We shall now show that the necessary conditions we have found for the existence of an invariant bilinear functional for $s_k$ not both $0, -1, \ldots$ $(k = 1, 2)$ are also sufficient. What we must show is that if $n_k$ are not both $0, 1, \ldots$, there exists in $D_\chi$ a bilinear functional $B(\varphi, \chi)$ which is invariant under $T_\chi(g)$ for all $\varphi, \psi \in D_\chi$ and for all $g \in G$.

In Section 4.3 we showed that if such a functional exists it is given by the expression

$$B(\varphi, \psi) = \left(\frac{i}{2}\right)^2 \int (z_1 - z_2)^{-n_1-1}(\bar{z}_1 - \bar{z}_2)^{-n_2-1} \varphi(z_1)\psi(z_2)\, dz_1\, d\bar{z}_1\, dz_2\, d\bar{z}_2 \quad (1)$$

for $\varphi(z)$ and $\psi(z)$ in $K$. We shall now prove the following assertion. *The functional* $B(\varphi, \psi)$ *defined for functions in* $K$ *by Eq.* (1) *can be extended to a continuous bilinear functional on all of* $D_\chi$, *and this extended functional is invariant under* $T_\chi(g)$.

The proof is simple in essence, but somewhat cumbersome. If he desires, the reader may skip it and go on to the complete statement in Theorem 1 at the end of the section.

**Proof.** We start with the following lemma.

**Lemma.** *If* $\varphi(z)$ *and* $\psi(z)$ *are in* $K$ *and* $g$ *is an element of* $G$ *such that* $T_\chi(g)\varphi(z)$ *and* $T_\chi(g)\psi(z)$ *are also in* $K$, *then for* $B(\varphi, \psi)$ *as defined by Eq.* (1),

$$B(\varphi, \psi) = B(T_\chi(g)\varphi, T_\chi(g)\psi). \quad (2)$$

More explicitly, Eq. (2) states that

$$\begin{aligned}
&\left(\frac{i}{2}\right)^2 \int (z_1 - z_2)^{-n_1-1}(\bar{z}_1 - \bar{z}_2)^{-n_2-1}\varphi(z_1)\psi(z_2)\, dz_1\, d\bar{z}_1\, dz_2\, d\bar{z}_2 \\
&\quad = \left(\frac{i}{2}\right)^2 \int (z_1 - z_2)^{-n_1-1}(\bar{z}_1 - \bar{z}_2)^{-n_2-1} \\
&\quad\quad \times (\beta z_1 + \delta)^{n_1-1}(\bar{\beta}\bar{z}_1 + \bar{\delta})^{n_2-1}\varphi\left(\frac{\alpha z_1 + \gamma}{\beta z_1 + \delta}\right) \\
&\quad\quad \times (\beta z_2 + \delta)^{n_1-1}(\bar{\beta}\bar{z}_2 + \bar{\delta})^{n_2-1}\psi\left(\frac{\alpha z_2 + \gamma}{\beta z_2 + \delta}\right) dz_1\, d\bar{z}_1\, dz_2\, d\bar{z}_2\,.
\end{aligned} \quad (3)$$

[14] This regulariation is discussed in Footnote 13.

Both sides of this equation are analytic in $n_1$ and $n_2$.[15] It is therefore sufficient to show that it holds for $\mathrm{Re}(n_1 + n_2) < 0$. But in this case the assumptions we have made concerning $\varphi(z)$, $\psi(z)$, and $g$ imply that the integrals on both sides of the equation converge absolutely. Now let us write

$$\frac{\alpha z_1 + \gamma}{\beta z_1 + \delta} = w_1\,, \qquad \frac{\alpha z_2 + \gamma}{\beta z_2 + \delta} = w_2$$

in the integrand on the right-hand side of (3). After some simple operations we find that the integral becomes

$$\left(\frac{i}{2}\right)^2 \int (w_1 - w_2)^{-n_1-1}(\bar{w}_1 - \bar{w}_2)^{-n_2-1}\, \varphi(w_1)\psi(w_2)\, dw_1\, d\bar{w}_1\, dw_2\, d\bar{w}_2\,,$$

which is the same as the integral on the left-hand side of (3). This proves the lemma.

Let us now extend $B(\varphi, \psi)$ to all of $D_\chi$, maintaining its invariance. Let $\varphi(z)$ and $\psi(z)$ be in $D_\chi$. Assume first that there exists an open set $\Omega$ on which both $\varphi(z)$ and $\psi(z)$ vanish. Then there exists an element $g_0 \in G$ such that $T_\chi(g_0)\varphi(z)$ and $T_\chi(g_0)\psi(z)$ have bounded support, for it is possible to choose

$$g_0 = \begin{Vmatrix} \alpha & \beta \\ \gamma & \delta \end{Vmatrix}$$

so that the linear-fractional transformation

$$w = \frac{\alpha z + \gamma}{\beta z + \delta}$$

corresponding to it maps some point of $\Omega$ onto the point at infinity. Then we define the bilinear functional for these functions by the expression

$$B(\varphi, \psi) = B(T_\chi(g_0)\varphi,\, T_\chi(g_0)\psi).$$

Moreover, this definition does not depend on the choice of $g_0$. Indeed, let $g_1$ be some other element of $G$ such that $T_\chi(g_1)\varphi(z)$ and $T_\chi(g_1)\psi(z)$ have bounded support. Then $T_\chi(g_0 g_1^{-1})$ transforms $T_\chi(g_1)\varphi$ and $T_\chi(g_1)\psi$, both of which have bounded support, into $T_\chi(g_0)\varphi$ and $T_\chi(g_0)\psi$, which also have bounded support, so that

$$B(T_\chi(g_1)\varphi,\, T_\chi(g_1)\psi) = B(T_\chi(g_0)\varphi,\, T_\chi(g_0)\psi).$$

[15] More exactly, they are analytic in $\lambda = n_1 + n_2$ for fixed integer $n_1 - n_2$.

In this way $B(\varphi, \psi)$ is extended to all $\varphi(z)$ and $\psi(z)$ in $D_\chi$ which vanish in a given open region. Obviously this extended functional is bilinear, for if $\varphi_1$, $\varphi_2$, and $\psi$ vanish in some region $\Omega$, then

$$B(a_1\varphi_1 + a_2\varphi_2, \psi) = a_1B(\varphi_1, \psi) + a_2B(\varphi_2, \psi).$$

Similarly, if $\varphi$, $\psi_1$, and $\psi_2$ vanish in some region $\Omega$, then

$$B(\varphi, b_1\psi_1 + b_2\psi_2) = b_1B(\varphi, \psi_1) + b_2B(\varphi, \psi_2).$$

It is also obvious that this functional is invariant under the entire representation $T(g)$.

What remains is to define $B(\varphi, \psi)$ for any $\varphi(z)$ and $\psi(z)$ in $D_\chi$. On the complex plane we choose three regions $\Omega_1$, $\Omega_2$, $\Omega_3$ such that (1) every pair of them has a nonempty intersection and (2) there exists no point belonging simultaneously to the closures of all three regions. Then obviously every $\varphi(z) \in D_\chi$ can be written in the form

$$\varphi(z) = \varphi_1(z) + \varphi_2(z) + \varphi_3(z),$$

where each $\varphi_i(z)$ vanishes on $\Omega_i$ ($i = 1, 2, 3$) and belongs to $D_\chi$.

Now let $\varphi$ and $\psi$ be any two functions in $D_\chi$. As above, we write

$$\varphi(z) = \varphi_1(z) + \varphi_2(z) + \varphi_3(z),$$

$$\psi(z) = \psi_1(z) + \psi_2(z) + \psi_3(z).$$

The functional $B(\varphi_i, \psi_j)$ is defined for any pair $\varphi_i$, $\psi_j$, since these functions vanish in a common region (namely the intersection of $\Omega_i$ and $\Omega_j$). We then define $B(\varphi, \psi)$ by the expression

$$B(\varphi, \psi) = \sum_{i,j=1}^{3} B(\varphi_i, \psi_j).$$

It is easily shown that this definition is independent of how $\varphi$ and $\psi$ are broken up into the $\varphi_i$ and $\psi_i$ and that $B(\varphi, \psi)$ so defined is an invariant bilinear functional.

This completes the proof of the sufficiency. Summarizing, we have the following theorem.

**Theorem 1.** *Assume the representations*

$$T_{\chi_1}(g)\varphi(z) = (\beta z + \delta)^{n_1-1}(\bar{\beta}\bar{z} + \bar{\delta})^{n_2-1}\varphi\left(\frac{\alpha z + \gamma}{\beta z + \delta}\right) \tag{4}$$

*and*

$$T_{\chi_2}(g)\psi(z) = (\beta z + \delta)^{m_1-1}(\bar{\beta}\bar{z} + \bar{\delta})^{m_2-1}\psi\left(\frac{\alpha z + \gamma}{\beta z + \delta}\right) \tag{4'}$$

*of G to be such that* $s_1 = -\frac{1}{2}(n_1 + m_1)$ *and* $s_2 = -\frac{1}{2}(n_2 + m_2)$ *are not both negative integers or zero. Then there exists a continuous bilinear functional* $B(\varphi, \psi)$ *invariant under* $T_{\chi_1}(g)$ *and* $T_{\chi_2}(g)$ *if and only if* $n_1 = m_1$, $n_2 = m_2$ *(or in other words* $\chi_1 = \chi_2 = \chi$*). When it exists,* $B(\varphi, \psi)$ *is defined for* $\varphi(z)$ *and* $\psi(z)$ *in K by the expression*

$$B(\varphi, \psi) = (\tfrac{1}{2}i)^2 \int (z_1 - z_2)^{-n_1-1}(\bar{z}_1 - \bar{z}_2)^{-n_2-1} \times \varphi(z_1)\psi(z_2)\, dz_1\, d\bar{z}_1\, dz_2\, d\bar{z}_2\,. \tag{5}$$

*[For* $\mathrm{Re}(n_1 + n_2) > 0$ *the integral is to be understood in the sense of its regularization.] For functions with unbounded support,* $B(\varphi, \psi)$ *is defined by bilinearity and invariance.*

It may be remarked that for $\mathrm{Re}(n_1 + n_2) < 0$, Eq. (5) can be used to calculate $B(\varphi, \psi)$ for *all* functions in $D_\chi$. This is because the integral in (5) then converges in the ordinary sense for all $\varphi, \psi \in D_\chi$.[16]

## 4.5. Conditions for the Existence of Invariant Bilinear Functionals (Singular Case)

In Section 4.3 and 4.4 we dealt with the nonsingular case, in which $s_1$ and $s_2$ are not both negative integers or zero. Let us now go on to the singular case, in which $s_1, s_2 = 0, -1, -2, \ldots$ .

In Section 4.2 it was found that a bilinear functional defined on the functions of bounded support, invariant under parallel translations and dilations in the complex plane, must be of the form

$$B(\varphi, \psi) = (\tfrac{1}{2}i)^2 \int \delta^{(-s_1, -s_2)}(z_1 - z_2)\varphi(z_1)\psi(z_2)\, dz_1\, d\bar{z}_1\, dz_2\, d\bar{z}_2\,, \tag{1}$$

where $\delta^{(-s_1,-s_2)}(z) = \partial^{-s_1-s_2}\,\delta(z)/\partial z^{-s_1}\,\partial\bar{z}^{-s_2}$. Let us now study the additional conditions imposed on $\chi_1$ and $\chi_2$ if $B(\varphi, \psi)$ is to be invariant under the inversion $z \to -1/z$. The matrix

$$g_0 = \begin{Vmatrix} 0, & 1 \\ -1, & 0 \end{Vmatrix}$$

corresponding to this inversion is represented by the operators

$$T_{\chi_1}(g_0)\varphi(z) = z^{n_1-1}\bar{z}^{n_2-1}\varphi(-z^{-1}),$$

$$T_{\chi_2}(g_0)\psi(z) = z^{m_1-1}\bar{z}^{m_2-1}\psi(-z^{-1}).$$

[16] This convergence is most easily seen if one writes out the expression for $B(\varphi, \psi)$ in homogeneous coordinates [see Eq. (11) of Section 4.2].

We shall assume that $\varphi(z)$ and $\psi(z)$ not only have bounded support, but vanish in a neighborhood of $z = 0$. Then $T_{\chi_1}(g_0)\varphi$ and $T_{\chi_2}(g_0)\psi$ also have bounded support, and

$$\begin{aligned} B(T_{\chi_1}(g_0)\varphi(z), T_{\chi_2}(g_0)\psi(z)) \\ = \left(\frac{i}{2}\right)^2 \int \delta^{(-s_1,-s_2)}(z_1 - z_2) z_1^{n_1-1}\bar{z}_1^{n_2-1}\varphi(-z_1^{-1}) \\ \times z_2^{m_1-1}\bar{z}_2^{m_2-1}\psi(-z_2^{-1})\, dz_1\, d\bar{z}_1\, dz_2\, d\bar{z}_2 . \end{aligned}$$

Thus invariance of $B(\varphi, \psi)$ under inversion may be written

$$\begin{aligned} \left(\frac{i}{2}\right)^2 \int \delta^{(-s_1,-s_2)}(z_1 - z_2)\varphi(z_1)\psi(z_2)\, dz_1\, d\bar{z}_1\, dz_2\, d\bar{z}_2 \\ = \left(\frac{i}{2}\right)^2 \int \delta^{(-s_1,-s_2)}(z_1 - z_2) z_1^{n_1-1}\bar{z}_1^{n_2-1}\varphi(-z_1^{-1}) \\ \times z_2^{m_1-1}\bar{z}_2^{m_2-1}\psi(-z_2^{-1})\, dz_1\, d\bar{z}_1\, dz_2\, d\bar{z}_2 , \end{aligned} \tag{2}$$

where $\varphi(z)$ and $\psi(z)$ are any two functions in $K$ which vanish in a neighborhood of $z = 0$. We now write $w_k = -1/z_k$ $(k = 1, 2)$ in the integral on the right-hand side, and bear in mind that

$$(-1)^{n_1-n_2+m_1-m_2} = (-1)^{-2(s_1-s_2)} = 1.$$

Then Eq. (2) becomes

$$\begin{aligned} \left(\frac{i}{2}\right)^2 \int \delta^{(-s_1,-s_2)}(z_1 - z_2)\varphi(z_1)\psi(z_2)\, dz_1\, d\bar{z}_1\, dz_2\, d\bar{z}_2 \\ = \left(\frac{i}{2}\right)^2 \int \delta^{(-s_1,-s_2)} \left(\frac{w_1 - w_2}{w_1 w_2}\right) w_1^{-n_1-1}\bar{w}_1^{-n_2-1} \\ \times w_2^{-m_1-1}\bar{w}_2^{-m_2-1}\varphi(w_1)\psi(w_2)\, dw_1\, d\bar{w}_1\, dw_2\, d\bar{w}_2 . \end{aligned} \tag{3}$$

The integral on the right-hand side can be simplified somewhat by using the properties of the $\delta$ function. Let $P(w) \equiv P(w_1, \ldots, w_n)$ be an entire analytic function of $n$ complex variables such that the manifold whose equation is $P = 0$ has no singular points, and let $a(w)$ be an infinitely differentiable function that vanishes nowhere. Then

$$\delta^{(k,l)}(aP) = a^{-k-1}\bar{a}^{-l-1}\delta^{(k,l)}(P) \tag{4}$$

(this is discussed in Appendix B of Volume 1). We now set $P = w_1 - w_2$ and $a = 1/w_1 w_2$. This might at first seem to be in violation of our requirements, since $a$ has singularities at $w_1 = 0$ and $w_2 = 0$. Recall, however, that the function $\varphi(w_1)\psi(w_2)$ in Eq. (3) vanishes by assumption in the neighborhood of $w_1 = 0$ and $w_2 = 0$, so that the behavior of $a$ in this neighborhood is irrelevant. Thus we may replace the derivative of the $\delta$ function in Eq. (3) by $(w_1 w_2)^{-s_1+1}(\bar{w}_1\bar{w}_2)^{-s_2+1}\delta^{(-s_1,-s_2)}(w_1 - w_2)$. This yields

$$\left(\frac{i}{2}\right)\int \varphi^{(-s_1,-s_2)}(z)\psi(z)\,dz\,d\bar{z} = \left(\frac{i}{2}\right)\int w^{-s_1-m_1}\bar{w}^{-s_2-m_2}\psi(w) \times [w^{-s_1-n_1}\bar{w}^{-s_2-n_2}\varphi(w)]^{(-s_1,-s_2)}\,dw\,d\bar{w}. \quad (5)$$

Obviously this can be true only if

$$\varphi^{(-s_1,-s_2)}(w) = w^{-s_1-m_1}\bar{w}^{-s_2-m_2}[w^{-s_1-n_1}\bar{w}^{-s_2-n_2}\varphi(w)]^{(-s_1,-s_2)}. \quad (6)$$

Let us find the exponents for which this may hold. The expression on the right-hand side of (6) can be written as a sum of the form

$$\sum_{k=0}^{-s_1}\sum_{l=0}^{-s_2} c_{k,l}w^{\alpha_k}\bar{w}^{\beta_l}\varphi^{(k,l)}(w)$$

and according to Eq. (6) this sum must consist of the single term $\varphi^{(-s_1,-s_2)}(w)$. For $s_1 \neq 0$ and $s_2 \neq 0$ this is obviously possible only if $-s_1 - n_1 = 0$ and $-s_2 - n_2 = 0$. If $s_1 = 0$ and $s_2 \neq 0$, it is possible only if $-s_2 - n_2 = 0$. Similarly, if $s_1 \neq 0$ and $s_2 = 0$, it is possible only if $-s_1 - n_1 = 0$. Thus at least one of the following conditions must be fulfilled:

*Case 1.* $s_1 = -n_1,\ s_2 = -n_2$;

*Case 2.* $s_1 = 0,\ s_2 = 0$;

*Case 3.* $s_1 = -n_1,\ s_2 = 0$;

*Case 4.* $s_1 = 0,\ s_2 = -n_2$.

Summarizing, we have the following result.

*Let* $\chi_1 = (n_1, n_2)$ *and* $\chi_2 = (m_1, m_2)$, *and assume that* $s_1 = -\frac{1}{2}(n_1 + m_1)$ *and* $s_2 = -\frac{1}{2}(n_2 + m_2)$ *are negative integers or zero. Then there exists a bilinear functional* $B(\varphi, \psi)$ *invariant under* $T_{\chi_1}(g)$ *and* $T_{\chi_2}(g)$ *only if one of the following conditions is fulfilled:*

*Case 1.* $n_1 = m_1,\ n_2 = m_2;\ n_1, n_2 = 0, 1, 2, \ldots$;

*Case 2.* $n_1 = -m_1,\ n_2 = -m_2$;

*Case 3.* $n_1 = m_1,\ n_2 = -m_2,\ n_1 = 1, 2, \ldots$;

*Case 4.* $n_1 = -m_1,\ n_2 = m_2,\ n_2 = 1, 2, \ldots$.

We have established also that if such an invariant bilinear functional exists, it is given in each of these cases by the following expressions:

*Case 1.*

$$B(\varphi, \psi) = \frac{i}{2} \int \varphi^{(n_1, n_2)}(z)\psi(z)\, dz\, d\bar{z}. \tag{7}$$

*Case 2.*

$$B(\varphi, \psi) = \frac{i}{2} \int \varphi(z)\psi(z)\, dz\, d\bar{z}. \tag{8}$$

*Case 3.*

$$B(\varphi, \psi) = \frac{i}{2} \int \varphi^{(n_1, 0)}(z)\psi(z)\, dz\, d\bar{z}. \tag{9}$$

*Case 4.*

$$B(\varphi, \psi) = \frac{i}{2} \int \varphi^{(0, n_2)}(z)\psi(z)\, dz\, d\bar{z}. \tag{10}$$

[It is assumed in all these equations that $\varphi(z)$ and $\psi(z)$ are in $K$.]

We have thus not only found the values of the $n_k$ and $m_k$ for which an invariant bilinear functional may exist, but also have found the form of this functional on functions in $K$ when it does exist. We shall now show that these conditions are not only necessary, but sufficient.

Let us first consider Case 2, the simplest, namely $n_k = -m_k$. For $\varphi(z)$ and $\psi(z)$ in $K$, Eq. (8) defines $B(\varphi, \psi)$. But the integral in (8) converges as well for all $\varphi(z)$ and $\psi(z)$ in $D_{\chi_1}$ and $D_{\chi_2}$, respectively. Indeed, we know that as $z \to \infty$, such functions behave asymptotically according to the two relations

$$\begin{aligned} \varphi(z) &\sim C_1 z^{n_1-1}\bar{z}^{n_2-1}, \\ \psi(z) &\sim C_2 z^{m_1-1}\bar{z}^{m_2-1} = C_2 z^{-n_1-1}\bar{z}^{-n_2-1}. \end{aligned} \tag{11}$$

Therefore $\varphi(z)\psi(z) \sim C_1 C_2 | z |^{-4}$, so that the integral in (8) converges. Further, Eq. (3) defines an *invariant* bilinear functional, as may be seen from the following. With $n_k = -m_k$ and for any element

$$g = \begin{Vmatrix} \alpha & \beta \\ \gamma & \delta \end{Vmatrix}$$

of $G$, we find that

$$\begin{aligned} &B(T_{\chi_1}(g)\varphi, T_{\chi_2}(g)\psi) \\ &\quad = \frac{i}{2} \int T_{\chi_1}(g)\varphi(z) T_{\chi_2}(g)\psi(z)\, dz\, d\bar{z} \\ &\quad = \frac{i}{2} \int (\beta z + \delta)^{-2}(\bar{\beta}\bar{z} + \bar{\delta})^{-2} \varphi\left(\frac{\alpha z + \gamma}{\beta z + \delta}\right) \psi\left(\frac{\alpha z + \gamma}{\beta z + \delta}\right) dz\, d\bar{z}. \end{aligned}$$

Now if we write $(\alpha z + \gamma)/(\beta z + \delta) = w$ in this integral it becomes $\frac{i}{2}\int \varphi(w)\psi(w)\, dw\, d\bar{w}$. Thus

$$B(T_{\chi_1}(g)\varphi, T_{\chi_2}(g)\psi) = B(\varphi, \psi),$$

as asserted.

Let us now turn to Case 1, in which $n_k = m_k$ are nonnegative integers. Then for $\varphi(z)$ and $\psi(z)$ in $K$ we have

$$\begin{aligned} B(\varphi, \psi) &= \left(\frac{i}{2}\right)^2 \int \delta^{(n_1, n_2)}(z_1 - z_2)\varphi(z_1)\psi(z_2)\, dz_1\, d\bar{z}_1\, dz_2\, d\bar{z}_2 \\ &= \frac{i}{2}\int \varphi^{(n_1, n_2)}(z)\psi(z)\, dz\, d\bar{z}. \end{aligned} \tag{12}$$

Now this integral converges for all $\varphi(z)$ and $\psi(z)$ in $D_\chi$, $\chi = (n_1, n_2)$. To see this we make use of the asymptotic series

$$\varphi(z) \sim z^{n_1-1}\bar{z}^{n_2-1} \sum_{j,k=0}^{\infty} a_{jk} z^{-j}\bar{z}^{-k}, \qquad z \to \infty$$

[see Eq. (6) of Section 2.3]. Now since the $n_k$ are nonnegative integers, the derivative appearing in (12) behaves asymptotically as

$$\varphi^{(n_1, n_2)}(z) \sim C_1 z^{-n_1-1}\bar{z}^{-n_2-1}.$$

Further, $\psi(z) \sim C_2 z^{n_1-1}\bar{z}^{n_2-1}$ so that $\varphi^{(n_1, n_2)}(z)\psi(z) \sim C_1C_2|\, z\,|^{-4}$ and the integral in Eq. (12) converges. Consequently Eq. (12) defines a functional $B(\varphi, \psi)$ for all $\varphi, \psi \in D_\chi$; furthermore this functional is invariant, that is,

$$B(T_\chi(g)\varphi, T_\chi(g)\psi) = B(\varphi, \psi). \tag{13}$$

Since this has already been established in Section 4.2 for parallel translations and dilations, we need only prove it for inversions,

$$T_\chi(g_0)\varphi(z) = z^{n_1-1}\bar{z}^{n_2-1}\varphi(-z^{-1}) \equiv \hat{\varphi}(z),$$

that is, we must show that

$$B(\hat{\varphi}, \hat{\psi}) \equiv B(T_\chi(g_0)\varphi, T_\chi(g_0)\psi) = B(\varphi, \psi). \tag{13$'$}$$

Now by definition, we have

$$\begin{aligned} B(\hat{\varphi}, \hat{\psi}) = (\tfrac{1}{2}i)^2 \int \delta^{(n_1, n_2)}(z_1 - z_2) z_1^{n_1-1}\bar{z}_1^{n_2-1}\varphi(-z_1^{-1}) \\ \times z_2^{n_1-1}\bar{z}_2^{n_2-1}\psi(-z_2^{-1})\, dz_1\, d\bar{z}_1\, dz_2\, d\bar{z}_2\,. \end{aligned} \tag{14}$$

Let us first assume that $\varphi$ and $\psi$ as well as their inversions $\hat{\varphi}$ and $\hat{\psi}$ are in $K$ (which means that $\varphi$ and $\psi$ must vanish in a neighborhood of $z = 0$). We then write $w_k = -1/z_k$ in (14), and as before, since $\varphi(w_1)\psi(w_2)$ vanishes in a neighborhood of $w_1 = 0$ and $w_2 = 0$ [see Eq. (4)], we arrive at

$$\begin{aligned} B(\hat{\varphi}, \hat{\psi}) &= \left(\frac{i}{2}\right)^2 \int \delta^{(n_1, n_2)}\left(\frac{w_1 - w_2}{w_1 w_2}\right) w_1^{-n_1-1}\bar{w}_1^{-n_2-1}\varphi(w_1) \\ &\quad \times w_2^{-n_1-1}\bar{w}_2^{-n_2-1}\psi(w_2)\, dw_1\, d\bar{w}_1\, dw_2\, d\bar{w}_2 \\ &= \left(\frac{i}{2}\right)^2 \int \delta^{(n_1, n_2)}(w_1 - w_2)\varphi(w_1)\psi(w_2)\, dw_1\, d\bar{w}_1\, dw_2\, d\bar{w}_2 \\ &= B(\varphi, \psi). \end{aligned} \tag{15}$$

This proves Eq. (13′) for $\varphi$, $\psi$, $\hat{\varphi}$, $\hat{\psi}$ in $K$.

The case of more general $\varphi$ and $\psi$ is easily reduced to this one. Functions in $K$ whose inversions are also in $K$ do not form an everywhere dense set in $D_\chi$. Nevertheless for every pair of functions $\varphi, \psi \in D_\chi$ one can find a pair of functions $\varphi_1, \psi_1 \in K$, whose inversions $\hat{\varphi}_1, \hat{\psi}_1 \in K$, such that $B(\varphi, \psi)$ and $B(\hat{\varphi}, \hat{\psi})$ simultaneously differ by an arbitrarily small amount from $B(\varphi_1, \psi_1)$ and $B(\hat{\varphi}_1, \hat{\psi}_1)$. Then since $B(\varphi_1, \psi_1) = B(\hat{\varphi}_1, \hat{\psi}_1)$, the assertion can be proved.

Remark. These approximating functions $\varphi_1$ and $\psi_1$ are obtained by "smoothing to zero" the initial functions $\varphi$ and $\psi$ in the neighborhood of $z = 0$ and $z = \infty$. Since Eq. (12) involves the derivative of $\varphi$, the smoothing must be performed where $\psi_1(z)$ is already equal to zero. #

Cases 3 and 4 can be treated quite analogously.

Summarizing, we have the following theorem.

**Theorem 2.** *Assume the representations*

$$T_{\chi_1}(g)\varphi(z) = (\beta z + \delta)^{n_1-1}(\bar{\beta}\bar{z} + \bar{\delta})^{n_2-1}\varphi\left(\frac{\alpha z + \gamma}{\beta z + \delta}\right)$$

*and*

$$T_{\chi_2}(g)\psi(z) = (\beta z + \delta)^{m_1-1}(\bar{\beta}\bar{z} + \bar{\delta})^{m_2-1}\psi\left(\frac{\alpha z + \gamma}{\beta z + \delta}\right)$$

*of $G$ to be such that $s_1 = -\frac{1}{2}(n_1 + m_1)$ and $s_2 = -\frac{1}{2}(n_2 + m_2)$ are negative integers or zero. Then there exists a continuous bilinear functional $B(\varphi, \psi)$ invariant under $T_{\chi_1}(g)$ and $T_{\chi_2}(g)$ if and only if one of the following conditions is fulfilled:*[17]

[17] Note that since $n_1 - n_2$ is an integer, *all* the $n_k$ and $m_k$ in Cases 3 and 4 are integers.

*Case 1.* $n_k = m_k = 0, 1, 2, \ldots;$

Case *2.* $n_k = -m_k;$

*Case 3.* $n_1 = m_1 = 1, 2, \ldots, n_2 = -m_2;$

*Case 4.* $n_1 = -m_1, n_2 = m_2 = 1, 2, \ldots .$

*When it exists,* $B(\varphi, \psi)$ *is defined for each of these cases by the following expressions.*

*Case 1.*

$$B(\varphi, \psi) = \frac{i}{2} \int \varphi^{(n_1, n_2)}(z)\psi(z)\, dz\, d\bar{z}. \tag{16}$$

*Case 2.*

$$B(\varphi, \psi) = \frac{i}{2} \int \varphi(z)\psi(z)\, dz\, d\bar{z}. \tag{17}$$

*Case 3.*

$$B(\varphi, \psi) = \frac{i}{2} \int \varphi^{(n_1, 0)}(z)\psi(z)\, dz\, d\bar{z}. \tag{18}$$

*Case 4.*

$$B(\varphi, \psi) = \frac{i}{2} \int \varphi^{(0, n_2)}(z)\psi(z)\, dz\, d\bar{z}. \tag{19}$$

## 4.6. Degeneracy of Invariant Bilinear Functionals[18]

A bilinear functional $B(\varphi, \psi)$ defined on a pair of spaces $D_1$ and $D_2$ is called *degenerate on a subspace* $E_1 \subset D_1$ if $B(\varphi, \psi) = 0$ for all $\varphi \in E_1$ and $\psi \in D_2$. Similarly, it may be degenerate on a subspace $E_2 \subset D_2$. We shall show in this section that the invariant bilinear functionals defined in Sections 4.4 and 4.5 are sometimes degenerate.

Consider first Case 1, in which $n_k = m_k = 0, 1, 2, \ldots$. Then it is clear from Eq. (16) of the previous section that $B(\varphi, \psi) = 0$ if $\varphi(z)$ is a polynomial of the form

$$\varphi(z) = \sum_{j=0}^{n_1-1} \sum_{k=0}^{n_2-1} a_{jk} z^j \bar{z}^k. \tag{1}$$

[We already saw in Section 3.1 that these polynomials form the invariant subspace $E_{\chi_1} \subset D_{\chi_1}, \chi_1 = (n_1, n_2)$.] Hence if $n_1$ and $n_2$ are nonnegative integers, the invariant bilinear functional $B(\varphi, \psi)$ defined on $D_{\chi_1}$ is degenerate on the finite-dimensional subspace $E_{\chi_1}$.

[18] This section and the following one will not be used later, and the reader may omit them if he so desires.

Now consider the case in which $n_k = m_k$, but the $n_k$ are not nonnegative integers. We then have

$$B(\varphi, \psi) = \left(\frac{i}{2}\right)^2 \int (z_1 - z_2)^{-n_1-1}(\bar{z}_1 - \bar{z}_2)^{-n_2-1}\varphi(z_1)\psi(z_2)\, dz_1\, d\bar{z}_1\, dz_2\, d\bar{z}_2 .$$

It is then obvious that if in addition $n_1$ and $n_2$ are *negative* integers, $B(\varphi, \psi)$ vanishes for all $\varphi(z) \in D_{\chi_1}$ whose moments

$$b_{jk} = \frac{i}{2} \int z^j \bar{z}^k \varphi(z)\, dz\, d\bar{z}$$

vanish for $j = 0, 1, ..., -n_1 - 1, k = 0, 1, ..., -n_2 - 1$. [We already saw in Section 3.1 that these functions form an invariant infinite-dimensional subspace $F_{\chi_1} \subset D_{\chi_1}$ such that the factor space $D_{\chi_1}/F_{\chi_1}$ has the finite dimension $n_1 n_2$.]

The bilinear functional will be degenerate also in Cases 3 and 4 if *both* $n_1$ and $n_2$ are nonnegative integers. In fact, for instance, Eq. (18) of Section 4.5 shows that then in Case 3 $B(\varphi, \psi)$ vanishes for $\varphi$ given by (1). As in Case 1, these functions form the finite-dimensional invariant subspace $E_{\chi_1} \subset D_{\chi_1}$. The situation is similar in Case 4.

## 4.7. Conditionally Invariant Bilinear Functionals

Consider a representation on $D_\chi$ for integer $\chi$ (i.e., $n_1$ and $n_2$ are nonzero integers of the same sign). As we have seen, there exists on $D_\chi$ an invariant bilinear functional, degenerate on an invariant subspace. It is possible to introduce another bilinear functional on this invariant subspace which will also be invariant under $T_\chi(g)$.

Assume first that $n_1$ and $n_2$ are negative integers. We then have

$$B(\varphi, \psi) = \frac{i}{2} \int (z_1 - z_2)^{-n_1-1}(\bar{z}_1 - \bar{z}_2)^{-n_2-1}\varphi(z_1)\psi(z_2)\, dz_1\, d\bar{z}_1\, dz_2\, d\bar{z}_2 , \quad (1)$$

which is degenerate on the subspace $F_\chi$ of functions $\varphi$ whose moments $b_{ks}$ vanish for $0 \leqslant k \leqslant -n_1 - 1, 0 \leqslant s \leqslant -n_2 - 1$. We wish to find an invariant bilinear functional on this subspace $F_\chi$. But first note the following. Early in our considerations (see Section 4.2) we used the homogeneity condition

$$B(\varphi(z), \psi(z)) = \alpha^{2s_1+2}\bar{\alpha}^{2s_2+2}B(\varphi(\alpha^2 z), \psi(\alpha^2 z)), \quad (2)$$

where, as always, $s_k = -\frac{1}{2}(n_k + m_k)$. From this we derived the result that $B(\varphi, \psi)$ must be of the form

$$B(\varphi, \psi) = C\left(\frac{i}{2}\right)^2 \int z^{s_1-1}\bar{z}^{s_2-1} \left[\int \varphi(z_1)\psi(z + z_1)\, dz_1\, d\bar{z}_1\right] dz\, d\bar{z} \tag{3}$$

(in our case $n_k = m_k$, so that $s_k = -n_k$, $k = 1, 2$). If, however, we restrict our considerations only to $\varphi(z)$ and $\psi(z)$ in $F_\chi$, Eq. (2) will be satisfied not only by (3), but also by

$$B_1(\varphi, \psi) = \left(\frac{i}{2}\right)^2 \int z^{-n_1-1}\bar{z}^{-n_2-1} \ln |z| \int \varphi(z_1)\psi(z + z_1)\, dz\, d\bar{z}\, dz_1\, d\bar{z}_1 . \tag{4}$$

This is because $z^{-n_1-1}\bar{z}^{-n_2-1} \ln|z|$ is an associated homogeneous generalized function (see Volume 1, Appendix B1.5), so that

$$B_1(\varphi(\alpha^2 z), \psi(\alpha^2 z))$$
$$= \alpha^{2n_1-2}\bar{\alpha}^{2n_2-2}[B_1(\varphi(z), \psi(z)) + \ln |\alpha| B(\varphi(z), \psi(z))].$$

But $\varphi(z), \psi(z) \in F_\chi$ implies that $B(\varphi, \psi) = 0$. Therefore

$$B_1(\varphi(\alpha^2 z), \psi(\alpha^2 z)) = \alpha^{-2s_1-2}\bar{\alpha}^{-2s_2-2}B_1(\varphi(z), \psi(z)),$$

so that $B_1(\varphi, \psi)$ is invariant under the dilations

$$T_{\chi_1}(g)\varphi(z) = \alpha^{-n_1+1}\bar{\alpha}^{-n_2+1}\varphi(\alpha^2 z),$$

$$T_{\chi_2}(g)\psi(z) = \alpha^{-m_1+1}\bar{\alpha}^{-m_2+1}\psi(\alpha^2 z).$$

One can then show by direct calculation that $B_1(\varphi, \psi)$ is an invariant bilinear functional on $F_\chi$.

In summary, then, if $n_1, n_2 = -1, -2, \ldots$, there exists an invariant bilinear functional on the subspace $F_\chi \subset D_\chi$, and this functional is given by the expression

$$B_1(\varphi, \psi) = \left(\frac{i}{2}\right)^2 \int (z_2 - z_1)^{-n_1-1}(\bar{z}_2 - \bar{z}_1)^{-n_2-1} \ln |z_1 - z_2|$$
$$\times \varphi(z_1)\psi(z_2)\, dz_1\, d\bar{z}_1\, dz_2\, d\bar{z}_2 . \tag{5}$$

Now let $n_1$ and $n_2$ be positive integers. We then have

$$B(\varphi, \psi) = \frac{i}{2} \int \varphi^{(n_1, n_2)}(z)\psi(z)\, dz\, d\bar{z},$$

which is degenerate on the subspace $E_\chi$ of polynomials of degree $n_1 - 1$ in $z$ and $n_2 - 1$ in $\bar{z}$. In this case there also exists an invariant bilinear functional $B_2(\varphi, \psi)$ on the subspace. Obviously this functional will be uniquely defined by the numbers

$$c_{ks,rt} = B_2(z^k \bar{z}^s, z^r \bar{z}^t),$$
$$0 \leqslant k, \quad r \leqslant n_1 - 1, \quad 0 \leqslant s, \quad t \leqslant n_2 - 1.$$

We first show that $c_{ks,rt} = 0$ if $k + r \neq n_1 - 1$ or $s + t \neq n_2 - 1$. Again we use the fact that $B_2$ must be invariant under dilations. This yields

$$c_{ks,rt} = \alpha^{-2(n_1-k-r-1)} \bar{\alpha}^{-2(n_2-s-t-1)} c_{ks,rt}.$$

If $c_{ks,rt} \neq 0$, this equation can be satisfied only if $k + r = n_1 - 1$ and $s + t = n_2 - 1$.

Thus we need find only the $c_{ks} \equiv c_{ks;n_1-k-1,n_2-s-1}$. This may be done by using invariance under parallel translations, which implies that

$$\begin{aligned} 0 &= B_2(z^k \bar{z}^s, z^{n_1-k} \bar{z}^{n_2-s-1}) \\ &= B_2([z + z_0]^k [\bar{z} + \bar{z}_0]^s, [z + z_0]^{n_1-k} [\bar{z} + \bar{z}_0]^{n_2-s-1}). \end{aligned}$$

We now expand the right-hand side in powers of $z_0$ and set the coefficients of $z_0$ equal to zero. We then obtain the recurrence relation

$$k c_{k-1,s} + (n_1 - k) c_{ks} = 0.$$

It is shown similarly that

$$s c_{k,s-1} + (n_2 - s) c_{ks} = 0.$$

These relations may be solved (for simplicity we set $c_{00} = 1$) to yield

$$c_{ks} = (-1)^{k+s} \frac{k!(n_1 - k - 1)! s!(n_2 - s - 1)!}{(n_1 - 1)!(n_2 - 1)!}.$$

Therefore for polynomials of the form

$$\varphi(z) = \sum_{k=0}^{n_1-1} \sum_{s=0}^{n_2-1} a_{ks}^{(1)} z^k \bar{z}^s$$

and

$$\psi(z) = \sum_{k=0}^{n_1-1} \sum_{s=0}^{n_2-1} a_{ks}^{(2)} z^k \bar{z}^s$$

we arrive at the functional

$$B_2(\varphi, \psi) = \sum_{k=0}^{n_1-1} \sum_{s=0}^{n_2-1} (-1)^{k+s} \frac{k!(n_1 - k - 1)!s!(n_2 - s - 1)!}{(n_1 - 1)!(n_2 - 1)!} a_{ks}^{(1)} a_{n_1-k-1, n_2-s-1}^{(2)} . \tag{6}$$

In summary, then, if $n_1, n_2 = 1, 2, \ldots$, there exists an invariant bilinear functional $B_2(\varphi, \psi)$ on the subspace $E_\chi$ of polynomials of degree $n_1 - 1$ in $z$ and $n_2 - 1$ in $\bar{z}$. This functional is defined by Eq. (6).

## 5. Equivalence of Representations of *G*

We now wish to study the conditions on $\chi_1 = (n_1, n_2)$ and $\chi_2 = (m_1, m_2)$ such that the representations

$$T_{\chi_1}(g)\varphi(z) = (\beta z + \delta)^{n_1-1}(\bar{\beta}\bar{z} + \bar{\delta})^{n_2-1}\varphi\left(\frac{\alpha z + \gamma}{\beta z + \delta}\right) \tag{1}$$

and

$$T_{\chi_2}(g)\psi(z) = (\beta z + \delta)^{m_1-1}(\bar{\beta}\bar{z} + \bar{\delta})^{m_2-1}\psi\left(\frac{\alpha z + \gamma}{\beta z + \delta}\right) \tag{2}$$

of $G$ are equivalent. In other words we shall we shall study the conditions under which there exists a linear operator $A$ which is a one-to-one mapping of $D_{\chi_1}$ onto $D_{\chi_2}$ and such that

$$AT_{\chi_1}(g) = T_{\chi_2}(g)A. \tag{3}$$

(We shall than say that $A$ *intertwines* $T_{\chi_1}$ and $T_{\chi_2}$.) In the process we shall discover which representations $T_{\chi_1}(g)$ and $T_{\chi_2}(g)$ are what is called "partially equivalent," that is, we shall determine when there exists a linear operator $A$ satisfying Eq. (3) and mapping $D_{\chi_1}$ into $D_{\chi_2}$ (but not isomorphically).

Also in the process it will be shown that any continuous operator on $D_\chi$ that commutes with the $T_\chi(g)$ is a multiple of the unit operator [which then establishes the operator irreducibility of $T(g)$].

### 5.1. Intertwining Operators

Let us find for which representations $T_{\chi_1}(g)$ and $T_{\chi_2}(g)$ there exists a continuous linear mapping $A \neq 0$ of $D_{\chi_1}$ into $D_{\chi_2}$ (not necessarily one-to-one and not necessarily onto all of $D_{\chi_2}$) such that

$$AT_{\chi_1}(g) = T_{\chi_2}(g)A. \tag{1}$$

If $A$ is an isomorphism, that is, if $A$ is a bicontinuous one-to-one mapping of $D_{\chi_1}$ *onto* $D_{\chi_2}$, then $T_{\chi_1}(g)$ and $T_{\chi_2}(g)$ are called equivalent.

We first establish a relation between such intertwining operators $A$ and invariant bilinear functionals. As was shown in Section 4.5, there exists on the pair of spaces $D_{-\chi_2}$ and $D_{\chi_2}$, where $\chi_2 = (m_1, m_2)$ and $-\chi_2 = (-m_1, -m_2)$, a nondegenerate invariant bilinear functional, and this functional is given by

$$(\psi, \varphi) = \frac{i}{2} \int \psi(z)\varphi(z)\, dz\, d\bar{z}. \tag{2}$$

Now let $A$ be a linear operator that maps $D_{\chi_1}$ into $D_{\chi_2}$. With $A$ we associate the bilinear functional

$$B(\varphi, \psi) = (\psi, A\varphi) \tag{3}$$

on the pair of spaces $D_{\chi_1}$ and $D_{-\chi_2}$. Here $\varphi \in D_{\chi_1}$ and $\psi \in D_{-\chi_2}$.

We then assert: *$A$ intertwines $T_{\chi_1}(g)$ and $T_{\chi_2}(g)$, that is, satisfies Eq.* (1), *if and only if $B(\varphi, \psi) = (\psi, A\varphi)$ is invariant under the representations $T_{\chi_1}(g)$ and $T_{-\chi_2}(g)$.*

**Proof.** Equation (1) is equivalent to the statement that the equation

$$(T_{-\chi_2}(g)\psi, AT_{\chi_1}(g)\varphi) = (T_{-\chi_2}(g)\psi, T_{\chi_2}(g)A\varphi) \tag{4a}$$

holds for all $\varphi \in D_{\chi_1}$ and $\psi \in D_{-\chi_2}$. The invariance of $B(\varphi, \psi)$, on the other hand, may be written

$$(T_{-\chi_2}(g)\psi, AT_{\chi_1}(g)\varphi) = (\psi, A\varphi). \tag{4b}$$

The left-hand sides of these equations are identical. Moreover, the right-hand sides are equal because $(\psi, \varphi)$ is invariant under $T_{-\chi_2}(g)$ and $T_{\chi_2}(g)$. Consequently (4a) is the same as (4b), or the commutation of $A$ through $T_{\chi_1}(g)$ and $T_{\chi_2}(g)$ and the invariance of $B(\varphi, \psi)$ under $T_{\chi_1}(g)$ and $T_{-\chi_2}(g)$ are both expressed by the same equation. This proves the assertion.

In Section 4 we found the possible $\chi_1$ and $\chi_2$ for which an invariant bilinear functional may exist, and hence to find the conditions under which the intertwining operator $A$ may exist we need only replace $\chi_2$ by $-\chi_2$ in the results of Section 4. In this way we arrive immediately at the following result.

*An operator $A \neq 0$ mapping $D_{\chi_1}$ continuously into $D_{\chi_1}$, for $\chi_1 = (n_1, n_2)$ and $\chi_2 = (m_1, m_2)$, and intertwining the representations, i.e., satisfying Eq.* (1), *exists if and only if one of the following four conditions is fulfilled*[19]:

[19] These cases are the same as those of Section 4, but are renumbered for convenience and because $\chi_2$ has been replaced by $-\chi_2$.

*Case 1.* $n_1 = m_1, n_2 = m_2$;

*Case 2.* $n_1 = -m_1, n_2 = -m_2, n_1, n_2$ not simultaneously nonnegative integers;

*Case 3.* $n_1 = -m_1, n_2 = -m_2, n_1, n_2$ nonnegative integers;

*Case 4.* $n_1 = -m_1 = 1, 2, \ldots, n_2 = m_2$;

*Case 4'.* $n_1 = m_1, n_2 = -m_2 = 1, 2, \ldots$.

Further, the results of Section 4 can be used to establish the form of the intertwining operator for each of these four cases. In fact if $A$ exists, then there exists the invariant bilinear functional

$$B(\varphi, \psi) = \tfrac{1}{2}i \int \psi(z) A\varphi(z)\, dz\, d\bar{z} \tag{5}$$

on the pair of spaces $D_{\chi_1}$ and $D_{-\chi_2}$. We saw in Section 4, on the other hand, that if it exists, such an invariant bilinear functional is determined uniquely up to a factor, and by replacing $\chi_2$ by $-\chi_2$ in the results of that chapter we arrive at the following results (in a somewhat different order).

*Case 1.*

$$B(\varphi, \psi) = c\,\frac{i}{2} \int \psi(z)\varphi(z)\, dz\, d\bar{z}, \tag{6}$$

*Case 2.*

$$B(\varphi, \psi) = c\left(\frac{i}{2}\right)^2 \int (z_2 - z_1)^{-n_1-1}(\bar{z}_2 - \bar{z}_1)^{-n_2-1} \times \varphi(z_1)\psi(z_2)\, dz_1\, d\bar{z}_1\, dz_2\, d\bar{z}_2 . \tag{7}$$

*Case 3.*

$$B(\varphi, \psi) = c\,\frac{i}{2} \int \varphi^{(n_1, n_2)}(z)\psi(z)\, dz\, d\bar{z}. \tag{8}$$

*Case 4.*

$$B(\varphi, \psi) = c\,\frac{i}{2} \int \varphi^{(n_1, 0)}(z)\psi(z)\, dz\, d\bar{z}. \tag{9}$$

*Case 4'.*

$$B(\varphi, \psi) = c\,\frac{i}{2} \int \varphi^{(0, n_2)}(z)\psi(z)\, dz\, d\bar{z}. \tag{10}$$

To obtain expressions for $A$, we need only compare Eqs. (6)–(10) with Eq. (5). Thus we may summarize the results of Section 5.1 as follows. Let $T_{\chi_1}(g)$ and $T_{\chi}(g_2)$ be two representations of $G$. An operator $A$

mapping $D_{\chi_1}$ continuously into $D_{\chi_2}$ and and intertwining these representations, that is, satisfying Eq. (1), exists in the following four cases.

*Case 1.* $\chi_1 = \chi_2$. Then $A$ is of the form

$$A\varphi(z) = c\varphi(z), \tag{11}$$

and is therefore a multiple of the unit operator.

*Case 2.* $\chi_1 = (n_1, n_2)$, and $\chi_2 = -\chi_1 = (-n_1, -n_2)$, where $n_1$ and $n_2$ are not simultaneously nonnegative integers. Then $A$ is of the form[20]

$$A\varphi(z) = c\frac{i}{2}\int (z - z_1)^{-n_1-1}(\bar{z} - \bar{z}_1)^{-n_2-1}\varphi(z_1)\, dz_1\, d\bar{z}_1 . \tag{12}$$

*Case 3.* $\chi_1 = -\chi_2$, where $n_1$ and $n_2$ are nonnegative integers. Then $A$ is of the form[21]

$$A\varphi(z) = c\frac{\partial^{n_1+n_2}\varphi(z)}{\partial z^{n_1}\partial \bar{z}^{n_2}} . \tag{13}$$

*Case 4.* $\chi_1 = (n_1, n_2)$, $\chi_2 = (-n_1, n_2)$, where $n_1$ is a positive integer. Then $A$ is of the form

$$A\varphi(z) = c\frac{\partial^{n_1}\varphi(z)}{\partial z^{n_1}} . \tag{14}$$

*Case 4′.* $\chi_1 = (n_1, n_2)$, $\chi_2 = (n_1, -n_2)$, where $n_2$ is a positive integer. Then $A$ is of the form

$$A\varphi(z) = c\frac{\partial^{n_2}\varphi(z)}{\partial \bar{z}^{n_2}} . \tag{14′}$$

We call particular attention to Case 1, in which $\chi_1 = \chi_2$, or $D_{\chi_1} = D_{\chi_2}$. This case establishes that *any continuous linear operator on* $D_\chi$ *that commutes with all the operators of* $T_\chi(g)$ *is a multiple of the unit operator.* We have thus, in fact, demonstrated the operator irreducibility of $T_\chi(g)$.

[20] The integral is to be understood in the sense of its regularization.

[21] Note that Eq. (13) may be thought of as a special case of (12). This may be seen by rewriting (12) with $c^{-1} = \Gamma(-\frac{1}{2}n_1 - \frac{1}{2}n_2 + \frac{1}{2}|\, n_1 - n_2\,|)$ and recalling that the generalized function

$$\frac{(z_1 - z)^{-n_1-1}(\bar{z}_1 - \bar{z})^{-n_2-1}}{\Gamma(-\frac{1}{2}n_1 - \frac{1}{2}n_2 + \frac{1}{2}|\, n_1 - n_2\,|)}$$

becomes $\delta^{(n_1,n_2)}(z_1 - z)$ for nonnegative integers $n_1$ and $n_2$.

### 5.2. Equivalence of Two Representations

Let us now find the necessary and sufficient conditions for the representations $T_{\chi_1}(g)$ and $T_{\chi_2}(g)$ to be equivalent. Recall that two representations are called equivalent if there exists an operator $A$ which is a bicontinuous one-to-one mapping of $D_{\chi_1}$ onto $D_{\chi_2}$ such that

$$AT_{\chi_1}(g) = T_{\chi_2}(g)A. \tag{1}$$

In Section 5.1 we found the necessary and sufficient conditions on $\chi_1$ and $\chi_2$ for there to exist such an operator $A$, and we found expressions for $A$ in several cases. To establish whether $T_{\chi_1}(g)$ and $T_{\chi_2}(g)$ are furthermore equivalent we need only find what additional conditions must be placed on $\chi_1$ and $\chi_2$ if $A$ is to be a bicontinuous one-to-one mapping of $D_{\chi_1}$ onto $D_{\chi_2}$.

Case 1 is of no interest, since then $D_{\chi_1}$ and $D_{\chi_2}$ simply coincide. Let us therefore turn immediately to Case 2.

We then assert that *if $n_1$ and $n_2$ are not integers of the same sign, the representations $T_{\chi_1}(g)$ and $T_{-\chi_1}(g)$ are equivalent.* For the proof, consider an operator $A$ mapping $D_{\chi_1}$ into $D_{\chi_2}$ (where $\chi_2 = -\chi_1$) and such that Eq. (1) is satisfied. As was shown in Section 5.1, this operator must be of the form

$$A\varphi(z) = \frac{i}{2}\int (z - z_1)^{-n_1-1}(\bar{z} - \bar{z}_1)^{-n_2-1}\varphi(z_1)\, dz_1\, d\bar{z}_1 .$$

Consider also the operator $A_1$ mapping $D_{\chi_2}$ into $D_{\chi_1}$ and intertwining the representations, which means in this case that

$$A_1 T_{\chi_2}(g) = T_{\chi_1}(g)A_1 .$$

Then similarly $A_1$ must be of the form

$$A_1\varphi(z) = \frac{i}{2}\int (z - z_1)^{n_1-1}(\bar{z} - \bar{z}_1)^{n_2-1}\varphi(z_1)\, dz_1\, d\bar{z}_1 .$$

Now the operator $A_1A$ maps $D_{\chi_1}$ into itself. Further, we have

$$A_1AT_{\chi_1}(g) = A_1T_{\chi_2}(g)A = T_{\chi_1}(g)A_1A.$$

It therefore follows from the operator irreducibility of $T_{\chi_1}(g)$ that $A_1A$ is a multiple of the unit operator $E$, and we write

$$A_1A = \mu_1 E.$$

Similarly

$$AA_1 = \mu_2 E.$$

If we can now that $\mu_1 \neq 0$ and $\mu_2 \neq 0$, we will have shown that $A$ and $A_1$ are isomorphisms of $D_{\chi_1}$ and $D_{\chi_2}$, and therefore that the representations are equivalent. Thus let $\varphi(z)$ be an infinitely differentiable rapidly decreasing function in $D_{\chi_1}$. Its Fourier transform $\tilde{\varphi}(w)$ is also an infinitely differentiable rapidly decreasing function. Let us make the further assumption that $\tilde{\varphi}(w)$ vanishes in a neighborhood of $w = 0$. Now consider $A_1A$ operating on $\varphi$. We first find the Fourier transform $F[A\varphi]$ of $A\varphi$. Since $A\varphi$ is the convolution of $\varphi(z)$ with the generalized function $z^{-n_1-1}\bar{z}^{-n_2-1}$, its Fourier transform is given by

$$F[A\varphi] = F[z^{-n_1-1}\bar{z}^{-n_2-1}]F(w).$$

But it is known that

$$F[z^{-n_1-1}\bar{z}^{-n_2-1}] = \frac{2^{-n_1-n_2}\pi i^{|n_1-n_2|}\Gamma(-\frac{1}{2}[n_1+n_2]+\frac{1}{2}|n_1-n_2|)}{\Gamma(\frac{1}{2}[n_1+n_2]+\frac{1}{2}|n_1-n_2|+1)} w^{n_1}\bar{w}^{n_2}. \quad (2)$$

Therefore by assumption concerning $\varphi(z)$, $F[A\varphi]$ is also an infinitely differentiable rapidly decreasing function vanishing in the neighborhood of $w = 0$. Let us now find the Fourier transform of $A_1A\varphi$. We have

$$F[A_1A\varphi] = F[z^{n_1-1}\bar{z}^{n_2-1}] \cdot F[A\varphi].$$

Now on the right-hand side we may insert for the first factor an expression similar to (2), but with $n_k$ replaced by $-n_k$, and for the second the appropriate expression in terms of (2), and then after canceling some factors and using the fact that $F[A_1A\varphi] = \mu_1\tilde{\varphi}$, we obtain

$$\mu_1 = (-1)^{n_1-n_2}4\pi^2(n_1+n_2+|n_1-n_2|)^{-1}$$
$$\times(-n_1-n_2+|n_1-n_2|)^{-1}. \quad (3)$$

The same expression will obviously be obtained for $AA_1$.

We thus find that $A_1A = AA_1 = \mu E$, where $\mu$ is given by Eq. (3). Consequently $\mu \neq 0$, and the operators $A$ and $A_1$ are isomorphisms of $D_{\chi_1}$ and $D_{\chi_2}$. This proves the equivalence of $T_{\chi_1}(g)$ and $T_{-\chi_1}(g)$, for $n_1$ and $n_2$ not nonzero integers of the same sign.

Now consider the case of integer $\chi_1 = -\chi_2$, i.e., $n_1$ and $n_2$ nonzero integers of the same sign.[22] We may then show that for this case $T_{\chi_1}(g)$ and $T_{-\chi_1}(g)$ are not equivalent.

[22] The case $n_1 = n_2 = 0$ is, of course, unnecessary to consider because in this case $D_{\chi_1}$ coincides with $D_{\chi_2}$.

Indeed, let us assume, without loss of generality, that $n_k \geqslant 0$ [otherwise we could merely interchange the roles of $T_{\chi_1}(g)$ and $T_{\chi_2}(g)$]. Then the operator $A$ mapping $D_{\chi_1}$ into $D_{\chi_2}$ and intertwining the representations is of the form

$$A = c \frac{\partial^{n_1+n_2}}{\partial z^{n_1} \partial \bar{z}^{n_2}}.$$

This operator is obviously singular. Indeed, it annihilates all polynomials in $D_{\chi_1}$ of the form

$$\varphi(z) = \sum_{j=0}^{n_1-1} \sum_{k=0}^{n_2-1} a_{jk} z^j \bar{z}^k.$$

Therefore $T_{\chi_1}(g)$ and $T_{\chi_2}(g)$ are not equivalent.

The only cases left are 4 and 4′, namely those in which $\chi_1 = (n_1, n_2)$, $\chi_2 = (-n_1, n_2)$ for positive integer $n_1$, and $\chi_1 = (n_1, n_2)$, $\chi_2 = (n_1, -n_2)$ for positive integer $n_2$. In these two cases $T_{\chi_1}(g)$ and $T_{\chi_2}(g)$ are not equivalent. Indeed, if they were equivalent, there would exist an operator $A_1$ mapping $D_{\chi_2}$ into $D_{\chi_1}$ and intertwining the representations. But we know that no such operator exists.[23]

In summary, we arrive at the following result.

*Two representations* $T_{\chi_1}(g)$ *and* $T_{\chi_2}(g)$, *where* $\chi_1 = (n_1, n_2) \neq \chi_2 = (m_1, m_2)$, *are equivalent if and only if*

$$n_1 = -m_1, \qquad n_2 = -m_2,$$

*and* $n_1$ *and* $n_2$ *are not integers of the same sign. When they are equivalent, the bicontinuous one-to-one mapping* $A$ *of* $D_{\chi_1}$ *onto* $D_{\chi_2}$ *such that*

$$AT_{\chi_1}(g) = T_{\chi_2}(g)A$$

*is given by*

$$A\varphi(z) = \frac{i}{2} \int (z - z_1)^{-n_1-1} (\bar{z} - \bar{z}_1)^{-n_2-1} \varphi(z_1)\, dz_1\, d\bar{z}_1 .$$

[*For* $\mathrm{Re}(n_1 + n_2) > 0$ *the integral is to be understood in the sense of its regularization.*]

### 5.3. Partially Equivalent Representations[24]

We have already seen that there may exist an operator $A$ intertwining a pair of representations without these representations being equiv-

[23] In any case, we know that any intertwining operator $A$ mapping $D_{\chi_1}$ into $D_{\chi_2}$ is a differential operator and that it therefore annihilates some subspace.

[24] This section is not needed for the rest of the book and may be omitted if so desired.

alent. Let us call two such representations *partially equivalent*; we shall analyze this phenomenon in some detail in this section.

Recall that two representations $T_{\chi_1}(g)$, $\chi_1 = (n_1, n_2)$, and $T_{\chi_2}(g)$, $\chi_2 = (m_1, m_2)$, possess an intertwining operator but are inequivalent in the following three cases.

*Case 1.* Integer $\chi_1 = -\chi_2$ (that is, $n_1$ and $n_2$ are integers of the same sign, or $n_1 n_2 > 0$).

*Case 2.* $n_1 = -m_1$, $n_2 = m_2$, where $n_1 = 1, 2, \ldots$ .

*Case 3.* $n_1 = m_1$, $n_2 = -m_2$, where $n_2 = 1, 2, \ldots$ .

*Case 1.* We may assume without loss of generality that $n_1$ and $n_2$ are positive integers. Then $D_{\chi_1}$ and $D_{\chi_2}$ have invariant subspaces. Specifically, the invariant subspace of $D_{\chi_1}$ is spanned by the polynomials

$$\sum_{k=0}^{n_1-1} \sum_{l=0}^{n_2-1} a_{jk} z^j \bar{z}^k, \tag{1}$$

and we have called it $E_{\chi_1}$. The invariant subspace of $D_{\chi_2}$ is spanned by all functions $\psi(z)$ whose moments

$$b_{jk} = \frac{i}{2} \int z^j \bar{z}^k \psi(z)\, dz\, d\bar{z} \tag{2}$$

vanish for $j = 0, 1, \ldots, n_1 - 1$ and $k = 0, 1, \ldots, n_2 - 1$. We have called this subspace $F_{\chi_2}$.

Thus $T_{\chi_1}(g)$ and $T_{\chi_2}(g)$ each induce two other representations of $G$, namely those on the invariant subspaces and that on the factor spaces. It turns and that the representation of $G$ on $D_{\chi_1}/E_{\chi_1}$ is equivalent to the representation on $F_{\chi_2}$. Similarly, the representations on $D_{\chi_2}/F_{\chi_2}$ and on $E_{\chi_1}$ are equivalent.

Let us first prove the first of these assertions. Consider the operator $A$ which maps $D_{\chi_1}$ continuously into $D_{\chi_2}$ and is such that

$$AT_{\chi_1}(g) = T_{\chi_2}(g)A.$$

We have seen that this operator is of the form

$$A = \frac{\partial^{n_1+n_2}}{\partial z^{n_1} \partial \bar{z}^{n_2}}.$$

Let us find the kernel and image under the mapping $A$. Obviously $A$ annihilates all the polynomials of Eq. (1) and nothing more, so that the kernel of $A$ is $E_{\chi_1}$.

We assert further that *the image of* $D_{\chi_1}$ *under* $A$ *is the subspace* $F_{\chi_2}$. Indeed, let $\varphi(z)$ be some function in $D_{\chi_2}$ with bounded support. Then

$$A\varphi(z) = \frac{\partial^{n_1+n_2}\varphi(z)}{\partial z^{n_1}\partial \bar{z}^{n_2}}$$

is in $F_{\chi_2}$, for integration by parts yields

$$b_{kl} = \frac{i}{2}\int z^k \bar{z}^l \frac{\partial^{n_1+n_2}\varphi(z)}{\partial z^{n_1}\partial \bar{z}^{n_2}} dz\, d\bar{z}$$

$$= (-1)^{n_1+n_2}\frac{i}{2}\int \frac{\partial^{n_1+n_2}(z^k \bar{z}^l)}{\partial z^{n_1}\partial \bar{z}^{n_2}} \varphi(z)\, dz\, d\bar{z},$$

which vanishes for nonnegative integers $k < n_1$ and $l < n_2$ [see Eq. (2)]. Further, $F_{\chi_2}$ contains also all functions of the form $\psi(z) = AT_{\chi_1}(g)\varphi(z)$, where $\varphi(z)$ has bounded support. Indeed,

$$\psi(z) = AT_{\chi_1}(g)\varphi(z) = T_{\chi_2}(g)A\varphi(z).$$

But we have seen that $A\varphi(z)$ belongs to $F_{\chi_2}$, and we know that $F_{\chi_2}$ is invariant under $T_{\chi_2}(g)$. Therefore $\psi(z)$ also belongs to $F_{\chi_2}$. To complete the first step of the proof, we remark that every function $\varphi(z) \in D_{\chi_1}$ can be written as a linear combination of functions of the form $T_{\chi_1}(g)\varphi_k(z)$ where $\varphi_k(z)$ has bounded support. This means that $\varphi(z) \in D_{\chi_1}$ implies that $A\varphi(z) \in F_{\chi_2}$, and therefore that $A$ maps $D_{\chi_1}$ into the subspace $F_{\chi_2}$ of $D_{\chi_2}$.

We assert finally that $A$ maps $D_{\chi_1}$ onto all of $F_{\chi_2}$. Let $D^{-1}$ and $\bar{D}^{-1}$ be operators on the functions $\psi(z)$ of bounded support, defined by the equations

$$\begin{aligned} D^{-1}\psi(z) &= \int \psi(z)\, dx + i\int \psi(z)\, dy, \\ \bar{D}^{-1}\psi(z) &= \int \psi(z)\, dx - i\int \psi(z)\, dy. \end{aligned} \tag{3}$$

(The integrals in these equations are understood as primitive functions vanishing as $x \to -\infty$ and $y \to -\infty$, respectively.) Now it is easily shown that if $\psi(z)$ has bounded support and is in $F_{\chi_2}$, then

$$\varphi(z) = D^{-n_1}\bar{D}^{-n_2}\psi(z)$$

also has bounded support and that $A\varphi(z) = \partial^{n_1+n_2}\varphi(z)/\partial z^{n_1}\, \partial \bar{z}^{n_2} = \psi(z)$. We have thus established that every $\psi(z) \in F_{\chi_2}$ with bounded support is the image under $A$ of some $\varphi(z)$ with bounded support. This implies that every function in $F_{\chi_2}$ of the form $\psi_1(z) = T_{\chi_2}(g)\psi(z)$, where $\psi(z) \in F_{\chi_2}$ has bounded support, is also the image of some function in $D_{\chi_1}$, for we have seen that there exists a $\varphi(z) \in D_{\chi_1}$ such that $\psi(z) = A\varphi(z)$, and therefore

$$\psi_1(z) = T_{\chi_2}(g)A\varphi(z) = AT_{\chi_1}(g)\varphi(z).$$

It is easily shown that every function in $F_{\chi_2}$ can be written as a linear combination of functions of the form $T_{\chi_2}(g)\psi_k(z)$, where $\psi_k(z) \in F_{\chi_2}$ has bounded support. This means that every function in $F_{\chi_2}$ is the image under $A$ of some function in $D_{\chi_1}$, and thus the image of $D_{\chi_1}$ under $A$ is the *entire* subspace $F_{\chi_2}$.

**Remark.** This last assertion could also have been proven differently. In fact consider the operator

$$B\psi(z) = \frac{2}{\pi(n_1 - 1)!(n_2 - 1)!} \times \frac{i}{2}\int (z - z_1)^{n_1-1}(\bar{z} - \bar{z}_1)^{n_2-1} \ln|z - z_1|\,\psi(z_1)\,dz_1\,d\bar{z}_1,$$

and recall that $z^{n_1-1}\bar{z}^{n_2-1}\ln|z|$ is an associated homogeneous generalized function of degree $(n_1 - 1, n_2 - 1)$. Direct calculation will show that this is a mapping which carries a function from $F_{\chi_2}$ into $D_{\chi_1}$. Further, the fact that

$$\frac{\partial^{n_1+n_2}}{\partial z^{n_1}\partial\bar{z}^{n_2}}[z^{n_1-1}\bar{z}^{n_2-1}\ln|z|] = \frac{\pi(n_1 - 1)!(n_2 - 1)!}{2}\delta(z)$$

implies that $AB\psi(z) = \psi(z)$ for all $\psi(z) \in F_{\chi_2}$. Therefore $A$ maps $D_{\chi_1}$ onto all of $F_{\chi_2}$. #

We have thus shown that the kernel of the mapping $A$ is the invariant subspace $E_{\chi_1} \subset D_{\chi_1}$ and that the image of $D_{\chi_1}$ under $A$ is the invariant subspace $F_{\chi_2} \subset D_{\chi_2}$. Therefore $A$ induces a one-to-one mapping of the factor space $D_{\chi_1}/E_{\chi_1}$ onto $F_{\chi_1}$. Since by assumption

$$AT_{\chi_1}(g) = T_{\chi_2}(g)A,$$

this proves the equivalence of the representations on $D_{\chi_1}/E_{\chi_1}$ and $F_{\chi_2}$.

Let us now analyze the representations induced by $T_{\chi_1}(g)$ and $T_{\chi_2}(g)$ on $E_{\chi_1}$ and $D_{\chi_2}/F_{\chi_2}$, respectively. These representations are also equivalent. To prove this we consider the operator $A_1$ mapping $D_{\chi_2}$ into $D_{\chi_1}$ and intertwining the representations, defined by[25]

$$A_1\psi(z) = \frac{i}{2}\int (z - z_1)^{n_1-1}(\bar{z} - \bar{z}_1)^{n_2-1}\psi(z_1)\, dz_1\, d\bar{z}_1 . \tag{4}$$

It is easily seen that $A_1$ transforms any function $\psi(z) \in D_{\chi_2}$ into a polynomial of degree $n_1 - 1$ in $z$ and $n_2 - 1$ in $\bar{z}$. In other words, $A_1$ maps $D_{\chi_2}$ into $E_{\chi_1} \subset D_{\chi_1}$. Obviously the kernel of this mapping consists of functions whose moments $b_{ks}$ vanish for $0 \leqslant k \leqslant n_1 - 1$, $0 \leqslant s \leqslant n_2 - 1$. Now it is always possible to find a function with bounded support having a given set of moments, and therefore $A_1$ maps $D_{\chi_2}$ onto all of $E_{\chi_1}$. This then proves the equivalence of the representations induced on $E_{\chi_1}$ and $D_{\chi_2}/F_{\chi_2}$.

We have thus established the following results for Case 1, in which $\chi_1 = (n_1, n_2) = -\chi_2$, where $n_1$ and $n_2$ are positive integers.

(i) The representation induced by $T_{\chi_1}(g)$ on the factor space $D_{\chi_1}/E_{\chi_1}$ is equivalent to the representation induced by $T_{\chi_2}(g)$ on $F_{\chi_2}$.

(ii) The representation induced by $T_{\chi_1}(g)$ on $E_{\chi_1}$ is equivalent to the representation induced by $T_{\chi_2}(g)$ on the factor space $D_{\chi_2}/F_{\chi_2}$.

We can now go on to Cases 2 and 3 listed at the beginning of this section. For these cases the operator $A$ that intertwines $T_{\chi_1}(g)$ and $T_{\chi_2}(g)$ is of the form $A = \partial^{n_1}/\partial z^{n_1}$ in Case 2 and of the form $A = \partial^{n_2}/\partial \bar{z}^{n_2}$ in Case 3. Then by considerations similar to those we have just gone through, we arrive at the following result. Let $n_1 = -m_1$, $n_2 = m_2$, where $n_1$ is a positive integer. If $n_2 > 0$, the operator $A = \partial^{n_1}/\partial z^{n_1}$ establishes an isomorphism between the factor space $D_{\chi_1}/E_{\chi_1}$ and $D_{\chi_2}$. If $n_2 < 0$, this same operator establishes an isomorphism between $D_{\chi_1}$ and $F_{\chi_2}$.

A similar assertion holds for the case in which $n_1 = m_1$, $n_2 = -m_2$, where $n_2$ is a positive integer.

Thus *for $n_1$ and $n_2$ positive integers, we have established the equivalence of the following four representations*:

(a) *The representation induced by $T_{\chi_1}(g)$, $\chi_1 = (n_1, n_2)$, on the factor space $D_{\chi_1}/E_{\chi_1}$.*

[25] In general if $n_1$ and $n_2$ are any numbers, the integral in (4) may diverge, and it must then be understood in the sense of its regularization. In our case, however, when $n_1$ and $n_2$ are positive integers, the integral converges. This follows immediately from the fact that as $|z_1| \to \infty$ the functions $\psi(z_1) \in D_{\chi_1}$ have the asymptotic behavior

$$\psi(z_1) \sim C z_1^{-n_2-1} \bar{z}_1^{-n_2-1}.$$

(b) *The representation induced by* $T_{\chi_2}(g)$, $\chi_2 = (-n_1, -n_2)$, on $F_{\chi_2}$.

(c) *The representation* $T_{\chi_3}(g)$, $\chi_3 = (-n_1, n_2)$, *on* $D_{\chi_3}$.

(d) *The representation* $T_{\chi_4}(g)$, $\chi_4 = (n_1, -n_2)$, *on* $D_{\chi_4}$.

*In addition, the representation induced by* $T_{\chi_1}(g)$ *on* $E_{\chi_1}$ *is equivalent to the representation induced by* $T_{\chi_2}(g)$ *in the factor space* $D_{\chi_2}/F_{\chi_2}$.

## 6. Unitary Representations of G

### 6.1. Invariant Hermitian Functionals on $D_\chi$

In this section we wish to establish the conditions under which there may exist an *invariant scalar product* $(\varphi, \psi)$ on a $D_\chi$. By invariant scalar product product we shall understand a positive definite Hermitian functional on $D_\chi$ such that $(\varphi, \psi) = (T_\chi(g)\varphi, T_\chi(g)\psi)$ for all $\varphi(z)$ and $\psi(z)$ in $D_\chi$ and for all $g$ in $G$. Let us first find the conditions under which there may exist an invariant, but not necessarily positive definite, Hermitian functional on $D_\chi$. We do this by finding a relation between invariant *Hermitian* functionals and the invariant *bilinear* functionals of Section 4.

With every Hermitian functional $(\varphi, \psi)$ on $D_\chi$, where $\chi = (n_1, n_2)$, we may associate the bilinear functional

$$B(\varphi, \psi) = (\varphi, \bar{\psi}).$$

This bilinear functional is defined for $\varphi(z) \in D_\chi$ and $\psi(z) \in D_{\bar{\chi}}$, where $\bar{\chi} = (\bar{n}_2, \bar{n}_1)$. Then $(\varphi, \psi)$ is invariant under $T_\chi(g)$ if and only if $B(\varphi, \psi) = (\varphi, \psi)$ is invariant under the representations $T_\chi(g)$ and $T_{\bar{\chi}}(g)$, as follows immediately from the obviously valid equation

$$B(T_\chi(g)\varphi, T_{\bar{\chi}}(g)\psi) = (T_\chi(g)\varphi, \overline{T_{\bar{\chi}}(g)\psi})$$
$$= (T_\chi(g)\varphi, T_\chi(g)\bar{\psi}).$$

Therefore the study of invariant Hermitian functionals on $D_\chi$ reduces to the study of bilinear functionals invariant under $T_\chi(g)$ and $T_{\bar{\chi}}(g)$. But in Section 4 we found the pairs $\chi_1, \chi_2$ for which such invariant bilinear functionals may exist, as well as their form when they do exist. In order, in fact, to find the conditions under which an invariant Hermitian functional exists in $D_\chi$ all we need do is write $\chi_1 = \chi$ and $\chi_2 = \bar{\chi}$ in the earlier development. Moreover, this functional will be given by the formula obtained by replacing $\psi$ by $\bar{\psi}$ in the corresponding formula for $B(\varphi, \psi)$. We may thus immediately state the result.

*An invariant Hermitian functional exists on a* $D_\chi$, *where* $\chi = (n_1, n_2)$, *if and only if* $n_1 = -\bar{n}_2$ *or* $n_1 = \bar{n}_2$.

*In the first case* ($n_1 = -\bar{n}_2$) *this functional is given by*

$$(\varphi, \psi) = \frac{i}{2} \int \varphi(z)\bar{\psi}(z)\, dz\, d\bar{z}. \tag{1}$$

*The second case* ($n_1 = \bar{n}_2$) *implies that* $n_1 = n_2 = \rho$ *is real, since* $n_1 - n_2$ *is an integer; then the invariant Hermitian functional is given by*

$$(\varphi, \psi) = c\left(\frac{i}{2}\right)^2 \int | z_1 - z_2 |^{-2\rho-2} \varphi(z_1)\bar{\psi}(z_2)\, dz_1\, d\bar{z}_1\, dz_2\, d\bar{z}_2, \tag{2}$$

*if* $\rho \neq 0, 1, \ldots$ (*and as usual the integral is understood in the sense of its regularization*), *and by*

$$(\varphi, \psi) = c\tfrac{1}{2}i \int \varphi^{(q,q)}(z)\bar{\psi}(z)\, dz\, d\bar{z}, \tag{2'}$$

*if* $\rho = q$ *is a nonnegative integer.*

These two equations may, incidentally, be unified by using the fact that

$$\left.\frac{| z |^{-2\rho-2}}{\Gamma(-\rho)}\right|_{\delta=q} = \frac{(-1)^q \pi}{q!}\, \delta^{(q,q)}(z).$$

We then have

$$(\varphi, \psi) = \frac{c}{\Gamma(-\rho)} (\tfrac{1}{2}i)^2 \int | z_1 - z_2 |^{-2\rho-2} \varphi(z_1)\bar{\psi}(z_2)\, dz_1\, d\bar{z}_1\, dz_2\, d\bar{z}_2. \tag{3}$$

## 6.2. Positive Definite Invariant Hermitian Functionals

We have now established that an invariant Hermitian functional exists on $D_\chi$ in the following two cases.

*Case 1.* $n_1 = -\bar{n}_2$, or $n_1 = \frac{1}{2}(n + i\rho)$, $n_2 = \frac{1}{2}(-n + i\rho)$, where $n$ is an integer and $\rho$ is any real number.

*Case 2.* $n_1 = n_2 = \rho$ *is a real number.*

In Case 1 the Hermitian functional is

$$(\varphi, \psi) = \frac{i}{2} \int \varphi(z)\bar{\psi}(z)\, dz\, d\bar{z} \tag{1}$$

and is obviously positive definite. We shall call these Case 1 representations, in which $\chi = (\frac{1}{2}[n + i\rho], \frac{1}{2}[-n + i\rho])$, *representations of the principal series.*

Let us now turn to Case 2 and find which values of $\rho$ yield Hermitian functionals which are positive definite. Assume first that $\rho < 0$. Then as we have seen,

$$(\varphi, \psi) = \frac{1}{\Gamma(-\rho)} \left(\frac{i}{2}\right)^2 \int | z_1 - z_2 |^{-2\rho-2} \varphi(z_1)\bar{\psi}(z_2) \, dz_1 \, d\bar{z}_1 \, dz_2 \, d\bar{z}_2 \,. \tag{2}$$

We wish to establish, consequently, the values of $\rho$ for which $(\varphi, \varphi) > 0$ for arbitrary nonvanishing $\varphi \in D_\chi$. For this purpose let us write $(\varphi, \varphi)$ in the form

$$(\varphi, \varphi) = \frac{1}{\Gamma(-\rho)} \frac{i}{2} \int | z |^{-2\rho-2} \theta(z) \, dz \, d\bar{z},$$

where

$$\theta(z) = \frac{i}{2} \int \varphi(z_1)\bar{\varphi}(z + z_1) \, dz_1 \, d\bar{z}_1 \,.$$

We now make use of the Fourier transform of $\varphi(z)$, which is[26]

$$\tilde{\varphi}(w) = \frac{i}{2} \int \varphi(z) e^{i\mathrm{Re}(zw)} \, dz \, d\bar{z}. \tag{3}$$

Since $\theta(z)$ is the convolution of $\varphi(z)$ and $\varphi^*(z) = \bar{\varphi}(-z)$, its Fourier transform is $| \tilde{\varphi}(w)|^2$. The Fourier transform of $| z |^{-2\rho-2}/\Gamma(-\rho)$ is the generalized function

$$\frac{\pi \, | w |^{2\rho}}{2^{2\rho}\Gamma(\rho + 1)}$$

(see Volume 1, Appendix B, Section 1.7). Consequently $(\varphi, \varphi)$ is given by

$$(\varphi, \varphi) = \frac{1}{2^{2\rho+2}\pi\Gamma(\rho + 1)} \frac{i}{2} \int | w |^{2\rho} \, | \tilde{\varphi}(w)|^2 \, dw \, d\bar{w}. \tag{4}$$

Of course this integral is to be understood in general in the sense of its regularization, but note that for $-1 < \rho < 0$ it converges in the ordinary sense and that then $(\varphi, \varphi) > 0$ for $\varphi \neq 0$.

[26] Note that asymptotically for large absolute values of $z$ (as $| z | \to \infty$) we have $| \varphi(z)| \sim C| z |^{2\rho-2}$, and recall that $\rho < 0$. It follows that $\varphi(z)$ is an absolutely integrable function. Thus the integral in Eq. (3) converges in the usual sense.

Hence *if* $n_1 = n_2 = \rho$, *where* $-1 < \rho < 0$, *the invariant Hermitian functional on* $D_\chi$ *is positive definite.*

If, on the other hand, $\rho$ lies in the interval $-q-1 < \rho < -q$, where $q$ is a positive integer, the regularization of (4) is given by

$$(\varphi, \varphi) = \frac{1}{2^{2\rho+2}\pi\Gamma(\rho+1)} \times \frac{i}{2}\int |w|^{2\rho}\left[|\tilde{\varphi}(w)|^2 - \sum_{j+k=0}^{2q-2} \frac{w^j \bar{w}^k}{j!k!} \frac{\partial^{j+k} |\tilde{\varphi}(0)|^2}{\partial w^j \, \partial \bar{w}^k}\right] dw \, d\bar{w}. \quad (5)$$

We shall study this functional in some detail in the next section, where it will be shown in particular that it is not of definite sign.

If $\rho = -q - 1$ is a negative integer, then

$$\left.\frac{|w|^{2\rho}}{\Gamma(\rho+1)}\right|_{\rho=-q-1} = \frac{(-1)^q \pi}{q!} \delta^{(q,q)}(w).$$

Accordingly, for these values of $\rho$ Eq. (4) becomes

$$(\varphi, \varphi) = \frac{(-1)^q}{2^{-2q} q!} \frac{\partial^{2q} |\tilde{\varphi}(0)|^2}{\partial w^q \, \partial \bar{w}^q}. \quad (6)$$

We shall study this functional in some detail in Section 6.4, where it will be shown in particular that it is not of definite sign.

We have now studied the invariant Hermitian functionals in the $D_\chi$, where $\chi = (\rho, \rho)$, for $\rho < 0$. These are, however, the only ones we need consider, as the problem for $\rho > 0$ can be reduced to this case. Indeed, we saw in Section 5.2 that $T_{\chi_1}(g)$, $\chi_1 = (\rho, \rho)$, and $T_{\chi_2}(g)$, $\chi_2 = (-\rho, -\rho)$, are equivalent representations if $\rho$ is not an integer (positive, negative, or zero). That is to say, there exists a bicontinuous one-to-one mapping $A$ of $D_{\chi_2}$ onto $D_{\chi_1}$ such that

$$AT_{\chi_2}(g) = T_{\chi_1}(g)A.$$

The operator $A$ is given, in fact, by

$$A\varphi(z) = \frac{1}{\Gamma(-\rho)} \frac{i}{2} \int |z - z_1|^{2\rho-2} \varphi(z_1) \, dz_1 \, d\bar{z}_1,$$

where the integral is to be understood in the sense of its regularization. Now assume that $(\varphi, \psi)$ is an invariant Hermitian functional defined on $D_{\chi_1}$, $\chi_1 = (\rho, \rho)$. Then clearly

$$(\varphi, \psi)_1 = (A\varphi, A\psi)$$

defines an invariant Hermitian functional on $D_{\chi_2}$, $\chi_2 = (-\rho, -\rho)$. Obviously $(\varphi, \psi)_1$ will be positive definite if and only if $(\varphi, \psi)$ is positive definite. But we have already established that the invariant Hermitian functionals on $D_{\chi_1}$ are positive definite if $-1 < \rho < 0$, and it therefore follows that they must also be positive definite if $0 < \rho < 1$.

It may be noted that $\rho = 0$ also corresponds to a positive definite invariant Hermitian functional, given in this case by Eq. (1).

Summarizing, then, we have the following result.

*A representation $T_\chi(g)$ on $D_\chi$, where $\chi = (n_1, n_2)$, possesses a Hermitian positive definite invariant functional $(\varphi, \psi)$ in the following two cases.*

*Case 1.* $n_1 = \frac{1}{2}(n + i\rho)$, $n_2 = \frac{1}{2}(-n + i\rho)$, *where $n$ is an integer and $\rho$ is any real number. In this case*

$$(\varphi, \psi) = \frac{i}{2} \int \varphi(z)\bar{\psi}(z)\, dz\, d\bar{z}.$$

*Such representations are called representations of the principal series.*

*Case 2.* $n_1 = n_2 = \rho$, *where* $-1 < \rho < 1$, $\rho \neq 0$. *In this case*

$$(\varphi, \psi) = \frac{1}{\Gamma(-\rho)} \left(\frac{i}{2}\right)^2 \int |z_1 - z_2|^{-2\rho-2} \varphi(z_1)\bar{\psi}(z_2)\, dz_1\, d\bar{z}_1\, dz_2\, d\bar{z}_2$$

*(for $\rho > 0$ this integral is to be understood in the sense of its regularization). Such representations are called representations of the supplementary series.*

## 6.3. Invariant Hermitian Functionals for Noninteger $\rho$, $|\rho| \geqslant 1$[27]

Let us return to the case in which $\rho$ is a real number such that $-q - 1 < \rho < -q$, $q = 1, 2, \ldots$ . As was shown in Section 6.2, the invariant Hermitian functional in this case is given by

$$(\varphi, \varphi) = \frac{1}{2^{2\rho+2}\pi\Gamma(\rho + 1)} \frac{i}{2} \int |w|^{2\rho} |\tilde{\varphi}(w)|^2\, dw\, d\bar{w}, \tag{1}$$

where $\tilde{\varphi}(w)$ is the Fourier transform of $\varphi(z)$, i.e., where

$$\tilde{\varphi}(w) = \frac{i}{2} \int \varphi(z) e^{i\operatorname{Re}(zw)}\, dz\, d\bar{z}.$$

[27] This section may be omitted if so desired.

We remark again that the integral is to be understood in the sense of its regularization, namely

$$(\varphi, \varphi) = \frac{1}{2^{2\rho+2}\pi\Gamma(\rho+1)}$$
$$\times \frac{i}{2}\int |w|^{2\rho}\left[|\tilde{\varphi}(w)|^2 - \sum_{j+k=0}^{2q-2} \frac{w^j\bar{w}^k}{j!k!}\frac{\partial^{j+k}|\tilde{\varphi}(0)|^2}{\partial w^j\,\partial\bar{w}^k}\right] dw\, d\bar{w}.$$

Consider the set $L_q$ of functions $\varphi(z) \in D_\chi$ such that

$$\frac{\partial^{j+k}|\tilde{\varphi}(0)|^2}{\partial w^j\,\partial\bar{w}^k} = 0 \quad \text{for} \quad j+k \leqslant 2q-2.$$

For such $\varphi(z)$ our functional is given by

$$(\varphi, \varphi) = \frac{1}{2^{2\rho+2}\pi\Gamma(\rho+1)}\frac{i}{2}\int |w|^{2\rho}|\tilde{\varphi}(w)|^2\, dw\, d\bar{w},$$

and is thus of definite sign on $L_q$. We assert that *the function* $\varphi(z)$ *is in* $L_q$ *if and only if its moments*

$$b_{jk} = \frac{i}{2}\int z^j\bar{z}^k\varphi(z)\, dz\, d\bar{z}$$

*vanish for* $j+k \leqslant q-1$ (*where* $j, k \geqslant 0$). This will then imply that $L_q$ is a linear subspace of $D_\chi$.

Proof. Obviously $b_{jk} = 0$ if and only if $\partial^{j+k}\tilde{\varphi}(0)/\partial w^j\,\partial\bar{w}^k = 0$, where $\tilde{\varphi}(w)$ is the Fourier transform of $\varphi(z)$. What we must therefore show is that

$$\frac{\partial^{j+k}|\tilde{\varphi}(0)|^2}{\partial w^j\,\partial\bar{w}^k} = 0 \quad \text{for} \quad j+k \leqslant 2q-2$$

if and only if

$$\frac{\partial^{j+k}\tilde{\varphi}(0)}{\partial w^j\,\partial\bar{w}^k} = 0 \quad \text{for} \quad j+k \leqslant q-1.$$

Obviously the first of these relations follows from the second. We shall use induction on $j+k$ to prove that the second follows from the first. Hence assume the first to be satisfied. Then if $j+k=0$, we have $|\tilde{\varphi}(0)|^2 = 0$, so that $\tilde{\varphi}(0) = 0$. Now assume the second condition to have been proven from the first for $j+k < s$ (where $s < q-1$); we shall then show it to be true also for $j + k = s$. Indeed, from the formula for the derivative of a product we have

$$\frac{\partial^{2s} |\tilde{\varphi}(0)|^2}{\partial w^s \, \partial \bar{w}^s} = \sum_{j,k=0}^{s} C_s^j C_s^k \frac{\partial^{j+k}\tilde{\varphi}(0)}{\partial w^j \, \partial \bar{w}^k} \overline{\frac{\partial^{2s-j-k}\tilde{\varphi}(0)}{\partial w^{s-k} \, \partial \bar{w}^{s-j}}} = 0.$$

When we drop all terms in this sum which vanish by the inductive assumption, we arrive at

$$\sum_{j+k=s} C_s^j C_s^k \left| \frac{\partial^{j+k}\tilde{\varphi}(0)}{\partial w^j \, \partial \bar{w}^k} \right|^2 = 0,$$

which means that the condition is true also for $j + k = s$. This completes the proof.

Having shown that the Hermitian functional is of definite sign on $L_q$, we proceed to show that it is not of definite sign on $D_\chi$. In fact we shall construct a subspace $M_q$ of $D_\chi$ complementary to $L_q$ such that on $M_q$ the functional $(\varphi, \varphi)$ takes on both positive and negative values. This subspace may be constructed as follows.

Consider discontinuous functions $\tilde{\varphi}(w)$ of the form

$$\tilde{\varphi}(w) = \begin{cases} P(w) & \text{for} \quad |w| \leqslant 1, \\ 0 & \text{for} \quad |w| > 1, \end{cases}$$

where $P(w) = \sum_{j+k=0}^{q-1} a_{jk} w^j \bar{w}^k$. For $|w| \leqslant 1$, we have

$$|\tilde{\varphi}(w)|^2 = \sum_{j+k=0}^{2q-2} \frac{w^j \bar{\omega}^k}{j!k!} \frac{\partial^{j+k} |P(0)|^2}{\partial w^j \, \partial \bar{w}^k}.$$

It then follows that a function $\varphi(z)$ whose Fourier transform is one of these $\tilde{\varphi}(w)$ has the property that

$$(\varphi, \varphi) = -\frac{1}{2^{2\rho+2}\pi\Gamma(\rho + 1)} \frac{i}{2} \int_{|w|>1} |w|^{2\rho} |P(w)|^2 \, dw \, d\bar{w},$$

which shows that for functions of this type the sign of $(\varphi, \varphi)$ is different than it is on $L_q$. But these $\varphi(z)$ functions do not belong to $D_\chi$, for although they are infinitely differentiable [since $\tilde{\varphi}(w)$ has bounded support] they do not have the proper asymptotic behavior as $|z| \to \infty$. However, the functions $\psi(z) = \exp(-\epsilon |z|^2)\varphi(z)$ are in $D_\chi$. It is easily shown that for sufficiently small $\epsilon$ the sign of $(\psi, \psi)$ is the same as that of $(\varphi, \varphi)$. We shall designate by $M_q$ the subspace of the $\psi(z)$ functions.

We wish to show also that $M_q$ is complementary to $L_q$. First, it may be noted that the only common element of $L_q$ and $M_q$ is $\varphi_0(z) \equiv 0$

(since the signs of $(\varphi, \varphi)$ differ for these two subspaces). Second, the dimension of $M_q$ is obviously $\frac{1}{2}q(q+1)$. But the dimension of the factor space $D_\chi/L_q$ is also $\frac{1}{2}q(q+1)$, since every coset of $L_q$ in $D_\chi$ is given by the values of the $b_{jk}$, with $j + k \leqslant q - 1$, and the number of these moments is $\frac{1}{2}q(q+1)$. This immediately implies that $L_q$ and $M_q$ are complementary.

We have thus shown that for $-q-1 < \rho < -q$, where $q$ is a positive integer, $D_\chi$, $\chi = (\rho, \rho)$ is the sum of two nonintersecting subspaces $L_q$ and $M_q$ on which $(\varphi, \varphi)$ takes on different signs and such that $M_q$ is of dimension $\frac{1}{2}q(q+1)$.

**Remark.** It is easily shown that this implies that $L_q$ is the maximum subspace of $D_\chi$ on which $(\varphi, \varphi)$ is of definite sign. In other words, if $L$ is any subspace of $D_\chi$ containing $L_q$ as a proper subspace, then there exists in $L$ and element $\psi$ such that the sign of $(\psi, \psi)$ is different from that of $(\varphi, \varphi)$ for $\varphi \in L_q$. #

Now since $T_\chi(g)$ and $T_{-\chi}(g)$, where $\chi = (\rho, \rho)$ are equivalent for noninteger $\rho$, the results we have obtained will hold also for $q < \rho < q+1$, where $q$ is a positive integer. In other words, for such $\rho$ it follows that $D_\chi$ is the sum of two nonintersecting subspaces on which the invariant Hermitian functional takes on opposite signs.

### 6.4. Invariant Hermitian Functionals in the Special Case of Integer $n_1 = n_2$

We shall now discuss the invariant Hermitian functional on $D_\chi$ for $\chi = (q, q)$, where $q$ is an integer.

Assume first that $q$ is positive. Then as was mentioned in Section 6.1,[28]

$$(\varphi, \psi) = (-1)^q \frac{i}{2} \int \varphi^{(q,q)}(z)\bar{\psi}(z)\, dz\, d\bar{z}. \tag{1}$$

This functional is obviously degenerate, since it vanishes for all $\psi(z) \in D_\chi$ if $\varphi(z)$ is a polynomial of the form

$$\varphi(z) = \sum_{i,j=0}^{q-1} a_{ij} z^i \bar{z}^j.$$

Recall that such polynomials form the finite-dimensional invariant subspace $E_\chi \subset D_\chi$. Thus in this case we may treat $(\varphi, \psi)$ as an invariant

[28] The factor $(-1)^q$ multiplying the integral is chosen for convenience.

Hermitian functional on the factor space $D_\chi/E_\chi$. We assert that *on $D_\chi/E_\chi$ the functional is nondegenerate and positive definite.*

Proof. Recall that it was shown in Section 5.3 that the representation $T_\chi(g)$, $\chi = (q, q)$, induces on $D_\chi/E_\chi$ a representation equivalent to $T_{\chi_1}(g)$, $\chi_1 = (-q, q)$. But $T_{\chi_1}(g)$ belongs to the principal series and therefore possesses an invariant positive definite functional, defined up to a factor. Therefore the functional $(\varphi, \psi)$ on the factor space $D_\chi/E_\chi$ is nondegenerate and of definite sign. In fact it is *positive* definite. Indeed, let $\varphi(z)$ be a function with bounded support. Then by integration by parts we obtain

$$\begin{aligned}(\varphi, \varphi) &= (-1)^q \frac{i}{2} \int \varphi^{(q,q)}(z)\bar{\varphi}(z)\, dz\, d\bar{z} \\ &= \frac{i}{2} \int |\varphi^{(q,0)}(z)|^2\, dz\, d\bar{z}.\end{aligned} \tag{2}$$

Thus $(\varphi, \varphi) > 0$ as asserted.

Now let $n_1 = n_2 = -q$, where $q$ is a positive integer. Then our invariant Hermitian functional becomes

$$\begin{aligned}(\varphi, \psi) &= \left(\frac{i}{2}\right)^2 \int | z_1 - z_2 |^{2q-2} \varphi(z_1)\bar{\psi}(z_2)\, dz_1\, d\bar{z}_1\, dz_2\, d\bar{z}_2 \\ &= \left(\frac{i}{2}\right)^2 \int (z_1 - z_2)^{q-1}(\bar{z}_1 - \bar{z}_2)^{q-1}\varphi(z_1)\bar{\psi}(z_2)\, dz_1\, d\bar{z}_1\, dz_2\, d\bar{z}_2 .\end{aligned} \tag{3}$$

Obviously this functional is degenerate on the functions $\varphi(z)$ whose moments

$$b_{jk} = \frac{i}{2} \int z^j \bar{z}^k \varphi(z)\, dz\, d\bar{z}$$

vanish for $j, k = 0, 1, \ldots, q - 1$. Recall that these functions form the invariant subspace $F_\chi \subset D_\chi$, so that we may consider $(\varphi, \psi)$ to be an invariant Hermitian functional on the factor space $D_\chi/F_\chi$. We assert that *on $D_\chi/F_\chi$ this functional is nondegenerate and that it is not of definite sign if $q > 1$.*

Proof. The representation $T_\chi(g)$, $\chi = (-q, -q)$, induces a representation on $D_\chi/E_\chi$ equivalent to that induced by $T_{\chi_1}(g)$, $\chi_1 = (q, q)$, on the subspace of the polynomials

$$\varphi(z) = \sum_{i,j=0}^{q-1} a_{ij} z^i \bar{z}^j$$

(see Section 5.3). Then we may use Eq. (6) of Section 4.7 to show that for such polynomials the invariant Hermitian functional is given by

$$(\varphi, \psi) = \sum_{i,j=0}^{q-1} (-1)^{i+j} \frac{i!(q-i-1)!j!(q-j-1)!}{[(q-1)!]^2} a_{ij}\bar{c}_{q-1-j,q-1-i}, \qquad (4)$$

where the $c_{ij}$ are the coefficients of the polynomial $\psi(z)$. Obviously this functional is nondegenerate and not of definite sign.

Note, however, that the functional given by (4) is positive definite on the subspace $E_\chi^+$ of polynomials whose coefficients satisfy the relation $a_{ij} = (-1)^{i+j}\bar{a}_{q-1-j,q-1-i}$. This subspace is of dimension $\frac{1}{2}q(q+1)$ for $q$ even, and of dimension $\frac{1}{2}q(q-1)$ for $q$ odd. On the other hand, the functional given by (4) is negative definite on the subspace $E_\chi^-$ of polynomials whose coefficients satisfy the relation $a_{ij} = -(-1)^{i+j}\bar{a}_{q-1-j,q-1-i}$. This subspace is of dimension $\frac{1}{2}q(q-1)$ for $q$ even, and of dimension $\frac{1}{2}q(q+1)$ for $q$ odd.

Remark. It is easily shown that $E_\chi^+$ and $E_\chi^-$ are orthogonal with respect to $(\varphi, \psi)$ and that their sum is $E_\chi$. #

Now there exists on $F_\chi$, $\chi = (-q, -q)$, a positive definite Hermitian functional invariant under $T_\chi(g)$. Indeed, this representation induces a representation on $F_\chi$ equivalent to $T_{\chi_1}(g)$ on $D_{\chi_1}$, where $\chi_1 = (-q, q)$ (see Section 5.3). But we know that the latter representation belongs to the principal series. Therefore on $D_{\chi_1}$, and therefore also on $F_\chi$, there exists an invariant positive definite Hermitian functional.

Recall that we have already studied the invariant Hermitian functional on $D_{\chi_1}$ and know the operator $A$ mapping $F_\chi$ into $D_{\chi_1}$. This means that it is a simple matter to obtain the expression for the invariant Hermitian functional on $F_\chi$. It is given, in fact, by

$$(\varphi, \psi) = \left(\frac{i}{2}\right)^2 \int |z_1 - z_2|^{2q-2} \ln |z_1 - z_2| \varphi(z_1)\bar{\psi}(z_2)\, dz_1\, d\bar{z}_1\, dz_2\, d\bar{z}_2. \qquad (5)$$

We leave the derivation of this equation as an exercise to the reader.

### 6.5. Unitary Representations of G by Operators on Hilbert Space

In Section 6.3 we found the conditions under which there exists a positive definite Hermitian functional $(\varphi, \psi)$ invariant under $T_\chi(g)$, that is such that

$$(\varphi, \psi) = (T_\chi(g)\varphi, T_\chi(g)\psi).$$

If such a Hermitian functional exists, it may be taken as the scalar product in $D_\chi$. Then we may complete $D_\chi$ with respect to the norm $\| \varphi \|$ defined by

$$\| \varphi \|^2 = (\varphi, \varphi),$$

to obtain a Hilbert space $H$ in which $D_\chi$ forms an everywhere dense subset.

Since the operators of $T_\chi(g)$ are isometric on $D_\chi$, they can be extended uniquely to unitary operators on $H$. We shall denote these operators as before by $T_\chi(g)$. Obviously these operators also possess the group property

$$T_\chi(g_1 g_2) = T_\chi(g_1) T_\chi(g_2).$$

and therefore form a representation of $G$.

Thus *to every representation $T_\chi(g)$ possessing an invariant positive definite Hermitian functional there corresponds a representation of $G$ by unitary operators on Hilbert space.* We shall show that in this correspondence *equivalent representations correspond to equivalent ones, and nonequivalent representations to nonequivalent ones.*[29]

Proof. Let $T_{\chi_1}(g)$ and $T_{\chi_2}(g)$ be equivalent representations, that is, assume that there exists a bicontinuous one-to-one mapping $A$ of $D_{\chi_1}$ onto $D_{\chi_2}$ such that $AT_{\chi_1}(g) = T_{\chi_2}(g)A$. Assume further that there exists an invariant positive definite Hermitian functional $(\varphi, \psi)_2$ on $D_{\chi_2}$. To it there corresponds the invariant positive definite Hermitian functional $(\varphi, \psi)_1$ on $D_{\chi_1}$ defined by

$$(\varphi, \psi)_1 = (A\varphi, A\psi)_2 . \tag{1}$$

Let us choose $(\varphi, \psi)_1$ and $(\varphi, \psi)_2$ as scalar products in $D_{\chi_1}$ and $D_{\chi_2}$, and let $H_1$ and $H_2$ be completions of these spaces with respect to these scalar products.

Now according to Eq. (1), $A$ is an isometric mapping of $D_{\chi_1}$ onto $D_{\chi_2}$. This mapping can be extended to an isometric mapping $\hat{A}$ of $H_1$ onto $H_2$ which will also intertwine the representations:

$$\hat{A} T_{\chi_1}(g) = T_{\chi_2}(g) \hat{A},$$

[29] Two unitary representations $T_1(g)$ and $T_2(g)$ of a group $G$ on Hilbert spaces $H_1$ and $H_2$ are called equivalent if there exists an isometric mapping $A$ of $H_1$ onto $H_2$ such that $AT_1(g) = T_2(g)A$.

This shows that if $T_{\chi_1}(g)$ and $T_{\chi_2}(g)$ are equivalent representations of $G$ on $D_{\chi_1}$ and $D_{\chi_2}$, and if they possess positive definite invariant Hermitian functionals, they can be extended to equivalent representations on $H_1$ and $H_2$.

Conversely, let $T_{\chi_1}(g)$, $\chi_1 = (n_1, n_2)$, and $T_{\chi_2}(g)$, $\chi_2 = (m_1, m_2)$, induce equivalent representations on the Hilbert space $H_1$ and $H_2$. Then by definition there exists an isometric operator $\hat{A}$ mapping $H_1$ onto $H_2$ such that $\hat{A}T_{\chi_1}(g) = T_{\chi_2}(g)\hat{A}$. Let $(\varphi, \psi)_2$ be an invariant scalar product on $H_2$. Consider the Hermitian functional

$$B(\varphi, \psi) = (\hat{A}\varphi, \psi)_2,$$

where $\varphi$ may be any function in $H_1$ and $\psi$ may be any function in $H_2$. This functional is invariant under the pair of representations $T_{\chi_1}(g)$ and $T_{\chi_2}(g)$. In particular, $B(\varphi, \psi)$ may be considered an invariant Hermitian functional defined on the pair of spaces $D_{\chi_1}$ and $D_{\chi_2}$. But such a functional can exist only in one of the following four cases.

*Case* 1. $n_1 = -\bar{m}_2$, $n_2 = -\bar{m}_1$.

*Case* 2. $n_1 = \bar{m}_2$, $n_2 = \bar{m}_1$.

*Case* 3. $n_1 = \bar{m}_2 = 1, 2, \ldots, n_2 = -\bar{m}_1$.

*Case* 4. $n_1 = -\bar{m}_1$, $n_2 = \bar{m}_2 = 1, 2, \ldots$.

Cases 3 and 4 may be immediately excluded, since for them $B(\varphi, \psi)$ is not of definite sign.

Thus if $T_{\chi_1}(g)$ is a representation of the principal series, that is, if $n_1 = \frac{1}{2}(-n + i\rho)$ and $n_2 = \frac{1}{2}(-n - i\rho)$, then either $m_1 = \frac{1}{2}(n - i\rho)$ and $m_2 = \frac{1}{2}(n + i\rho)$, or $m_1 = \frac{1}{2}(-n + i\rho)$ and $m_2 = \frac{1}{2}(-n - i\rho)$. In both cases $T_{\chi_1}(g)$ is equivalent to $T_{\chi_2}(g)$. If, on the other hand, $T_{\chi_1}(g)$ is a representation of the supplementary series, that is, if $n_1 = n_2 = \rho$ is a real number such that $-1 < \rho < 1$, $\rho \neq 0$, then either $m_1 = m_2 = -\rho$ or $m_1 = m_2 = \rho$. Again in both cases $T_{\chi_1}(g)$ is equivalent to $T_{\chi_2}(g)$.

This proves that if the representations on $D_{\chi_1}$ and $D_{\chi_2}$ can be extended to equivalent representations on Hilbert spaces, the representations on $D_{\chi_1}$ and $D_{\chi_2}$ are themselves equivalent.

### 6.6. Subspace Irreducibility of the Unitary Representations

In Section 6.5 we have constructed representations of $G$ by unitary operators on Hilbert space, and now we shall show that these representations are subspace irreducible, that is, that *there exists no closed proper subspace in* $H$ *invariant under* $T_{\chi'}(g)$.

**Proof.** Let $H_1$ be some closed invariant subspace of the Hilbert space $H$. Consider the projection operator $P$ associated with this subspace (that is, such that $P\varphi = \varphi$ for all $\varphi \in H_1$ and $P\varphi = 0$ for all $\varphi \in H_2$ where $H_2$ is the subspace orthogonal to $H_1$). This operator commutes with all operators of $T_\chi(g)$. Indeed, every element $\varphi \in H$ may be written $\varphi = \varphi_1 + \varphi_2$, where $\varphi_1 \in H_1$ and $\varphi_2 \in H_2$. Then from the unitarity of $T_\chi(g)$ and the invariance of $H_1$ it follows that $T_\chi(g)\varphi_1$ belongs to $H_1$ and that $T_\chi(g)\varphi_2$ is orthogonal to $H_1$. Therefore

$$PT_\chi(g)\varphi = T_\chi(g)\varphi_1 = T_\chi(g)P\varphi.$$

Now consider the Hermitian functional

$$(\varphi, \psi)_1 = (P\varphi, \psi),$$

where $(\varphi, \psi)$ is the invariant scalar product in $H$. By the above, $(\varphi, \psi)_1$ is invariant under $T_\chi(g)$. We shall now consider only the restriction of this representation to $D_\chi$. Recall that if an invariant Hermitian functional on $D_\chi$ exists, it is defined uniquely up to a constant factor. Therefore there exists a number $\lambda$ such that for any two elements $\varphi(z)$ and $\psi(z)$ in $D_\chi$ we have

$$(P\varphi, \psi) = \lambda(\varphi, \psi),$$

or

$$(P\varphi - \lambda\varphi, \psi) = 0.$$

But $D_\chi$ is an everywhere dense set in $H$, so that $P\varphi = \lambda\varphi$ on all of $H$. This means that $P$ is either the unit operator or the null operator and thus $H_1$ is either the null space or $H$ itself. This proves the subspace irreducibility of $T_\chi(g)$.

CHAPTER IV

# HARMONIC ANALYSIS ON THE GROUP OF COMPLEX UNIMODULAR MATRICES IN TWO DIMENSIONS

In this chapter we shall study the Fourier transform on the group $G$ of complex unimodular matrices in two dimensions. Our treatment will be modeled on the Fourier transform on the line (or in Euclidean $n$-space). Fourier transforms on Lie groups have many unique properties not possessed by the ordinary Fourier transform, and they therefore deserve special consideration.[1] The group $G$ we have chosen is a typical example of such properties.

Integral geometry on a hyperboloid in four complex dimensions, discussed in Chapter II, Section 2, is very closely related to the problems of the present chapter. (This relation is discussed in Section 3.1.)

We defer until Section 1.5, when we will have introduced all the necessary basic concepts, a summary of the content of this chapter.

## 1. Definition of the Fourier Transform on a Group. Statement of the Problems and Summary of the Results

### 1.1. Fourier Transform on the Line

Let us start by listing some elementary facts concerning the Fourier transform on the line.

[1] A somewhat more useful model for the definition of the Fourier transform on a group would be a Lobachevskian rather than a Euclidean space, because in a certain sense Euclidean space is degenerate: it is the limit of a Lobachevskian space as the curvature $k$ approaches zero. The principal difference between a Lobachevskian and a Euclidean space is that in the former the volume of a sphere depends on its radius exponentially (that is, goes as $e^{|k|r}$), while in a Euclidean space it depends on some power or $r$. This difference is reflected in the groups of motions associated with these spaces. Let $G$ be such a group of motions, and $U$ a compact neighborhood of the identity, and consider the invariant measure $\mu(U^n)$ of $U^n$ as a function of $n$. The group of Lobachevskian motions is locally isomorphic to the group of complex unimodular matrices in two dimensions. For this group $\mu(U^n)$ is an exponential function of $n$. For the Euclidean motions, however, $\mu(U^n)$ goes as a power of $n$. In semisimple Lie groups the behavior is similar to our group of matrices. Nilpotent groups, on the other hand, have properties more like those of the Euclidean motions. We shall discuss the Fourier transform in a Lobachevskian space in Chapter VI.

*The Fourier transform of a summable function $f(x)$ defined on the line is given by*

$$F(\lambda) = \int_{-\infty}^{\infty} f(x)e^{i\lambda x}\, dx, \tag{1}$$

*where $\lambda$ is a real or complex number.* This integral converges for all real values of $\lambda$. Under certain additional requirements on $f(x)$, for instance if $|f(x)| \leqslant Ce^{-a|x|}$ for all $a > 0$, the Fourier transform $F(\lambda)$ is defined for all complex $\lambda$ and is an entire analytic function of $\lambda$. The basic results of harmonic analysis on the line are the following.

1. *Behavior of the Fourier transform under translation, differentiation, and convolution.* When $f(x)$ is replaced by $f(x - a)$ (that is, under translation) $F(\lambda)$ is multiplied by $e^{i\lambda a}$.

When $f(x)$ is differentiated, $F(\lambda)$ is multiplied by $-i\lambda$, and when $f(x)$ is multiplied by $ix$, $F(\lambda)$ is transformed to its derivative $F'(\lambda)$.

If $F_1(\lambda)$ and $F_2(\lambda)$ are the Fourier transforms of $f_1(x)$ and $f_2(x)$, respectively, the Fourier transform of the convolution

$$f_1 * f_2(x) = \int_{-\infty}^{\infty} f_1(y)f_2(x - y)\, dy$$

is $F_1(\lambda)F_2(\lambda)$.

The Fourier transform of $f(-x)$ is $F(-\lambda)$, and that of $\bar{f}(x)$ is $\bar{F}(-\bar{\lambda})$. The analogs of these results for $G$ will be obtained in Section 2.

2. *Inverse Fourier transform.* In terms of its Fourier transform, $f(x)$ is given by

$$f(x) = (2\pi)^{-1} \int_{-\infty}^{\infty} F(\lambda)e^{-i\lambda x}\, d\lambda \tag{2}$$

(assuming that $F(\lambda)$ is also a summable function). The analog of this result for $G$ will be obtained in Section 3.

3. *Plancherel's theorem.* If the square modulus of $f(x)$ is integrable, then so is that of $F(\lambda)$. In fact

$$\int_{-\infty}^{\infty} |f(x)|^2\, dx = (2\pi)^{-1} \int_{-\infty}^{\infty} |F(\lambda)|^2\, d\lambda. \tag{3}$$

Equation (2) then holds if the equality is understood in the sense of convergence in the mean.

The analog of Plancherel's theorem for $G$ will be obtained in Section 3. We may remark that for the additive group of real numbers the domain of definition of the Fourier transform is the same for square integrable and summable functions, namely the real axis. It can be shown (see Section 1.4) that these domains differ for $G$. One of them is the analog of the real axis, and the other the analog of a strip.

4. *The Paley-Wiener theorem.* The function $F(\lambda)$ of a complex variable $\lambda$ is the Fourier transform of an infinitely differentiable function $f(x)$ with bounded support if and only if $F(\lambda)$ is in $Z$, that is, if it is an entire analytic function of $\lambda$ such that

$$|\lambda^m F(\lambda)| \leqslant C_m e^{a|\mathrm{Im}\,\lambda|}, \qquad m = 0, 1, \ldots . \tag{4}$$

The analog of the Paley-Wiener theorem for $G$ will be obtained in Section 5. We may remark that the situation is somewhat more complicated than for the real line. In addition to analyticity and inequalities such as (4), the Fourier transform $F(\lambda)$ of a differentiable function $f(g)$ with compact support must satisfy certain algebraic conditions (cf. also Chapter II, Section 2.7).

5. *Bochner's theorem.* Every positive definite continuous function $f(x)$ is the Fourier transform of a finite positive measure.

Remark. In the case of $G$, the set of basis functions involved in Bochner's theorem differs from those of Plancherel's theorem, although for the real line both theorems involve the $e^{i\lambda x}$, where $\lambda$ is a real number. This difference lies in the fact that the square integrable functions are expanded in the unitary representations only of the principal series, while the summable positive definite functions are expanded in unitary representations of both the principal and supplementary series. #

We shall not deal with Bochner's theorem for $G$ in this book.

### 1.2. Functions on $G$

In this section we discuss the basic properties of functions on $G$. A function $f(g)$ on $G$, the group of complex unimodular matrices

$$g = \begin{Vmatrix} \alpha & \beta \\ \gamma & \delta \end{Vmatrix}, \tag{1}$$

is called *rapidly decreasing* if

$$|f(g)| < C\,|g|^{-n}, \tag{2}$$

for every $n > 0$, where

$$|g|^2 = |\alpha|^2 + |\beta|^2 + |\gamma|^2 + |\delta|^2.$$

In order now to introduce the concept of a summable function, we must define the element of volume on the group. It turns out that the element of volume $dg$ can be defined so that it is invariant under left and right translations on the group, as well as under inversions (passing to the inverse), that is, so that

$$dg = d(gg_0) = d(g_0g) = d(g^{-1})$$

for every $g_0 \in G$. (The proof will be found in the Appendix to this section.)

In terms of the $\alpha$, $\beta$, and $\delta$ appearing in (1),

$$dg = \left(\frac{i}{2}\right)^3 d\alpha\, d\bar{\alpha}\, d\beta\, d\bar{\beta}\, d\delta\, d\bar{\delta}\, |\beta|^{-2}. \tag{3}$$

A function $f(g)$ is called *summable* on $G$ if the integral

$$\int |f(g)|\, dg$$

converges. It can be shown that every rapidly decreasing function is summable (see the Appendix to this section).

Finally, we define differentiability on $G$. With each element $g \in G$, whose matrix elements are given by (1), we may associate a point on the surface

$$\alpha\delta - \beta\gamma = 1 \tag{4}$$

in the space of four complex dimensions. We may choose any three of the parameters $\alpha$, $\beta$, $\gamma$, $\delta$ to be local coordinates in a neighborhood of any such point. [Assume, for instance, that $\beta \neq 0$ in this neighborhood. Then (4) implies that $\gamma$ is a continuous function of $\alpha$, $\beta$, $\delta$, and therefore that the points in this neighborhood can be specified by giving the values of these three parameters.] We shall call $f(g)$ *infinitely differentiable* in a neighborhood of $g_0$ if, considered as a function of the three coordinates chosen in this neighborhood, it has derivatives of all orders in this neighborhood. It can be shown that this definition is independent of the choice of coordinate system.

### 1.3. Fourier Transform on G

We shall define the Fourier transform on $G$ by first defining those functions on $G$ which are the analogs of the exponentials. The ordinary exponentials $e^{i\lambda x}$ are solutions of the functional equation

$$f(x_1 + x_2) = f(x_1)f(x_2).$$

Similarly, the analogs of the $e^{i\lambda x}$ for $G$ will be solutions of the functional equation

$$f(g_1 g_2) = f(g_1) f(g_2). \tag{1}$$

However, the only scalar function satisfying this equation is $f(g) \equiv 1$. One is therefore led naturally to seek solutions of (1) in operator functions, that is, functions of operators acting on some space. [Every solution of (1) is a representation of $G$.]

The simplest (namely, operator irreducible) solutions of Eq. (1) have already been found in Chapter III. These were the operators $T_\chi(g)$, $\chi = (n_1, n_2)$, acting in the function space $D_\chi$ according to

$$T_\chi(g)\varphi(z) = (\beta z + \delta)^{n_1 - 1}(\bar{\beta}\bar{z} + \bar{\delta})^{n_2 - 1} \varphi\left(\frac{\alpha z + \gamma}{\beta z + \delta}\right). \tag{2}$$

Consequently these $T_\chi(g)$ shall be our analogs of the exponentials $e^{i\lambda x}$ for $G$. In this analog $g$ plays the role of $x$, while $\chi$ plays the role of $i\lambda$.

Remark. Such "exponentials," that is, solutions of equations of the form $T(g_1, g_2) = T(g_1)T(g_2)$, give rise to almost all special functions. More accurately, special functions are matrix elements of operators belonging to representations of various groups. The only exceptions among the special functions are the Lamé and Mathieu functions, which, as far as is known at present, seem to be unrelated to group representations. #

We shall call *the Fourier transform of a function $f(g)$ on $G$* the operator function $F(\chi)$ defined by

$$F(\chi) = \int f(g) T_\chi(g)\, dg. \tag{3}$$

Thus for every $\chi$ such that $F(\chi)$ is defined, this function is an operator on $D_\chi$ whose action is given by

$$F(\chi)\varphi(z) = \int f(g) T_\chi(g)\varphi(z)\, dg. \tag{4}$$

What we wish to study is the properties of such Fourier transforms. We shall analyze their behavior under translation and convolution and under differentiation of $f(g)$. As we have said before, we shall obtain the inverse Fourier transform formula, giving $f(g)$ in terms of its Fourier transform $F(\chi)$, in Section 3, where we shall also obtain the analog of Plancherel's theorem. In Section 5 we shall obtain the analog of the Paley-Wiener theorem.

## 1.4. Domain of Definition of $F(\chi)$

We must first establish the values of $\chi$ for which $F(\chi)$ is defined. Consider first the Fourier transforms of rapidly decreasing functions $f(g)$. We shall show that if $f(g)$ is a continuous rapidly decreasing function, the integral

$$F(\chi) = \int f(g) T_\chi(g)\, dg$$

converges for all $\chi = (n_1, n_2)$, or in other words that the integral

$$F(\chi)\varphi(z) = \int f(g) T_\chi(g) \varphi(z)\, dg$$

converges for every $\varphi(z) \in D_\chi$, and moreover that $F(\chi)\varphi(z) \in D_\chi$. Further, $F(\chi)$ is continuous in the topology of $D_\chi$.

Proof. We choose the realization of $D_\chi$ as the space of infinitely differentiable homogeneous functions $\varphi(z_1, z_2)$ of degree $(n_1 - 1, n_2 - 1)$. In this space $T_\chi(g)$ is given by

$$T_\chi(g)\varphi(z_1, z_2) = \varphi(\alpha z_1 + \gamma z_2, \beta z_1 + \delta z_2)$$

(see Chapter III, Section 2.4).

We must first show that the integral

$$\begin{aligned} F(\chi)\varphi(z_1, z_2) &\equiv \int f(g) T_\chi(g)\varphi(z_1, z_2)\, dg \\ &= \int f(g)\varphi(\alpha z_1 + \gamma z_2, \beta z_1 + \delta z_2)\, dg \end{aligned} \tag{1}$$

converges. Note first that because $\varphi(z_1, z_2)$ is homogeneous, we have

$$|\varphi(\alpha z_1 + \gamma z_2, \beta z_1 + \delta z_2)| \leqslant M[|\alpha z_1 + \gamma z_2|^2 + |\beta z_1 + \delta z_2|^2]^k,$$

where $k = \frac{1}{2}\operatorname{Re}(n_1 + n_2) - 1$, and $M$ is the maximum value of $\varphi(z_1, z_2)$ on the sphere whose equation is $|z_1|^2 + |z_2|^2 = 1$. From the Cauchy-Bunyakovskii inequality we have

$$\begin{aligned} &|\alpha z_1 + \gamma z_2|^2 + |\beta z_1 + \delta z_2|^2 \\ &\qquad \leqslant (|\alpha|^2 + |\gamma|^2 + |\beta|^2 + |\delta|^2)(|z_1|^2 + |z_2|^2) \\ &\qquad = |g|^2(|z_1|^2 + |z_2|^2). \end{aligned}$$

Thus for $k \geqslant 0$,

$$\int |f(g)T_\chi(g)\varphi(z_1, z_2)|\, dg \leqslant M(|z_1|^2 + |z_2|^2)^k \int |g|^{2k} |f(g)|\, dg. \tag{2}$$

This last integral converges because $f(g)$ is a rapidly decreasing function. If, however, $k \leqslant 0$, we make use of the inequality

$$|\delta z_1' - \gamma z_2'|^2 + |-\beta z_1' + \alpha z_2'|^2 \leqslant |g|^2(|z_1'|^2 + |z_2'|^2).$$

When we write $z_1' = \alpha z_1 + \gamma z_2$, $z_2' = \beta z_1 + \delta z_2$, this becomes

$$|\alpha z_1 + \gamma z_2|^2 + |\beta z_1 + \delta z_2|^2 \geqslant |z_1|^2 + |z_2|^2 |g|^{-2}.$$

Thus for $k \leqslant 0$,

$$\int |f(g)T_\chi(g)\varphi(z_1, z_2)|\, dg \leqslant M(|z_1|^2 + |z_2|^2)^k \int |g|^{-2k} |f(g)|\, dg, \tag{3}$$

which again implies the convergence of (1).

It is obvious that $F(\chi)\varphi(z_1, z_2)$ is homogeneous of the same degree as $\varphi(z_1, z_2)$ and that it is infinitely differentiable for $(z_1, z_2) \neq (0, 0)$. Thus this function also belongs to $D_\chi$.

Finally, Eqs. (2) and (3) imply immediately that $F(\chi)\varphi(z_1, z_2)$ depends continuously on $\varphi(z_1, z_2)$ in the topology of $D_\chi$. This completes the proof.

Consider now the Fourier transform of a summable function on $G$. Recall that the Fourier transform $F(\chi)$ has the same domain of definition for summable and square integrable functions on the line, namely the real axis. For $G$ the situation is somewhat different. It can be shown, in fact, that if $f(g)$ is a *summable* function on $G$, its Fourier transform $F(\chi)$ is defined "on the strip" $-2 \leqslant \mathrm{Re}(n_1 + n_2) \leqslant 2$. We shall see in Section 3, on the other hand, that the Fourier transform of a *square integrable* function on $G$ is defined in general only on the "real axis" $\mathrm{Re}(n_1 + n_2) = 0$. Thus for $G$ the Fourier transform has different domains of definition for summable and square integrable functions.

We wish to prove a somewhat different assertion about summable $f(g)$ [see Ref. (17), p. 504]. *Let $\chi = (n_1, n_2)$ lie on the strip $-2 \leqslant \mathrm{Re}(n_1+n_2) \leqslant 2$. Then $D_\chi$ can be imbedded as an everywhere dense subset in a certain Banach space $L_\chi$ such that the $T_\chi(g)$ are isometric operators on $L_\chi$. Moreover, the integral*

$$F(\chi) = \int f(g)T_\chi(g)\, dg$$

*will converge in the norm of $L_\chi$ for every summable function $f(g)$.*

Proof. We introduce the norm $\|\varphi\|_p$ in $D_\chi$, where $p = 4[2 - \mathrm{Re}(n_1 + n_2)]^{-1}$ (so that $1 \leqslant p \leqslant \infty$), defined by

$$\|\varphi\|_p = \left[\frac{i}{2}\int |\varphi(z)|^p \, dz \, d\bar{z}\right]^{1/p},$$

for $p < \infty$, and $\|\varphi\|_\infty = \max |\varphi(z)|$. It is easily shown that $\|\varphi\|_p < \infty$ for every $\varphi \in D_\chi$. If we now complete $D_\chi$ with respect to the norm $\|\varphi\|_p$, we obtain a Banach space which we shall denote by $L_\chi$. What we must show is that the $T_\chi(g)$ are isometric in $L_\chi$. The case in which $\mathrm{Re}(n_1 + n_2) = 2$ is trivial. Assume, therefore, that $\mathrm{Re}(n_1 + n_2) < 2$. We then have

$$\begin{aligned}\| T_\chi(g)\varphi(z)\|_p^p &= \frac{i}{2}\int | T_\chi(g)\varphi(z)|^p \, dz \, d\bar{z} \\ &= \frac{i}{2}\int \left|(\beta z + \delta)^{n_1 - 1}(\bar{\beta}\bar{z} + \bar{\delta})^{n_2 - 1}\varphi\left(\frac{\alpha z + \gamma}{\beta z + \delta}\right)\right|^p dz \, d\bar{z}.\end{aligned}$$

By writing $(\alpha z + \gamma)/(\beta z + \delta) = w$ we transform this integral to the form

$$\begin{aligned}&\int |(-\beta w + \alpha)^{(-n_1 - n_2 + 2)p - 4}| \, |\varphi(w)|^p \, dw \, d\bar{w} \\ &\quad = \frac{i}{2}\int | -\beta w + \alpha |^{[2 - \mathrm{Re}(n_1 + n_2)]p - 4} \, |\varphi(w)|^p \, dw \, d\bar{w}.\end{aligned}$$

Now inserting the expression for $p$ in terms of $n_1$ and $n_2$, we arrive at

$$\| T_\chi(g)\varphi(z)\|_p^p = \frac{i}{2}\int |\varphi(w)|^p \, dw \, d\bar{w} = \|\varphi\|_p^p,$$

which shows that $T_\chi(g)$ is an isometric mapping.

The fact that it is isometric implies that

$$\left\| \int f(g) T_\chi(g) \, dg \right\|_p \leqslant \int |f(g)| \, dg,$$

and therefore that $\int f(g) T_\chi(g) \, dg$ converges in this norm for every summable $f(g)$.

## 1.5. Summary of the Results of Chapter IV

This summary will be unavoidably brief, for we cannot yet assume that the reader has mastered all the concepts involving the Fourier transform on the group. The summary will therefore not replace the text, which is of a form perhaps somewhat unfamiliar to most analysts.

We have defined the Fourier transform of a function on $G$ as an operator $F(\chi)$ depending on a pair of complex numbers $\chi = (n_1, n_2)$ whose difference is an integer. We shall see later in Section 2 that $F(\chi)$ is an integral operator with a certain kernel $K(z_1, z_2; \chi)$, and that therefore the Fourier transform of a function on a group is a function of the form $K(z_1, z_2; \chi)$, where $z_1$ and $z_2$ are complex variables.

The function $K(z_1, z_2; \chi)$ is closely related to the discussion in Chapter II, Section 2 of integral transforms on a hyperboloid in a space of four complex dimensions. This relation will be established in Section 2.3 of the present chapter. Later the reader will see that Chapter II, Section 2 is essentially a different, or geometric, version of the present chapter. (More accurately, the content of Chapter II, Section 2 is related to that of Chapter IV in the same way that the Radon transform in $n$ dimensions is related to the Fourier transform in $n$ dimensions.)

It can be shown that the more rapidly $f(g)$ decreases, the wider the strip in which its Fourier transform $F(\chi)$ is defined. Moreover, if $f(g)$ is what we have called *rapidly decreasing*, $F(\chi)$ is defined in the entire "complex $\chi$ plane," as will be shown in Section 2.4. (For simplicity we call it the complex $\chi$ plane, although what is involved is actually a countable number of complex planes.) It can be shown that if $f(g)$ is rapidly decreasing, then $F(\chi)$ is an analytic function of $\chi$.

This chapter will be concerned essentially with two basic questions: Which $F(\chi)$ correspond to square integrable functions (Section 3), and which to rapidly decreasing functions (Section 5)? The fundamental result of Section 3 is the following. Let $f(g)$ be a square integrable function. Then its Fourier transform $F(\chi)$ is defined only for "real" $\chi$, that is, for $\chi = (n_1, n_2)$ such that $n_2 = -\bar{n}_1$, which may therefore be written $\chi = (\frac{1}{2}[n + i\rho], \frac{1}{2}[-n + i\rho])$, where $n$ is an integer and $\rho$ is a real number. (We saw in Chapter III, Section 6 that such $\chi$ correspond to representations of the principal series.) In addition, Plancherel's theorem is obtained in the form

$$\int |f(g)|^2 \, dg = -(8\pi^4)^{-1} \int n_1 n_2 \operatorname{Tr}[F(\chi)F^*(\chi)] \, d\chi \qquad (1)$$

[where $F^*(\chi)$ is defined at the end of Section 2.1]. Here the integral over $\chi$ is to be understood as the integral over $\rho$ and sum over $n$; $\operatorname{Tr}(FF^*)$ denotes the trace of the operator $FF^*$. [$F(\chi)$ is a Hilbert-Schmidt operator.] The interesting group-theoretical meaning of the weight factor $n_1 n_2$ will be mentioned toward the end of Section 3.4.

Plancherel's theorem, that is, Eq. (1), can also be written in terms of the kernel $K$:

$$\int |f(g)|^2\, dg = -(8\pi^4)^{-1}\left(\frac{i}{2}\right)^2 \int n_1 n_2 \,|\, K(z_1, z_2; \chi)|^2\, dz_1\, d\bar{z}_1\, dz_2\, d\bar{z}_2\, d\chi. \quad (2)$$

Note that the integral does not involve all the irreducible unitary representations, but only those of the principal series. This is a peculiarity of the Fourier transform on the group.

One may ask then whether every $F(\chi)$ operator whose kernel is such that the right-hand side of Eq. (2) converges corresponds to a square integrable $f(g)$. It is found that this is not always so: the Fourier transform $F(\chi)$ of a square integrable function $f(g)$ has an additional symmetry property relating $F(\chi)$ to $F(-\chi)$, where $-\chi = (-n_1, -n_2)$. Specifically, there exists an operator $A(\chi)$ such that

$$F(\chi) = A^{-1}(\chi)F(-\chi)A(\chi).$$

This operator is of the form

$$A\varphi(z) = \frac{i}{2}\int (z - z_1)^{-n_1-1}(\bar{z} - \bar{z}_1)^{-n_2-1}\varphi(z_1)\, dz_1\, d\bar{z}_1,$$

where the integral is to be understood in the sense of its regularization. This additional symmetry requirement arises from the fact that every representation is counted twice, for the representations on $D_\chi$ and $D_{-\chi}$ are equivalent. These conditions are then found to be sufficient for the operator function $F(\chi)$ to be the Fourier transform of some square integrable $f(g)$. Further, when these conditions are fulfilled, it is possible to write the inverse Fourier transform formula giving $f(g)$ in terms of $F(\chi)$ (see Section 3.4).

Fourier transforms of rapidly decreasing functions are also of considerable interest (Section 5), and we go into the question of the necessary and sufficient conditions that $F(\chi)$ be the Fourier transform of some rapidly decreasing $f(g)$. We shall not here describe these conditions in all detail. Essentially it is a matter of stating the degree of smoothness of $K(z_1, z_2; \chi)$ and its rate of decrease as $|z_1| \to \infty$ and $|z_2| \to \infty$. In addition, $F(\chi)$ satisfies certain symmetry conditions. But even this, however, is not sufficient. It turns out that $F(\chi)$ satisfies an additional set of conditions at integer $\chi$. (Recall that integer $\chi$ occurs when $n_1$ and $n_2$ are integers of the same sign. In general integer $\chi$ play a special role in representation theory.) From the group-theoretical point of view these conditions arise from the fact that at integer $\chi$ there exists in $D_\chi$ either a finite-dimensional invariant subspace or an invariant subspace such that the factor space is finite dimensional. The conditions of which we are speaking may be written

$$\frac{\partial^{n_1}}{\partial z_1^{n_1}} K(z_1, z_2; n_1, n_2) = (-1)^{n_1} \frac{\partial^{n_1}}{\partial z_2^{n_1}} K(z_1, z_2; -n_1, n_2)$$

for $n_1 = 1, 2, \ldots$, and

$$\frac{\partial^{n_1}}{\partial \bar{z}_1^{n_2}} K(z_1, z_2; n_1, n_2) = (-1)^{n_2} \frac{\partial^{n_2}}{\partial \bar{z}_2^{n_2}} K(z_1, z_2; n_1, -n_2)$$

for $n_2 = 1, 2, \ldots$ .

No such conditions occur in the case of square integrable functions. The more rapidly $f(g)$ decreases the greater the number of conditions its Fourier transform satisfies. For instance, if the rate of decrease is only slightly greater than that required for summability, only the first of these relations (that is, for $n_1 = n_2 = 1$) need hold.

**Remark.** In view of the great interest at present in the zeta-function we might perhaps compare its typical properties with those of $F(\chi)$, which we think of as an analytic function of $\chi$. These properties are the following. *1.* The special role of the critical strip $-2 \leqslant \mathrm{Re}(n_1 + n_2) \leqslant 2$. For instance, for any $\chi$ in this strip (and, seemingly, only in this strip), the $T_\chi(g)$ are bounded as functions of $g$. *2.* The functional equation

$$F(\chi) = A^{-1}(\chi)F(-\chi)A(\chi).$$

*3.* Singularities at integer points.[2] #

We may mention another fact closely related to such properties of $F(\chi)$. If $f(g)$ is a rapidly decreasing function, it possesses moments, that is, integrals of the form

$$\int f(g)a(g)\, dg,$$

where $a(g)$ is the matrix element of a finite-dimensional representation of $G$. It is remarkable that these moments can be written explicitly in terms of the kernel $K$ of the Fourier transform of $f(g)$ (see Section 5.5). In particular, we find that

$$\int f(g)\, dg = \frac{i}{2} \int K(z_1, z_2; 1, 1)\, dz_2\, d\bar{z}_2$$

(the integral on the right-hand side is independent of $z_1$).

[2] Perhaps without actually having realized it Selberg [Ref. (40)] was led by just this relation to a certain class of zeta-functions which are in fact related to group representations [see also Gel'fand and Pyatetskii-Shapiro, Ref. (18)].

## Appendix. Functions on $G$

### 1. Rapidly Decreasing Functions on $G$

We shall now study in more detail the fundamental concepts concerning functions on the group of complex unimodular matrices in two dimensions. We start by defining functions with bounded support and rapidly decreasing functions.

Let us call

$$|g| = (|\alpha|^2 + |\beta|^2 + |\gamma|^2 + |\delta|^2)^{\frac{1}{2}}$$

the *norm*[3] of the matrix

$$g = \begin{Vmatrix} \alpha & \beta \\ \gamma & \delta \end{Vmatrix}.$$

We shall say that a function $f(g)$ on $G$ *has bounded support* if there exists a number $A$ such that $f(g) = 0$ for $|g| > A$. Further, $f(g)$ is called *continuous* on $G$ if $\lim_{n\to\infty} |g_n - g| = 0$ implies $\lim_{n\to\infty} f(g_n) = f(g)$.

The continuous functions with bounded support on $G$ form a linear space $C$, and a topology can be introduced on this space in a natural way. Specifically, we shall say that a sequence $\{f_n(g)\}$ of continuous functions with bounded support converges to zero if:

1. There exists an $A$ such that all the $f_n(g)$ vanish for $|g| > A$.
2. For $|g| \leqslant A$, the sequence $\{f_n(g)\}$ converges uniformly to zero.

A function $f(g)$ is called *rapidly decreasing* if for every $n$

$$\lim_{|g|\to\infty} |g|^n f(g) = 0.$$

Rapidly decreasing continuous functions also form a linear space. A topology is introduced on this space by means of the countable sequence of norms

$$\|f\|_n = \sup_g |g|^n |f(g)|.$$

Remark. The rapidly decreasing functions on $G$ have properties similar not so much to the rapidly decreasing functions on the line as to

[3] The following properties of this norm are easily verified.

$$|g_1 + g_2| \leqslant |g_1| + |g_2|;$$
$$|g_1 g_2| \leqslant |g_1|\,|g_2|;$$
$$|ag| = |a|\,|g|.$$

where $a$ is any number.

those decreasing more rapidly than any function of the form $e^{-a|x|}$, where $a > 0$. This will become clearer later, when we develop the properties of the Fourier transform. The analog of the rapidly decreasing functions on the line would more properly be $f(g)$ such that $\lim_{|g|\to\infty}(\ln|g|)^n f(g) = 0$ for all $n$ (cf. Footnote 1 at the beginning of this chapter). #

## 2. INVARIANT INTEGRATION ON $G$

In order to define integration on $G$ we must first define the element of volume $dg$. We shall define it so that it remains invariant under left and right translations on the group, i.e., so that when $g$ is replaced by $g_0 g$ or $gg_0$ we have

$$dg = d(g_0 g) = d(gg_0).$$

When the volume element is invariant, the integral also becomes invariant:

$$\int f(g_0 g)\, dg = \int f(gg_0)\, dg = \int f(g)\, dg \tag{1}$$

(assuming the integrals to converge). In other words the integral is invariant under *left and right translation* of $f(g)$.

We wish to obtain an expression for the invariant element of volume on $G$. Consider first the set of *all* complex matrices in two dimensions. With each such matrix

$$g = \begin{Vmatrix} \alpha & \beta \\ \gamma & \delta \end{Vmatrix}$$

we associate a point $(\alpha, \beta, \gamma, \delta)$ in a space of four complex dimensions. The unimodular matrices form the quadratic surface $\alpha\delta - \beta\gamma = 1$ in this space. With this surface we associate the differential form $\omega$ defined by

$$d\alpha\, d\beta\, d\gamma\, d\delta = d(\alpha\delta - \beta\gamma) \cdot \omega. \tag{2}$$

Then in terms of the parameters of $g$ this form becomes

$$\omega = \delta^{-1}\, d\beta\, d\gamma\, d\delta = \gamma^{-1}\, d\alpha\, d\gamma\, d\delta = -\beta^{-1}\, d\alpha\, d\beta\, d\delta = -\alpha^{-1}\, d\alpha\, d\beta\, d\gamma. \tag{3}$$

Since $\alpha$, $\beta$, $\gamma$, and $\delta$ cannot all simultaneously vanish, $\omega$ has no singularities on $G$. Moreover, $\omega$ is invariant under translations. Indeed, under translations $\alpha$, $\beta$, $\gamma$, $\delta$ undergo a linear transformation whose determinant is 1, so that $d\alpha\, d\beta\, d\gamma\, d\delta$ is invariant under such transformations. On the other hand, the determinant $\alpha\delta - \beta\gamma$ of the matrix $g$ is invariant under such transformations. Hence $d(\alpha\delta - \beta\gamma)$ is also invariant. This then implies that $\omega$ as defined by Eq. (2) is invariant under translations.

We now define the element of volume on $G$ to be

$$dg = \left(\frac{i}{2}\right)^3 \omega\bar{\omega}. \tag{4}$$

It is seen from what has gone before that this element of volume is invariant under left and right translations.[4] The volume element can be written in several ways in terms of the parameters of $g$. Using Eq. (3), we have

$$\begin{aligned} dg &= \left(\frac{i}{2}\right)^3 \frac{d\beta\, d\gamma\, d\delta\, d\bar{\beta}\, d\bar{\gamma}\, d\bar{\delta}}{|\delta|^2} = \left(\frac{i}{2}\right)^3 \frac{d\alpha\, d\gamma\, d\delta\, d\bar{\alpha}\, d\bar{\gamma}\, d\bar{\delta}}{|\gamma|^2} \\ &= \left(\frac{i}{2}\right)^3 \frac{d\alpha\, d\beta\, d\delta\, d\bar{\alpha}\, d\bar{\beta}\, d\bar{\delta}}{|\beta|^2} = \left(\frac{i}{2}\right)^3 \frac{d\alpha\, d\beta\, d\gamma\, d\bar{\alpha}\, d\bar{\beta}\, d\bar{\gamma}}{|\alpha|^2}. \end{aligned} \tag{5}$$

It is noteworthy that $dg$ is invariant not only under left and right translations but also under inversions, i.e. that

$$d(g^{-1}) = dg.$$

Indeed, if $g$ has determinant 1, then

$$g^{-1} = \begin{Vmatrix} \delta & -\beta \\ -\gamma & \alpha \end{Vmatrix}.$$

Thus we may replace $\alpha$, $\beta$, $\gamma$, and $\delta$ in the expressions of Eq. (5) by $\delta$, $-\beta$, $-\gamma$, and $\alpha$, respectively; under this replacement Eqs. (5) remain invariant. Consequently

$$\int f(g^{-1})\, dg = \int f(g)\, dg.$$

Remark. The invariance of $dg$ could have been established directly by calculating the Jacobian of the transformation from $g$ to $g_0 g$ or $g g_0$. #

We now assert that *the invariant integral converges absolutely for every continuous rapidly decreasing function on $G$* (and in particular for every continuous function with bounded support).

Consider the region $\Omega_\alpha$ of the group $G$ consisting of matrices $g$ such that $|\alpha| > \frac{1}{2}$. Similarly, we may define the three other regions $\Omega_\beta$, $\Omega_\gamma$, $\Omega_\delta$. The union of $\Omega_\alpha$, $\Omega_\beta$, $\Omega_\gamma$, $\Omega_\delta$ is the entire group $G$,[5] and it is

[4] It can be shown that invariance determines $dg$ uniquely up to a constant factor.

[5] This is because $\alpha\delta - \beta\gamma = 1$ implies that the inequalities $|\alpha| \leqslant \frac{1}{2}$, $|\beta| \leqslant \frac{1}{2}$, $|\gamma| \leqslant \frac{1}{2}$, $|\delta| \leqslant \frac{1}{2}$ cannot hold simultaneously.

therefore sufficient to establish the absolute convergence of the integral on each of these regions.

Consider, for instance, $\Omega_\beta$. Let $f(g)$ be a rapidly decreasing continuous function on $G$. Then for every $n$ we have $|f(g)| \leqslant C_n |g|^{-n}$, where $|g|$ is the norm defined in Section 1 of this Appendix. Therefore

$$\int_{\Omega_\beta} |f(g)|\, dg$$
$$\leqslant C\left(\frac{i}{2}\right)^3 \int_{|\beta| > \frac{1}{2}} (|\alpha|^2 + |\beta|^2 + |\gamma|^2 + |\delta|^2)^{-\frac{1}{2}n} |\beta|^{-2}\, d\alpha\, d\beta\, d\delta\, d\bar{\alpha}\, d\bar{\beta}\, d\bar{\delta}.$$

It is a simple matter to show that the integral on the right-hand side converges for sufficiently large values of $n$ (specifically, for $n > 4$). Therefore $\int_{\Omega_\beta} |f(g)|\, dg$ also converges.

Henceforth in speaking of summable functions, square integrable functions, etc., we shall always mean integrability with respect to $dg$ (the *invariant measure*).

## 2. Properties of the Fourier Transform on $G$

### 2.1. Simplest Properties

In this section we shall deal with some of the simplest properties of the Fourier transform on $G$, all of which are easily obtained from the functional equation

$$T_\chi(g_1)T_\chi(g_2) = T_\chi(g_1 g_2)$$

satisfied by the "exponentials" $T_\chi(g)$.

Let us first study the behavior of the Fourier transform $F(\chi)$ under translation of $f(g)$. The Fourier transform of $f(gg_0)$ is the operator function

$$\int f(gg_0)T_\chi(g)\, dg.$$

Since the integral is invariant under the translation $g \to gg_0$, we have

$$\int f(gg_0)T_\chi(g)\, dg = \int f(g)T_\chi(gg_0^{-1})\, dg$$
$$= \left[\int f(g)T_\chi(g)\, dg\right] T_\chi^{-1}(g_0).$$

Consequently

$$\int f(gg_0)T_\chi(g)\, dg = F(\chi)T_\chi^{-1}(g_0). \tag{1}$$

Similarly

$$\int f(g_0^{-1}g)T_\chi(g)\,dg = T_\chi(g_0)F(\chi). \tag{1'}$$

Thus *if $F(\chi)$ is the Fourier transform of $f(g)$, the Fourier transforms of $f(gg_0)$ and $f(g_0^{-1}g)$ are $F(\chi)T_\chi^{-1}(g_0)$ and $T_\chi(g_0)F(\chi)$, respectively.* Note that $T_\chi^{-1}(g_0)$ and $T_\chi(g_0)$, as well as $F(\chi)$, are operator functions of $\chi$; multiplication of operator functions is understood as multiplication of operators on $D_\chi$ for each value of $\chi$.

We now turn to the Fourier transform of the *convolution*

$$f_1 * f_2(g) = \int f_1(g_1)f_2(g_1^{-1}g)\,dg_1$$

of two rapidly decreasing continuous functions $f_1(g)$ and $f_2(g)$ with Fourier transforms $F_1(\chi)$ and $F_2(\chi)$, respectively. Then the Fourier transform of $f_1 * f_2(g)$ is given by

$$\int f_1 * f_2(g)T_\chi(g)\,dg = \int\int f_1(g_1)f_2(g_1^{-1}g)T_\chi(g)\,dg_1\,dg.$$

Let us change the order of integration and then replace $g$ by $g_1g_2$.[6] Then invariance under translation leads to

$$\int f_1 * f_2(g)T_\chi(g)\,dg = \int\int f_1(g_1)f_2(g_2)T_{\chi_1}(g_1)T_{\chi_2}(g_2)\,dg_1\,dg_2 = F_1(\chi)F_2(\chi). \tag{2}$$

Thus *the Fourier transform of the convolution of two functions on $G$ is the product of their Fourier transforms.* This result has a well-known analog for Fourier transforms of functions of a real variable.

Let us now find the Fourier transform of $f(g^{-1})$,[7] defined by the integral

$$\int f(g^{-1})T_\chi(g)\,dg = \int f(g)T_\chi(g^{-1})\,dg$$

(where we have used invariance under inversion). Now recall that $D_\chi$ can be imbedded in a natural way into the space $D'_{-\chi}$ which is the dual of $D_{-\chi}$ (see Chapter III, Section 2.6). Under this operation it is found that $T_\chi(g^{-1}) = T'_{-\chi}(g)$, where $T'_{-\chi}(g)$ is the dual of $T_{-\chi}(g)$. Thus if $F(\chi)$ is the Fourier transform of $f(g)$, then

$$\int f(g^{-1})T_\chi(g)\,dg = \int f(g)T'_{-\chi}(g)\,dg = F'(-\chi). \tag{3}$$

[6] The order of integration can be changed because the functions are rapidly decreasing.

[7] To go from $f(g)$ to $f(g^{-1})$ is analogous to going from $f(x)$ to $f(-x)$ on the real line.

Here $F'(-\chi)$ is the operator dual to $F(-\chi)$, that is, the operator such that

$$\frac{i}{2}\int [F(-\chi)\varphi(z)]\psi(z)\,dz\,d\bar{z} = \frac{i}{2}\int \varphi(z)F'(-\chi)\psi(z)\,dz\,d\bar{z}$$

for every $\varphi(z) \in D_{-\chi}$ and $\psi(z) \in D_\chi$. Note that $F'(-\chi)$ and $F(\chi)$ both act on the same space $D_\chi$.

Thus *if $F(\chi)$ is the Fourier transform of $f(g)$, the Fourier transform of $f(g^{-1})$ is $F'(-\chi)$, where $F'(\chi)$ is the dual of $F(\chi)$.*

We now turn to the Fourier transform of $\bar{f}(g)$, defined by

$$\int \bar{f}(g)T_\chi(g)\,dg = \int \overline{f(g)\bar{T}_\chi(g)\,dg}.$$

But $\bar{T}_\chi(g) = T_{\bar{\chi}}(g)$, where $\bar{\chi} = (\bar{n}_2, \bar{n}_1)$. We thus obtain

$$\int \bar{f}(g)T_\chi(g)\,dg = \int \overline{f(g)T_{\bar{\chi}}(g)\,dg} = \bar{F}(\bar{\chi}). \tag{4}$$

Here $\bar{F}(\bar{\chi})$ operates on the functions of $D_\chi$ according to $\bar{F}(\bar{\chi})\varphi(z) = \overline{F(\bar{\chi})\bar{\varphi}(z)}$.[8]

Thus *if $F(\chi)$ is the Fourier transform of $f(g)$, the Fourier transform of $\bar{f}(g)$ is $\bar{F}(\bar{\chi})$.*

We may combine the last two results to obtain the Fourier transform of $f^*(g) = \bar{f}(g^{-1})$. We then find that *if $F(\chi)$ is the Fourier transform of $f(g)$, the Fourier transform of $f^*(g)$ is $F^*(\chi^*)$*, where we have written $F^* = \bar{F}'$, and $\chi^* = (-\bar{n}_2, -\bar{n}_1)$.

## 2.2. Fourier Transform as Integral Operator

We shall now prove the following assertion. The Fourier transform of any rapidly decreasing continuous function $f(g)$ is an integral operator of the form

$$F(\chi)\varphi(z_1) = \frac{i}{2}\int K(z_1, z_2; \chi)\varphi(z_2)\,dz_2\,d\bar{z}_2 \tag{1}$$

for each $\chi$. [We shall show moreover in Section 2.5 that the kernel $K(z_1, z_2; \chi)$ is a continuous function of $z_1$ and $z_2$, and shall discuss its

[8] Let $F$ be an operator on the functions of $D_\chi$. Then by $\bar{F}$ we denote the operator on $D_{\bar{\chi}}$ defined by

$$\bar{F}\varphi(z) = \overline{F\bar{\varphi}(z)}$$

for every $\varphi(z) \in D_\chi$.

geometric meaning. Later, in Section 5, we shall discuss the analytic properties of this kernel.]

Proof. By definition of the Fourier transform we have

$$F(\chi)\varphi(z_1) = \int f(g)T_\chi(g)\varphi(z_1)\, dg \tag{2}$$

for every $\varphi(z) \in D_\chi$. This integral converges absolutely for every $\chi$ if $f(g)$ is a rapidly decreasing continuous function (see Section 1.4). We now insert

$$T_\chi(g)\varphi(z_1) = (\beta z_1 + \delta)^{n_1-1}(\bar{\beta}\bar{z}_1 + \bar{\delta})^{n_2-1}\varphi\left(\frac{\alpha z_1 + \gamma}{\beta z_1 + \delta}\right)$$

and the expression for $dg$ in terms of $\alpha$, $\beta$, and $\delta$ into Eq. (2). This yields

$$F(\chi)\varphi(z_1) = \left(\frac{i}{2}\right)^3 \int f(\alpha, \beta, \delta)(\beta z_1 + \delta)^{n_1-1}(\bar{\beta}\bar{z}_1 + \bar{\delta})^{n_2-1}\varphi\left(\frac{\alpha z_1 + \gamma}{\beta z_1 + \delta}\right) | \beta |^{-2}$$
$$\times\, d\alpha\, d\bar{\alpha}\, d\beta\, d\bar{\beta}\, d\delta\, d\bar{\delta}. \tag{3}$$

Let us now change variables according to

$$\lambda = \beta z_1 + \delta$$

and

$$z_2 = (\alpha z_1 + \gamma)/(\beta z_1 + \delta).$$

These equations can be inverted to give[9]

$$\delta = \lambda - \beta z_1, \qquad \alpha = \lambda^{-1} + \beta z_2.$$

Then the Jacobian of the transformation is

$$D(\alpha, \beta, \delta)/D(z_2, \beta, \lambda) = \beta.$$

In terms of these new variables, Eq. (3) becomes

$$F(\chi)\varphi(z_1) = \left(\frac{i}{2}\right)^3 \int f(\lambda^{-1} + \beta z_2, \beta, \lambda - \beta z_1)\varphi(z_2)\lambda^{n_1-1}\bar{\lambda}^{n_2-1}$$
$$\times\, d\lambda\, d\bar{\lambda}\, d\beta\, d\bar{\beta}\, dz_2\, d\bar{z}_2. \tag{4}$$

[9] Recall that $\gamma = (\alpha\delta - 1)\beta^{-1}$.

This proves the assertion. We have in addition obtained an expression for the kernel, namely[10]

$$K(z_1, z_2; \chi) = \left(\frac{i}{2}\right)^2 \int f(\lambda^{-1} + \beta z_2, \beta, \lambda - \beta z_1)\lambda^{n_1-1}\bar{\lambda}^{n_2-1}\, d\lambda\, d\bar{\lambda}\, d\beta\, d\bar{\beta}. \quad (5)$$

### 2.3. Geometric Interpretation of $K(z_1, z_2; \chi)$. The functions $\varphi(z_1, z_2; \lambda)$ and $\Phi(u, v; u', v')$.

The kernel $K$ can be calculated in two steps. First we use $f(g) = f(\alpha, \beta, \delta)$ to calculate the new function

$$\varphi(z_1, z_2; \lambda) = \frac{i}{2} \int f(\lambda^{-1} + \beta z_2, \beta, \lambda - \beta z_1)\, d\beta\, d\bar{\beta}, \quad (1)$$

and then the kernel is given in terms of this function by

$$K(z_1, z_2; \chi) = \frac{i}{2} \int \varphi(z_1, z_2; \lambda)\lambda^{n_1-1}\bar{\lambda}^{n_2-1}\, d\lambda\, d\bar{\lambda}. \quad (2)$$

The geometric meaning of $\varphi(z_1, z_2; \lambda)$ is quite simple. Let us think of $f(g)$ as a function on the hypersurface whose equation is $\alpha\delta - \beta\gamma = 1$. Then for each fixed $z_1, z_2, \lambda$ the equations

$$\begin{aligned} \alpha &= \lambda^{-1} + \beta z_2, & \beta &= \beta, \\ \gamma &= (\alpha\delta - 1)\beta^{-1} = \lambda z_2 - \lambda^{-1} z_1 - \beta z_1 z_2, & \delta &= \lambda - \beta z_1 \end{aligned} \quad (3)$$

determine a (complex) line generator of this surface. It is easily shown that by changing the values of $z_1, z_2, \lambda$ we obtain all of the linear generators of the surface that are not parallel to the $\beta = 0$ plane.[11] Thus $\varphi(z_1, z_2; \lambda)$ is the integral of $f(g)$, $g = (\alpha, \beta, \gamma, \delta)$ along the linear generator

$$\frac{\alpha - \lambda^{-1}}{z_2} = \frac{\beta}{1} = \frac{\gamma - \lambda z_2 + \lambda^{-1} z_1}{-z_1 z_2} = \frac{\delta - \lambda}{-z_1}.$$

Obviously $z_1$ and $z_2$ determine the direction of this generator.

[10] Since the integral in Eq. (2) converges absolutely, it follows from Fubini's theorem that the integral in Eq. (5) converges absolutely for all $z_1$ and for almost all $z_2$ and that the integral of Eq. (1) converges absolutely.

[11] Indeed, any straight line on the $\alpha\delta - \beta\gamma = 1$ surface which is not parallel to the $\beta = 0$ is given by equations of the form $\alpha = \alpha_0 + \beta z_2$, $\delta = \lambda - \beta z_1$, where $\alpha_0$, $\lambda$, $z_1$, and $z_2$ are constants. By setting $\beta = 0$ and using the equation of the surface, we find that $\alpha_0 \lambda = 1$. Then the expression for $\gamma$ in Eq. (3) follows immediately from the fact that $\gamma = (\alpha\delta - 1)\beta^{-1}$.

In this interpretation $K(z_1, z_2; \lambda)$ is, for given $z_1$ and $z_2$, the average of $\varphi(z_1, z_2; \lambda)$ with weight factor $\lambda^{n_1-1}\bar{\lambda}^{n_2-1}$ over the set of linear generators corresponding to these fixed values of $z_1$ and $z_2$, that is, over a set of mutually parallel linear generators.

**Remark.** The expression for the kernel in terms of $\varphi(z_1, z_2; \lambda)$ is analogous to the usual formula for the Mellin transform. For this reason we shall call $K(z_1, z_2; \lambda)$ the *Mellin transform of* $\varphi(z_1, z_2; \lambda)$. #

It is sometimes convenient to designate the lines on the $\alpha\delta - \beta\gamma = 1$ surface not in terms of $z_1, z_2, \lambda$, but in terms of homogeneous coordinates $u, v; u', v'$ (as in Section 2 of Chapter II). Recall that we showed that every line on the $\alpha\delta - \beta\gamma = 1$ surface is given by the set of linear equations[12]

$$\begin{aligned} u\alpha + v\gamma &= u', \\ u\beta + v\delta &= v', \end{aligned} \tag{4}$$

where $(u, v) \neq (0, 0)$ and $(u', v') \neq (0, 0)$. The geometric meaning of these homogeneous coordinates is now evident. If $(u, v)$ and $(u', v')$ are the coordinates of two points in the plane, the first is carried into the second by the transformation whose matrix is

$$g = \begin{Vmatrix} \alpha & \beta \\ \gamma & \delta \end{Vmatrix},$$

and what we have been calling a line is the set of all transformations $g$ that carry $(u, v)$ into $(u', v')$.

We may define the integral of $f(g)$ along the line whose homogeneous coordinates are $(u, v; u', v')$ by

$$\Phi(u, v; u', v') = \frac{i}{2}\int f(g)\omega\bar{\omega}, \tag{5}$$

where $\omega = d\alpha/vu' = d\beta/vv' = -d\gamma/uu' = -d\delta/uv'$. Then $\Phi$ is a homogeneous function of $u, v, u', v'$ of degree $(-2, -2)$; that is,

$$\Phi(tu, tv; tu', tv') = t^{-2}\bar{t}^{-2}\Phi(u, v; u', v')$$

for every $t \neq 0$.

The relation between the homogeneous coordinates and $z_1, z_2, \lambda$, and thereby also the relation between $\Phi(u, v; u', v')$ and $\varphi(z_1, z_2; \lambda)$, is

[12] In Section 2 of Chapter II we denoted the coordinates of the points not by $\alpha, \beta, \gamma, \delta$ but by $z_1, z_2, z_3, z_4$.

easily established. From Eq. (3) we find that the line on the $\alpha\delta - \beta\gamma = 1$ surface is given by

$$\begin{aligned} z_1\alpha + \gamma &= \lambda z_2 , \\ z_1\beta + \delta &= \lambda. \end{aligned} \tag{6}$$

When we compare this with Eq. (4), we find that the relations between the two sets of coordinates may be written

$$z_1 = u/v, \qquad z_2 = u'/v', \qquad \lambda = v'/v. \tag{7}$$

Further, $\varphi(z_1 , z_2 ; \lambda)$ is related to $\Phi(u, v; u', v')$, the integral of $f(g)$ over this line, by

$$\varphi(z_1 , z_2; \lambda) = \lambda\bar{\lambda}\Phi(z_1 , 1; \lambda z_2 , \lambda). \tag{8}$$

The inverse of this relationship, obtained in a simple way from (8), is

$$\Phi(u, v; u', v') = v^{-1}\bar{v}^{-1}v'^{-1}\bar{v}'^{-1}\varphi(u/v, u'/v'; v'/v). \tag{9}$$

## 2.4. Properties of $K(z_1 , z_2 ; \chi)$

In Section 2.1 we studied how the Fourier transform of a function on $G$ behaves when the function $f(g)$ is replaced by $f(g^{-1})$, $\bar{f}(g)$, or $\bar{f}(g^{-1})$. These properties can obviously also be stated in terms of the kernel $K$ of the Fourier transform operator $F(\chi)$. The results obtained are then the following.

Let $f(g)$ be a rapidly decreasing continuous function on $G$, and let $K(z_1 , z_2 ; \chi)$ be the kernel of its Fourier transform $F(\chi)$. Then the Fourier transform of $f(gg_0)$ has kernel $T_{-\chi}(g_0)K(z_1 , z_2 ; \chi)$, where $T_{-\chi}(g_0)$ acts on $K$ as a function of $z_2$ . Explicitly this new kernel is

$$T_{-\chi}(g_0)K(z_1 , z_2; \chi) = (\beta_0 z_2 + \delta_0)^{-n_1-1}(\bar{\beta}_0\bar{z}_2 + \bar{\delta}_0)^{-n_2-1}K\left(z_1 , \frac{\alpha_0 z_2 + \gamma_0}{\beta_0 z_2 + \delta_0}; \chi\right). \tag{1}$$

The Fourier transform of $f(g_0^{-1}g)$ has kernel $T_\chi(g_0)K(z_1 , z_2 ; \chi)$, where $T_\chi(g_0)$ acts on $K$ as a function of $z_1$ . Explicitly this new kernel is

$$T_\chi(g_0)K(z_1 , z_2; \chi) = (\beta_0 z_1 + \delta_0)^{n_1-1}(\bar{\beta}_0\bar{z}_1 + \bar{\delta}_0)^{n_2-1}K\left(\frac{\alpha_0 z_1 + \gamma_0}{\beta_0 z_1 + \delta_0}, z_2; \chi\right). \tag{2}$$

The Fourier transform of $f(g^{-1})$ has kernel

$$K'(z_1 , z_2; \chi) = K(z_2 , z_1; -\chi). \tag{3}$$

The Fourier transform of $\bar{f}(g)$ has kernel

$$\bar{\bar{K}}(z_1, z_2; \chi) = \bar{K}(z_1, z_2; \bar{\chi}). \tag{4}$$

The Fourier transform of $f^*(g) = \bar{f}(g^{-1})$ has kernel

$$K^*(z_1, z_2; \chi) = \bar{K}(z_2, z_1; \chi^*), \tag{5}$$

where $\chi^* = -\bar{\chi} = (-\bar{n}_2, -\bar{n}_1)$.

Let $K_1$ and $K_2$ be the kernels of the Fourier transforms of $f_1$ and $f_2$, respectively. Then the Fourier transform of the convolution $f_1 * f_2(g)$ has kernel

$$K(z_1, z_2; \chi) = \frac{i}{2} \int K_1(z_1, z_3; \chi) K_2(z_3, z_2; \chi)\, dz_3\, d\bar{z}_3 . \tag{6}$$

All of these assertions can be proven simply on the basis of the results of Section 2.1 [for instance, Eq. (3) follows from the fact that the dual operator is obtained from the transposed kernel].

## 2.5. Continuity of $K(z_1, z_2; \chi)$

*Let $f(g)$ be a continuous rapidly decreasing function. Then the corresponding kernel $K(z_1, z_2; \chi)$ is a continuous function of $z_1$ and $z_2$ for all values of $\chi$.*

To prove this all we need show is that the defining integral

$$K(z_1, z_2; \chi) = \left(\frac{i}{2}\right)^2 \int f(\lambda^{-1} + \beta z_2, \beta, \lambda - \beta z_1) \lambda^{n_1 - 1} \bar{\lambda}^{n_2 - 1}\, d\lambda\, d\bar{\lambda}\, d\beta\, d\bar{\beta} \tag{1}$$

converges uniformly in every bounded region of the complex $z_1$, $z_2$ plane.

Now since $f(g)$ is rapidly decreasing, the inequality

$$| f(g) \lambda^{n_1 - 1} \bar{\lambda}^{n_2 - 1} | \leqslant C | \lambda |^{2k} | g |^{-2m} \tag{2}$$

holds for all $m$, where, as before, the norm of $g$ is given by $g = (| \alpha |^2 + | \beta |^2 + | \gamma |^2 + | \delta |^2)^{\frac{1}{2}}$, and $k = \frac{1}{2} \operatorname{Re}(n_1 + n_2) - 1$. Let us derive some estimates for $| \lambda |$. Recall that according to Eq. (6) of Section 2.3

$$\lambda = \beta z_1 + \delta, \qquad \lambda z_2 = \alpha z_1 + \gamma. \tag{3}$$

Thus

$$| \lambda |^2 = \frac{| \alpha z_1 + \gamma |^2 + | \beta z_1 + \delta |^2}{1 + | z_2 |^2} .$$

Then from the Cauchy-Bunyakovskii inequality we arrive easily at

$$|\lambda|^2 \leqslant |g|^2 \frac{1 + |z_1|^2}{1 + |z_2|^2}. \tag{4}$$

Let us now obtain a lower bound for $\lambda$. We first rewrite Eq. (3) in the form[13]

$$\lambda(\alpha - \beta z_2) = 1, \qquad \lambda(\delta z_2 - \gamma) = z_1, \tag{3'}$$

from which we obtain

$$|\lambda|^2 = \frac{1 + |z_1|^2}{|\alpha - \beta z_2|^2 + |\delta z_2 - \gamma|^2}.$$

We may then use the Cauchy-Bunyakovskii inequality again to obtain

$$|\lambda|^2 \leqslant \frac{1}{|g|^2} \frac{1 + |z_1|^2}{1 + |z_2|^2}. \tag{4'}$$

Now we apply (2), (4), and (4′) to be integral of Eq. (1). If $k \geqslant 0$, we find that

$$\left(\frac{i}{2}\right)^2 \int_{|\beta|^2+|\lambda|^2 \geqslant A} |f(\lambda^{-1} + \beta z_2, \beta, \lambda - \beta z_1)\lambda^{n_1-1}\bar{\lambda}^{n_2-1}| \, d\lambda \, d\bar{\lambda} \, d\beta \, d\bar{\beta}$$

$$\leqslant C \left(\frac{1 + |z_1|^2}{1 + |z_2|^2}\right)^k \left(\frac{i}{2}\right)^2 \int_{|\beta|^2+|\lambda|^2 \geqslant A} |g|^{2k-2m} \, d\lambda \, d\bar{\lambda} \, d\beta \, d\bar{\beta}.$$

We have thus to show that in every bounded region of the complex $z_1$, $z_2$ plane the integral

$$J(A) = \left(\frac{i}{2}\right)^2 \int_{|\beta|^2+|\lambda|^2 \geqslant A} |g|^{2k-2m} \, d\lambda \, d\bar{\lambda} \, d\beta \, d\bar{\beta}$$

converges uniformly to zero as $A \to \infty$ for sufficiently large $m$. Let us write $\lambda = \beta z_1 + \delta$. Then since $|\beta|^2 + |\lambda|^2 \leqslant (|\beta|^2 + |\delta|^2)(1 + |z_1|^2)$, it follows that

$$J(A) \leqslant \left(\frac{i}{2}\right)^2 \int_{|\beta|^2+|\delta|^2 \geqslant \frac{1}{2}A(1+|z_1|^2)^{-1}} (|\beta|^2 + |\delta|^2)^{k-m} \, d\delta \, d\bar{\delta} \, d\beta \, d\bar{\beta}.$$

From this it follows immediately that if $m > k + 2$, then $J(A) \to 0$ uniformly as $A \to 0$ in every bounded region in $z_1$.

The case of $k \leqslant 0$ can be treated similarly.

[13] Multiply the first of Eqs. (3) by $\alpha$ and the second by $-\beta$ and add. This gives the first of Eqs. (3′). The second is obtained similarly.

## 2.6. Asymptotic Behavior of $K(z_1, z_2; \chi)$

We wish now to study the behavior of the kernel as $|z_1| \to \infty$ or $|z_2| \to \infty$. Let $f(g)$ be a continuous rapidly decreasing function. We have seen that then $K(z_1, z_2; \chi)$ depends continuously on $z_1$ and $z_2$. Since $f(g_0^{-1}g)$ is also a continuous rapidly decreasing function, the kernel $K_1$ corresponding to it is also continuous in $z_1$ and $z_2$. Now let

$$g_0 = \begin{Vmatrix} 0 & 1 \\ -1 & 0 \end{Vmatrix},$$

and use Eq. (2) of Section 2.4. We find that then

$$K_1(z_1, z_2; \chi) \equiv z_1^{n_1-1}\bar{z}_1^{n_2-1}K(-1/z_1, z_2; \chi)$$

is continuous in $z_1$ and $z_2$ for all finite values of these variables. This equation shows that asymptotically as $|z_1| \to \infty$ the kernel behaves according to

$$K(z_1, z_2; \chi) \sim C_1 z_1^{n_1-1}\bar{z}_1^{n_2-1}, \tag{1}$$

where $C_1 = (-1)^{n_1-n_2}K_1(0, z_2; \chi)$.

Similarly, by considering right translations of $f(g)$, we find that

$$K_2(z_1, z_2; \chi) = z_2^{-n_1-1}\bar{z}_2^{-n_2-1}K(z_1, -1/z_2; \chi)$$

is continuous in $z_1$ and $z_2$ for all finite values of the variables. This equation shows that asymptotically as $|z_2| \to \infty$ the kernel behaves according to

$$K(z_1, z_2; \chi) \sim C_2 z_2^{-n_1-1}\bar{z}_2^{-n_2-1}, \tag{2}$$

where $C_2 = (-1)^{n_1-n_2}K_2(z_1, 0; \chi)$.

Finally, by considering simultaneously left and right translations of $f(g)$, we find that

$$K_3(z_1, z_2; \chi) = z_1^{n_1-1}\bar{z}_1^{n_2-1}z_2^{-n_1-1}\bar{z}_2^{-n_2-1}K(-1/z_1, -1/z_2; \chi)$$

is continuous in $z_1$ and $z_2$ for all finite values of these variables. This equation shows that as $|z_1| \to \infty$ *and* $|z_2| \to \infty$ the kernel behaves asymptotically according to

$$K(z_1, z_2; \chi) \sim C_3 z_1^{n_1-1}\bar{z}_1^{n_2-1}z_2^{-n_1-1}\bar{z}_2^{-n_2-1}, \tag{3}$$

where $C_3 = K_3(0, 0; \chi)$.

### 2.7. Trace of the Fourier Transform

From Section 2.6 we may derive the following property of the Fourier transform.

*Let $f(g)$ be a continuous rapidly decreasing function and let $K(z_1, z_2; \chi)$ be the kernel of its Fourier transform. Then the integral*

$$\frac{i}{2}\int K(z, z; \chi)\, dz\, d\bar{z} \tag{1}$$

*converges absolutely for every $\chi$.*

**Proof.** Equation (3) of Section 2.6 implies that as $|z| \to \infty$,

$$K(z, z; \chi) \sim C\, |z|^{-4}.$$

Since $\frac{i}{2}\int_{|z|\geq 1} |z|^{-4}\, dz\, d\bar{z}$ converges, (1) converges absolutely.

**Remark.** In Section 5 of this chapter we shall introduce the concept of an infinitely differentiable function $f(g)$ on the group, decreasing rapidly with all of its derivatives. The kernel of the Fourier transform of such a function is not only continuous but also infinitely differentiable in $z_1$ and $z_2$. If this kernel also had bounded support, we could conclude that $F(\chi)$ has a trace for every $\chi$, and that this trace is equal to

$$\frac{i}{2}\int K(z, z; \chi)\, dz\, d\bar{z}.$$

In fact, however, $K(z_1, z_2; \chi)$ does not have bounded support and may even have a singularity at $z_1 = \infty$ or $z_2 = \infty$. Nevertheless the Fourier transform operator $F(\chi)$ of an infinitely differentiable function $f(g)$ decreasing rapidly with all of its derivatives possesses a trace given by this integral. To prove this one must realize $D_\chi$ not as the space of functions of $z$, but as the space of functions defined on a compact manifold [e.g., the functions $f(z_1, z_2)$ defined on the sphere whose equation is $|z_1|^2 + |z_2|^2 = 1$]. In this realization the kernel of $F(\chi)$ is an infinitely differentiable function on this compact manifold, and consequently difficulties associated with the behavior of the kernel at infinity will not arise. #

We may thus assert that if $f(g)$ is an infinitely differentiable function decreasing rapidly with all of its derivatives, the operator

$$F(\chi) = \int f(g) T_\chi(g)\, dg \tag{2}$$

possesses a trace. This trace is given in terms of the kernel $K(z_1, z_2; \chi)$ of $F(\chi)$ by

$$\operatorname{Tr} F(\chi) = \frac{i}{2} \int K(z, z; \chi)\, dz\, d\bar{z}. \tag{3}$$

This trace can also be expressed directly in terms of $f(g)$ by expressing the kernel in Eq. (3) in terms of $f(g)$. The resulting equation (whose detailed derivation we leave as an exercise to the reader) is then

$$\operatorname{Tr} F(\chi) = \int f(g)(\lambda^{n_1}\bar{\lambda}^{n_2} + \lambda^{-n_1}\bar{\lambda}^{-n_2}) | \lambda - \bar{\lambda} |^{-2}\, dg, \tag{4}$$

where $\lambda$ and $\lambda^{-1}$ are the eigenvalues of $g$.

By comparing Eqs. (2) and (4), we see that $\operatorname{Tr} F(\chi)$ can be obtained from the expression for $F(\chi)$ by replacing $T_\chi(g)$ in that expression by $(\lambda^{n_1}\bar{\lambda}^{n_2} + \lambda^{-n_1}\bar{\lambda}^{-n_2}) | \lambda - \bar{\lambda} |^{-2}$ (where $\lambda$ and $\bar{\lambda}$ are the eigenvalues of the $g$ matrix), which we may therefore think of as the generalized trace of $T_\chi(g)$. [Note that $T_\chi(g)$ has no trace in the ordinary sense.] It would be interesting to obtain the formula for the generalized trace of $T_\chi(g)$ by a more straightforward calculation.

## 3. Inverse Fourier Transform and Plancherel's Theorem for G

### 3.1. Statement of the Problem

In the preceding section we have defined the Fourier transform of a rapidly decreasing function on $G$. In the present one, we shall obtain the inverse Fourier transform formula, which expresses the function in terms of its Fourier transform. Further, we shall find the class of operator functions $F(\chi)$ that are the Fourier transforms of square integrable functions (i.e., functions whose square modulus is integrable) and shall obtain the analog of Plancherel's theorem. We shall use as a model the analogous formulas for the Fourier transforms of functions defined on the real line (or in $n$-dimensional Euclidean space). As is well known, if $f(x)$ is a square integrable function on the line, it is given in terms of its Fourier transform by

$$f(x) = (2\pi)^{-1} \int_{-\infty}^{\infty} F(\lambda) e^{-i\lambda x}\, d\lambda.$$

In addition, Plancherel's theorem for this case is

$$\int_{-\infty}^{\infty} | f(x) |^2\, dx = (2\pi)^{-1} \int_{-\infty}^{\infty} | F(\lambda) |^2\, d\lambda.$$

We shall see that there is a significant difference between these formulas and their analogs for $G$. Specifically, on the line we must know $F(\lambda)$ for all *real* values of $\lambda$. These are precisely those values of $\lambda$ for which multiplication by $e^{i\lambda x}$ is a unitary operation on the line. An analogous situation obtains in $n$-dimensional Euclidean space. One might have supposed that the inverse transform formula and Plancherel's theorem for $G$ would require knowing $F(\chi)$ at those values $\chi$ for which $T_\chi(g)$ is a unitary operator. It is found, however, that if $f(g)$ is a square integrable function on $G$, its expansion requires not all unitary representations of $G$, but only in those of the principal series (i.e., those for which $n_2 = -\bar{n}_1$). Recall that in addition to these, the representations of the supplementary series, for which

$$n_1 = n_2 = \rho, \qquad -1 < \rho < 1, \quad \rho \neq 0,$$

are unitary.

**Remark.** The case in which the square integrable functions are expanded in all the unitary representations is in a certain sense degenerate. It occurs for compact groups and Euclidean spaces (as well as for all spaces on which, for instance, a nilpotent Lie group operates). For semisimple Lie groups, however, there always exist many unitary representations that do not enter into the expansion of a square integrable function. #

Let us proceed therefore to formulate our problem. Let $F(\chi)$ be the Fourier transform of a square integrable function $f(g) = f(\alpha, \beta, \delta)$ on $G$. Then $F(\chi)$ is defined on the "real axis," that is, for those $\chi = (n_1, n_2)$ which correspond to unitary representations of the principal series. We have seen in Chapter III that the points of the "real axis" are characterized by the fact that $n_2 = -\bar{n}_1$, or that

$$n_1 = \tfrac{1}{2}(n + i\rho), \qquad n_2 = \tfrac{1}{2}(-n + i\rho),$$

where $\rho$ is a real number and $n$ is an integer.

Our goal is to express $f(g)$ in terms of $F(\chi)$, for $\chi$ on the "real axis."

We may state this problem also in terms of the kernel of $F(\chi)$. Recall that this kernel is given by

$$K(z_1, z_2; \chi) = \left(\frac{i}{2}\right)^2 \int f(\lambda^{-1} + \beta z_2, \beta, \lambda - \beta z_1)\lambda^{n_1-1}\bar{\lambda}^{n_2-1}\, d\lambda\, d\bar{\lambda}\, d\beta\, d\bar{\beta}. \quad (1)$$

Thus our problem is to invert Eq. (1), or to express $f(g)$ in terms of $K(z_1, z_2; \chi)$, with $\chi$ on the real axis.

We shall solve this problem in two steps. First we shall consider the function $\varphi(z_1, z_2; \lambda)$ defined by

$$\varphi(z_1, z_2; \lambda) = \frac{i}{2} \int f(\lambda^{-1} + \beta z_2, \beta, \lambda - \beta z_1)\, d\beta\, d\bar{\beta}, \tag{2}$$

in terms of which $K(z_1, z_2; \chi)$ is given by

$$K(z_1, z_2; \chi) = \frac{i}{2} \int \varphi(z_1, z_2; \lambda)\lambda^{n_1-1}\bar{\lambda}^{n_2-1}\, d\lambda\, d\bar{\lambda}. \tag{3}$$

(Elsewhere we have called the integral on the right-hand side the Mellin transform of $\varphi$.) Then Eq. (1) may be inverted in two steps by the following procedure.

1. Invert Eq. (3), that is, express $\varphi(z_1, z_2; \lambda)$ in terms of $K(z_1, z_2; \chi)$.
2. Invert Eq. (2), that is, express $f(g)$ in terms of $\varphi(z_1, z_2; \lambda)$.

The first of these steps is relatively simple. Essentially all it involves is to write a function in terms of its ordinary Mellin transform. We shall do this in Section 3.2.

It is the second step which presents the real problem. We shall solve this one in Section 3.3, discussing here only its rather simple geometric meaning. Recall that $f(g)$ may be considered a function on the quadratic surface $\alpha\delta - \beta\gamma = 1$ in a linear space of four complex dimensions. Then $\varphi(z_1, z_2; \lambda)$ is the integral of $f(g)$ along a line generator of this surface, as we have seen in Section 2.3. The problem is therefore to express a function over the $\alpha\delta - \beta\gamma = 1$ surface in terms of its integrals along all the line generators of this surface. Stated in this way the problem is one of integral geometry.

We have solved this problem in Section 2 of Chapter II. The solution we obtained there may be written

$$f(g) = -(8\pi^2)^{-1}\left(\frac{i}{2}\right) \int_\Gamma L\bar{L}\Phi(u, v; u\alpha + v\gamma, u\beta + v\delta)$$
$$\times (u\, dv - v\, du)(\bar{u}\, d\bar{v} - \bar{v}\, d\bar{u}), \tag{4}$$

where $\Phi(u, v; u', v')$ is the integral along a linear generator expressed in terms of the homogeneous coordinates $u, v; u', v'$. In this equation the integral is along any contour $\Gamma$ on the $(u, v)$ plane which intersects every complex line passing through the origin just once, and $L$ and $\bar{L}$ are the operators

$$L = -u\frac{\partial}{\partial u} - v\frac{\partial}{\partial v} + u'\frac{\partial}{\partial u'} + v'\frac{\partial}{\partial v'}, \tag{5}$$

$$\bar{L} = -\bar{u}\frac{\partial}{\partial \bar{u}} - \bar{v}\frac{\partial}{\partial \bar{v}} + \bar{u}'\frac{\partial}{\partial \bar{u}'} + \bar{v}'\frac{\partial}{\partial \bar{v}'}. \tag{5'}$$

Geometrically, Eq. (4) means that in order to obtain $f(g)$ we must first apply the infinitesimal parallel translation operators $L$ and $\bar{L}$ to $\Phi(u, v; u', v')$ and then average $L\bar{L}\Phi$ over the set of all the linear generators that pass through $g$.

Now let us rewrite Eq. (4) in terms of the inhomogeneous coordinates $z_1, z_2, \lambda$ by transforming from $\Phi(u, v; u', v')$ to $\varphi(z_1, z_2; \lambda)$. According to Eq. (9) of Section 2.3, we have

$$\Phi(u, v; u', v') = v^{-1}\bar{v}^{-1}v'^{-1}\bar{v}'^{-1}\varphi(u/v, u'/v'; v'/v). \tag{6}$$

We choose the contour $\Gamma$ in Eq. (4) to be the line $v = 1$, and then after some elementary operations we arrive at

$$f(g) = -\frac{1}{2\pi^2}\frac{i}{2}\int \varphi_{\lambda\bar{\lambda}}\left(z, \frac{\alpha z + \gamma}{\beta z + \delta}; \beta z + \delta\right) dz\, d\bar{z}. \tag{7}$$

In this chapter, however, we shall not rely on the results of Chapter II, and in Section 3.3 we shall go through an independent derivation of Eq. (7) based on the Fourier transform. This new derivation exhibits explicitly the one-to-one nature of the mapping of $f$ to $\varphi$.

In order to avoid difficulties involving convergence of integrals and parameterization of $G$, we shall assume that $f(g) = f(\alpha, \beta, \delta)$ has bounded support on the group and that it vanishes for sufficiently small values of $|\beta|$. This makes it possible to treat $f(g)$ as a function of bounded support in the parameters $\alpha, \beta, \delta$. We shall further assume that $f$ is infinitely differentiable in these parameters. Later we shall drop these requirements.

### 3.2. Expression for $\varphi(z_1, z_2; \lambda)$ in Terms of $K(z_1, z_2; \chi)$

In this section we shall find the inversion formula for the Mellin transform

$$K(z_1, z_2; \chi) = \frac{i}{2}\int \varphi(z_1, z_2; \lambda)\lambda^{n_1-1}\bar{\lambda}^{n_2-1}\, d\lambda\, d\bar{\lambda}. \tag{1}$$

The following result will be obtained. *The inversion formula is*

$$\varphi(z_1, z_2; \lambda) = (4\pi^2)^{-1}\int_{X_0} K(z_1, z_2; \chi)\lambda^{-n_1}\bar{\lambda}^{-n_2}\, d\chi, \tag{2}$$

*where $X_0$ is the "real axis," that is, the set of points $\chi = (\frac{1}{2}[n + i\rho], \frac{1}{2}[-n + i\rho])$, where $\rho$ is real and $n$ is an integer.* Integration with respect to $\chi$ is understood as integration over $\rho$ and summation over $n$.

**Proof.** The proof is based on the observation that Eq. (1) reduces to the ordinary Fourier transform. Indeed, on writing $\lambda = e^{\tau+i\alpha}$, we see that

$$K(z_1, z_2; \chi)$$

$$= \int_0^{2\pi}\int_{-\infty}^{\infty} \varphi(z_1, z_2; e^{\tau+i\alpha}) \exp[\tau(n_1 + n_2)] \exp[i(n_1 - n_2)\alpha]\, d\tau\, d\alpha. \quad (3)$$

If $\text{Re}(n_1 + n_2) = 0$, that is, if $\chi$ lies on the "real axis," this becomes

$$K(z_1, z_2; \chi) = \int_0^{2\pi}\int_{-\infty}^{\infty} \varphi(z_1, z_2; e^{\tau+i\alpha}) e^{i\rho\tau} e^{in\alpha}\, d\tau\, d\alpha. \quad (4)$$

Thus $K(z_1, z_2; \chi)$ is the Fourier transform of $\varphi(z_1, z_2; e^{\tau+i\alpha})$ with respect to $\tau$ and $\alpha$ (more accurately, it is a Fourier *transform* with respect to $\tau$ and a Fourier *coefficient* with respect to $\alpha$). We may now use the ordinary inverse Fourier transform formula to obtain[14]

$$\varphi(z_1, z_2; e^{\tau+i\alpha}) = \frac{1}{4\pi^2} \sum_{n=-\infty}^{\infty} \int_{-\infty}^{\infty} K\left(z_1, z_2; \frac{n+i\rho}{2}, \frac{-n+i\rho}{2}\right) e^{-i\rho\tau} e^{-in\alpha}\, d\rho. \quad (5)$$

This equation is more conveniently written

$$\varphi(z_1, z_2; \lambda) = (4\pi^2)^{-1} \int_{X_0} K(z_1, z_2; \chi) \lambda^{-n_1} \bar{\lambda}^{-n_2}\, d\chi,$$

where $X_0$ represents the "real axis" of $\chi$ and the integral is understood as integration over $\rho$ and summation over $n$.

**Remark.** It can be shown that the Fourier transform of an infinitely differentiable function $f(g)$ with bounded support is an analytic function of $\chi$. In general, therefore, one may replace integration along the "contour" $X_0$ (the analog of the real axis) by integration along some other contour. This means that $\varphi(z_1, z_2; \lambda)$ can be expressed in terms of the kernel $K(z_1, z_2; \chi)$ for values of $\chi$ corresponding to nonunitary representations of $G$. However, we shall not go into this interesting question here. #

[14] We may apply the inverse Fourier transform formula because we have assumed that $f(g)$ has bounded support. Specifically, under the assumptions we have made, $\varphi(z_1, z_2; \lambda)$ as defined by Eq. (2) of Section 3.1 has bounded support in $\lambda$ for fixed $z_1$ and $z_2$ and is infinitely differentiable in $\lambda$ and $\bar{\lambda}$.

### 3.3. Expression for $f(g)$ in Terms of $\varphi(z_1, z_2; \lambda)$

We now wish to invert the integral transform

$$\varphi(z_1, z_2; \lambda) = \frac{i}{2} \int f(\lambda^{-1} + \beta z_2, \beta, \lambda - \beta z_1)\, d\beta\, d\bar{\beta}. \tag{1}$$

Let us first introduce the new parameters

$$p_1 = -\delta/\beta, \qquad p_2 = \alpha/\beta, \qquad p_3 = 1/\beta$$

in $G$. We shall denote by $f_1(p_1, p_2, p_3)$ our original function $f(g) = f(\alpha, \beta, \delta)$ expressed in terms of the $p_i$. Explicitly,[15]

$$f(\alpha, \beta, \delta) = f_1(-\delta/\beta, \alpha/\beta, 1/\beta). \tag{2}$$

In terms of these new parameters Eq. (1) becomes

$$\varphi(z_1, z_2; \lambda) = \frac{i}{2} \int f_1(z_1 - \lambda p_3, z_2 + \lambda^{-1} p_3, p_3) | p_3 |^{-4}\, dp_3\, d\bar{p}_3 . \tag{3}$$

Our problem is now to invert this integral transform.

Let us now obtain the Fourier transform $\Phi(w_1, w_2; \lambda)$ with respect to $z_1$ and $z_2$ of $\varphi(z_1, z_2; \lambda)$:[16]

$$\Phi(w_1, w_2; \lambda) = \left(\frac{i}{2}\right)^2 \int \varphi(z_1, z_2; \lambda) \exp[i \operatorname{Re}(z_1 \bar{w}_1 + z_2 \bar{w}_2)]\, dz_1\, d\bar{z}_1\, dz_2\, d\bar{z}_2 . \tag{4}$$

In terms of the $p_i$ this becomes

$$\Phi(w_1, w_2; \lambda) = \left(\frac{i}{2}\right)^3 \int f_1(z_1 - \lambda p_3, z_2 + \lambda^{-1} p_3, p_3) \exp[i \operatorname{Re}(z_1 \bar{w}_1 + z_2 \bar{w}_2)] \times | p_3 |^{-4}\, dz_1\, d\bar{z}_1\, dz_2\, d\bar{z}_2\, dp_3\, d\bar{p}_3 . \tag{5}$$

Changing variables in the integrand, we write

$$p_1 = z_1 - \lambda p_3, \qquad p_2 = z_2 + \lambda^{-1} p_3 .$$

[15] Note that by the assumptions made concerning $f(g)$ at the end of Section 3.1, our new function $f_1$ is an infinitely differentiable function of the $p_i$ with bounded support and vanishes for sufficiently small values of $| p_3 |$.

[16] Note that under the assumptions we have made concerning $f(g)$, the integral in (4) converges, so that the usual inverse Fourier transform formula may be used. Indeed, $\varphi(z_1, z_2; \lambda)$ is an infinitely differentiable function of bounded support in $z_1$ and $z_2$ for fixed $\lambda$.

This yields

$$\Phi(w_1, w_2; \lambda) = \left(\frac{i}{2}\right)^3 \int f_1(p_1, p_2, p_3) | p_3 |^{-4} \times \exp\{i \operatorname{Re}[p_1\bar{w}_1 + p_2\bar{w}_2 + p_3(\lambda\bar{w}_1 - \lambda^{-1}\bar{w}_2)]\} \times dp_1\, d\bar{p}_1\, dp_2\, d\bar{p}_2\, dp_3\, d\bar{p}_3 . \tag{6}$$

We see from this that $\Phi(w_1, w_2; \lambda)$ is simply related to the Fourier transform $F(w_1, w_2, w_3)$ of $f_1(p_1, p_2, p_3)| p_3 |^{-4}$, whose formula is

$$F(w_1, w_2, w_3) = \left(\frac{i}{2}\right)^3 \int f_1(p_1, p_2, p_3)| p_3 |^{-4} \exp[i \operatorname{Re}(p_1\bar{w}_1 + p_2\bar{w}_2 + p_3\bar{w}_3)] \times dp_1\, d\bar{p}_1\, dp_2\, d\bar{p}_2\, dp_3\, d\bar{p}_3 . \tag{7}$$

Specifically,

$$\Phi(w_1, w_2; \lambda) = F(w_1, w_2, \bar{\lambda} w_1 - \bar{\lambda}^{-1} w_2). \tag{8}$$

This equation is the basic one relating $f$ and $\varphi$. It is obvious that it can be used to express $F(w_1, w_2, w_3)$ and therefore also $f_1(p_1, p_2, p_3)$ in terms of $\Phi(w_1, w_2; \lambda)$. We may use the inverse Fourier transform to write

$$f_1(p_1, p_2, p_3)| p_3 |^{-4} = (2\pi)^{-6}\left(\frac{i}{2}\right)^3 \int F(w_1, w_2, w_3) \exp[-i \operatorname{Re}(p_1\bar{w}_1 + p_2\bar{w}_2 + p_3\bar{w}_3)] \times dw_1\, d\bar{w}_1\, dw_2\, d\bar{w}_2\, dw_3\, d\bar{w}_3 . \tag{9}$$

Now writing $w_3 = \bar{\lambda} w_1 - \bar{\lambda}^{-1} w_2$ in order to transform in the integrand from $F$ to $\Phi$, we arrive at[17]

$$f_1(p_1, p_2, p_3)| p_3 |^{-4} = \tfrac{1}{2}(2\pi)^{-6}\left(\frac{i}{2}\right)^3 \int \Phi(w_1, w_2; \lambda)| \bar{w}_1 + \lambda^{-2}\bar{w}_2 |^2 \times \exp[-i \operatorname{Re}\{p_1\bar{w}_1 + p_2\bar{w}_2 + p_3(\lambda\bar{w}_1 - \lambda^{-1}\bar{w}_2)\}] \times dw_1\, d\bar{w}_1\, dw_2\, d\bar{w}_2\, d\lambda\, d\bar{\lambda}. \tag{10}$$

[17] The extra factor of $\frac{1}{2}$ arises because when $\lambda$ runs once over the complex plane, $w_3$ runs over it twice.

This gives $f_1$ in terms of the Fourier transform of $\varphi(z_1, z_2; \lambda)$ with respect to $z_1$ and $z_2$. The next step is to insert Eq. (4) into Eq. (10), which then yields

$$f_1(p_1, p_2, p_3)|p_3|^{-4} = \tfrac{1}{2}(2\pi)^{-6}\int \varphi(z_1, z_2; \lambda) \times \exp[i\,\mathrm{Re}\{\bar{w}_1(z_1 - p_1 - p_3\lambda) + \bar{w}_2(z_2 - p_2 + p_3\lambda^{-1})\}] \times |\bar{w}_1 + \lambda^{-2}\bar{w}_2|^2\, dv, \tag{11}$$

where $dv = (i/2)^5\, dw_1\, d\bar{w}_1\, dw_2\, d\bar{w}_2\, d\lambda\, d\bar{\lambda}\, dz_1\, d\bar{z}_1\, dz_2\, d\bar{z}_2$.

This formula can be simplified by integrating over $w_1$ and $w_2$. We make use of the fact that

$$(2\pi)^{-4}\left(\frac{i}{2}\right)^2 \int \exp\{i\,\mathrm{Re}\,[\bar{w}_1(z_1 - p_1 - p_3\lambda) + \bar{w}_2(z_2 - p_2 + p_3\lambda^{-1})]\} \times dw_1\, d\bar{w}_1\, dw_2\, d\bar{w}_2 = \delta(z_1 - p_1 - p_3\lambda, z_2 - p_2 + p_3\lambda^{-1}).$$

Differentiating this equation with respect to $\lambda$ and $\bar{\lambda}$, we obtain

$$\tfrac{1}{4}(2\pi)^{-4}|p_3|^2\left(\frac{i}{2}\right)^2 \int \exp\{i\,\mathrm{Re}\,[\bar{w}_1(z_1 - p_1 - p_3\lambda) + \bar{w}_2(z_2 - p_2 + p_3\lambda^{-1})]\} \times |\bar{w}_1 + \lambda^{-2}\bar{w}_2|^2\, dw_1\, d\bar{w}_1\, dw_2\, d\bar{w}_2 = \frac{\partial^2}{\partial\lambda\,\partial\bar{\lambda}}\delta(z_1 - p_1 - p_3\lambda, z_2 - p_2 + p_3\lambda^{-1}).$$

Thus after integrating over $w_1$ and $w_2$ we reduce Eq. (11) to the form

$$\begin{aligned} f_1(p_1, p_2, p_3) &= -(2\pi^2)^{-1}|p_3|^2\left(\frac{i}{2}\right)^3 \int \varphi(z_1, z_2; \lambda) \\ &\quad \times \frac{\partial^2}{\partial\lambda\,\partial\bar{\lambda}}\delta(z_1 - p_1 - p_3\lambda, z_2 - p_2 + p_3\lambda^{-1})\, dz_1\, d\bar{z}_1\, dz_2\, d\bar{z}_2\, d\lambda\, d\bar{\lambda} \\ &= -(2\pi^2)^{-1}|p_3|^2\left(\frac{i}{2}\right) \int \varphi_{\lambda\bar{\lambda}}(p_1 + p_3\lambda, p_2 - p_3\lambda^{-1}; \lambda)\, d\lambda\, d\bar{\lambda}, \end{aligned} \tag{12}$$

where

$$\varphi_{\lambda\bar{\lambda}}(p_1 + p_3\lambda, p_2 - p_3\lambda^{-1}; \lambda) = \left.\frac{\partial^2\varphi(z_1, z_2; \lambda)}{\partial\lambda\,\partial\bar{\lambda}}\right|_{\substack{z_1 = p_1 + p_3\lambda \\ z_2 = p_2 - p_3\lambda^{-1}}}. \tag{13}$$

This is the desired inversion formula.

All we need now do is state this in terms of our original parameters $\alpha$, $\beta$, $\delta$. This is simply

$$f(\alpha, \beta, \delta) = -\frac{1}{2\pi^2}\frac{1}{|\beta|^2}\frac{i}{2}\int \varphi_{\lambda\bar{\lambda}}\left(\frac{\lambda - \delta}{\beta}, \frac{\alpha - \lambda^{-1}}{\beta}; \lambda\right) d\lambda\, d\bar{\lambda}.$$

Now writing $z = (\lambda - \delta)\beta^{-1}$, we finally arrive at

$$f(\alpha, \beta, \delta) = -\frac{1}{2\pi^2}\frac{i}{2}\int \varphi_{\lambda\bar{\lambda}}\left(z, \frac{\alpha z + \gamma}{\beta z + \delta}; \beta z + \delta\right) dz\, d\bar{z}. \tag{14}$$

Thus *let* $f(g) = f(\alpha, \beta, \delta)$ *be an infinitely differentiable function of bounded support on G, and let* $\varphi(z_1, z_2; \lambda)$ *be defined by Eq.* (1). *Then f is given in terms of* $\varphi$ *by Eq.* (14).

Recall that we have required that $f(\alpha, \beta, \delta)$ vanish for sufficiently small values of $|\beta|$. Actually this requirement is not necessary (see Section 3.4). It will become clear in Section 3.5 that Eq. (14) remains valid for any square integrable function $f(g)$ so long as the integrals of Eqs. (1) and (14) converge.

### 3.4. Expression for $f(g)$ in Terms of Its Fourier Transform $F(\chi)$

We may now combine the results of Sections 3.2 and 3.3 to express $f(g) = f(\alpha, \beta, \delta)$ in terms of the kernel of its Fourier transform. We shall use Eq. (14) of Section 3.3, namely

$$f(\alpha, \beta, \delta) = -\frac{1}{2\pi^2}\frac{i}{2}\int \varphi_{\lambda\bar{\lambda}}\left(z, \frac{\alpha z + \gamma}{\beta z + \delta}; \beta z + \delta\right) dz\, d\bar{z}, \tag{1}$$

and Eq. (2) of Section 3.2, namely

$$\varphi(z_1, z_2; \lambda) = (4\pi^2)^{-1}\int_{X_0} K(z_1, z_2; \chi)\lambda^{-n_1}\bar{\lambda}^{-n_2}\, d\chi \tag{2}$$

(recall that $X_0$ is the "real axis" of $\chi$),[18] to derive the desired formula.

Let us differentiate Eq. (2) with respect to $\lambda$ and $\bar{\lambda}$ and insert the result into (1). Then we arrive at the following result. *Let* $f(g) = f(\alpha, \beta, \delta)$ *be*

[18] Recall also that the integral over $\chi$ is to be understood as integration over $\rho$ and summation over $n$.

*an infinitely differentiable function with bounded support on G. Then in terms of the kernel* $K(z_1, z_2; \chi)$ *of its Fourier transform it may be written*

$$f(\alpha, \beta, \delta) = -\frac{1}{8\pi^4}\frac{i}{2}\int_{X_0}\int n_1 n_2 K\left(z, \frac{\alpha z + \gamma}{\beta z + \delta}; \chi\right) \times (\beta z + \delta)^{-n_1-1}(\bar{\beta}\bar{z} + \bar{\delta})^{-n_2-1}\, dz\, d\bar{z}\, d\chi. \tag{3}$$

*Here* $\chi = (n_1, n_2) = (\frac{1}{2}[n + i\rho], \frac{1}{2}[-n + i\rho])$, *where n is an integer and* $\rho$ *is real; the integral over* $\chi$ *is understood in the sense of integration over* $\rho$ *and summation over n.* In a more explicit notation Eq. (3) may be written

$$f(\alpha, \beta, \delta) = \frac{1}{32\pi^4}\sum_{n=-\infty}^{\infty}\frac{i}{2}\int_{-\infty}^{\infty}(n^2 + \rho^2)\, d\rho \times \int K\left(z, \frac{\alpha z + \gamma}{\beta z + \delta}; \frac{n + i\rho}{2}, \frac{-n + i\rho}{2}\right) \times (\beta z + \delta)^{-\frac{1}{2}(n+i\rho)-1}(\bar{\beta}\bar{z} + \bar{\delta})^{\frac{1}{2}(n-i\rho)-1}\, dz\, d\bar{z}. \tag{4}$$

We remark that we could have dropped the assumption that $f(\alpha, \beta, \delta)$ has bounded support. This will be seen in Section 3.5.

As a final step, let us now express $f(g)$ directly in terms of its Fourier transform $F(\chi)$. Note first that

$$K\left(z_1, \frac{\alpha z_2 + \gamma}{\beta z_2 + \delta}; \chi\right)(\beta z_2 + \delta)^{-n_1-1}(\bar{\beta}\bar{z}_2 + \bar{\delta})^{-n_2-1}$$

is the kernel of $F(\chi)T_\chi^{-1}(g)$ (see Section 2.4). Hence

$$\frac{i}{2}\int K\left(z_1, \frac{\alpha z + \gamma}{\beta z + \delta}; \chi\right)(\beta z + \delta)^{-n_1-1}(\bar{\beta}\bar{z} + \bar{\delta})^{-n_2-1}\, dz\, d\bar{z} = \mathrm{Tr}[F(\chi)T_\chi^{-1}(g)]. \tag{5}$$

Inserting this into (3), we obtain[19]

$$f(g) = -(8\pi^4)^{-1}\int_{X_0}\mathrm{Tr}[F(\chi)T_\chi^{-1}(g)]c(\chi)\, d\chi, \qquad c(\chi) = n_1 n_2. \tag{6}$$

[19] When one is dealing with compact groups, the $D_\chi$ spaces are replaced by finite-dimensional spaces. In that case the coefficient $c(\chi)$ is the dimension of this carrier space of the representation. This means that in the present case $c(\chi) = n_1 n_2$ is the analog of the dimension of a finite-dimensional representation. [Note, incidentally, that finite-dimensional representations of $G$ are realized on the space of polynomials of $z$ and $\bar{z}$ of degree $(n_1 - 1, n_2 - 1)$ and that for these $c(\chi) = n_1 n_2$ is indeed the dimension of the carrier space.]

Remark. The inversion formula (6) has been proved for infinitely differentiable functions with bounded support that vanish for sufficiently small values of $|\beta|$. In fact, however, this last requirement is unnecessary, as is immediately evident if we can prove the following assertion. If Eq. (6) is valid for some $f(g)$ it is also valid for $f_1(g) = f(gg_0)$, where $g_0$ is any element of $G$. That this is true is seen when we recall that it was shown in Section 2.1 that if $F(\chi)$ is the Fourier transform of $f(g)$, the Fourier transform of $f_1(g)$ is $F_1(\chi) = F(\chi)T_\chi^{-1}(g_0)$. Now since Eq. (6) holds for $f(g)$, we have [using the fact that $T_\chi^{-1}(g) = T_\chi(g^{-1})$]

$$\begin{aligned} f_1(g) = f(gg_0) &= (8\pi^4)^{-1}\int_{X_0} \mathrm{Tr}[F(\chi)T_\chi(g_0^{-1}g^{-1})]c(\chi)\, d\chi \\ &= (8\pi^4)^{-1}\int_{X_0} \mathrm{Tr}[F(\chi)T_\chi(g_0^{-1})T_\chi(g^{-1})]c(\chi)\, d\chi \\ &= -(8\pi^4)^{-1}\int_{X_0} \mathrm{Tr}[F_1(\chi)T_\chi(g^{-1})]c(\chi)\, d\chi, \end{aligned}$$

which proves the assertion. #

We shall show in Section 3.5 that Eq. (6) will hold for any square integrable $f(g)$ if Eq. (3) is understood in the sense of convergence in the mean.

## 3.5. Analog of Plancherel's Theorem for $G$

We have constructed the Fourier transform for functions on $G$ and have shown that infinitely differentiable functions with bounded support can be expressed in terms of the values of their Fourier transforms on the "real axis" $X_0$ of the set $X$ of representations, which is given by the condition $n_2 = -\bar{n}_1$. We shall now extend the relation between functions $f(g)$ and operators $F(\chi)$ to all square integrable functions. Later, in Section 3.6, we shall define the class of operators $F(\chi)$ corresponding to such functions.

Let us start by establishing that the mapping $f(g) \to K(z_1, z_2; \chi)$ can be made isometric by suitably choosing the scalar product in the space of the $f(g)$ functions and the space of the kernels. Specifically, we choose the scalar product

$$(f_1, f_2) = \int f_1(g)\bar{f}_2(g)\, dg \tag{1}$$

for the functions on $G$, and

$$(K_1, K_2) = -(8\pi^4)^{-1}\left(\frac{i}{2}\right)^2 \int_{X_0} K_1(z_1, z_2; \chi)\bar{K}_2(z_1, z_2; \chi)c(\chi)\, dz_1\, d\bar{z}_1\, dz_2\, d\bar{z}_2\, d\chi \tag{2}$$

for the kernels, where as before $c(\chi) = n_1 n_2 = -| n_1 |^2$. (As usual, the integral with respect to $\chi$ is understood as integration over $\rho$ and summation over $n$.)

*Let $f_1(g)$ and $f_2(g)$ be infinitely differentiable functions with bounded support, and let $K_1(z_1, z_2; \chi)$ and $K_2(z_1, z_2; \chi)$ be the kernels of their Fourier transforms. Then*

$$(f_1, f_2) = (K_1, K_2),$$

*or more explicitly,*

$$\int f_1(g)\bar{f}_2(g)\, dg = -(8\pi^4)^{-1}\left(\frac{i}{2}\right)^2 \int_{X_0}\int K_1(z_1, z_2; \chi)$$

$$\times\, \bar{K}_2(z_1, z_2; \chi)c(\chi)\, dz_1\, d\bar{z}_1\, dz_2\, d\bar{z}_2\, d\chi, \tag{3}$$

*which is then the analog of Plancherel's theorem.*

**Proof.** Consider the convolution $f(g) = f_1 * f_2^*(g)$ of $f_1(g)$ and $f_2^*(g) = \bar{f}_2(g^{-1})$. This convolution is also an infinitely differentiable function with bounded support. Therefore the inverse Fourier transform formula is valid for it, and we may write, in particular,

$$f(e) = -(8\pi^4)^{-1}\left(\frac{i}{2}\right) \int_{X_0}\int K(z, z; \chi)c(\chi)\, d\chi\, dz\, d\bar{z}, \tag{4}$$

where $K(z_1, z_2; \chi)$ is the kernel of the Fourier transform of $f(g)$. As was shown in Section 2.4, this kernel may be written

$$K(z_1, z_2; \chi) = \frac{i}{2}\int K_1(z_1, z_3; \chi)\bar{K}_2(z_2, z_3; \chi)\, dz_3\, d\bar{z}_3 .$$

Therefore

$$\frac{i}{2}\int K(z_1, z_1; \chi)\, dz_1\, d\bar{z}_1$$

$$= \left(\frac{i}{2}\right)^2 \int K_1(z_1, z_2; \chi)\bar{K}_2(z_1, z_2; \chi)\, dz_1\, d\bar{z}_1\, dz_2\, d\bar{z}_2 .$$

On the other hand, from the definition of the convolution

$$f(g) = f_1 * f_2^*(g) = \int f_1(g_1) f_2^*(g_1^{-1} g)\, dg_1$$

$$= \int f_1(g_1) \bar{f}_2(g^{-1} g_1)\, dg_1 ,$$

we obtain

$$f(e) = \int f_1(g_1) \bar{f}_2(g_1)\, dg_1 = (f_1 , f_2).$$

Thus it follows from Eq. (4) that

$$(f_1 , f_2) = -(8\pi^4)^{-1} \left(\frac{i}{2}\right)^2 \int_{X_0} \int K_1(z_1 , z_2 ; \chi) \bar{K}_2(z_1 , z_2 ; \chi)$$

$$\times\, c(\chi)\, dz_1\, dz_2\, d\bar{z}_1\, d\bar{z}_2\, d\chi = (K_1 , K_2). \tag{5}$$

This completes the proof of Plancherel's theorem for infinitely differentiable functions of bounded support.

Consider now the space $\mathfrak{H}_f$ of all square integrable functions (i.e., those whose square modulus is integrable) on $G$. The infinitely differentiable functions with bounded support form an everywhere dense set in this space. This implies that the mapping $f(g) \to K(z_1 , z_2 ; \chi)$, so far defined only for infinitely differentiable functions with bounded support, can be extended to an isometric mapping of $\mathfrak{H}_f$ into the space $\mathfrak{H}_K$ of all kernels $K$ such that

$$(K, K) < \infty.$$

This establishes the following analog of Plancherel's theorem for square integrable functions on $G$:

$$\int |f(g)|^2\, dg = -(8\pi^4)^{-1} \left(\frac{i}{2}\right)^2 \int_{X_0} \int |K(z_1 , z_2; \chi)|^2 c(\chi)\, dz_1\, d\bar{z}_1\, dz_2\, d\bar{z}_2\, d\chi. \tag{6}$$

Remark. Equation (6) may also be rewritten for convenience by putting the right-hand side in terms of $F(\chi)$. Now $F(\chi)$ is a Hilbert-Schmidt operator, since its kernel satisfies

$$\left(\frac{i}{2}\right)^2 \int |K(z_1 , z_2; \chi)|^2\, dz_1\, d\bar{z}_1\, dz_2\, d\bar{z}_2 < \infty. \tag{7}$$

Thus the operator $F(\chi)F^*(\chi)$ possesses a trace, which is given by (7). It follows then that Eq. (6) can be rewritten in the form

$$\int |f(g)|^2\, dg = -(8\pi^4)^{-1} \int_{X_0} c(\chi)\, \mathrm{Tr}[F(\chi)F^*(\chi)]\, d\chi. \quad \# \tag{8}$$

### 3.6. Symmetry Properties of $F(\chi)$

We wish now to establish which functions of the form $K(z_1, z_2; \chi)$ are the kernels of Fourier transforms of square integrable functions on $G$. Recall for this purpose how we developed the transition from $f(g) = f(\alpha, \beta, \delta)$ to $K(z_1, z_2; \chi)$. This was done by the sequence of mappings

$$\begin{aligned} &f(\alpha, \beta, \delta) \to f_1(p_1, p_2, p_3) \to F(w_1, w_2, w_3) \\ &\qquad \to \Phi(w_1, w_2; \lambda) \to \varphi(z_1, z_2; \lambda) \to K(z_1, z_2; \chi). \end{aligned} \tag{1}$$

Explicitly, these mappings are given by the following formulas:

$$f_1(-\delta/\beta, \alpha/\beta, 1/\beta) = f(\alpha, \beta, \delta), \tag{2}$$

$$\begin{aligned} F(w_1, w_2, w_3) = \left(\frac{i}{2}\right)^3 \int & f_1(p_1, p_2, p_3) |p_3|^{-4} \\ & \times \exp\{i \operatorname{Re}(p_1\bar{w}_1 + p_2\bar{w}_2 + p_3\bar{w}_3)\}\, dp_1\, d\bar{p}_1\, dp_2\, d\bar{p}_2\, dp_3\, d\bar{p}_3, \end{aligned} \tag{3}$$

$$\Phi(w_1, w_2; \lambda) = F(w_1, w_2, \bar{\lambda} w_1 - \bar{\lambda}^{-1} w_2), \tag{4}$$

$$\begin{aligned} \varphi(z_1, z_2; \lambda) = (2\pi)^{-4} \left(\frac{i}{2}\right)^2 \int & \Phi(w_1, w_2; \lambda) \exp\{i \operatorname{Re}(z_1\bar{w}_1 + z_2\bar{w}_2)\} \\ & \times dw_1\, d\bar{w}_1\, dw_2\, d\bar{w}_2, \end{aligned} \tag{5}$$

$$K(z_1, z_2; \chi) = \frac{i}{2} \int \varphi(z_1, z_2; \lambda) \lambda^{n_1 - 1} \bar{\lambda}^{n_2 - 1}\, d\lambda\, d\bar{\lambda}. \tag{6}$$

Elementary calculation will show that these mappings are isometric with respect to the scalar products

$$\begin{aligned} (f_1, f_2) = \int f_1(g) \bar{f}_2(g)\, dg \equiv \left(\frac{i}{2}\right)^3 \int & f_1(\alpha, \beta, \delta) \bar{f}_2(\alpha, \beta, \delta) |\beta|^{-2} \\ & \times d\alpha\, d\bar{\alpha}\, d\beta\, d\bar{\beta}\, d\delta\, d\bar{\delta}, \end{aligned} \tag{7}$$

$$\begin{aligned} (f_{1,1}, f_{1,2}) = \left(\frac{i}{2}\right)^3 \int & \{f_{1,1}(p_1, p_2, p_3) |p_3|^{-4}\}\{\bar{f}_{1,2}(p_1, p_2, p_3) |p_3|^{-4}\} |p_3|^2 \\ & \times dp_1\, d\bar{p}_1\, dp_2\, d\bar{p}_2\, dp_3\, d\bar{p}_3, \end{aligned} \tag{8}$$

$$\begin{aligned} (F_1, F_2) = -4(2\pi)^{-6} \left(\frac{i}{2}\right)^3 \int & F_1(w_1, w_2, w_3) \frac{\partial^2 \bar{F}_2(w_1, w_2, w_3)}{\partial w_3\, \partial \bar{w}_3} \\ & \times dw_1\, d\bar{w}_1\, dw_2\, d\bar{w}_2\, dw_3\, d\bar{w}_3, \end{aligned} \tag{9}$$

$$(\Phi_1, \Phi_2) = -2(2\pi)^{-6}\left(\frac{i}{2}\right)^3 \int \Phi_1(w_1, w_2; \lambda) \frac{\partial^2 \bar{\Phi}_2(w_1, w_2; \lambda)}{\partial\lambda\, \partial\bar{\lambda}} \times dw_1\, d\bar{w}_1\, dw_2\, d\bar{w}_2\, d\lambda\, d\bar{\lambda}, \tag{10}$$

$$(\varphi_1, \varphi_2) = -(2\pi^2)^{-1}\left(\frac{i}{2}\right)^3 \int \varphi_1(z_1, z_2; \lambda) \frac{\partial^2 \bar{\varphi}_2(z_1, z_2; \lambda)}{\partial\lambda\, \partial\bar{\lambda}} \times dz_1\, d\bar{z}_1\, dz_2\, d\bar{z}_2\, d\lambda\, d\bar{\lambda}, \tag{11}$$

$$(K_1, K_2) = -(8\pi^4)^{-1}\left(\frac{i}{2}\right)^2 \int_{X_0} \int K_1(z_1, z_2; \chi)\bar{K}_2(z_1, z_2; \chi) \times c(\chi)\, dz_1\, d\bar{z}_1\, dz_2\, d\bar{z}_2\, d\chi. \tag{12}$$

In this way we obtain the chain of mappings

$$\mathfrak{H}_f \to \mathfrak{H}_{f_1} \to \mathfrak{H}_F \to \mathfrak{H}_\Phi \to \mathfrak{H}_\varphi \to \mathfrak{H}_K,$$

where the $\mathfrak{H}_f, \mathfrak{H}_F, \ldots$ are Hilbert spaces of functions in which the scalar products are given by Eqs. (7)–(12). Now of these mappings $\mathfrak{H}_{f_1} \to \mathfrak{H}_F$, $\mathfrak{H}_\Phi \to \mathfrak{H}_\varphi$, and $\mathfrak{H}_\varphi \to \mathfrak{H}_K$ can be reduced to the ordinary Fourier transform, and are therefore one-to-one mappings onto the entire Hilbert space. Obviously $\mathfrak{H}_f \to \mathfrak{H}_{f_1}$ as defined by Eq. (2) is also one-to-one.

Now consider $\mathfrak{H}_F \to \mathfrak{H}_\Phi$, the mapping defined by Eq. (4). This is not a mapping onto all of $\mathfrak{H}_\Phi$, for the functions that can be obtained from it are only those satisfying the *symmetry condition*

$$\Phi(w_1, w_2; \lambda) = \Phi(w_1, w_2; -\bar{w}_2/\lambda\bar{w}_1). \tag{13}$$

The converse is easily established: every $\Phi(w_1, w_2; \lambda)$ satisfying this equation can be obtained from some $F(w_1, w_2, w_3)$ in $\mathfrak{H}_F$.

Thus we have proven that *the kernels of the Fourier transforms of square integrable functions on the group are all functions $K(z_1, z_2; \chi)$ such that*

$$-(8\pi^4)^{-1}\left(\frac{i}{2}\right)^2 \int_{X_0} \int |K(z_1, z_2; \chi)|^2 c(\chi)\, dz_1\, d\bar{z}_1\, dz_2\, d\bar{z}_2\, d\chi < \infty$$

*and such that the function*

$$\Phi(w_1, w_2; \lambda) = (2\pi)^{-2}\left(\frac{i}{2}\right)^2 \int_{X_0} \int K(z_1, z_2; \chi) \exp\{-i \operatorname{Re}(z_1\bar{w}_1 + z_2\bar{w}_2)\}\lambda^{-n_1}\bar{\lambda}^{-n_2} \times dz_1\, d\bar{z}_1\, dz_2\, d\bar{z}_2\, d\chi \tag{14}$$

*satisfies Eq.* (13), *i.e., remains invariant under replacement of $\lambda$ by $-\bar{w}_2/(\lambda\bar{w}_1)$.*

This symmetry condition can be stated in terms of $K(z_1, z_2; \chi)$. Without going through the derivation, we merely give the result:

$$\frac{i}{2}\int (z_1 - z)^{-n_1-1}(\bar{z}_1 - \bar{z})^{-n_2-1}K(z, z_2; \chi)\,dz\,d\bar{z}$$

$$= (-1)^{n_1-n_2}\left(\frac{i}{2}\right)\int (z_2 - z)^{-n_1-1}(\bar{z}_2 - \bar{z})^{-n_2-1}K(z_1, z; -\chi)\,dz\,d\bar{z}$$

$$(n_2 = -\bar{n}_1).$$

In Section 5 we shall discuss the group-theoretical content of this equation. It is related to the fact, established in Section 5.2 of Chapter III, that the representations on $D_\chi$ and $D_{-\chi}$ are equivalent. The geometric meaning has already been discussed in Section 2.7 of Chapter II, where we gave a geometric derivation of Eq. (13).

### 3.7. Fourier Integral and the Decomposition of the Regular Representation of the Lorentz Group into Irreducible Representations

We wish now to establish the relation between the Fourier integral expansion of a function $f(g)$ on the Lorentz group and the expansion of the regular representation of this group in irreducible representations. The discussion in this section is not related to the material that follows.

We start with some definitions. The *regular* (more accurately, *right regular*) representation of any locally compact group $G$ is constructed on the Hilbert space $H$ of functions $f(g)$ on the group such that the integral

$$\|f\|^2 = \int |f(g)|^2\,dg$$

converges (where $dg$ is a right invariant measure on the group). With every element $g_0 \in G$ we associate the operator $R(g_0)$, called the right translation operator on $H$, which transforms all the $f(g)$ according to

$$R(g_0)f(g) = f(gg_0).$$

Obviously

$$R(g_1g_2) = R(g_1)R(g_2),$$

so that $R(g)$ is a representation of $G$. This is the so-called *regular representation*. Because $dg$ is invariant under right translations, the regular representation is unitary. One may then ask for the decomposition of this representation into irreducible representations.

In proceeding, we shall need the concept of the continuous direct sum (or *direct integral*) of Hilbert spaces.[20] Consider a set $X$ with some positive measure $\mu$. With every $x \in X$ let there be associated a separable Hilbert space $H(x)$ of dimension $n(x)$, where $n(x)$ is either a positive integer or infinite, and assume $n(x)$ to be measurable with respect to $\mu$. We define the direct integral first for the case in which all of the $H(x)$ have the same dimension $n$ (a positive integer or infinity). In this case all the $H(x)$ can be thought of as a single Hilbert space $H$ of dimension $n$. For this space we construct the space $\mathfrak{H}$ of *vector functions* (i.e., functions with values in $H$) $\xi = h(x)$ over $X$ such that

1. For every $h_0 \in H$ the ordinary function $(h_0, h(x))$ is measurable with respect to $\mu$ [where $(h_1, h_2)$ is the scalar product in $H$];
2. The ordinary function $\| h(x)\|^2$ is integrable with respect to $\mu$:

$$\int_X \| h(x)\|^2 \, d\mu < +\infty.$$

The linear operations in $\mathfrak{H}$ are defined as follows. Let $\xi = h(x)$ and $\eta = g(x)$ be two vector functions. Then $\xi + \eta = h(x) + g(x)$ and $a\xi = ah(x)$; the scalar product as defined by the equation

$$(\xi, \eta) = \int_X (h(x), g(x)) \, d\mu.$$

It can be shown that $\mathfrak{H}$ is then a Hilbert space, and we shall call it the direct integral of the Hilbert spaces $H(x)$ with respect to $\mu$, and shall write

$$\mathfrak{H} = \int \oplus \, H(x) \, d\mu.$$

Assume now that the $H(x)$ are of different dimensions. We then break up $H$ into measurable subsets $X_1, \ldots, X_n, \ldots$, on each of which $n(x) = n$ is a constant. Then we use the above definition to construct the Hilbert spaces

$$\mathfrak{H}_n = \int_{X_n} \oplus \, H(x) \, d\mu$$

and then take their direct sum

$$\mathfrak{H} = \sum_{n=1}^{\infty} \oplus \, \mathfrak{H}_n .$$

[20] For more details, see Volume 4, Chapter I, Section 4.4.

This is called the continuous direct sum or direct integral of the $H(x)$ with respect to the measure $\mu$, and is denoted by

$$\mathfrak{H} = \int_X \oplus H(x)\, d\mu.$$

A unitary representation $T(g)$ of a group $G$ can be *decomposed by a direct integral.* Let us define what we mean by this. Assume that $T(g)$ is a representation of $G$ on a *nuclear* space $\Phi$ (whose definition and properties will be found in Volume 4, p. 62). This representation is called unitary with respect to the scalar product $(\varphi, \psi)$ in $\Phi$, which we assume to be continuous with respect to both arguments, if for every $\varphi, \psi \in \Phi$ and for every $g \in G$ we have

$$(\varphi, \psi) = (T(g)\varphi, T(g)\psi).$$

We now complete $\Phi$ with respect to this scalar product, which yields a Hilbert space $\mathfrak{H}$ on which $T(g)$ is extended to a unitary representation of $G$ on $\mathfrak{H}$. Let us assume that $\mathfrak{H}$ can be written as, or, as we shall say, expanded in a direct integral of Hilbert spaces $H(x)$ with respect to some measure $\mu$. Then the following result is obtained (the proof is similar to the proof of Theorem 1′ in Volume 4, p. 117).

*For every $x \in X$ there exists a nuclear operator $P(x)$ mapping $\Phi$ into $H(x)$ such that for every $\varphi \in \Phi$ the functions $\varphi(x)$ and $P(x)\varphi$ differ only on a set of measure zero* [*where $\varphi(x)$ is the vector function corresponding to the element $\varphi$ when $\mathfrak{H}$ is realized in the form of the direct integral*]. *$P(x)$ maps $\Phi$ on an everywhere dense subset of $H(x)$.*

Now assume that there exist unitary representations $T_x(g)$ of the group $G$ on Hilbert spaces $H(x)$ such that for every $\varphi \in \Phi$

$$T_x(g)P(x)\varphi = P(x)T(g)\varphi.$$

*We then say that $T(g)$ is the direct integral of the $T_x(g)$, and we write*

$$T(g) = \int_X \oplus T_x(g)\, d\mu.$$

Each $T_x(g)$ is called the *component* of $T(g)$ in $H(x)$. As a rule we shall be interested in the case in which the $T_x(g)$ are irreducible, and we then say that $T(g)$ is decomposed by the direct integral into its irreducible components.

The simplest example of the decomposition of a representation by a direct integral of irreducible representations is the decomposition of the regular representation of the additive group of real numbers by means of the Fourier integral. The regular representation of this group is constructed on $L^2$, the space of square integrable functions $f(t)$, and is defined by

$$R(t_0)f(t) = f(t + t_0).$$

Let $F(\lambda)$ be the Fourier transform of $f(t)$, or

$$F(\lambda) = \int f(t)e^{i\lambda t}\, dt.$$

Plancherel's theorem implies that the correspondence $f(t) \to F(\lambda)$ can be thought of as an expansion of $L^2$ in the direct integral of one-dimensional Hilbert spaces $H(\lambda)$. Under the translation $f(t) \to f(t + t_0)$ the Fourier transform changes from $F(\lambda)$ to $\exp(-i\lambda t_0)F(\lambda)$. Thus in each one-dimensional $H(\lambda)$ the translation operator $R(t)$ reduces to multiplication by $e^{-i\lambda t}$. These operators then form (one-dimensional) representations of the additive group of real numbers, and we have thus decomposed the regular representation $R(t)$ of this group into irreducible representations.

In the same way, the expression for $f(g)$ in a Fourier integral on the Lorentz group, derived in Section 3.4, gives rise to a decomposition of the regular representation of this group, as we shall now demonstrate. Let $K(z_1, z; \chi)$ be the kernel of the Fourier transform of a square integrable function $f(g)$ on the Lorentz group $G$. This kernel may be treated as a function of $z$ depending on the parameters $z_1$ and $\chi$. Let $X$ be the set of ordered pairs $x = (z_1, \chi)$, where

$$\chi = (\tfrac{1}{2}[n + i\rho], \tfrac{1}{2}[-n + i\rho]),$$

in which $\rho$ is a real number and $n$ is an integer, and let the measure $\mu$ be defined in this set by

$$\int_X \Phi(z_1, \chi)\, d\mu = -(8\pi^4)^{-1} \left(\frac{i}{2}\right) \int_{X_0} \int c(\chi)\Phi(z_1, \chi)\, dz_1\, d\bar{z}_1\, d\chi,$$

where $X_0$ is the set of admissible $\chi$ (that is, what we have called the "real axis") and the integral over $\chi$ is understood, as usual, as integration over $\rho$ and summation over $n$. As before, $c(\chi) = -\frac{1}{4}(n^2 + \rho^2)$.

Now with each $x \in X$ we associate the Hilbert space $H_{z_1,\chi}$, consisting of functions $\varphi(z)$ such that

$$\| \varphi \|^2_{z_1,\chi} = \frac{i}{2} \int |\varphi(z)|^2\, dz\, d\bar{z} < \infty.$$

Obviously each $K(z_1, z; \chi)$ can then be thought of as a vector function over $X$ taking on values from $H_{z_1,\chi}$. Then the analog of Plancherel's theorem [see Eq. (3) of Section 3.6] implies that[21]

$$\int |f(g)|^2 \, dg = \int \| K(z_1, z; \chi)\|^2_{z_1,\chi} \, d\mu.$$

This shows that the correspondence $f(g) \to K(z_1, z; \chi)$ is an expansion of $H$, the space of the $f(g)$ functions, in a direct integral of the form

$$H = \int_X \oplus H_{z_1,\chi} \, d\mu.$$

This expansion corresponds to a decomposition of the regular representation of $G$ by a direct integral into the irreducible unitary representations $T_\chi(g)$ of the principal series. To see this one need only recall that under the translation $f(g) \to f(gg_0)$ the kernel transforms according to

$$K(z_1, z; \chi) \to (\beta_0 z + \delta_0)^{-n_1-1}(\bar{\beta}_0 \bar{z} + \bar{\delta}_0)^{-n_2-1} K\left(z_1, \frac{\alpha_0 z + \gamma_0}{\beta_0 z + \delta_0}; \chi\right).$$

Thus as a function of $z$ the kernel transforms under such translations according to the representation $T_{-\chi}(g_0)$ of the principal series. This shows that the expansion of the functions on the Lorentz group $G$ in the Fourier integral defines the decomposition of its regular representation $R(g)$ by a direct integral into irreducible representations of the principal series.

To each $\chi$ there corresponds a set of component spaces depending on the other parameter $z_1$, on each of which the same representation $T_{-\chi}(g)$ is induced.

Actually the correspondence $f(g) \to K(z_1, z; \chi)$ defines also the decomposition of the *left regular* representation $L(g)$ of the Lorentz group into irreducible components. This representation is defined by

$$L(g_0) f(g) = f(g_0^{-1} g).$$

Equation (2) of Section 2.4 shows that to $L(g_0)$ corresponds the transformation of the kernel given by

$$K(z_1, z; \chi) \to (\beta_0 z_1 + \delta_0)^{n_1-1}(\bar{\beta}_0 \bar{z}_1 + \bar{\delta}_0)^{n_2-1} K\left(\frac{\alpha_0 z_1 + \gamma_0}{\beta_0 z_1 + \delta_0}, z; \chi\right).$$

[21] Here $\| K(z_1, z; \chi)\|_{z_1,\chi}$ denotes the norm of $K(z_1, z; \chi)$ treated as a function of $z$, which is defined by

$$\| K(z_1, z; \chi)\|^2_{z_1,\chi} = \frac{i}{2} \int | K(z_1, z; \chi)|^2 \, dz \, d\bar{z}.$$

Thus under left translations the kernel $K(z_1, z; \chi)$ transforms as a function of $z_1$ under the unitary representation $T_\chi(g)$ of the principal series. This means that the Fourier transform of $f(g)$ defines also the decomposition of the left regular representation into irreducible components.

## 4. Differential Operators on G

In Section 1 infinitely differentiable functions on $G$ were defined by choosing local coordinates in $G$. We should like now to give a more convenient definition of such functions, basing it now on what we shall call *differential operators on G*. This definition was given previously in different terms in Section 2.7 of Chapter II.

### 4.1. Tangent Space to G

Consider the point

$$e = \begin{Vmatrix} 1 & 0 \\ 0 & 1 \end{Vmatrix}$$

in $G$ and "curves" of the form

$$h(t) = \begin{Vmatrix} a(t) & b(t) \\ c(t) & d(t) \end{Vmatrix}$$

that pass through it. Here $t$ is a complex parameter that we shall choose to be zero at $e$, that is, such that $a(0) = d(0) = 1$ and $b(0) = c(0) = 0$. Assume $a(t)$, $b(t)$, $c(t)$, and $d(t)$ to be analytic in $t$ for small values of $|t|$.

We shall call the matrix

$$h = h'(0) = \begin{Vmatrix} a'(0) & b'(0) \\ c'(0) & d'(0) \end{Vmatrix}$$

the *tangent vector* of the curve $h(t)$ at $e = h(0)$. The vectors at $e$ tangent to all possible curves passing through this point form a complex linear space[22] which we shall call the *tangent space* to the group at $e$ and shall denote by $\Lambda$. Obviously the tangent space is of the same complex dimension as $G$ itself, namely of dimension three.

[22] Indeed, if $h_1$ and $h_2$ are vectors tangent at $e$ to the curves $h_1(t)$ and $h_2(t)$, then $\lambda_1 h_1 + \lambda_2 h_2$ is the vector tangent to $h(t) = h_1(\lambda_1 t) h_2(\lambda_2 t)$.

It is easily shown that a matrix belongs to $\Lambda$ if and only if its trace vanishes. Indeed, the matrix elements of $h(t)$ are related by

$$a(t)d(t) - b(t)c(t) = 1.$$

When this equation is differentiated with respect to $t$ and evaluated at $t = 0$, we find that $a'(0) + d'(0) = 0$, which means that the trace of the tangent vector vanishes. Conversely, the matrices with trace zero form a complex linear space of the same dimension as $\Lambda$, namely three. Therefore they all belong to $\Lambda$.

**Remark.** It can be shown that to every matrix $h$ of the tangent space corresponds a unique one-parameter subgroup $h(t)$ of $G$ whose tangent vector at $e$ is $h$. [A one-parameter subgroup is a curve $h(t)$ such that $h(t_1 + t_2) = h(t_1)h(t_2)$ for all complex numbers $t_1$ and $t_2$.] #

## 4.2. Lie Operators

To each matrix $h \in \Lambda$ correspond two linear differential operators $A_h$ and $\bar{A}_h$ on $G$. These are defined as follows. Let $h(t)$ be a curve passing through $e$ with tangent vector $h$. Then we make the definitions

$$A_h f(g) = \left.\frac{\partial f[h(t)g]}{\partial t}\right|_{t=0}, \qquad \bar{A}_h f(g) = \left.\frac{\partial f[h(t)g]}{\partial \bar{t}}\right|_{t=0}.$$

It is easily shown that $A_h$ and $\bar{A}_h$ do not depend on the choice of $h(t)$ so long long as it has the proper tangent vector, and they are therefore completely determined by the tangent vector $h$.[23] We shall call $A_h$ and $\bar{A}_h$ the *left derivatives along the vector* $h \in \Lambda$, or simply *left Lie operators*.

An obvious calculation shows that

$$A_{\lambda_1 h_1 + \lambda_2 h_2} = \lambda_1 A_{h_1} + \lambda_2 A_{h_2}, \qquad \bar{A}_{\lambda_1 h_1 + \lambda_2 h_2} = \bar{\lambda}_1 \bar{A}_{h_1} + \bar{\lambda}_2 \bar{A}_{h_2}.$$

[23] To prove this let us parameterize the elements

$$g = \begin{Vmatrix} \alpha & \beta \\ \gamma & \delta \end{Vmatrix}$$

of $G$ by $\alpha$, $\beta$, and $\delta$. Then a simple calculation shows that

$$A_h f(g) = a'(0)\left[\alpha \frac{\partial f}{\partial \alpha} + \beta \frac{\partial f}{\partial \beta} - \delta \frac{\partial f}{\partial \delta}\right] + b'(0)\left[\gamma \frac{\partial f}{\partial \alpha} + \delta \frac{\partial f}{\partial \beta}\right] + c'(0)\beta \frac{\partial f}{\partial \delta}.$$

A similar expression is obtained for $\bar{A}_h$.

Thus the $A_h$ themselves form a linear space, and the same may be said of the $\bar{A}_h$. It is convenient in these spaces to set up the following bases. In $\Lambda$ we take as a basis the three matrices

$$a_0 = \begin{Vmatrix} \frac{1}{2} & 0 \\ 0 & -\frac{1}{2} \end{Vmatrix} \qquad a_+ = \begin{Vmatrix} 0 & 1 \\ 0 & 0 \end{Vmatrix}, \qquad a_- = \begin{Vmatrix} 0 & 0 \\ 1 & 0 \end{Vmatrix}.$$

These are tangent vectors to the one-parameter subgroups

$$a_0(t) = \begin{Vmatrix} e^{\frac{1}{2}t} & 0 \\ 0 & e^{-\frac{1}{2}t} \end{Vmatrix}, \qquad a_+(t) = \begin{Vmatrix} 1 & t \\ 0 & 1 \end{Vmatrix}, \qquad a_-(t) = \begin{Vmatrix} 1 & 0 \\ t & 1 \end{Vmatrix}.$$

We then choose as bases in the two left-derivative operator spaces those operators $A_h$ and $\bar{A}_h$ for which $h$ is $a_0$, $a_+$, or $a_-$, denoting them by $A_0$, $A_+$, $A_-$ and $\bar{A}_0$, $\bar{A}_+$, $\bar{A}_-$, respectively.

In addition to the left derivative operators we can define *right derivative operators* $B_h$ and $\bar{B}_h$ for $G$ by the equations

$$B_h f(g) = \left.\frac{\partial f[gh(t)]}{\partial t}\right|_{t=0}, \qquad \bar{B}_h f(g) = \left.\frac{\partial f[gh(t)]}{\partial \bar{t}}\right|_{t=0},$$

where $h(t)$ is any curve for which $h(0) = e$ and $h'(0) = h$. As is true for the left derivative operators, these also form three-dimensional linear spaces, and in these spaces we choose our bases similarly: $B_0$, $B_+$, $B_-$ and $\bar{B}_0$, $\bar{B}_+$, $\bar{B}_-$ correspond to $a_0$, $a_+$, and $a_-$, respectively.

In terms of $\alpha$, $\beta$, $\delta$, and $\gamma$ (we choose the first three as coordinates) these operators are given by

$$A_0 = \frac{1}{2}\left(\alpha\frac{\partial}{\partial\alpha} + \beta\frac{\partial}{\partial\beta} - \delta\frac{\partial}{\partial\delta}\right),$$

$$A_+ = \gamma\frac{\partial}{\partial\alpha} + \delta\frac{\partial}{\partial\beta},$$

$$A_- = \beta\frac{\partial}{\partial\delta},$$

$$B_0 = \frac{1}{2}\left(\alpha\frac{\partial}{\partial\alpha} - \beta\frac{\partial}{\partial\beta} - \delta\frac{\partial}{\partial\delta}\right),$$

$$B_+ = \alpha\frac{\partial}{\partial\beta} + \gamma\frac{\partial}{\partial\delta},$$

$$B_- = \beta\frac{\partial}{\partial\alpha}.$$

The expressions for the barred operators are obtained from these by replacing $\alpha, \beta, \gamma, \delta$ and differentiation with respect to them by their complex conjugates and by differentiation with respect to their complex conjugates.

### 4.3. Relation between Left and Right Derivative Operators

Since the elements of $G$ are given by six real parameters, there exist on $G$ exactly six first-order linear differential operators that are linearly independent over the reals. For this reason there must exist linear relations between the left and right derivative operators (in which the coefficients depend on $g$). These relations are conveniently written in matrix form in terms of the matrices

$$A = \begin{Vmatrix} A_0 & A_+ \\ A_- & -A_0 \end{Vmatrix}, \qquad B = \begin{Vmatrix} B_0 & B_+ \\ B_- & -B_0 \end{Vmatrix}$$

whose elements are Lie operators. We then have

$$g'A = Bg', \tag{1}$$

where $g'$ is the transpose of $g$. (In this notation $Bg'$ denotes symbolic matrix multiplication; that is, the elements of $g'$ are not differentiated.) A similar relation exists for the barred operators.

*Derivation of Eq.* (1). Consider the general linear group (i.e., of all non-singular matrices)

$$g = \begin{Vmatrix} \alpha & \beta \\ \gamma & \delta \end{Vmatrix}.$$

The tangent space at the identity of this group consists of all $2 \times 2$ matrices, and we choose our basis in this tangent space to be

$$a_{11} = \begin{Vmatrix} 1 & 0 \\ 0 & 0 \end{Vmatrix}, \qquad a_{12} = \begin{Vmatrix} 0 & 1 \\ 0 & 0 \end{Vmatrix}, \qquad a_{21} = \begin{Vmatrix} 0 & 0 \\ 1 & 0 \end{Vmatrix}; \qquad a_{22} = \begin{Vmatrix} 0 & 0 \\ 0 & 1 \end{Vmatrix}.$$

Let $A_{ij}$ and $B_{ij}$ $(i, j = 1, 2)$ be the corresponding left and right derivative operators. Then a simple calculation gives

$$A_{11} = \alpha \frac{\partial}{\partial \alpha} + \beta \frac{\partial}{\partial \beta}, \qquad A_{12} = \gamma \frac{\partial}{\partial \alpha} + \delta \frac{\partial}{\partial \beta},$$

$$A_{21} = \alpha \frac{\partial}{\partial \gamma} + \beta \frac{\partial}{\partial \delta}, \qquad A_{22} = \gamma \frac{\partial}{\partial \gamma} + \delta \frac{\partial}{\partial \delta}.$$

In terms of the symbolic matrix

$$\frac{\partial}{\partial g} = \begin{Vmatrix} \partial/\partial\alpha & \partial/\partial\beta \\ \partial/\partial\gamma & \partial/\partial\delta \end{Vmatrix}$$

we may write

$$\begin{Vmatrix} A_{11} & A_{12} \\ A_{21} & A_{22} \end{Vmatrix} = \frac{\partial}{\partial g} g'$$

and the similar expression

$$\begin{Vmatrix} B_{11} & B_{12} \\ B_{21} & B_{22} \end{Vmatrix} = g' \frac{\partial}{\partial g}.$$

Then it follows immediately that

$$g' \begin{Vmatrix} A_{11} & A_{12} \\ A_{21} & A_{22} \end{Vmatrix} = \begin{Vmatrix} B_{11} & B_{12} \\ B_{21} & B_{22} \end{Vmatrix} g'. \tag{2}$$

Note that in terms of the $A_{ij}$ the left Lie operators on the unimodular group $G$ may be written

$$A_0 = \tfrac{1}{2}(A_{11} - A_{22}), \qquad A_+ = A_{12}, \qquad A_- = A_{21}.$$

We thus have

$$\begin{Vmatrix} A_{11} & A_{12} \\ A_{21} & A_{22} \end{Vmatrix} = \begin{Vmatrix} A_0 & A_+ \\ A_- & -A_0 \end{Vmatrix} + \tfrac{1}{2}(A_{11} + A_{22}) \begin{Vmatrix} E & 0 \\ 0 & E \end{Vmatrix},$$

where $E$ is the unit operator. In a similar way,

$$\begin{Vmatrix} B_{11} & B_{12} \\ B_{21} & B_{22} \end{Vmatrix} = \begin{Vmatrix} B_0 & B_+ \\ B_- & -B_0 \end{Vmatrix} + \tfrac{1}{2}(B_{11} + B_{22}) \begin{Vmatrix} E & 0 \\ 0 & E \end{Vmatrix}.$$

Note now that

$$A_{11} + A_{22} = B_{11} + B_{22} = \alpha \frac{\partial}{\partial\alpha} + \beta \frac{\partial}{\partial\beta} + \gamma \frac{\partial}{\partial\gamma} + \delta \frac{\partial}{\partial\delta},$$

which, together with Eq. (2), yields

$$g' \begin{Vmatrix} A_0 & A_+ \\ A_- & -A_0 \end{Vmatrix} = \begin{Vmatrix} B_0 & B_+ \\ B_- & -B_0 \end{Vmatrix} g'.$$

### 4.4. Commutation Relations for the Lie Operators

Note first that the left Lie operators commute with the right Lie operators. This follows immediately from the fact that left translations commute with right translations on $G$.

It is helpful to introduce the *commutator* $[A, B]$ of two operators $A$ and $B$, defined in the usual way by

$$[A, B] = AB - BA.$$

Then we have immediately

$$[A_{h_1}, B_{h_2}] = 0, \qquad [\bar{A}_{h_1}, \bar{B}_{h_2}] = 0, \qquad [A_{h_1}, \bar{B}_{h_2}] = 0,$$
$$[\bar{A}_{h_1}, B_{h_2}] = 0.$$

Further, because differentiation with respect to $t$ commutes with differentiation with respect to $\bar{t}$, we have also

$$[A_{h_1}, \bar{A}_{h_2}] = 0, \qquad [B_{h_1}, \bar{B}_{h_2}] = 0.$$

It may be shown for the left derivative operators that

$$\begin{aligned} [A_+, A_0] &= A_+, \\ [A_-, A_0] &= -A_-, \\ [A_+, A_-] &= -2A_0. \end{aligned} \tag{1}$$

Similarly, for the right derivative operators we have

$$\begin{aligned} [B_+, B_0] &= -B_+, \\ [B_-, B_0] &= B_-, \\ [B_+, B_-] &= 2B_0. \end{aligned} \tag{1'}$$

*Derivation of Eqs.* (1) *and* (1′). These relations can be obtained directly by making use of the definitions of $A_h$ and $B_h$ in Section 4.2. It is simpler, however, to obtain them from more general considerations. Let $h_1$ and $h_2$ be matrices in the tangent space $\Lambda$. Then $[h_1, h_2]$ is also in $\Lambda$, for

$$\mathrm{Tr}(h_1h_2 - h_2h_1) = \mathrm{Tr}(h_1h_2) - \mathrm{Tr}(h_2h_1) = 0.$$

Now it can be shown[24] that

$$[A_{h_1}, A_{h_2}] = -A_{[h_1,h_2]}, \qquad [B_{h_1}, B_{h_2}] = B_{[h_1,h_2]}.$$

[24] See, for instance, Ref. (16).

Thus Eqs. (1) and (1′) can be obtained by first finding $[a_+, a_0]$, $[a_-, a_0]$, and $[a_+, a_-]$. A simple direct calculation yields

$$[a_+, a_0] = -a_+, \qquad [a_-, a_0] = a_-, \qquad [a_+, a_-] = -2a_0,$$

from which (1) and (1′) follow immediately.

The commutation relations can now be used to write any polynomial of Lie operators with constant coefficients as a linear combination of ordered products of the form

$$A_0^{k_1} A_+^{k_2} A_-^{k_3} B_0^{l_1} B_+^{l_2} B_-^{l_3} \bar{A}_0^{m_1} \bar{A}_+^{m_2} \bar{A}_-^{m_3} \bar{B}_0^{n_1} \bar{B}_+^{n_2} \bar{B}_-^{n_3}. \tag{2}$$

## 4.5. Laplacian Operators

Operators that commute with infinitesimal translations (that is, with Lie operators) are of particular importance to the study of functions on a given space. In Euclidean space, for instance, this property is possessed by the ordinary Laplacian.[25]

Let us then turn our attention to those operators on $G$ which commute with all Lie operators. It is seen from the commutation relations of Section 4.4 that there exist no first-order operators with this property. On the other hand, we shall show that the quadratic operator

$$\Delta = 2A_0^2 + A_+A_- + A_-A_+$$

has this commutation property.

We shall call $\Delta$ the *Laplacian operator* for $G$. It is clear from its definition that

$$\Delta = \operatorname{Tr} A^2, \qquad \text{where} \quad A = \begin{Vmatrix} A_0 & A_+ \\ A_- & -A_0 \end{Vmatrix}.$$

This equation can be used to express $\Delta$ in terms of the right Lie operators. We have seen in Section 4.3 that the left and right Lie operators are related by

$$A = g'^{-1} B g'$$

and it therefore follows that

$$\operatorname{Tr} A^2 = \operatorname{Tr} B^2,$$

[25] For this particular case we mean translation any transformation preserving the distance between two points.

and therefore that

$$\Delta = \operatorname{Tr} B^2, \qquad \text{where} \quad B = \begin{Vmatrix} B_0 & B_+ \\ B_- & -B_0 \end{Vmatrix}.$$

It is now easy to shows that $\Delta$ commutes with all the Lie operators. Indeed, since it can be written as a polynomial of the $A_h$ with constant coefficients, it commutes with all the $B_h$. On the other hand, it can also be written as a polynomial of the $B_h$ with constant coefficients and must therefore commute with all the $A_h$. It therefore commutes with all the Lie operators.

The operator $\bar{\Delta}$ on $G$ defined by

$$\bar{\Delta} = \operatorname{Tr} \bar{A}^2 = 2\bar{A}_0^2 + \bar{A}_+\bar{A}_- + \bar{A}_-\bar{A}_+$$

also commutes with all the Lie operators. It can be shown, in fact, *that every differential operator commuting with all the Lie operators is a polynomial* (*with constant coefficients*) *in* $\Delta$ *and* $\bar{\Delta}$.

## 4.6. Functions on G with Rapidly Decreasing Derivatives

In Section 1.2 we defined infinitely differentiable functions on $G$. This definition can obviously be stated in terms of Lie operators: if every operator which is a polynomial of Lie operators can be applied to a function $f(g)$, then $f(g)$ is called an infinitely differentiable function on $G$. We shall call expressions of the form $P(X)f(g)$ derivatives of $f(g)$ for all constant-coefficient polynomials $P(X)$ of Lie operators. The order of the derivative is defined as the degree of the polynomial when all the terms are written in the canonical form of Eq. (2), Section 4.4.

With $G$ we may associate the space $S$ of functions with rapidly decreasing derivatives. This space consists of the infinitely differentiable functions on $G$, all of whose derivatives decrease rapidly asymptotically as $|g| \to \infty$. This means that for every $P(X)$ and for any $n$ we must have

$$\lim_{|g|\to\infty} |g|^n P(X) f(g) = 0.$$

We introduce a topology on $S$ by the sequence of norms

$$\|f\|_n = \sup_{g,P} |g|^n \, |P(X)f(g)|,$$

where $P(X)$ runs through all Lie operator polynomials of degree no greater than $n$, whose coefficients, in the canonical form, are of modulus no greater than one.

Among the properties of $S$ are the following.

1. $S$ is invariant under left and right translations and under inversion [i.e., under $f(g) \rightarrow f(g^{-1})$].

2. $S$ is invariant under multiplication of $f(g)$ by any polynomial $P(g)$ of the matrix elements of $g$ or their complex conjugates.

3. The convolution $f_1 * f_2(g)$ of two functions in $S$ also belongs to $S$.

An example of a function in $S$ is $\exp\{-|g|^2\}$. Also in $S$ is $P \exp\{-|g|^2\}$, where $P$ is any polynomial (with constant coefficients) of matrix elements of $g$, of their complex conjugates, or of Lie operators on $G$.

## 4.7. Fourier Transforms of Lie Operators

In Section 1.3 we established the correspondence between every function $f(g)$ and its Fourier transform, an operator function $F(\chi)$. Let us now see how Lie operators transform under this correspondence. As was shown in Section 2.1, under the left translation $f(g) \rightarrow f(g_0^{-1}g)$, the Fourier transform $F(\chi)$ of $f(g)$ is transformed to $T_\chi(g_0)F(\chi)$. It then follows that the Fourier transform of $f[h(t)g]$ is the operator function $T_\chi^{-1}[h(t)]F(\chi)$. Therefore the Fourier transform of

$$A_h f(g) = \left. \frac{\partial f[h(t)g]}{\partial t} \right|_{t=0}$$

is the operator function $\tilde{A}_h(\chi)F(\chi)$, where

$$\tilde{A}_h(\chi) = \left. \frac{\partial T_\chi^{-1}[h(t)]}{\partial t} \right|_{t=0}$$

We thus see that the left derivative along $h$ of a function $f(g)$ induces multiplication of its Fourier transform $F(\chi)$ on the left by the operator $\tilde{A}_h(\chi)$ defined by Eq. (1). In the same way it can be shown that the right derivative along $h$ of $f(g)$ induces multiplication of its Fourier transform on the right by $\tilde{A}_h(\chi)$.

**Remark.** The Fourier transforms of the left Lie operators corresponding to the basis $a_0$, $a_+$, $a_-$ in the tangent space are

$$\tilde{A}_0(\chi) = \frac{n_1 - 1}{2} E - z \frac{\partial}{\partial z},$$

where $E$ is the unit operator,

$$\tilde{A}_+(\chi) = -(n_1 - 1)z + z^2 \frac{\partial}{\partial z},$$

$$\tilde{A}_-(\chi) = -\frac{\partial}{\partial z}. \quad \#$$

We now find the Fourier transforms of the Laplacians $\Delta$ and $\bar{\Delta}$.

**Theorem.** *When the $f(g)$ are replaced by their Fourier transforms $F(\chi)$, $\chi = (n_1, n_2)$, the Laplacian operators are replaced by multiples of the unit operator. Specifically, $\Delta$ goes over into $n_1^2 - 1$ and $\bar{\Delta}$ goes over into $n_2^2 - 1$.*

Proof. Let us write $\tilde{\Delta}(\chi)$ for the Fourier transform of $\Delta$. It is easily shown that because $\Delta$ commutes with all the Lie operators, it commutes with all the translation operators.[26] But the Fourier transforms of the translation operators are the $T_\chi(g)$. Therefore $\tilde{\Delta}(\chi)$ must commute with all the $T_\chi(g)$. But it was shown in Chapter III, Section 5.1 that all of the $T_\chi(g)$ are operator irreducible, so that $\tilde{\Delta}(\chi)$ must be a multiple of the unit operator. We thus have

$$\tilde{\Delta}(\chi) = k(\chi)E,$$

where $k(\chi)$ is a scalar. Similar considerations apply to the Fourier transform of $\bar{\Delta}$.

A more direct calculation, in which one inserts into the relation

$$\tilde{\Delta}(\chi) = 2\tilde{A}_0^2 + \tilde{A}_+\tilde{A}_- + \tilde{A}_-\tilde{A}_+$$

the expressions for the operators on the right-hand side, gives the value of $k(\chi)$. In this way we obtain

$$k(\chi) = n_1^2 - 1.$$

## 5. The Paley-Wiener Theorem for the Fourier Transform on $G$

In Section 3.6 we established the necessary and sufficient conditions for $K(z_1, z_2; \chi)$ to be the kernel of the Fourier transform of a square integrable function of $g$. We shall now establish similar conditions, but for $f(g) \in S$, that is, for infinitely differentiable rapidly decreasing func-

[26] This is because the translation operators are obtained as integrals of the Lie operators.

tions on $G$. Necessary conditions are obtained in Sections 5.3 and 5.4, and sufficient conditions in Section 5.6, where the results are summarized in an analog of the Paley-Wiener theorem.

## 5.1. Integrals of $f(g)$ along "Line Generators"

Recall that the transition from $f(g)$ to the kernel $K(z_1, z_2; \chi)$ of its Fourier transform may be broken up into two steps. First we transform from $f(g)$ to $\varphi(z_1, z_2; \lambda)$, defined by

$$\varphi(z_1, z_2; \lambda) = \frac{i}{2} \int f(\lambda^{-1} + \beta z_2, \beta, \lambda - \beta z_1)\, d\beta\, d\bar{\beta}, \tag{1}$$

and then from this function to the kernel according to

$$K(z_1, z_2; \chi) = \frac{i}{2} \int \varphi(z_1, z_2; \lambda) \lambda^{n_1 - 1} \bar{\lambda}^{n_2 - 1}\, d\lambda\, d\bar{\lambda}, \qquad \chi = (n_1, n_2). \tag{2}$$

Now the second of these steps is essentially an ordinary Mellin transform, so that for our purposes we need only study the transition from $f(g)$ to $\varphi(z_1, z_2; \lambda)$. We have seen in Section 2.3 that $\varphi(z_1, z_2; \lambda)$ has a clear geometric meaning which may be described as follows. If $f(g)$ is considered as a function defined on the quadratic surface $\alpha\delta - \beta\gamma = 1$ in four complex dimensions, then $\varphi(z_1, z_2; \lambda)$ is the integral of $f(g)$ along a "line generator" of this surface. The variables $z_1, z_2, \lambda$, are then the coordinates or parameters specifying the generator.

As elsewhere, it will be convenient to use the homogeneous coordinates of these generators. We choose them again as the coefficients of the set of linear equations

$$u\alpha + v\gamma = u', \qquad u\beta + v\delta = v', \tag{3}$$

which determine a line on the quadratic surface.

The integral of $f(g)$ along the line generator specified by (3) is then defined as

$$\Phi(u, v; u', v') = \frac{i}{2} \int f(g) \omega \bar{\omega}, \tag{4}$$

where

$$\omega = d\alpha / vu' = d\beta / vv' = -d\gamma / uu' = -d\delta / uv'.$$

This defines $\Phi(u, v; u', v')$ for all values of $u, v, u', v'$ other than $u = v = 0$ and $u' = v' = 0$, and the function so defined is homogeneous of degree $(-2, -2)$, i.e., satisfies the equation

$$\Phi(\alpha u, \alpha v; \alpha u', \alpha v') = \alpha^{-2}\bar{\alpha}^{-2}\Phi(u, v; u', v')$$

for all $\alpha \neq 0$. Recall in addition that $z_1, z_2, \lambda$ are related to the homogeneous coordinates by

$$z_1 = u/v, \qquad z_2 = u'/v', \qquad \lambda = v'/v.$$

Further, $\varphi$ and $\Phi$ are related by

$$\varphi(z_1, z_2; \lambda) = \lambda\bar{\lambda}\Phi(z_1, 1; \lambda z_2, \lambda), \tag{5}$$

$$\Phi(u, v; u', v') = |v|^{-2} |v'|^{-2}\varphi(u/v, u'/v'; v'/v), \tag{6}$$

as has been shown in Section 2.3.

In Sections 5.2–5.4 we study the properties of $\Phi(u, v; u', v')$.

### 5.2. Behavior of $\Phi(u, v; u', v')$ under Translation and Differentiation of $f(g)$

Consider the function $f_1(g) = f(g_0^{-1}g)$, where

$$g_0 = \begin{Vmatrix} \alpha_0 & \beta_0 \\ \gamma_0 & \delta_0 \end{Vmatrix},$$

and consider also the function

$$\Phi_1(u, v; u', v') = \frac{i}{2}\int f(g_0^{-1}g)\omega\bar{\omega} \tag{1}$$

associated with it. We wish to obtain a relation between $\Phi_1$ and $\Phi$.

Let us write Eq. (3) of Section 5.1 in the abbreviated from

$$(u, v)g = (u', v'),$$

where $g$ has matrix elements $\alpha, \beta, \gamma, \delta$, and $(u, v)$ is understood as a row vector. According to this equation, then, the line consists of all the matrices $g$ which carry $(u, v)$ into $(u', v')$. We wish to show that the integral in (1) is the integral of $f(g)$ along the line

$$(u, v)g_0 g = (u', v'),$$

that is, the line whose homogeneous coordinates are

$$(u\alpha_0 + v\gamma_0, u\beta_0 + v\delta_0; u', v'). \tag{2}$$

To do this we replace $g_0 g$ by $g$ in the integrand, which transforms the differential form $\omega$ into

$$\omega' = d\alpha'/vu' = d\beta'/vv' = -d\gamma'/uu' = -d\delta'/uv',$$

where $\alpha', \beta', \gamma', \delta'$ are the matrix elements of $g_0 g$. It is a simple matter to use these expressions to show that

$$\omega' = \frac{d(\delta_0\alpha' - \beta_0\gamma')}{(u\beta_0 + v\delta_0)u'} \equiv \frac{d\alpha}{(u\beta_0 + v\delta_0)u'}\,.$$

This demonstrates that $\omega'$ is precisely that differential form which defines integration along the line whose homogeneous coordinates are those given in Eq. (2), as asserted.

This establishes that *under the left translation* $f(g) \to f(g_0^{-1}g)$ *the function* $\Phi(u, v; u', v')$ *is transformed into*

$$\Phi_1(u, v; u', v') \equiv \Phi(u\alpha_0 + v\gamma_0\,,\, u\beta_0 + v\delta_0; u', v').$$

Similar considerations show that *under the right translation* $f(g) \to f(gg_0)$, *the function* $\Phi(u, v; u', v')$ *is transformed into*

$$\Phi_2(u, v; u', v') \equiv \Phi(u, v; u'\alpha_0 + v'\gamma_0\,,\, u'\beta_0 + v'\delta_0).$$

With these results it can be found how $\Phi$ behaves also when some Lie operator is applied to $f(g)$. The corresponding operators acting on the $\Phi$ functions shall also be called Lie operators and shall be denoted by the same symbols $A_0$, $A_+$, $A_-$, etc. The equations of Section 4.2 can be used to obtain explicit expressions for these operators. One then finds that

$$A_0 = -\tfrac{1}{2}\left(u\frac{\partial}{\partial u} - v\frac{\partial}{\partial v}\right),$$

$$A_+ = -u\frac{\partial}{\partial v},$$

$$A_- = -v\frac{\partial}{\partial u};$$

$$B_0 = \tfrac{1}{2}\left(u'\frac{\partial}{\partial u'} - v\frac{\partial}{\partial v'}\right),$$

$$B_+ = u'\frac{\partial}{\partial v'},$$

$$B_- = v'\frac{\partial}{\partial u'}\,.$$

The expressions for the corresponding barred operators are obtained from these by replacing $u, v, u', v'$ by their complex conjugates and the derivatives with respect to these variables by the derivatives with respect to their complex conjugates.

**Remark.** Note that in the space of the $\Phi$ functions the left derivative operators depend only on the unprimed variables $u$ and $v$ and their complex conjugates and on the derivatives with respect to these, while the right translation operators depend similarly only on the primed variables. #

### 5.3. Differentiability and Asymptotic Behavior of $\Phi(u, v; u', v')$

Let us now assume that $f(g)$ is a rapidly decreasing infinitely differentiable function. This means that for every $n > 0$ and for every polynomial $P(X)$ of Lie operators, the function

$$|g|^n P(X) f(g)$$

is bounded on $G$. Then the corresponding $\Phi(u, v; u', v')$ will also satisfy certain differentiability and asymptotic conditions. We now turn to the problem of finding these conditions.

We assert first that *$\Phi(u, v; u', v')$ is infinitely differentiable with respect to $u, v, u', v'$ and their complex conjugates everywhere except at $u = v = 0$ and $u' = v' = 0$* (where the function is not defined).

**Proof.** Consider a region of $(u, v; u', v')$-space such that one of the variables $(u, v)$ and one of the variables $(u', v')$ fails to vanish in this region. We shall assume, to be specific, that $u \neq 0$ and $v' \neq 0$. Let us write the integral along a line $(u, v; u', v')$, one whose coordinates are given by

$$\begin{aligned} \alpha\delta - \beta\gamma &= 1, \\ u\alpha + v\gamma &= u', \\ u\beta + v\delta &= v', \end{aligned} \tag{1}$$

using $\delta$ for the variable of integration. From the equations of the line we find that

$$\Phi(u, v; u', v')$$
$$= \frac{i}{2} \int f\left(\frac{uv + u'v' - vu'\delta}{uv'}, \frac{v' - v\delta}{u}, \frac{u'\delta - u}{v'}, \delta\right) \frac{d\delta\, d\bar{\delta}}{|u|^2 |v'|^2}. \tag{2}$$

This equation shows immediately that if $f(g) = f(\alpha, \beta, \gamma, \delta)$ is an infinitely differentiable rapidly decreasing function, then $\Phi$ is infinitely differentiable in the region we have chosen. The same considerations hold for regions in which $u \neq 0$, $u' \neq 0$; regions in which $v \neq 0$, $u' \neq 0$; and, finally, regions in which $v \neq 0$, $v' \neq 0$. This completes the proof.

We now assert that $\Phi(u, v; u', v')$ has the following property. *For any number k the function*

$$\left(\frac{|u'|^2 + |v'|^2}{|u|^2 + |v|^2}\right)^k (|u|^2 + |v|^2)(|u'|^2 + |v'|^2)\Phi(u, v; u', v') \tag{3}$$

*is a bounded function of $u, v, u', v'$. The same is true of any derivative $P(X)\Phi$, where $P(X)$ is any polynomial of Lie operators with constant coefficients.*

**Proof.** By assumption, for all $k > 0$ we may write

$$|f(g)| < C\,|g|^{-2k}.$$

Hence

$$|\Phi(u, v; u', v')| < C\frac{i}{2}\int \frac{\omega\bar{\omega}}{|g|^{2k}}, \tag{4}$$

where the integral is along the line of Eq. (1), on which the differential form $\omega$ is

$$\omega = d\alpha/vu' = d\beta/vv' = -d\gamma/uu' = -d\delta/uv'.$$

Let us calculate the integral on the right-hand side of (4). Again, to be specific, we choose $u \neq 0$, $v' \neq 0$. Then we may choose $\delta$ to be the variable of integration, and the remaining variables are given by

$$\alpha = \frac{uv + u'v' - u'v\delta}{uv'}, \qquad \beta = \frac{v' - v\delta}{u}, \qquad \gamma = \frac{u'\delta - u}{v'}. \tag{5}$$

With these expressions we may write $|g|^2$ in the form

$$|g|^2 = |a\delta + b|^2 + |g_0|^2,$$

where

$$|a|^2 = \frac{(|u|^2 + |v|^2)(|u'|^2 + |v'|^2)}{|u|^2\,|v'|^2},$$

and $|g_0|$ is the minimum value of $|g|$ on the line. We then have

$$\frac{i}{2}\int \frac{\omega\bar{\omega}}{|g|^{2k}} = \frac{1}{|u|^2\,|v'|^2}\frac{i}{2}\int \frac{d\delta\, d\bar{\delta}}{(|a\delta + b|^2 + |g_0|^2)^k}$$

$$= \frac{\pi}{k-1}\frac{|g_0|^{-2(k-1)}}{|u|^2\,|v'|^2\,|a|^2} = \frac{\pi}{k-1}\frac{|g_0|^{-2(k-1)}}{(|u|^2 + |v|^2)(|u'|^2 + |v'|^2)}.$$

We thus establish that for every $k > 0$

$$| \Phi(u, v; u', v')| < C(| u |^2 + | v |^2)^{-1}(| u' |^2 + | v' |^2)^{-1} | g_0 |^{-2(k-1)}. \tag{6}$$

Let us now find $| g_0 |$, the minimum value of $| g |$ on the line. From Eq. (1) we have

$$| u' |^2 + | v' |^2 \leqslant (| u |^2 + | v |^2) | g |^2,$$

so that

$$| g_0 |^2 \geqslant \frac{| u' |^2 + | v' |^2}{| u |^2 + | v |^2} . \tag{7}$$

On the other hand, the line of Eq. (1) can be given by the equivalent set of equations

$$\begin{aligned} \alpha\delta - \beta\gamma &= 1, \\ u'\delta - v'\gamma &= u, \\ -u'\beta + v'\alpha &= v. \end{aligned} \tag{1'}$$

from which in a similar way we obtain

$$| u |^2 + | v |^2 \leqslant (| u' |^2 + | v' |^2) | g |^2,$$

and hence

$$| g_0 |^2 \geqslant \frac{| u |^2 + | v |^2}{| u' |^2 + | v' |^2} . \tag{7'}$$

By combining Eqs. (6), (7), and (7′) we find indeed that for every $k$ the function of Eq. (3) is bounded in $u$, $v$, $u'$, $v'$, as asserted.

Now consider $P(X)f(g)$, where $P(X)$ is a polynomial of Lie operators. We have already seen that to this function corresponds $P(X)\Phi(u, v; u', v')$. Since $P(X)f$ is by assumption a rapidly decreasing function on the group, the result we have obtained for $\Phi$ must hold also for $P(X)\Phi$.

### 5.4. Conditions on $K(z_1, z_2; \chi)$

It turns out that in addition to the differentiability and boundedness conditions established in Section 5.3, $\Phi(u, v; u', v')$ satisfies also certain additional symmetry conditions.[27] These conditions are most conveniently

[27] This is similar to the situation for the Radon transform $\check{f}(\zeta, s)$ of a rapidly decreasing function in an affine space of $n$ complex dimensions, for which the integral

$$i/2 \int \check{f}(\zeta, s) s^k \bar{s}^l \, ds \, d\bar{s}$$

is a homogeneous polygonomial of degree $(k, l)$ in $\zeta$, $\bar{\zeta}$, where $k, l = 0, 1, \ldots$ (see Section 3.4 of Chapter II).

formulated in terms of the kernel $K(z_1, z_2; \chi)$ rather than in terms of $\Phi$. Recall that the kernel is defined by

$$K(z_1, z_2; \chi) = \frac{i}{2}\int \Phi(z_1, 1; \lambda z_2, \lambda)\lambda^{n_1}\bar{\lambda}^{n_2}\, d\lambda\, d\bar{\lambda}, \qquad \chi = (n_1, n_2)$$

[see Eqs. (2) and (5) of Section 5.1]. Recall also that $K$ is the kernel of the operator function

$$F(\chi) = \int f(g)T_\chi(g)\, dg,$$

acting on $D_\chi$. We have seen in Section 1.4 that $F(\chi)$ is defined for all values of $\chi$.

The relations satisfied by $K$ are connected with the existence of intertwining operators for certain pairs of representations $T_{\chi_1}(g)$ and $T_{\chi_2}(g)$. We have found such pairs of representations and the corresponding intertwining operators in Section 5.1 of Chapter III. There we first established that for $\chi = (n_1, n_2)$ and $-\chi = (-n_1, -n_2)$, where $n_1$ and $n_2$ are not integers of the same sign, $T_\chi(g)$ and $T_{-\chi}(g)$ are equivalent. This means that there exists a one-to-one bicontinuous mapping $A$ of the carrier space of $T_\chi(g)$ onto the carrier space of $T_{-\chi}(g)$ such that

$$AT_\chi(g) = T_{-\chi}(g)A.$$

From this it follows that the operator function $F(\chi)$ satisfies the relation

$$AF(\chi) = F(-\chi)A. \tag{1}$$

This can be written as a condition on the kernel $K$ of $F(\chi)$. As was shown in Chapter III, $A$ is of the form

$$A\varphi(z) = \frac{i}{2}\int (z_1 - z)^{-n_1-1}(\bar{z}_1 - \bar{z})^{-n_2-1}\varphi(z_1)\, dz_1\, d\bar{z}_1, \tag{2}$$

where the integral is to be understood in the sense of its regularization.

Thus Eq. (1) can be written for the kernel of $F(\chi)$ in the form

$$\frac{i}{2}\int (z - z_1)^{-n_1-1}(\bar{z} - \bar{z}_1)^{-n_2-1}K(z, z_2; n_1, n_2)\, dz\, d\bar{z}$$
$$= \frac{i}{2}\int (z_2 - z)^{-n_1-1}(\bar{z}_2 - \bar{z})^{-n_2-1}K(z_1, z; -n_1, -n_2)\, dz\, d\bar{z} \tag{3}$$

for $n_1, n_2$ not integers of the same sign. As before, the integral of (3) is to be understood in the sense of its regularization.[28]

[28] At points where $n_1$ and $n_2$ are integers of the same sign, the integrals of Eq. (3), treated as analytic functions of $n_1$ and $n_2$, have singularities (see the Footnote 6 of Section 4.4, Chapter III).

We have thus obtained the conditions on the kernel $K(z_1, z_2; \chi)$. As was stated, these relations have to do with the fact that the representations defined by $(n_1, n_2)$ and $(-n_1, -n_2)$ are equivalent. In fact we already obtained these relations when we were studying the Fourier transforms of square integrable functions on the group. We established them there, however, only for $\chi = (n_1, n_2)$ such that $n_2 = -\bar{n}_1$ (the analog of the real axis), since the Fourier transform of a square integrable function is defined only on the "real axis." It can be shown that the more rapidly $f(g)$ decreases, the broader the strip in which its Fourier transform is defined and the greater the number of relations it then satisfies.

Recall now that intertwining operators may exist not only for equivalent representations, but also for partially equivalent ones with $\chi_1 = (n_1, n_2)$ and $\chi_2 = (-n_1, n_2)$, where $n_1$ is a positive integer, as well as for representations with $\chi_1 = (n_1, n_2)$ and $\chi_2 = (n_1, -n_2)$, where $n_2$ is a positive integer. In the first of these cases the intertwining operator is of the form $A = \partial^{n_1}/\partial z^{n_1}$ and in the second $A = \partial^{n_2}/\partial \bar{z}^{n_2}$ (see Section 5.3 of Chapter III).

Remark. The existence of intertwining operators $A$ is related to the degeneracy of the representations at integer points $\chi$. Specifically, let $\chi = (n_1, n_2)$ be an integer point (that is, let $n_1$ and $n_2$ be integers of the same sign, which we shall choose in this instance to be positive). Then we have seen in Chapter III that $D_\chi$ contains the finite-dimensional invariant subspace $E_\chi$, and $D_{-\chi}$ contains the infinite-dimensional invariant subspace $F_{-\chi}$. Further, $D_\chi/E_\chi \simeq D_{-n_1, n_2} \simeq D_{n_1, -n_2} \simeq F_{-\chi}$. #

Thus in addition to Eq. (1), $F(\chi) \equiv F(n_1, n_2)$ satisfies the further relations

$$A_1 F(n_1, n_2) = F(-n_1, n_2) A_1,$$

and

$$A_2 F(n_1, n_2) = F(n_1, -n_2) A_2,$$

where $A_1 = \partial^{n_1}/\partial z^{n_1}$ and $A_2 = \partial^{n_2}/\partial \bar{z}^{n_2}$ for $n_1, n_2 = 1, 2, \ldots$. In terms of $K(z_1, z_2; \chi)$, these relations are

$$\frac{\partial^{n_1}}{\partial z_1^{n_1}} K(z_1, z_2; n_1, n_2) = (-1)^{n_1} \frac{\partial^{n_1}}{\partial z_2^{n_1}} K(z_1, z_2; -n_1, n_2) \quad \text{for} \quad n_1 = 1, 2, \ldots; \tag{4}$$

$$\frac{\partial^{n_2}}{\partial \bar{z}_1^{n_2}} K(z_1, z_2; n_1, n_2) = (-1)^{n_2} \frac{\partial^{n_2}}{\partial \bar{z}_2^{n_2}} K(z_1, z_2; n_1, -n_2) \quad \text{for} \quad n_2 = 1, 2, \ldots. \tag{4'}$$

As we have mentioned, these relations occur as a result of the degeneracy of $T_\chi(g)$ at integer points.

### 5.5. Moments of $f(g)$ and Their Expression in Terms of the Kernel

We wish to make a brief aside concerning what we shall call the *moments* of $f(g)$. We define the moments of $f(g)$ as integrals of the form

$$\int f(g)a(g)\,dg,$$

where $a(g)$ is a matrix element of a finite-dimensional irreducible representation of $G$. It will be shown that if $f(g)$ is a rapidly decreasing function, all of its moments are easily expressed in terms of $K(z_1, z_2; n_1, n_2)$.

First we shall need to calculate the $a(g)$. We have constructed the irreducible finite-dimensional representations of $G$ in Section 3.1 of Chapter III. Each such representation is specified by a pair of positive integers $n_1, n_2$, and is realized on the space $E_{n_1,n_2}$ of polynomials of degree $n_1 - 1$ and lower in $z$ and of degree $n_2 - 1$ and lower in $\bar{z}$. Then the representation of

$$g = \begin{Vmatrix} \alpha & \beta \\ \gamma & \delta \end{Vmatrix}$$

is

$$T_\chi(g)\varphi(z) = (\beta z + \delta)^{n_1-1}(\bar{\beta}\bar{z} + \bar{\delta})^{n_2-1}\varphi\left(\frac{\alpha z + \gamma}{\beta z + \delta}\right).$$

It can be shown that up to equivalence these are all the irreducible finite-dimensional representations of $G$.

We shall now introduce a basis into $E_{n_1,n_2}$. Let the elements of this basis be the monomials $z^{k_1}\bar{z}^{k_2}$, where $k_j = 0, 1, ..., n_j - 1$. Then to each operator of the representation will correspond an $n_1 n_2$-dimensional matrix whose elements are conveniently denoted by index pairs $k_1, k_2$ and $l_1, l_2$, where the $l_j$ take on the same values as the $k_j$. Then $a_{k_1,k_2;l_1,l_2}(g)$ is the coefficient of $z^{l_1}\bar{z}^{l_2}$ in the polynomial

$$T_\chi(g)z^{k_1}\bar{z}^{k_2} = (\beta z + \delta)^{n_1-k_1-1}(\bar{\beta}\bar{z} + \bar{\delta})^{n_2-k_2-1}(\alpha z + \gamma)^{k_1}(\bar{\alpha}\bar{z} + \bar{\gamma})^{k_2}.$$

A simple calculation then leads to

$$_{k_1,k_2;l_1,l_2}(g) = \frac{1}{l_1!l_2!}$$

$$\times \frac{\partial^{l_1+l_2}}{\partial z^{l_1}\,\partial \bar{z}^{l_2}}[(\beta z + \delta)^{n_1-k_1-1}(\bar{\beta}\bar{z} + \bar{\delta})^{n_2-k_2-1}(\alpha z + \gamma)^{k_1}(\bar{\alpha}\bar{z} + \bar{\gamma})^{k_2}]_{z=0}. \quad (1)$$

We may now go on to express the moment

$$\int f(g)a_{k_1,k_2;l_1,l_2}(g)\,dg$$

in terms of $K(z_1, z_2; n_1, n_2)$, which is defined, as before, by

$$\int f(g)T_\chi(g)\varphi(z_1)\, dg = \frac{i}{2}\int K(z_1, z_2; n_1, n_2)\varphi(z_2)\, dz_2\, d\bar{z}_2 .$$

In this defining integral we now write $\varphi(z_1) = z_1^{k_1}\bar{z}_1^{k_2}$. This yields

$$\int f(g)(\beta z_1 + \delta)^{n_1-k_1-1}(\bar{\beta}\bar{z}_1 + \bar{\delta})^{n_2-k_2-1}(\alpha z_1 + \gamma)^{k_1}(\bar{\alpha}\bar{z}_1 + \bar{\gamma})^{k_2}\, dg$$
$$= \frac{i}{2}\int K(z_1, z_2; n_1, n_2)z_2^{k_1}\bar{z}_2^{k_2}\, dz_2\, d\bar{z}_2 . \tag{2}$$

With this and Eq. (1) we arrive at

$$\int f(g)a_{k_1,k_2;l_1,l_2}(g)\, dg$$
$$= \frac{1}{l_1!l_2!}\frac{i}{2}\frac{\partial^{l_1+l_2}}{\partial z_1^{l_1}\,\partial\bar{z}_1^{l_2}}\int K(z_1, z_2; n_1, n_2)z_2^{k_1}\bar{z}_2^{k_2}\, dz_2\, d\bar{z}_2\,|_{z_1=0} . \tag{3}$$

In particular, for $n_1 = n_2 = 1$ this becomes

$$\int f(g)\, dg = \frac{i}{2}\int K(z_1, z_2; 1, 1)\, dz_2\, d\bar{z}_2$$

(note that the integral on the right-hand is side independent of $z_1$).

It can be shown also that the moments of $f(g)$ have similar expressions in terms of $K(z_1, z_2; -n_1, -n_2)$.

As a by-product we have obtained the following interesting result. The integral

$$\frac{i}{2}\int K(z_1, z_2; n_1, n_2)z_2^{k_1}\bar{z}_2^{k_2}\, dz_2\, d\bar{z}_2 ,$$

(where $n_1$ and $n_2$ are positive integers and the $k_j = 0, 1, ..., n_j$) is a polynomial of degree no higher than $n_1 - 1$ in $z_1$ and no higher than $n_2 - 1$ in $\bar{z}_1$. This result could also have been easily obtained from Eqs. (4) and (4′) of Section 5.4. Note that from those symmetry relations we could have deduced also that

$$\frac{i}{2}\int K(z_1, z_2; -n_1, -n_2)z_1^{k_1}\bar{z}_1^{k_2}\, dz_1\, d\bar{z}_1$$

is a polynomial of degree no higher than $n_1 - 1$ in $z_2$ and no higher than $n_2 - 1$ in $\bar{z}_2$.

## 5.6. The Paley-Wiener Theorem for the Fourier Transform on $G$

In the preceding few sections we established the correspondence between the infinitely differentiable rapidly decreasing functions on $G$ and their integrals over all possible "line generators." If such a generator is given by the set of equations

$$\alpha\beta - \gamma\delta = 1,$$
$$u\alpha + v\gamma = u',$$
$$u\beta + v\delta = v',$$

then the integral along this generator is defined by

$$\Phi(u, v; u', v') = \frac{i}{2}\int f(g)\omega\bar{\omega}, \tag{1}$$

where $\omega = d\alpha/vu' = d\beta/vv' = -d\gamma/uu' = -d\delta/uv'$ [see also the equivalent statement in terms of $\varphi(z_1, z_2; \lambda)$ in Eq. (1) of Section 5.1]. Now recall that in Sections 3 and 4 we showed that $\Phi(u, v; u', v')$ is infinitely differentiable, found its asymptotic behavior, and showed that it satisfies certain additional conditions [which we formulated in terms of $K(z_1, z_2; \chi)$]. It turns out that all of these properties together are not only necessary, but also are sufficient conditions in order that a function $\Phi(u, v; u', v')$ be representable by Eq. (1) where $f(g) \in S$. In other words we may assert the following.

**Theorem.** (*Analog of the Paley-Wiener theorem*). *The necessary and sufficient conditions that a function $\Phi(u, v; u', v')$ have an integral representation of the form of Eq.* (1) *where $f(g)$ is an infinitely differentiable rapidly decreasing function on $G$, are the following.*

1. *$\Phi(u, v; u', v')$ is homogeneous of degree $(-2, -2)$, or in other words*

$$\Phi(\alpha u, \alpha v; \alpha u', \alpha v') = \alpha^{-2}\bar{\alpha}^{-2}\Phi(u, v; u', v')$$

*for all $\alpha \neq 0$.*

2. *$\Phi(u, v; u', v')$ is infinitely differentiable in $u, v; u', v'$ and their complex conjugates everywhere except at $u = v = 0$ and $u' = v' = 0$.*

3. *For every number $k$ the function*

$$\left(\frac{|u'|^2 + |v'|^2}{|u|^2 + |v|^2}\right)^k (|u|^2 + |v|^2)(|u'|^2 + |v'|^2)\Phi(u, v; u', v')$$

*is bounded in $u$, $v$; $u'$, $v'$, and so are all of its derivatives of the form $P(X)\Phi$ where $P(X)$ is a polynomial of Lie operators with constant coefficients.*[29]

In addition, $\Phi(u, v; u', v')$ must satisfy certain symmetry relations which are more conveniently formulated in terms of the kernel

$$K(z_1, z_2; n_1, n_2) = \frac{i}{2} \int \Phi(z_1, 1; \lambda z_2, \lambda) \lambda^{n_1} \bar{\lambda}^{n_2} \, d\lambda \, d\bar{\lambda}. \tag{2}$$

4. *$K(z_1, z_2; \chi)$, where $\chi = (n_1, n_2)$, satisfies the symmetry relation*

$$\frac{i}{2} \int (z_1 - z)^{-n_1 - 1} (\bar{z}_1 - \bar{z})^{-n_2 - 1} K(z, z_2; n_1, n_2) \, dz \, d\bar{z}$$
$$= \frac{i}{2} \int (z - z_2)^{-n_1 - 1} (\bar{z} - \bar{z}_2)^{-n_2 - 1} K(z_1, z; -n_1, -n_2) \, dz \, d\bar{z}. \tag{3}$$

5. *If $n_1$ is a positive integer, then*

$$\frac{\partial^{n_1}}{\partial z_1^{n_1}} K(z_1, z_2; n_1, n_2) = (-1)^{n_1} \frac{\partial^{n_1}}{\partial z_2^{n_1}} K(z_1, z_2; -n_1, n_2). \tag{4}$$

5′. *If $n_2$ is a positive integer, then*

$$\frac{\partial^{n_2}}{\partial \bar{z}_1^{n_2}} K(z_1, z_2; n_1, n_2) = (-1)^{n_2} \frac{\partial^{n_2}}{\partial \bar{z}_2^{n_2}} K(z_1, z_2; n_1, -n_2). \tag{4'}$$

*The symmetry relations expressed by Eqs.* (3), (4), *and* (4′) *are satisfied also by the kernels corresponding to derivatives of the form $P(X)\Phi$.*

**Proof.** The *necessity* has already been established in Sections 5.3 and 5.4. We now prove the *sufficiency*. Let $\Phi(u, v; u', v')$ satisfy the conditions of the theorem.

We show first that a square integrable function of $g$ exists such that its integrals along the linear generators give $\Phi(u, v; u', v')$. As elsewhere, we transform to the homogeneous coordinates $z_1$, $z_2$, $\lambda$ and write

$$\varphi(z_1, z_2; \lambda) = \lambda \bar{\lambda} \Phi(z_1, 1; \lambda z_2, \lambda). \tag{5}$$

Then Condition 3 implies that the integral

$$\left(\frac{i}{2}\right)^3 \int |\varphi'_\lambda(z_1, z_2; \lambda)|^2 \, dz_1 \, d\bar{z}_1 \, dz_2 \, d\bar{z}_2 \, d\lambda \, d\bar{\lambda} \tag{6}$$

[29] Recall that these Lie operators are

$$-\tfrac{1}{2}\left(u \frac{\partial}{\partial u} - v \frac{\partial}{\partial v}\right), \quad -u \frac{\partial}{\partial v}, \quad -v \frac{\partial}{\partial u}$$

and similar operators in $u'$ and $v'$ and the complex conjugates of all these variables (see Section 5.2).

converges. Indeed, it is easily seen that

$$\partial\varphi/\partial\lambda = \bar{\lambda}\Phi_1(z_1, 1; \lambda z_2, \lambda),$$

where $\Phi_1$ also satisfies Condition 3. Now consider

$$\begin{aligned} \int |\varphi_\lambda'(z_1, z_2; \lambda)|^2 \, dv &= \int |\lambda|^2 \, |\Phi_1(z_1, 1; \lambda z_2, \lambda)|^2 \, dv \\ &= \int |\Phi_1(z_1, 1; z_2, \lambda)|^2 \, dv, \end{aligned} \tag{7}$$

where $dv = (i/2)^3 \, dz_1 \, d\bar{z}_1 \, dz_2 \, d\bar{z}_2 \, d\lambda \, d\bar{\lambda}$. We break up the region of integration into a region I where $1 + |z_1|^2 < |z_2|^2 + |\lambda|^2$ and another II where $|z_2|^2 + |\lambda|^2 < 1 + |z_1|^2$. Then Condition 3 implies that

$$\int_{\text{I}} |\Phi_1(z_1, 1; z_2, \lambda)|^2 \, dv < C_1 \int_{\text{I}} \frac{dv}{(|z_2|^2 + |\lambda|^2)^4},$$

$$\int_{\text{II}} |\Phi_1(z_1, 1; z_2, \lambda)|^2 \, dv < C_2 \int_{\text{II}} \frac{dv}{(|z_1|^2 + 1)^4}.$$

Obviously the integrals on the right-hand sides converge.

To proceed we make use of the results of Section 3.5, where we found the conditions under which $\varphi(z_1, z_2; \lambda)$ is the integral of a square integrable function $f(g)$ along a line generator. These conditions were two. The first is that the kernel defined in Eq. (2), which we now write in the form

$$K(z_1, z_2; n_1, n_2) = \frac{i}{2} \int \varphi(z_1, z_2; \lambda) \lambda^{n_1 - 1} \bar{\lambda}^{n_2 - 1} \, d\lambda \, d\bar{\lambda},$$

be such that the integral

$$\sum_{n=-\infty}^{\infty} \int_{-\infty}^{\infty} (n^2 + \rho^2) \times \left\{ \left(\frac{i}{2}\right)^2 \int \left| K\left(z_1, z_2; \frac{n + i\rho}{2}, \frac{-n + i\rho}{2}\right) \right|^2 dz_1 \, d\bar{z}_1 \, dz_2 \, d\bar{z}_2 \right\} d\rho \tag{8}$$

converge. In our case this condition is fulfilled since this integral, as is easily shown, differs from Eq. (7) only by a factor. The second condition is that $\varphi(z_1, z_2; \lambda)$ have a certain symmetry property which, when stated in terms of $K(z_1, z_2; \chi)$, is just Eq. (3) (see the end of Section 3.6).

We thus see that $\varphi(z_1, z_2; \lambda)$ is the integral along a line generator of some square integrable $f(g)$. According to Section 3.3, $f(g)$ is given in terms of $\varphi(z_1, z_2; \lambda)$ by

$$f(g) = -\frac{1}{2\pi^2}\frac{i}{2}\int \varphi''_{\bar{\lambda}\lambda}\left(z, \frac{\alpha z + \gamma}{\beta z + \delta}; \beta z + \delta\right) dz\, d\bar{z}. \tag{9}$$

This formula may conveniently be written in the homogeneous coordinates. It then becomes[30]

$$f(g) = -(2\pi^2)^{-1}\left(\frac{i}{2}\right)\int_{\Gamma} \Psi(z_1, z_2; \alpha z_1 + \gamma z_2, \beta z_1 + \delta z_2)$$
$$\times (z_1\, dz_2 - z_2\, dz_1)(\bar{z}_1\, d\bar{z}_2 - \bar{z}_2\, d\bar{z}_1), \tag{10}$$

where

$$\Psi(u, v; u', v') = \frac{\partial^2}{\partial\lambda\, \partial\bar{\lambda}} [\lambda\bar{\lambda}\Phi(u, v; u', v')]_{\lambda=1}.$$

The integral here is along any contour $\Gamma$ in the $(z_1, z_2)$ plane that crosses each line passing through the origin exactly once (see Volume 1, Section B2.5). Equation (9) is obtained from (10) if $\Gamma$ is chosen to be $z_2 = 1$.

The integral in Eq. (10) converges. This is because it is independent of the choice of $\Gamma$, which may then be chosen to lie within some bounded region. Then the integral is taken over a compact set, and its convergence follows immediately from the fact that the integrand is bounded on this set.

What remains is to show that $f(g)$ as defined by Eq. (10) is infinitely differentiable and rapidly decreasing (with all of its derivatives). The infinite differentiability follows immediately from Conditions 3 and 4 imposed on $\Phi(u, v; u', v')$. Let us then show that $f(g)$ decreases rapidly.

[30] Equation (10) can also be written in the form

$$f(g) = -(8\pi^2)^{-1}\left(\frac{i}{2}\right)\int_{\Gamma} L\bar{L}\Phi(z_1, z_2; \alpha z_1 + \gamma z_2, \beta z_1 + \delta z_2)$$
$$\times (z_1\, dz_2 - z_2\, dz_1)(\bar{z}_1\, d\bar{z}_2 - z_2\, d\bar{z}_1),$$

where

$$L = u\frac{\partial}{\partial u} - v\frac{\partial}{\partial v} + u'\frac{\partial}{\partial u'} + v'\frac{\partial}{\partial v'}.$$

This is the form in which we obtained it in Section 2.4 of Chapter II, where it is Eq. (9).

We first write $g$ in the form $g = k_1 g_\epsilon k_2$, where $k_1$ and $k_2$ are unitary matrices,[31] and

$$g_\epsilon = \begin{Vmatrix} 0 & -\epsilon^{-1} \\ \epsilon & 0 \end{Vmatrix}.$$

We then have $|g|^2 = |\epsilon|^2 + |\epsilon|^{-1}$, so that as $|g| \to \infty$, either $\epsilon \to 0$ or $|\epsilon| \to \infty$. Without loss of generality we may assume that $\epsilon \to 0$, so that what we must show is that in this limit the function $f(k_1 g_\epsilon k_2)$ decreases faster than any power of $|\epsilon|$. Now the properties of $\Phi(u, v; u', v')$, and therefore also those of $f(g)$, are preserved under right and left translations. Therefore we need only show that $f(g_\epsilon)$ decreases rapidly as $\epsilon \to 0$.

Let us now apply (9) to $g_\epsilon$. We then have

$$\begin{aligned} f(g_\epsilon) &= -(2\pi^2)^{-1}\left(\frac{i}{2}\right)\int \varphi_{\lambda\tilde{\lambda}}(z, -\epsilon^2 z^{-1}; -\epsilon^{-1} z)\, dz\, d\bar{z} \\ &= -(2\pi^2)^{-1}\epsilon\bar{\epsilon}\left(\frac{i}{2}\right)\int \varphi_{\lambda\tilde{\lambda}}(-\epsilon z, \epsilon z^{-1}; z)\, dz\, d\bar{z}. \end{aligned} \tag{11}$$

It is easily shown by using Conditions 2 and 3 of the theorem that this integral is an infinitely differentiable function of $\epsilon$ in the neighborhood of $\epsilon = 0$. We may therefore expand $f(g_\epsilon)$ in an asymptotic Taylor's series about this point, writing

$$f(g_\epsilon) \sim -\frac{1}{2\pi^2}\sum_{k,l=0}^{\infty} a_{kl}\frac{\epsilon^{k+1}\bar{\epsilon}^{l+1}}{k!\,l!}, \tag{12}$$

where

$$\begin{aligned} a_{kl} = \frac{i}{2}\int \left(-z\frac{\partial}{\partial z_1} + z^{-1}\frac{\partial}{\partial z_2}\right)^k \left(-\bar{z}\frac{\partial}{\partial \bar{z}_1} + \bar{z}^{-1}\frac{\partial}{\partial \bar{z}_2}\right)^l \\ \times \varphi_{\lambda\tilde{\lambda}}(0, 0; z)\, dz\, d\bar{z}. \end{aligned} \tag{13}$$

We shall now show that all the $a_{kl}$ vanish, which is then a proof that $f(g_\epsilon)$ is rapidly decreasing. For this purpose we expand out the

[31] It is always possible to write $g$ in this form. This is because any matrix can be written $g = ka$, where $k$ is unitary and $a$ is positive definite. Further, there exists a unitary matrix $k_2$ which diagonalizes $a$, that is, such that $a = k_2^{-1}\delta k_2$, where $\delta$ is diagonal. Then $g = (kk_2^{-1}s^{-1})(s\delta)k_2$, where

$$s = \begin{Vmatrix} 0 & -1 \\ 1 & 0 \end{Vmatrix}.$$

This is obviously the required form.

differential operators in the integrand and integrate by parts. This leads to

$$a_{kl} = \sum_{p+q=k,\, p_1+q_1=l} (-1)^{p+p_1}(p-q)(p_1-q_1)C_k^p C_l^{p_1}$$

$$\times \frac{i}{2} \int z^{p-q-1} \bar{z}^{p_1-q_1-1} \frac{\partial^{p+q+p_1+q_1} \varphi(0,0;z)}{\partial z_1^p \, \partial z_2^q \, \partial \bar{z}_1^{p_1} \, \partial \bar{z}_2^{q_1}} \, dz \, d\bar{z}, \tag{14}$$

where $C_k^p$ and $C_l^{p_1}$ are binomial coefficients. In terms of $K(z_1, z_2; n_1, n_2)$ this may be written [see the relation between the kernel and $\varphi$ just above Eq. (8)]

$$a_{kl} = \sum_{p+q=k,\, p_1+q_1=l} (-1)^{p+p_1}(p-q)(p_1-q_1)C_k^p C_l^{p_1}$$

$$\times \frac{\partial^{p+q+p_1+q_1}}{\partial z_1^p \, \partial z_2^q \, \partial \bar{z}_1^{p_1} \, \partial \bar{z}_2^{q_1}} K(0,0; p-q, p_1-q_1).$$

We now combine each term that has a given $p > q$ and $p_1 > q_1$ with all those terms obtained by interchanging the indices $p$ and $q$ and the indices $p_1$ and $q_1$. The terms come in groups of four:

$$c_{pq} \frac{\partial^{2q+2q_1}}{\partial z_1^q \, dz_2^q \, d\bar{z}_1^{q_1} \, \partial \bar{z}_2^{q_1}} \Bigg[ (-1)^{n_1+n_2} \frac{\partial^{n_1+n_2}}{\partial z_1^{n_1} \, \partial \bar{z}_1^{n_2}} K(0,0; n_1, n_2)$$

$$-(-1)^{n_2} \frac{\partial^{n_1+n_2}}{\partial z_2^{n_1} \, \partial \bar{z}_1^{n_2}} K(0,0; -n_1, n_2)$$

$$-(-1)^{n_1} \frac{\partial^{n_1+n_2}}{\partial z_1^{n_1} \, \partial \bar{z}_2^{n_2}} K(0,0; n_1, -n_2)$$

$$+ \frac{\partial^{n_1+n_2}}{\partial z_2^{n_1} \, \partial \bar{z}_2^{n_2}}, K(0,0; -n_1, -n_2) \Bigg],$$

where $c_{pq} = (-1)^{q+q_1}(p-q)(p_1-q_1)C_k^p C_l^{p_1}$, $n_1 = p - q$, $n_2 = p_1 - q_1$. Then from Conditions 5 and 5′ of the theorem it follows immediately that the expression in square brackets vanishes, and therefore that $a_{kl} = 0$. This completes the proof of the rapid decrease of $f(g)$. That the derivatives $P(X)f$ are also rapidly decreasing may be seen as follows. Recall that $P(X)f$ corresponds to $P(X)\Phi$, and the latter function also satisfies the conditions of the theorem. Therefore $P(X)f(g)$ is also a rapidly decreasing function. This completes the proof of the analog of the Paley-Wiener theorem.

CHAPTER V

# INTEGRAL GEOMETRY IN A SPACE OF CONSTANT CURVATURE

In this chapter we shall study the analogs for a Lobachevskian and imaginary Lobachevskian space of the Radon problem for a Euclidean space, which was discussed in Chapter I.

Whereas previously with every $f(x)$ we associated its integrals on all possible hyperplanes [the Radon transform of $f(x)$] we shall now associate with every $f(x)$ defined on a Lobachevskian or imaginary Lobachevskian space its integrals over all possible hypersurfaces that are analog of hyperplanes, which we shall call horospheres. The geometry on the horospheres is Euclidean, and it is for this reason that they can be treated as analogs of hyperplanes in a Euclidean space (another analog would be actual hyperplanes in the Lobachevskian space). It is in this way that we extend the Radon transform to Lobachevskian and imaginary Lobachevskian spaces (Sections 2.1–3.5). We call the reader's attention to our method for averaging a functional over a noncompact manifold (hyperboloid) even when the formal averaging procedure gives rise to divergent integrals. The method of analytic continuation in the coordinates developed for this purpose is described in Section 3.2. It is our feeling that this averaging procedure will be fruitful also in very many other applications (for instance, in the theory of the representation of noncompact groups).

In the same way as the Radon transform is closely related to the Fourier transform in a Euclidean space, the analogous transform in a Lobachevskian or imaginary Lobachevskian space is related to the expansion of functions defined on this space in Fourier integrals. We shall use the results of Chapter V to develop these Fourier integral expansions in the next chapter.

## 1. Spaces of Constant Curvature[1]

### 1.1. Spherical and Lobachevskian Spaces

We shall start by stating the basic facts concerning spaces of constant curvature.

The simplest model of an $n$-dimensional space of constant *positive* curvature $k$ is the $n$-dimensional sphere

$$x_0^2 + x^2 + \cdots + x_n^2 = R^2, \qquad R = 1/k,$$

imbedded in an $(n + 1)$-dimensional Euclidean space $E_{n+1}$. The distance $\rho$ between points on the sphere will be defined by $\rho = R\alpha$, where $\alpha$ is the angle subtended at the center of the sphere by the arc of the great circle connecting these two points. In other words the distance between points on the sphere is defined as the arc length of the great circle connecting the points.

It is convenient to identify diametrically opposite points on the sphere and to define the distance between two points then as the shorter arc length. This defines a new space of constant positive curvature $k = 1/R$, and this space is called an *elliptic space* or a *Riemannian space.*

Now instead of pairs of diametrically opposite points on the sphere, let us consider the lines passing through these points. Then a Riemannian space can be defined as the space of all lines passing through the origin in $E_{n+1}$. The coordinate $(x_0, x_1, \ldots, x_n)$ of the points on these lines may then be treated as homogeneous coordinates in the Riemannian space. The distance $r$ between two lines, one of which passes through a point $M(x)$ and the other through a point $N(y)$, will be defined by

$$\cos^2 kr = \frac{(x, y)^2}{(x, x)(y, y)}, \qquad 0 \leqslant kr \leqslant \frac{\pi}{2}, \tag{1}$$

where

$$(x, y) = x_0 y_0 + x_1 y_1 + \cdots + x_n y_n \tag{2}$$

is the scalar product in $E_{n+1}$.

Mappings of a space onto itself which preserve the distances between points (and do not change orientations if the space is orientable) are called *motions* of the space. It is easily verified that the motions of the

[1] In what follows we shall use the generally accepted terminology "a space of constant positive (or negative) curvature." We shall, however, not have to make use of the concept of curvature itself, but use the term as a synonym for a spherical (or eliptical) space and a Lobachevskian space.

$n$-dimensional sphere and the $n$-dimensional Riemannian space are induced by orthogonal transformations of determinant one in $E_{n+1}$. (The transformations $g$ and $-g$ induce the same motion in a Riemannian space of odd dimension, and must therefore be identified.)

Let us now define a space of *negative* curvature by analogy with one of positive curvature. For this purpose consider the $(n+1)$-dimensional linear space $E_{n+1}$ with the bilinear form

$$[x, y] \equiv x_0 y_0 - x_1 y_1 - \cdots - x_n y_n \tag{3}$$

defined on it. Consider further the set of all lines passing through the origin of $E_{n+1}$ and lying inside the cone whose equation is

$$[x, x] \equiv x_0^2 - x_1^2 - \cdots - x_n^2 = 0, \tag{4}$$

that is, all lines whose points satisfy the inequality $[x, x] > 0$. We shall define the distance $r$ between two lines one of which passes through a point $M(x)$ and the other through a point $N(y)$ by

$$\cosh^2 kr = \frac{[x, y]^2}{[x, x][y, y]}, \tag{5}$$

the analog of Eq. (1). It is easily shown that for lines in the interior of the $[x, x] = 0$ cone the right-hand side of this equation is greater than or equal to one, being equal to one if and only if the two lines coincide. This means that as defined by Eq. (5), $r$ is a real nonnegative number equal to zero if and only if the two lines coincide. A direct calculation shows simply that this distance $r$ satisfies the usual axioms required for a distance function, namely symmetry and the triangular inequality.

The metric space obtained in this way is called an $n$-dimensional *hyperbolic space* or a *Lobachevskian space*. Then $k$ is called the *curvature* of the space.

Note that as a line approaches the $[x, x] = 0$ cone the distance between it and any fixed line increases without limit. Thus the lines on the cone are the points at infinity in a Lobachevskian space. The set of all points at infinity is called *absolute* of the space. Thus the absolute of a Lobachevskian space is the set of all lines on the $[x, x] = 0$ cone.

We shall denote by *hyperbolic rotations* the linear transformations of unit determinant in $E_{n+1}$ which preserve the quadratic form $[x, x]$ and which transform the absolute into itself. Obviously hyperbolic rotations do not change the distance between lines, and therefore induce motions of the Lobachevskian space. It can be shown that every motion of a Lobachevskian space is given by a hyperbolic rotation of $E_{n+1}$ and that the group of motions of a Lobachevskian space is *transitive*, that is, every point of the space can be transformed into any other by some motion.

### 1.2. Some Models of Lobachevskian Spaces

We have defined Lobachevskian $n$-space by identifying its point with the lines passing through the origin of $E_{n+1}$ and lying inside the $[x, x] = 0$ cone. Let us now go on to discuss some other models of these spaces.

Consider any surface inside the upper half of the cone such that every line passing through the origin intersects this surface at one and only one point, and associate with every such line its point of intersection with this surface. This gives rise to a model of the Lobachevskian space as the set of points on the surface. The two most common models obtained in this way are the following.

1. The set of points on the upper sheet of the two-sheeted hyperboloid

$$[x, x] = 1.$$

The distance $r$ between any two points and $y$ is then given by

$$\cosh kr = [x, y], \tag{1}$$

where we have used the fact that on the upper sheet of the hyperboloid $[x, y] \geqslant 1$.

From this formula we immediately obtain the expression

$$k^2\, ds^2 = -dx_0^2 + dx_1^2 + \cdots + dx_n^2 \tag{2}$$

for the differential element of arc length, where the differentials $dx_0, \dots dx_n$ are related, in view of the fact that $[x, x] = 1$, by the equation

$$x_0\, dx_0 - x_1\, dx_1 - \cdots - x_n\, dx_n = 0.$$

Although at first glance Eq. (2) does not seem to be positive definite, the relation between the differentials guarantees that it is in fact positive definite.

2. The set of those points on the $x_0 = 1$ plane that lie within the $[x, x] = 0$ cone. Obviously this plane intersects the cone in a (hyper-) sphere of unit radius, so that the Lobachevskian space may be interpreted as the interior of an $n$-dimensional sphere of unit radius (the ball of unit radius). Since a linear transformation on the space of the lines corresponds to a projective transformation of the $x_0 = 1$ plane, the motions of the Lobachevskian space in this model are realized as projective transformations in $n$-dimensions that preserve the unit ball.

Note that for this case the set of points at infinity (the absolute) is the unit sphere. Equation (5) of Section 1.1 can be used to show that in this model the distance $r$ between two points $M$ and $N$ is given by

$$r = \frac{1}{2k} \ln \left( \frac{MA}{NA} : \frac{MB}{NB} \right),$$

where $A$ and $B$ are the points of intersection of the chord $MN$ and the sphere of radius $k^{-1}$.

### 1.3. Imaginary Lobachevskian Spaces

We have been discussing the lines passing through the origin and lying within the $[x, x] = 0$ cone in $E_{n+1}$. Let us now consider *all* lines passing through the origin in $E_{n+1}$. We shall define the "distance" $r$ between any two such lines (assuming they are not on the $[x, x] = 0$ cone by the same formula used in the Lobachevskian space, namely by

$$\cosh^2 kr = \frac{[x, y]^2}{[x, x][y, y]}. \tag{1}$$

This "distance" is an invariant of a pair of points under linear transformation in $E_{n+1}$ which preserve the bilinear form

$$[x, y] = x_0 y_0 - x_1 y_1 - \cdots - x_n y_n .$$

The set of all lines passing through the origin breaks up into three families, on each of which these transformations are transitive. The first of these consists of those lines for which $[x, x] > 0$, which form the Lobachevskian space discussed above. The second consists of lines such that $[x, x] = 0$, which form the absolute of the Lobachevskian space. There is, finally, the third family consisting of lines such that $[x, x] < 0$, and we shall call this space, with the given definition of distance, an *imaginary Lobachevskian space.* In other words, an imaginary Lobachevskian is the set of lines in $E_{n+1}$ that pass through the origin and such that $[x, x] < 0$ (i.e., those lying outside the $[x, x] = 0$ cone). As before, the distance $r$ between two lines in this space is defined by Eq. (1).

Unlike the situation in a Lobachevskian space, in an imaginary Lobachevskian space the distance $r$ is no longer necessarily positive. In fact Eq. (1) shows that $\cosh kr$ can take on any value from zero to infinity. Therefore the distance between two points in an imaginary Lobachevskian space can be either real and nonnegative (if $1 \leqslant \cosh kr < \infty$) or imaginary in the interval $[0, \pi i/2k]$ (if $0 \leqslant \cosh kr \leqslant 1$).

Let us find the motions of an imaginary Lobachevskian space. Consider the linear transformation of unit determinant in $E_{n+1}$ which preserve $[x, x]$ and transform each sheet of the $[x, x] = 0$ cone into itself. These transformations obviously induce distance-preserving transformations in the imaginary Lobachevskian space, and we shall call them the motions of this space. The group of motions of this space is isomorphic to the group of motions of ordinary Lobachevskian space. According to our definition of an imaginary Lobachevskian space, the coordinates $x$ (that is, $x_0, \ldots, x_n$) of the lines are homogeneous coordinates in the space, and we shall later find occasion to specialize to particular cases by normalizing in one way or another. Different normalizations will give different models of the imaginary Lobachevskian space. The most convenient of these are the following.

1. If we normalize the coordinates according to $[x, x] = -1$, the model is the surface of a single-sheeted hyperboloid whose diametrically opposite points are identified. The distance $r$ between two points $x$ and $y$ of the hyperboloid is given by

$$\cosh^2 kr = [x, y]^2. \tag{2}$$

2. If we set $x_0 = 1$, the points of our imaginary Lobachevskian space are given by $n$ coordinates such that $x_1^2 + \cdots + x_n^2 > 1$. This is a realization of an imaginary Lobachevskian space on the exterior of the unit sphere in $n$ dimensions. In this model the motions are the projective transformations of the entire $n$-dimensional space that transform the exterior of the sphere into itself.

### 1.3a. Isotropic Lines of an Imaginary Lobachevskian Space

We now wish to define what we shall call *isotropic lines* in an imaginary Lobachevskian space. Since a point in such a space is defined as a line passing through the origin in $E_{n+1}$, it would be natural to call a "line" in such a space a two-dimensional plane passing through the origin of $E_{n+1}$. Hence a line in an imaginary Lobachevskian space is the set of of points of the form $x = sa + tb$, where $a$ and $b$ are fixed vectors and $s$ and $t$ are arbitrary numbers. We may without loss of generality assume that the basis vectors $a$ and $b$ are normalized according to $[a, a] = [b, b] = -1$. A line in an imaginary Lobachevskian space is called *isotropic* if the distance between any two points on it vanishes. We may then ask for the equation of an isotropic line.

Let an isotropic line be given by $x = sa + tb$, where $a, b$ are fixed vectors normalized as above, and $-\infty < s, t < \infty$. Since the distance between $a$ and $b$ must vanish, it follows from the fact that $\cosh kr = |[a, b]|$ that $|[a, b]| = 1$. We may now asume that $[a, b] = -1$, changing, if necessary, the direction of $b$. Recall also that $[a, a] = [b, b] = -1$.

It is more convenient to use the basis vectors $a$ and $\xi = b - a$, rather than $a$ and $b$. For these vectors we have $[\xi, \xi] = [a, \xi] = 0$. Thus an isotropic line is a set of points $x = sa + t\xi$ such that $-\infty < s, t < \infty$, and $a$, $\xi$ are fixed vectors, for which

$$[\xi, \xi] = [a, \xi] = 0, \qquad [a, a] = -1. \tag{1}$$

The converse is obvious: every set of points $x = sa + t\xi$ such that Eq. (1) is satisfied forms an isotropic line[2] (since the distance between any two such points vanishes). We shall call $\xi$ the *direction vector* of the isotropic line.

Since $[\xi, \xi] = 0$, the direction vector lies along one of the generators of the $[x, x] = 0$ cone. The vector $a$, on the other hand, is seen from the relation $[a, \xi] = 0$ to lie in a tangent plane to the cone. Thus in imaginary Lobachevskian space the isotropic lines correspond to, and in fact are two-dimensional planes tangent to, the $[x, x] = 0$ cone. The converse is easily verified; namely each such plane is an isotropic line of imaginary Lobachevskian space.

The various models we have discussed which arise from specializations of the $x$ coordinates lead to particularly helpful geometrical interpretations of isotropic lines. Consider, for instance, $x_0 = 1$. Obviously the $x_0 = 1$ hyperplane intersects the $[x, x] = 0$ cone in a sphere (the absolute) and therefore intersects two-dimensional planes tangent to the cone along the lines tangent to the sphere. Thus in this model on the $x_0 = 1$ hyperplane, isotropic lines are represented by lines tangent to the unit sphere (the absolute) in $n$ dimensions. Consider now the model on the hyperboloid whose equation is $[x, x] = -1$. If a point of the form $x = sa + t\xi$ belongs to this hyperboloid, then $[sa + t\xi, sa + t\xi] = -1$ and then Eq. (1) implies that $s^2 = 1$. Without loss of generality we may take $s = 1$. Consequently on a single-sheeted hyperboloid an isotropic line is simply a line generator given by

$$x = a + t\xi, \qquad \text{where} \quad [a, a] = -1, \quad [\xi, \xi] = [a, \xi] = 0.$$

It can be shown conversely that every line generator of the $[x, x] = -1$ hyperboloid is an isotropic line of the imaginary Lobachevskian space.

[2] This implies, in particular, that through any two points of an imaginary Lobachevskian space such that the distance between them is zero there passes one and only one isotropic line.

### 1.4. Spheres and Horospheres in a Lobachevskian Space

We shall call the set of all points $x$ at a distance $r$ from some point $a$ of Lobachevskian space the *sphere* of radius $r$ with center at $a$. Equation (1) of Section 1.3 then implies that

$$[x, a]^2 = c[a, a][x, x] \tag{1}$$

is the equation of this sphere[3] where we have written $c = \cosh^2 kr$, and where $[a, a] > 0$. We remark that if the Lobachevskian space is realized on one sheet of the hyperboloid whose equation is $[x, x] = 1$, Eq. (1) simplifies to

$$[x, a] = \cosh kr. \tag{2}$$

Let us now allow the center of the sphere to move off to infinity, still requiring the sphere to pass through a given fixed point $b$. Then in the limit the sphere becomes a sphere of infinite radius, which we shall call a *horosphere*. Obviously this horosphere is uniquely determined by $b$ and by the point $\xi$ on the absolute which is the limit of the center point.

Remark. Horospheres in a Lobachevskian space are interesting for many reasons. First, the intrinsic geometry of a horosphere in an $n$-dimensional Lobachevskian space is identical with the intrinsic geometry of an $(n - 1)$-dimensional Euclidean space. Thus horospheres in Lobachevskian spaces are the analogs of planes in Euclidean spaces. Further, the set of horospheres itself forms an interesting and a relatively simple space which will later be of fundamental importance. Horospheres can be defined also in another way. Consider the set of lines of a Lobachevskian space. If a point on such a line is allowed to move off to infinity, remaining on the line, it will in the limit become some point $\xi$ on the absolute. We then say that the line passes through the point $\xi$ of the absolute. We define the set of lines parallel to a given line in a Lobachevskian space as all lines passing through the same point of the absolute as the given line. It turns out that each horosphere is a surface orthogonal to one such set. Two horospheres may be called parallel if they are orthogonal to the same set of parallel lines. It can then be shown that two parallel horospheres cut segments of equal length on all parallel lines orthogonal to them. The length of these segments may naturally be called the distance between the two parallel horospheres. #

Let us now find the equation of a horosphere. To do this we first realize the Lobachevskian space as the pencil of lines passing through the

[3] Recall that $x = (x_0, x_1, ..., x_n)$ are homogeneous coordinates in the Lobachevskian space.

origin and lying inside the $[x, x] = 0$ cone of $E_{n+1}$. Again we use the homogeneous coordinates $x = (x_0, \ldots, x_n)$, with $[x, x] > 0$. As we have seen, in these homogeneous coordinates the equation of a sphere is

$$[x, a]^2 = c[a, a][x, x],$$

where $a = (a_0, \ldots, a_n)$ are the homogeneous coordinates of its center. Now let this center move off to infinity. In our realization this means that $a$ approaches some vector $\xi$ of the cone, that is, such that $[\xi, \xi] = 0$. We shall approach this limit keeping $c[a, a]$ constant. Then in the limit we obtain the equation

$$[x, \xi]^2 = c_1[x, x] \tag{3}$$

of the horosphere. We assert that in this equation $c_1 > 0$. This is because $[x, \xi]^2 \geqslant 0$ and $[x, x] > 0$, so that $c_1 \geqslant 0$. But in addition $c_1$ can not vanish, since there exists no point $x$ in a Lobachevskian space such that $[x, \xi] = 0$.[4]

Thus in homogeneous coordinates the equation of a horosphere in a Lobachevskian space is given by Eq. (3) with $c_1 > 0$, and with $\xi$ a point on the absolute. We shall call the *direction of this horosphere* the generator of the cone that passes through $\xi$.

Let us now turn to the equation of the horosphere in the model on the surface of the $[x, x] = 1$ hyperboloid. In this realization Eq. (3) becomes

$$[x, \xi]^2 = \lambda^2, \tag{4}$$

where again $[\xi, \xi] = 0$ and $\lambda \neq 0$. This equation remains invariant under replacement of $\xi$ by $-\xi$, and we shall therefore consider $\xi$ to lie on the *upper or positive sheet* of the $[\xi, \xi] = 0$ cone. Now note that if $x$ belongs to the positive sheet of the hyperboloid and $\xi$ to the positive cone, then $[x, \xi] > 0$, and hence Eq.(4) can be written

$$[x, \xi] = \lambda, \tag{5}$$

where $\lambda > 0$.

[4] If $[x, \xi] = 0$ for some $x$, then it would follow that

$$x_0\xi_0 = x_1\xi_1 + \cdots + x_n\xi_n .$$

According to the Cauchy-Bunyakovskii inequality we then have

$$x_0^2\xi_0^2 \leqslant (x_1^2 + \cdots + x_n^2)(\xi_1^2 + \cdots + \xi_n^2).$$

Now $[\xi, \xi] = 0$ implies $\xi_0^2 = \xi_1^2 + \cdots + \xi_n^2$, so that

$$x_0^2 \leqslant x_1^2 + \cdots + x_n^2,$$

which contradicts $[x, x] > 0$.

Now it is obvious that if we replace $\xi$ by $\alpha\xi$ and $\lambda$ by $\alpha\lambda$ in Eq. (5), with $\alpha > 0$, we obtain the same horosphere. This means that $(\xi, \lambda)$ may be thought of as homogeneous coordinates of the horosphere. In particular, the equation can be normalized by setting $\lambda = 1$. Then it becomes

$$[x, \xi] = 1, \tag{6}$$

where $\xi$ is a point on the positive $[\xi, \xi] = 0$ cone. Consequently every horosphere of a Lobachevskian space is given by a point $\xi$ on the upper sheet ($\xi_0 > 0$) of the cone.

Remark. Another important class of surfaces in a Lobachevskian space is the class of *planes*. These can be defined as follows. Let us consider again the realization in terms of the pencil of lines inside the $[x, x] = 0$ cone passing through the origin of $E_{n+1}$. Now consider a hyperplane passing through the origin of $E_{n+1}$ and intersecting the cone. We shall say that the lines of the Lobachevskian space that lie on this hyperplane form a *plane* in the Lobachevskian space. Consider the equation of such a plane. Every hyperplane in $E_{n+1}$ passing through the origin and intersecting the cone is given by an equation of the form

$$[x, \xi] = 0,$$

where $[\xi, \xi] < 0$. Thus this equation is also the equation of a plane in the Lobachevskian space, given in terms of homogeneous coordinates. Note that $\xi$ in this equation is arbitrary up to an arbitrary multiplicative factor. This means that the manifold of planes of a Lobachevskian space is identical to the set of lines of $E_{n+1}$ passing through the origin and lying outside the $[\xi, \xi] = 0$ cone, or to the set of points of an imaginary Lobachevskian space. The planes of a Lobachevskian space can, incidentally, be defined without using any particular realization. In fact they are the surfaces that may be characterized as follows: if a geodesic line has two points on the surface then it lies entirely in the surface. In this sense the planes of a Lobachevskian space are the analogs of the $(n - 1)$-dimensional hyperplanes of an $n$-dimensional Euclidean space. #

### 1.5. Spheres and Horospheres in an Imaginary Lobachevskian Space

A *sphere* of radius $r$ and with center point $a$ in an imaginary Lobachevskian space shall be the set of points $x$ at a distance $r$ from $a$. Recall that in an *imaginary* Lobachevskian space the distance may be a positive real number, an imaginary number, or zero. Therefore there are three types of spheres in such a space, namely those with real radius, zero radius, and imaginary radius. We remark that in the latter case, the radius lies in the interval $(0, \pi i/2k]$.

In homogeneous coordinates the equation of a sphere of radius $r$ with center at $a$ is of the same form as it is for the case of an ordinary Lobachevskian space, namely

$$[x, a]^2 = c[a, a][x, x], \qquad [a, a] < 0, \tag{1}$$

where $c = \cosh^2 kr$. The difference from the ordinary Lobachevskian space is that now $c$ need not be greater than one, but can take on any positive value. If $c > 1$, the sphere has real radius, if $c < 1$ the sphere has imaginary radius, and if $c = 1$ the sphere has zero radius. By definition the sphere of radius zero with center at $a$ is the set of points whose distance from $a$ vanishes. We have seen in Section 1.3$a$ that through any two points separated by a distance zero there passes an isotropic line. Thus the sphere of radius zero with center at $a$ is the surface generated by the isotropic lines passing through $a$. This surface may also be called an isotropic cone of an imaginary Lobachevskian space.

We now wish to introduce the concept of a horosphere in an imaginary Lobachevskian space. As in the ordinary case, we allow the center to move off to infinity, requiring that the sphere continue to pass through a given fixed point $b$, and we call the limit obtained in this way a *horosphere in an imaginary Lobachevskian space.* The equation of such a horosphere is obtained by allowing $a$ in Eq. (1) to approach a point $\xi$ on the cone $[\xi, \xi] = 0$ while the product $c[a, a]$ remains equal to some constant $c_1$. Thus in the limit we arrive at the equation for a horosphere in an imaginary Lobachevskian space,

$$[x, \xi]^2 = c_1[x, x], \tag{2}$$

where $[\xi, \xi] = 0$, $c_1 \leqslant 0$. Recall that the equation of a horosphere in an ordinary Lobachevskian space was of the same form. The difference is that in the ordinary case $c_1 \neq 0$, whereas now $c_1 = 0$ is allowed. We shall say that if $c_1 < 0$ the horosphere is of the *first kind*, and that the equation

$$[x, \xi] = 0 \tag{3}$$

describes a horosphere of the *second kind.*

A clear picture of horospheres of the second kind can be obtained in the model on the $x_0 = 1$ hyperplane, so that the imaginary

Lobachevskian space becomes the set of points outside the unit sphere in $n$ dimensions. Then Eq. (3) becomes

$$\xi_0 - \xi_1 x_1 - \cdots - \xi_n x_n = 0. \tag{4}$$

This is the equation of a hyperplane tangent to the absolute at the point $(\xi_1/\xi_0, ..., \xi_n/\xi_0)$. In other words, in this model a horosphere of the second kind is a hyperplane tangent to the absolute. But we already know that all the lines tangent to the absolute are isotropic lines. Thus a horosphere of the second kind in an imaginary Lobachevskian space is composed of isotropic lines.

We shall later be dealing with integral geometry in three-dimensional imaginary Lobachevskian space. For this case we shall deal not with the horospheres of the second kind but with the lines of which they are composed. For certain rather general reasons it is more convenient to call not the hyperplanes, but these isotropic lines the horospheres of the second kind.[5]

Let us turn to horospheres of the first kind. We wish to put the equation of such a horosphere, namely Eq. (2) with $c_1 < 0$, in canonical form. Now this equation is invariant under replacement of $\xi$ by $\alpha\xi$ and $c_1$ by $\alpha^2 c_1$, where $\alpha \neq 0$. Thus it is possible to normalize to $c_1 = -1$. On the other hand, since the equation remains invariant also when $\xi$ is replaced by $-\xi$, we may consider $\xi$ to be a point on the upper sheet of the $[\xi, \xi] = 0$ cone, and hence the canonical form of the equation becomes

$$[x, \xi]^2 = -[x, x], \tag{5}$$

so that such a horosphere is uniquely determined by a point on the positive cone whose equation may be written $[\xi, \xi] = 0$, $\xi_0 > 0$.

In particular, consider the realization of the imaginary Lobachevskian

[5] The general definition of a horosphere in the homogeneous space $X$ whose transformation group $G$ is the group of complex unimodular $2 \times 2$ matrices (see Chapter VI, Section 1.6) is the following. *A horosphere of $X$ is the trajectory of any point under the subgroup whose elements are matrices of the form*

$$\begin{Vmatrix} 1 & \zeta \\ 0 & 1 \end{Vmatrix}$$

(or any conjugate subgroup). It can be shown that in three-dimensional imaginary Lobachevskian space these trajectories are just the surfaces which we have here called the horospheres of the first kind and the isotropic lines. In Section 1.6 of Chapter VI we shall give a definition of a horosphere for the case of any complex semisimple Lie group of transformations.

space on the hyperboloid whose equation $[x, x] = -1$. Then Eq. (5) becomes

$$|[x, \xi]| = 1, \tag{6}$$

again with $[\xi, \xi] = 0$, $\xi_0 > 0$.

It is easily shown that there exists a hyperbolic rotation of $E_{n+1}$ which will carry any given point $\xi$ of the positive sheet of the $[\xi, \xi] = 0$ cone into any other given point of the same sheet. But this then implies that every horosphere of the first kind can be carried by a hyperbolic rotation into any other horosphere of the first kind. Similar considerations show that horospheres of the second kind, given by $[x, \xi] = 0$, also have the property that any one can be transformed into any other by some motion of the imaginary Lobachevskian space. As we have pointed out, these horospheres of the second kind are composed of isotropic lines, and it is moreover true that the isotropic lines themselves have this property.

## 1.6. Invariant Integration in a Space of Constant Curvature

We now wish to discuss the integrals of functions over a space of constant curvature. We shall require, as we did for group integrals, that these integrals have the invariance property

$$\int f(x)\, dx = \int f(xg)\, dx. \tag{1}$$

Here $xg$ designates the point into which $x$ is carried under the motion $g$. Thus our problem is to define a volume element which is invariant under the motions of the space.

It is well known that on the $n$-dimensional sphere $x_0^2 + \cdots + x_n^2 = R^2$ this invariant element of volume (invariant measure) is given by

$$dx = \frac{R\, dx_1 \cdots dx_n}{|x_0|} = \frac{R\, dx_1 \cdots dx_n}{[R^2 - (x_1^2 + \cdots + x_n^2)]^{\frac{1}{2}}}\,. \tag{2}$$

In order to deduce the analogous expression for the Lobachevskian space, let us take the model on the hyperboloid

$$[x, x] \equiv x_0^2 - x_1^2 - \cdots - x_n^2 = 1, \qquad x_0 > x,$$

in $E_{n+1}$. Now the element of volume $dv = dx_0 \cdots dx_n$ in $E_{n+1}$ remains invariant under all linear transformations with unit determinant, among which are the hyperbolic rotations. We may introduce the new set of coordinates $x_1, ..., x_n, r = [x, x]^{\frac{1}{2}}$ in the region $[x, x] > 0$, $x_0 > 0$. In these coordinates $dv$ becomes

$$dv = \frac{r\,dr\,dx_1 \cdots dx_n}{x_0} = \frac{r\,dr\,dx_1 \cdots dx_n}{(1 + x_1^2 + \cdots + x_n^2)^{\frac{1}{2}}}.$$

Hyperbolic rotations in $E_{n+1}$ leave both $dv$ and $r$ invariant, and therefore also

$$dx = \frac{dx_1 \cdots dx_n}{x_0} = \frac{dx_1 \cdots dx_n}{(1 + x_1^2 + \cdots + x_n^2)^{\frac{1}{2}}}. \tag{3}$$

This gives an invariant measure in a Lobachevskian space. In addition to this one, we may use any of the similar expressions

$$dx = \frac{dx_0 \cdots dx_{k-1}\,dx_{k+1} \cdots dx_n}{|x_k|}, \qquad k = 1, 2, ..., n. \tag{3'}$$

The invariant integral over the hyperboloid $[x, x] = 1$ can be written in terms of the generalized function $\delta([x, x] - 1)$. Specifically, it is easily shown[6] that

$$\int f(x)\,dx = 2 \int f(x)\delta([x, x] - 1)\,dv. \tag{4}$$

The invariance of the integral on the right-hand side of this equation follows immediately from the invariance of $dv$ and $[x, x]$ under hyperbolic rotations in $E_{n+1}$.

Let us now turn to an imaginary Lobachevskian space. This time we take the model on the single-sheeted hyperboloid $[x, x] = -1$. Then the invariant element of volume will be given by

$$dx = \frac{dx_1 \cdots dx_n}{|x_0|} = \frac{dx_1 \cdots dx_n}{(x_1^2 + \cdots + x_n^2 - 1)^{\frac{1}{2}}}, \tag{5}$$

whose derivation is quite analogous to that of Eq. (3). It is easily shown also that

$$\int f(x)\,dx = 2 \int f(x)\delta([x, x] + 1)\,dv, \tag{6}$$

where the integral on the left-hand side is taken over the entire $[x, x] = -1$ hyperboloid; in taking the integral one should bear in mind that $f(-x) = f(x)$.

[6] The generalized function $\delta(P)$ is discussed in Volume I, Chapter III, Section 1.3. It is self-evident that the $f(x)$ on the right-hand side of Eq. (4) represents a continuous extension to all of $E_{n+1}$ of the function defined on $[x, x] = 1$.

Later we shall need to integrate functions not only over Lobachevskian and imaginary Lobachevskian spaces (the upper sheet of the $[x, x] = 1$ hyperboloid and the $[x, x] = -1$ hyperboloid, respectively) but also over the $[x, x] = 0$ cone. The measure on the cone invariant under hyperbolic rotations of $E_{n+1}$ is given by

$$dx = \frac{dx_1 \cdots dx_n}{| x_0 |} = \frac{dx_1 \cdots dx_n}{(x_1^2 + \cdots + x_n^2)^{\frac{1}{2}}}, \tag{7}$$

whose derivation is analogous to that of (3) and (5).

## 1.7. Integration over a Horosphere

We now go on to discuss integration over a horosphere in a Lobachevskian and imaginary Lobachevskian space. Recall that for the realization of an ordinary Lobachevskian space on the upper sheet of the $[x, x] = 1$ hyperboloid, the equation of a horosphere is $[x, \xi] = 1$, where $[\xi, \xi] = 0$ (see Section 1.4).

We define the integral of $f(x)$ over the horosphere $\omega$ whose equation is $[x, \xi] = 1$ by

$$\int_\omega f(x)\, d\sigma = \int f(x)\delta([x, \xi] - 1)\, dx, \tag{1}$$

where $dx$ is the invariant measure. Since both $dx$ and $[x, \xi]$ remain invariant under simultaneous motion of $x$ and $\xi$, the integral defined in this way is invariant under displacement of the horosphere. This means that under a motion $g$ of the Lobachevskian space mapping $\omega$ into the new horosphere $\omega g$, we obtain

$$\int_\omega f(xg)\, d\sigma = \int_{\omega g} f(x)\, d\sigma_g, \tag{2}$$

where $d\sigma_g$ is the measure on $\omega g$. If, in particular, $g$ transforms $\omega$ into itself, then

$$\int_\omega f(xg)\, d\sigma = \int_\omega f(x)\, d\sigma.$$

We can define integration over a horosphere of the first kind in an imaginary Lobachevskian space by analogy with Eq. (1), namely by

$$\int_\omega f(x)\, d\sigma = \int f(x)\delta(|[x, \xi]| - 1)\, dx. \tag{3}$$

Here $dx$ is the invariant measure in the imaginary Lobachevskian space. Integrals on horospheres of the first kind also remain invariant under motions of the imaginary Lobachevskian space.

We now define integrals along isotropic lines in an imaginary Lobachevskian space. For this purpose we use the realization of the space on the hyperboloid

$$[x, x] = -1. \tag{4}$$

Then an isotropic line $l$ (a generator of the hyperboloid) is given by

$$x = \xi t + b, \tag{5}$$

where $b$ is some fixed point on the generator, and $\xi$ is its direction vector ($[\xi, \xi] = 0$). We then define the integral along the line of Eq. (5) by

$$\varphi(\xi, b) \equiv \int_{-\infty}^{\infty} f(\xi t + b)\, dt. \tag{6}$$

Obviously this definition depends not only on $l$, but also on the normalization of $\xi$.

It is notable that unlike the integral on a horosphere of the first kind, the integral along an isotropic line is not in general invariant, but is multiplied by some constant factor under the motions of the space. This is because there exist hyperbolic rotations of $E_{n+1}$ which carry a vector $\xi$ into another vector of the form $\lambda\xi$, where $\lambda \neq 0$.

## 1.8. Measures on the Absolute

We wish now to discuss measures on the absolute of a Lobachevskian space, namely on the set of linear generators of the cone $[\xi, \xi] = 0$. Since the generators of this cone are uniquely determined by any single point on them other than the origin, the definition of a measure on the absolute requires only the definition of a measure $d\sigma$ on any manifold $\mathfrak{M}$ on the cone that intersects each generator at one and only one point.

It can be shown that there exists no measure on the absolute invariant under all Lobachevskian motions. It is easy, however, to construct one which is invariant under rotations about some point $M$ (that is, under hyperbolic rotations leaving this point fixed). Recall that a point of a Lobachevskian space is a line passing through the origin of $E_{n+1}$ and lying inside our cone. Consider some line $M$ and let $x$ be a point on this line (other than the origin). Consider further the points $\xi$ on the cone such that $[x, \xi] = 1$ (the intersection of the cone and hyperplane).

It is easily seen that these points lie on one sheet of the cone and intersect each of its generators at one and only one point. Therefore this is the kind of set we need to define the measure on the absolute. We make this definition according to

$$\int f(\xi)\, d\sigma = \int f(\xi)\delta([x, \xi] - 1)\, d\xi, \tag{1}$$

where $d\xi$ is the invariant measure on the cone and $f(\xi)$ is a function on the absolute, that is, a function on the cone, constant on each generator.

We assert that $d\xi$ is invariant under rotations about $M$, i.e., hyperbolic rotations of $E_{n+1}$ leaving $M$ fixed. Such rotations will also leave fixed the point $x$ on the line, and therefore $[xg, \xi g] = [x, \xi g]$ for every rotation $g$ about $M$. Since $d\xi$ is invariant under Lobachevskian motions, it then follows that $d\sigma$ as defined by Eq. (1) is invariant under rotations about $M$. In other words, for every $f(\xi)$ defined on the absolute and for every rotation $g$ about $M$ we have

$$\int f(\xi g)\, d\sigma = \int f(\xi)\, d\sigma.$$

Obviously if we had chosen some point other than $x$ on $M$, we would have obtained a measure on the absolute differing from this one by a constant factor. All measures on the absolute invariant under Lobachevskian rotations about $M$ are of this type. Indeed, such rotations form a transitive group of transformations (of the absolute), and therefore the measure on the absolute invariant under this group is uniquely defined up to a constant factor

Remark. A measure on the absolute invariant under rotations about $M$ is called a *harmonic* measure. This is related to the fact that for the realization of the space in the unit ball (see Section 1.2) the measure of a fixed set on the absolute invariant under rotations about $M$ is a harmonic function of the coordinates of $M$. #

We now wish to repeat what we have just done, but for an imaginary Lobachevskian space. By definition, the points of such a space are the lines passing through the origin of $E_{n+1}$ and lying outside the $[\xi, \xi] = 0$ cone. Consider a line $M$ and let $x$ be a point on this line. We define a measure $d\sigma$ on the absolute by the equation

$$\int f(\xi)\, d\sigma = c_1 \int_{V_+} f(\xi)\delta([x, \xi] - 1)\, d\xi + c_2 \int_{V_-} f(\xi)\delta([x, \xi] - 1)\, d\xi, \tag{2}$$

where $V_+$ is the upper sheet of the $[\xi, \xi] = 0$ cone, and $V_-$ is the lower sheet. Ordinarily we shall write $c_1 = c_2 = 1$. This measure $d\sigma$ is invariant under rotations about $M$, for such rotations leave invariant both $d\xi$ and the expression $[x, \xi]$, and transform each sheet of the cone into itself.

The presence of two terms instead of one in Eq. (2) is related to the fact that the generators of the cone that intersect the $[x, \xi] = 1$ hyperplane break up into two subsets $\Omega_+$ and $\Omega_-$ on each of which the group is transitive. The first of these consists of the generators intersecting the hyperplane at points of $V_+$ and the other is defined similarly.

Actually the set of generators of the cone contains a third subset $\Omega_0$ on which the group of rotations about $M$ is transitive. This consists of those generators lying in the $[x, \xi] = 0$ plane, i.e., those parallel to the $[x, \xi] = 1$ plane. There exists no measure on $\Omega_0$ which is invariant under rotations about $M$. If, however, one chooses some point $N$ of the Lobachevskian space (not the imaginary Lobachevskian space), one can construct a measure invariant under all motions which leave $M$ and $N$ fixed. This measure is defined by the equation

$$\int f(\xi)\, d\sigma_0 = \int f(\xi)\delta([x, \xi], [y, \xi] - 1)\, d\xi,$$

where $d\xi$ is the invariant measure on the $[\xi, \xi] = 0$ cone, and $y$ is a point of $E_{n+1}$ lying on the line $N$.[7]

In order to visualize $\Omega_+$, $\Omega_-$, and $\Omega_0$ more clearly, let us turn to the realization of the imaginary Lobachevskian space outside of the unit sphere $\Omega$. Consider the isotropic cone whose vertex is at $M$. In this realization the isotropic lines are the lines tangent to $\Omega$(that is, tangent to the absolute). Therefore the isotropic cone with vertex at $M$ is tangent to $\Omega$ along a certain "circle of constant latitude." This "circle of constant latitude" is $\Omega_0$. It divides $\Omega$ into two regions, $\Omega_+$ and $\Omega_-$. Obviously under rotations of the imaginary Lobachevskian space about $M$ the absolute is transformed into itself, as is each sheet of the isotropic cone with vertex at $M$. At the same time each of $\Omega_+$, $\Omega_-$, and $\Omega_0$ is also transformed into itself.

## 2. Integral Transform Associated with Horospheres in a Lobachevskian Space

In Chapter I we discussed the Radon transform on a Euclidean space. This transform involved associating with every $f(x)$ of bounded support

[7] Recall again that the points of a Lobachevskian space are the lines through the origin lying *inside* the cone.

its integrals over all possible planes. We shall now turn to the analog of the Radon transform for Lobachevskian spaces, in which we associate with every $f(x)$ of bounded support its integrals over all possible horospheres.[8]

Just as in Euclidean space the Radon transform is closely related to the Fourier integral expansion of a function, so in Lobachevskian space the integral transform with which we shall be dealing is related to the Fourier integral. This relationship will be established in Section 3.2 of the next chapter. In general, it is often convenient to relate the Fourier integral of functions defined on homogeneous spaces to a preliminary transition from the function to its integrals over the horospheres of the space.

## 2.1. Integral Transform Associated with Horospheres

Let $f(x)$ be a function of bounded support on a Lobachevskian space, and let us associate with every such function its integrals over horospheres

$$h(\omega) = \int_{\omega} f(x)\, d\sigma, \tag{1}$$

where $d\sigma$ is the measure on $\omega$ defined in Section 1.7. This then associates with every $f(x)$ a function $h(\omega)$ defined on the set of horospheres of the Lobachevskian space. Let us call this association the *integral transform associated with horospheres*; that is, we say that $h(\omega)$ is the integral transform of $f(x)$.

Now we may think of $h(\omega)$ as a function defined on the upper sheet of the $[\xi, \xi] = 0$ cone by realizing the Lobachevskian space on the upper sheet of the $[x, x] = 1$ hyperboloid. In this realization each horosphere is uniquely specified by some vector $\xi$ on this positive cone through an equation of the form $[x, \xi] = 1$. Therefore we may write $h(\xi)$ in place of $h(\omega)$ and use Eq. (1) of Section 1.7 to write

$$h(\xi) = \int f(x)\delta([x, \xi] - 1)\, dx, \tag{2}$$

where $dx$ is the invariant measure in the Lobachevskian space.

[8] Since the intrinsic geometry of the horospheres in a Lobachevskian space is Euclidean, they form one of the analogs of hyperplanes of a Euclidean space. Other analogs of such hyperplanes are the actual hyperplanes of a Lobachevskian space. The integral transform corresponding to such hyperplanes is also of some interest, but it is not related to the analog of the Fourier integral on Lobachevskian space (see Chapter VI, Section 4.6). We shall not deal with this transform, but call the reader's attention to an interesting paper by Helgason [see Ref. (24)].

We now assert that if $f(x)$ is an infinitely differentiable function with bounded support on the $[x, x] = 1$ hyperboloid, then $h(\xi)$ is infinitely differentiable, has bounded support (on the cone), and vanishes in some neighborhood of the vertex.

Proof. That $h(\xi)$ is infinitely differentiable is obvious. To show that it vanishes in a neighborhood of the vertex, consider the inequality

$$|[x, \xi]| < |x_0\xi_0| + (x_1^2 + \cdots + x_n^2)^{\frac{1}{2}}(\xi_1^2 + \cdots + \xi_n^2)^{\frac{1}{2}}.$$

If $x$ lies on the $[x, x] = 1$ hyperboloid and $\xi$ on the $[\xi, \xi] = 0$ cone, this inequality becomes

$$|[x, \xi]| < |\xi_0|(|x_0| + [x_0^2 - 1]^{\frac{1}{2}}). \tag{3}$$

Now recall that $f(x)$ has bounded support. There exists, therefore, an $N > 1$ such that $f(x) = 0$ for $|x_0| > N$. Therefore the only nonzero contribution to the integral in Eq. (2) comes from the region in which $|x_0| < N$. But Eq. (3) shows that in this region $|[x, \xi]| < 1$ as long as $|\xi_0| < (N + [N^2 - 1]^{\frac{1}{2}})^{-1}$, so that for small enough $|\xi_0|$ the integral will vanish. This proves that $h(\xi)$ vanishes in a neighborhood of the vertex of the cone. That its support is bounded can be proven similarly with the aid of the inequality

$$|[x, \xi]| > |\xi_0|(|x_0| - [x_0^2 - 1]^{\frac{1}{2}}),$$

which can be obtained in the same way as was (3).

The main problem we shall be concerned with in these sections will be to derive the inversion formula for the integral transform of Eq. (2) [Eq. (17) of Section 2.2 for dimension $n = 3$ and Eqs.(7) and (8) of Section 2.3 for general $n$]. The method by which this will be done is closely related to the plane-wave expansion of the $\delta$ function in a Euclidean space (Volume 1, Chapter I, Section 3.10) in which the $\delta$ function is written in terms of functions that remain constant on planes. In outline, the procedure is the following.

To obtain an expression for $f(a)$, we first multiply both sides of Eq. (2) by a certain function $\varphi(a, \xi; \mu)$ depending on $a$, the point $\xi$ on the cone (or, equivalently, the horosphere $\omega$), and a complex parameter $\mu$. Then integrating both sides of the equation so obtained with respect to the invariant measure $d\xi$ on the cone, we obtain

$$\int \varphi(a, \xi; \mu)h(\xi)\, d\xi = \int \Phi(x, a; \mu)f(x)\, dx, \tag{4}$$

where

$$\Phi(x, a; \mu) = \int \varphi(a, \xi; \mu)\delta([x, \xi] - 1)\, d\xi. \tag{5}$$

Note that the integrand in (5) is concentrated on those points $\xi$ of the $[\xi, \xi] = 0$ cone for which $[x, \xi] = 1$; these points correspond to the horospheres passing through the point $x$. This means that $\Phi(x, a; \mu)$ is the integral of $\varphi(a, \xi; \mu)$ over the set of horospheres passing through $x$. Further, the measure $\delta([x, \xi] - 1)\, d\xi$ with respect to which this integration is taken is, as we have seen in Section 1.8, invariant under those motions $g$ which leave $x$ fixed (i.e., under rotations about $x$).

Now let $\varphi(a, \xi; \mu)$ be invariant under simultaneous motion of $a$ and $\xi$, i.e., let

$$\varphi(a, \xi; \mu) = \varphi(ag, \xi g; \mu).$$

Then the kernel $\Phi(x, a; \mu)$ is invariant under simultaneous motion of $a$ and $x$. But the only invariant of a pair of points in a Lobachevskian space is the distance between them, so that $\Phi(x, a; \mu)$ must be a function of $\mu$ and of the distance $r$ between $a$ and $x$.

Let us assume further that $\varphi(a, \xi; \mu)$ as a function of $\xi$ has the same singularities as does $|[a, \xi] - 1|^{\mu}$. It can then be shown that $\Phi(x, a; \mu)$ as a function of $r$ behaves in the neighborhood of $a$ as $Cr^{\mu}$. Thus for $\mu = -n$, this kernel has a delta function singularity concentrated at $a$. But then Eq. (4) becomes the desired inversion formula at $\mu = -n$, namely an expression for $f(a)$ in terms of $h(\xi)$.

We shall go through this derivation in detail first for $n = 3$ in Section 2.2 (this special case will be needed in the next chapter). The case of general $n$ will be discussed in Section 2.3.

## 2.2. Inversion Formula for $n = 3$

Let $f(x)$ be an infinitely differentiable function with bounded support on a Lobachevskian space (realized on the upper sheet of the $[x, x] = 1$ hyperboloid). We define $h(\xi)$ by

$$h(\xi) = \int f(x)\delta([x, \xi] - 1)\, dx, \tag{1}$$

where $dx$ is the invariant measure in the Lobachevskian space, and wish to obtain an expression for $f(x)$ in terms of $h(\xi)$. Recall that $h(\xi)$ is defined by Eq. (1) on the $[\xi, \xi] = 0$ cone, that it is infinitely differentiable, has bounded support, and vanishes in a neighborhood of the vertex of the cone. Therefore if Re $\mu > -1$, the integral

$$\int |[a, \xi] - 1|^{\mu} h(\xi)\, d\xi$$

converges, where $d\xi$ is the invariant measure on the $[\xi, \xi] = 0$ cone.

Clearly for Re $\mu > -1$ the formula

$$(\Phi, f(x)) = \int |[a, \xi] - 1|^{\mu} h(\xi)\, d\xi \tag{2}$$

defines a linear functional on the infinitely differentiable functions of bounded support. Our first step is to obtain an explicit expression for this functional, that is, to find a function $\Phi(x, a; \mu)$ such that

$$\int |[a, \xi] - 1|^{\mu} h(\xi)\, d\xi = \int \Phi(x, a; \mu) f(x)\, dx. \tag{3}$$

Equation (1) implies that this function is given by

$$\Phi(x, a; \mu) = \int |[a, \xi] - 1|^{\mu} \delta([x, \xi] - 1)\, d\xi, \tag{4}$$

or in other words that it is the integral of $|[a, \xi] - 1|^{\mu}$ over the set of horospheres passing through $x$.

Remark. Note that $\delta([x, \xi] - 1)\, d\xi$ is a harmonic measure on the absolute, invariant under rotations about $x$ (see Section 1.8). It is thus seen that the same factor $\delta([x, \xi] - 1)$ that is used to define the measure on the $[x, \xi] = 1$ horosphere leads also to the correct measure on the set of horospheres passing through $x$. #

In order to evaluate the integral of Eq. (4), note that the integrand does not change under simultaneous motion of $a$ and $x$, since such transformations leave invariant both the measure $d\xi$ and the expressions $[a, \xi]$ and $[x, \xi]$. We may thus write

$$\Phi(xg, ag; \mu) = \Phi(x, a; \mu),$$

where $g$ is any Lobachevskian motion. But every pair of points $x$, $a$ can be moved by Lobachevskian motions to the points $x'$ with coordinates $(1, 0, 0, 0)$ and $a'$ with coordinates $(\cosh kr, \sinh kr, 0, 0)$, where $r$ is the distance between $a$ and $x$.[9] Thus we need calculate $\Phi(x, a; \mu)$ only for $a = a'$ and $x = x'$.

[9] Indeed, there certainly exists a motion $g$ which carries $x$ to the point $x'$ whose coordinates are $(1, 0, 0, 0)$. Let this motion carry $a$ into some point $a^*$. Since Lobachevskian motions leave distances invariant, the distance between $x'$ and $a^*$ is $r$. Now rotate the $[x, x] = 1$ hyperboloid about the $x_0$ axis until $a^*$ is on the $x_0$, $x_1$ plane. Its coordinates are then obviously $(\cosh kr, \sinh kr, 0, 0)$, which are the coordinates of $a'$. This proves the assertion.

We thus find that

$$\begin{aligned}\Phi(x, a; \mu) &= \Phi(x', a'; \mu) \\ &= \int \mid \xi_0 \cosh kr - \xi_1 \sinh kr - 1 \mid^{\mu} \delta(\xi_0 - 1)\, d\xi,\end{aligned} \tag{5}$$

where $\xi_0, \xi_1, \xi_2, \xi_3$ are the coordinates of $\xi$. Now recall (Section 1.6) that the invariant measure on the $[\xi, \xi] = 0$ cone is given by

$$d\xi = \frac{d\xi_0\, d\xi_1\, d\xi_2}{\mid \xi_3 \mid},$$

where $\xi_3 = \pm (\xi_0^2 - \xi_1^2 - \xi_2^2)^{\frac{1}{2}}$. Thus Eq. (5) may be written

$$\Phi(x, a; \mu) = 2 \int_{-1}^{1} \mid \cosh kr - \xi_1 \sinh kr - 1 \mid^{\mu} \int_{-(1-\xi_1^2)^{\frac{1}{2}}}^{(1-\xi_1^2)^{\frac{1}{2}}} \frac{d\xi_2\, d\xi_1}{(1 - \xi_1^2 - \xi_2^2)^{\frac{1}{2}}}. \tag{6}$$

This integral converges for Re $\mu > -1$. A simple calculation shows that

$$\Phi(x, a; \mu) = \frac{4\pi \cosh \frac{1}{2}(\mu + 1)kr}{(\mu + 1) \cosh^{\mu+1} \frac{1}{2}kr} \sinh kr. \tag{7}$$

Thus we have shown that for Re $\mu > -1$ we may write Eq. (3) with kernel $\Phi(x, a; \mu)$ given by Eq. (7) if $f(x)$ is an infinitely differentiable function with bounded support. The definition of $(\Phi, f)$ by means of Eq. (2) is meaningless for Re $\mu < -1$. But since $h(\xi)$ is of bounded support and vanishes in a neighborhood of the vertex of $[\xi, \xi] = 0$ cone, the right-hand side of Eq. (2) is an analytic function of $\mu$ for Re $\mu > -1$, which makes it possible to continue $(\Phi, f)$ analytically into the region Re $\mu < -1$. This analytic continuation yields, since the $[a, \xi] = 1$ surface has no singular points, a functional which has only simple poles at $\mu = -1, -3, -5, \ldots$ (see Volume 1, Chapter III, Section 4). Then uniqueness of the analytic continuation preserves Eq. (3), in which the right-hand side is also understood in terms of its regularization.

We can now proceed to solve the fundamental problem of this section, namely to derive the inversion formula for the transform of Eq. (1), that is, to express $f(x)$ in terms of $h(\xi)$. To do this we consider Eq. (3) as $\mu \to -3$. As we have already mentioned, for this value of $\mu$ both sides of the equation have simple poles. It is by studying the residues at these poles that we shall obtain the desired inversion formula.

Consider the left-hand side of Eq. (3), namely the integral

$$\int \mid [a, \xi] - 1 \mid^{\mu} h(\xi)\, d\xi.$$

At $\mu = -3$ this integral has residue

$$\operatorname*{res}_{\mu=-3} \int |[a, \xi] - 1|^{\mu} h(\xi)\, d\xi = \int \delta''([a, \xi] - 1) h(\xi)\, d\xi. \tag{8}$$

This result may be obtained as follows. Recall that the generalized function $|t|^{\mu}$ of the single variable $t$ has a simple pole at $\mu = -3$ with residue

$$\operatorname*{res}_{\mu=-3} |t|^{\mu} = \delta''(t). \tag{9}$$

Now the intersection of the $[\xi, \xi] = 0$ cone and the $[a, \xi] = 1$ plane is a compact manifold with no singular points. This fact may be used with Eq. (9) to arrive at

$$\operatorname*{res}_{\mu=-3} |[a, \xi] - 1|^{\mu} = \delta''([a, \xi] - 1), \tag{10}$$

from which (8) follows immediately.

Let us now consider the right-hand side of (3), namely the integral

$$\int \Phi(x, a; \mu) f(x)\, dx,$$

where $\Phi(x, a; \mu)$ is given by Eq. (7). We assert that at $\mu = -3$ the generalized function $\Phi(x, a; \mu)$ has a simple pole with residue

$$\operatorname*{res}_{\mu=-3} \Phi(x; a; \mu) = -8\pi^2 \delta_a(x),$$

where $\delta_a(x)$ is the $\delta$ function concentrated at $a$, i.e. where

$$(\delta_a(x), f(x)) = f(a). \tag{11}$$

Now the only factor in Eq. (7) that is singular at $\mu = -3$ is $\sinh^{\mu} kr$, so that we need find the residue only of this function. In obtaining it, further, we may use the fact that this function is invariant under Lobachevskian motions, and thus without loss of generality we may take the coordinates of $a$ to be $(1, 0, 0, 0)$. With this in mind, we see that $r$ is determined by the equation

$$\cosh^2 kr = [a, x]^2 = x_0^2\,. \tag{12}$$

Because $x$ lies on the $[x, x] = 1$ hyperboloid, this means that

$$\sinh^2 kr = x_1^2 + x_2^2 + x_3^2\,. \tag{13}$$

Thus $\sinh^{\mu} kr$ becomes $(x_1^2 + x_2^2 + x_3^2)^{\frac{1}{2}\mu}$.

To find the residue of this generalized function at $\mu = -3$, we recall that the invariant measure on the Lobachevskian space is

$$dx = \frac{dx_1\, dx_2\, dx_3}{x_0} = \frac{dx_1\, dx_2\, dx_3}{(1 + x_1^2 + x_2^2 + x_3^2)^{\frac{1}{2}}}\,.$$

We then have

$$((x_1^2 + x_2^2 + x_3^2)^{\frac{1}{2}\mu}, f(x))$$

$$= \int (x_1^2 + x_2^2 + x_3^2)^{\frac{1}{2}\mu} \frac{f(x_1, x_2, x_3)\, dx_1\, dx_2\, dx_3}{(1 + x_1^2 + x_2^2 + x_3^2)^{\frac{1}{2}}}\,. \tag{14}$$

Now it was established in Volume 1, Chapter I, Section 3.9 that the generalized function $(x_1^2 + x_2^2 + x_3^2)^{\frac{1}{2}\mu}$ in Euclidean 3-space has a simple pole at $\mu = -3$ with residue given by

$$\operatorname*{res}_{\mu=-3} (x_1^2 + x_2^2 + x_3^2)^{\frac{1}{2}\mu} = 4\pi\delta(x_1, x_2, x_3).$$

Inserting this into (14), we arrive at

$$\operatorname*{res}_{\mu=-3} ((x_1^2 + x_2^2 + x_3^2)^{\frac{1}{2}\mu}, f(x)) = 4\pi f(0, 0, 0) = 4\pi f(a),$$

from which it follows that for any $a$

$$\operatorname*{res}_{\mu=-3} \sinh^\mu kr = 4\pi\delta_a(x),$$

where $\delta_a(x)$ is the generalized function in Lobachevskian space defined in Eq. (11).

Recall that $\sinh^\mu kr$ is the only factor in Eq. (7) that has a singularity at $\mu = -3$. We may thus set $r = 0$ and $\mu = -3$ in all the other factors, to arrive at the result

$$\operatorname*{res}_{\mu=-3} \Phi(x, a; \mu) = -8\pi^2\delta_a(x) \tag{15}$$

or

$$\operatorname*{res}_{\mu=-3} \int \Phi(x, a; \mu) f(x)\, dx = -8\pi^2 f(a). \tag{16}$$

This expression was obtained for $a$ with coordinates (1, 0, 0, 0). But since $\Phi(x, a; \mu)$ is invariant under simultaneous motion of $x$ and $a$, the equation holds for all $a$.

This essentially completes the derivation of the inversion formula. By combining (3), (8), and (16) we arrive at

$$f(a) = -(8\pi^2)^{-1} \int \delta''([a, \xi] - 1) h(\xi)\, d\xi, \tag{17}$$

which gives $f(a)$ in terms of $h(\xi)$. Summing up, we have established the following result.

*Let $f(x)$ be an infinitely differentiable function with bounded support on a three-dimensional Lobachevskian space, and let*

$$h(\xi) = \int f(x)\delta([x, \xi] - 1)\, dx$$

*be the integral of this function over the horosphere whose equation is $[x, \xi] = 1$. Then $f(a)$ is given in terms of $h(\xi)$ by Eq.* (17).

Equivalently, we have

$$\delta_a(x) = -(8\pi^2)^{-1} \int \delta''([a, \xi] - 1)\delta([x, \xi] - 1)\, d\xi.$$

Because this inversion formula is so important, we should like to write it in still another form. Equation (17) may be written as the double integral

$$f(a) = -(8\pi^2)^{-1} \int_0^\infty \int \delta(t - 1)\delta''([a, \xi] - t)h(\xi)\, d\xi\, dt.$$

Integrating over $t$ twice by parts, we arrive at

$$f(a) = -(8\pi^2)^{-1} \int_0^\infty \delta''(t - 1)H(a, t)\, dt = -(8\pi^2)^{-1}H_t''(a, 1) \tag{17'}$$

where

$$H(a, t) = \int h(\xi)\delta([a, \xi] - t)\, d\xi. \tag{18}$$

Thus $H(a, t)$ is the integral of $h(\xi)$ over the intersection of the $[\xi, \xi] = 0$ cone and the plane whose equation is $[a, \xi] = t$. The measure $\delta([a, \xi] - t)\, d\xi$ used in this integral is invariant under Lobachevskian rotations about $a$. In other words, it is a harmonic measure on the absolute invariant under rotations about $a$. It is easily established that the measure of the entire absolute is $4\pi t$.

In order to establish the geometric meaning of $H(a, t)$, let us find the distance $\tau$ from $a$ to the horosphere $\omega$ whose equation is $[x, \xi] = 1$.[10]

[10] By the distance from the point to the horosphere we mean the shortest distance from $a$ to any $x$ on the horosphere. It is equal to the length of the perpendicular dropped from $a$ onto the horosphere.

This distance is an invariant of the points $a$ and $\xi$, and must vanish when $a$ lies on the horosphere, that is, when $[a, \xi] = 1$. We assert that this distance is given by

$$\tau = k^{-1} | \ln [a, \xi]|. \tag{19}$$

Proof. The value of $\tau$ does not change under simultaneous motion of $a$ and the horophere $\omega$. But the equation of the transformed horosphere $\omega g$ is $[x, \xi g] = 1$; thus $\tau$ is invariant under simultaneous transformation of $a$ and $\xi$. If $a$ lies on the positive $[x, x] = 1$ hyperboloid and $\xi$ lies on the positive $[\xi, \xi] = 0$ cone, there exists a motion such that these points are carried into $a'$ and $\xi'$, where the coordinates of $a'$ are $(1, 0, 0, 0)$ and the coordinates of $\xi'$ are $(t, t, 0, 0)$, with $t = [a, \xi]$. Thus we need only calculate the distance between $a'$ and the horosphere $\omega'$ whose equation is $[x, \xi'] = 1 = tx_0 - tx_1$. Now from the definition of the distance between a point and a horosphere it follows that this distance is the radius of a sphere centered at $a'$ and tangent to $\omega'$. But every sphere centered at $a'$ is the intersection of the $[x, x] = 1$ hyperboloid with some plane of the form $x_0 = C$. Such an intersection is tangent to the $tx_0 - tx_1 = 1$ horosphere at the point where $x_0$ takes on its minimum value on the horosphere. It is easily seen that this point $y$ has coordinates $(\cosh k\tau, \sinh k\tau, 0, 0)$, where $\tau = k^{-1} | \ln t |$. Since the distance from $a'$ to $y$ is

$$\tau = k^{-1} | \ln t | = k^{-1} | \ln [a, \xi]|,$$

the distance from $a'$ to the horosphere $\omega'$ is also $\tau$, as is therefore also the distance from $a$ to the horosphere $\omega$ whose equation is $[x, \xi] = 1$. This proves the assertion.

This result shows that the horospheres corresponding to points $\xi$ on the cone such that $[a, \xi] = t$ are separated from $a$ by the distance $\tau$. In other words the integration in (18) is over the set of horospheres whose distance from $a$ is $\tau$, or in other words $H(a, t)$ is the integral of $h(\xi)$ over this set of horospheres. Since all these horospheres are tangent to the sphere of radius $\tau$ centered at $a$, we may also say that the integral is taken over this sphere. Then the measure with respect to which the integration is performed is invariant under rotations about $a$ and has the property that the total measure of the entire sphere is $4\pi t$. This result can be restated in the following way.

*Let $f(x)$ be a function of bounded support on Lobachevskian 3-space, and let the integrals*

$$h(\xi) = \int f(x)\delta([x, \xi] - 1)\, dx$$

*of this function over the horospheres* $[x, \xi] = 1$ *be known. Then the value of* $f(x)$ *at some point* $a$ *is given by*

$$f(a) = -(8\pi^2)^{-1} H_t''(a, 1),$$

*where* $H(a, t)$ *is the integral of* $h(\xi)$ *over the set of horospheres tangent to the sphere* $\Omega(a, \tau)$ *of radius* $\tau = k^{-1} \mid \ln t \mid$ *centered at* $a$. This integral reduces to the integral over $\Omega(a, \tau)$ with respect to a measure invariant under rotations about $a$ and such that the total measure of the entire sphere is $4\pi t$.[11]

In conclusion, we wish to establish the fact that $H(a, t)$ has the symmetry property

$$H(a, t) = H(a, t^{-1}). \tag{20}$$

This can be seen simply from the fact that, according to Eqs. (1) and (18),

$$\begin{aligned} H(a, t) &= \int h(\xi)\delta([a, \xi] - t)\, d\xi \\ &= \int f(x)\delta([x, \xi] - 1)\delta([a, \xi] - t)\, d\xi\, dx. \end{aligned} \tag{21}$$

Obviously the kernel $\int \delta([x, \xi] - 1)\, \delta([a, \xi] - t)\, d\xi$ is invariant under simultaneous motion of $a$ and $x$; in particular, it is invariant under interchange of $a$ and $x$. This implies that it is invariant also under interchange of $t$ and $t^{-1}$ (specifically, this requires a change of variables in the integral from $\xi$ to $t\xi$). Then Eq. (20) follows immediately.

It can be shown that the kernel formed by the integral of the two $\delta$ functions is the characteristic function of the region $[a, x] > \frac{1}{2}(t + t^{-1})$. This fact also exhibits the symmetry property of the kernel and hence of $H(a, t)$.

### 2.3. Inversion Formula for Arbitrary Dimension

So far we have dealt in detail with the case in which the dimension of the Lobachevskian space is 3. We now briefly describe the case of arbitrary dimension. In general outlines the derivation of the inversion formula for the integral transform,

$$h(\xi) = \int f(x)\delta([x, \xi] - 1)\, dx \tag{1}$$

[11] Since the horospheres tangent to this sphere are determined by their point of tangency with the absolute, the integral can be treated also as the integral over the absolute with respect to the harmonic measure invariant under rotations about $a$.

remains the same. It is only the calculation of $\Phi(x, a; \mu)$ which will be somewhat more complicated.

In this case the kernel is given by

$$\Phi(x, a; \mu) \equiv \int |[a, \xi] - 1|^{\mu}\delta([x, \xi] - 1)\, d\xi$$

$$= 2\int |\cosh kr - \xi_1 \sinh kr - 1|^{\mu} \frac{d\xi_1 \cdots d\xi_{n-1}}{(1 - \xi_1^2 - \cdots - \xi_{n-1}^2)^{\frac{1}{2}}}. \quad (2)$$

This integral can be expressed in terms of the hypergeometric function. Specifically, for $\operatorname{Re}\mu > -1$ we obtain

$$\Phi(x, a; \mu) = \frac{2\pi^{\frac{1}{2}(n-1)}\Gamma(\frac{1}{2}\mu + \frac{1}{2})}{\Gamma(\frac{1}{2}\mu + \frac{1}{2}n)} \sinh kr\, F(-\tfrac{1}{2}\mu, 1 - \tfrac{1}{2}[\mu + n]; \tfrac{1}{2}; \tanh^2 \tfrac{1}{2}kr), \quad (3)$$

where $r$ is the distance between $a$ and $x$, and $F(\alpha, \beta; \gamma; x)$ is the hypergeometric function.

The rest of the derivation of the inversion formula is the same as for the case $n = 3$. The following results concerning residues of generalized functions from Volume 1 are used. At $\mu = -n$ the generalized function $(x^2 + \cdots + x_n^2)^{\frac{1}{2}\mu}$ has a simple pole with residue

$$\operatorname*{res}_{\mu=-n} (x_1^2 + \cdots + x_n^2)^{\frac{1}{2}\mu} = \frac{2\pi^{\frac{1}{2}n}}{\Gamma(n/2)} \delta(x_1, \ldots, x_n). \quad (4)$$

The generalized function $|t|^{\mu}$ of a single variable has a simple pole at $\mu = -n$ for $n = 2m + 1$ odd. The residue is

$$\operatorname*{res}_{\mu=-2m-1} |t|^{\mu} = \frac{2\delta^{(2m)}(t)}{\Gamma(2m+1)}. \quad (5)$$

For $n = 2m$ even, the generalized function $|t|^{\mu}$ is regular and equal to $t^{-2m}$ at $\mu = -n$. With these facts we arrive at the following result.

*The inversion formula. Let $f(x)$ be a function with bounded support on an n-dimensional Lobachevskian space, and let*

$$h(\xi) = \int f(x)\delta([x, \xi] - 1)\, dx \quad (6)$$

*be integrals of $f(x)$ over horospheres $[x, \xi] = 1$ in this space. If the dimension $n = 2m + 1$ of the space is odd, the inverse of (6) is*

$$f(a) = \frac{(-1)^m}{2(2\pi)^{2m}} \int \delta^{(2m)}([a, \xi] - 1)h(\xi)\, d\xi. \quad (7)$$

*If $n = 2m$ is even, the inverse of* (6) *is*

$$f(a) = \frac{(-1)^m \Gamma(2m)}{(2\pi)^{2m}} \int ([a, \xi] - 1)^{-2m} h(\xi)\, d\xi. \tag{8}$$

*Here the integral is understood in terms of its regularization, specifically*

$$\int ([a, \xi] - 1)^{-2m} h(\xi)\, d\xi = \left[\int |[a, \xi] - 1\,|^{\mu} h(\xi)\, d\xi\right]_{\mu=-2m}.$$

Remark (*added in proof*). If $\varphi(a, \xi; \mu)$ is chosen as

$$\varphi(a, \xi; \mu) = [a, \xi]^{-\frac{1}{2}(\mu+n)}\, |[a, \xi] - 1\,|^{\mu},$$

rather than equal to $|[a, \xi] - 1\,|^{\mu}$, the kernel $\Phi(x, a; \mu)$ can be written in terms of elementary functions:

$$\Phi(x, a; \mu) = \frac{2\pi^{\frac{1}{2}(n-1)}\Gamma(\frac{1}{2}\mu + \frac{1}{2}) \sinh^{\mu} kr}{\Gamma(\frac{1}{2}\mu + \frac{1}{2}n) \cosh^{\mu+n-2} \frac{1}{2}kr}. \quad \#$$

## 2.4. Functions Depending on the Distance from a Point to a Horosphere, and Their Averages

The kernel $\Phi(x, a; \mu)$ introduced in the last few sections is obtained when $|[a, \xi] - 1\,|^{\mu}$ is integrated over the set of all horospheres whose equations are $[x, \xi] = 1$; these horospheres all pass through the given point $x$. In Section 2.2 it was shown that $[a, \xi]$ is simply related to the distance from $a$ to the horosphere $\omega$ whose equation is $[x, \xi] = 1$. Specifically, this distance is given by

$$\tau(a, \omega) = k^{-1}\, |\ln [a, \xi]|.$$

It follows then that $|[a, \xi] - 1\,|^{\mu}$ is a function of $\tau$. We now wish to turn to the more general case of an arbitrary function $\varphi(a, \omega)$ of the distance from $a$ to $\omega$, and to study the integral of such a function over the set of horospheres passing through a given point $x$.

Thus let $\varphi(a, \omega)$ depend only on the distance $\tau$ from $a$ to the horosphere $\omega$ whose equation is $[x, \xi] = 1$. This function may also be written as a function of $[a, \xi]$ in the form

$$\varphi(a, \omega) \equiv \varphi(\tau(a, \omega)) = \psi([a, \xi]). \tag{1}$$

Now consider any function $x$ of our Lobachevskian space, and let us integrate $\psi$ over the set of horospheres passing through $x$. We shall choose the measure on this set to be invariant under rotations about $x$. Then obviously the desired integral will be given by

$$\Phi(x, a) = \int \psi([a, \xi])\, d\sigma = \int \psi([a, \xi])\delta([x, \xi] - 1)\, d\xi, \tag{2}$$

where $d\xi$ is the invariant measure on the cone whose equation is $[\xi, \xi] = 0$ (see Section 1.6). As in the special case in which $\varphi(a, \xi)$ was equal to $|[a, \xi] - 1|^{\mu}$, it is easily verified that $\Phi(x, a)$ is invariant under simultaneous motion of $a$ and $x$, and therefore depends only on the distance $r$ between them.

Let us write $\Phi(r) \equiv \Phi(x, a)$ in the form of a single, noniterated integral. For this we choose the coordinates of $x$ to be $(1, 0, ..., 0)$ and those of $a$ to be $(\cosh kr, \sinh kr, 0, ..., 0)$; the distance between $x$ and $a$ is then $r$. For this special case we obtain

$$\Phi(r) = \frac{2\pi^{\frac{1}{2}(n-1)}}{\Gamma(\frac{1}{2}n - \frac{1}{2})} \int_{-1}^{1} \psi(\cosh kr - \xi_1 \sinh kr)(1 - \xi_1^2)^{\frac{1}{2}(n-3)}\, d\xi_1 .$$

In order to find the kernel of this integral transform, we write $\cosh kr - \xi_1 \sinh kr = \lambda$, obtaining

$$\Phi(r) = \frac{2\pi^{\frac{1}{2}(n-1)}}{\Gamma(\frac{1}{2}n - \frac{1}{2}) \sinh^{n-2} kr} \int_{e^{-kr}}^{e^{kr}} \psi(\lambda)(\lambda - e^{-kr})^{\frac{1}{2}(n-3)}(e^{kr} - \lambda)^{\frac{1}{2}(n-3)}\, d\lambda. \tag{3}$$

This implies that the kernel of the integral transform from $\psi(\lambda)$ to $\Phi(r)$ vanishes if $kr < |\ln \lambda|$, while for $kr \geqslant |\ln \lambda|$ it is given by

$$\frac{2\pi^{\frac{1}{2}(n-1)}}{\Gamma(\frac{1}{2}n - \frac{1}{2}) \sinh^{n-2} kr} (\lambda - e^{-kr})^{\frac{1}{2}(n-3)}(e^{kr} - \lambda)^{\frac{1}{2}(n-3)}. \tag{4}$$

Equation (3) is particularly simple when $n = 3$. It then becomes

$$\Phi(r) = \frac{2\pi}{\sinh kr} \int_{e^{-kr}}^{e^{kr}} \psi(\lambda)\, d\lambda. \tag{3'}$$

It is a simple matter, using Eq. (3), to obtain the form of the integral transform from $\varphi(\tau)$to $\Phi(r)$. One need only use Eq. (1) in the definition of $\psi$. Then setting $\lambda = e^{k\tau}$ in (3), we arrive at

$$\Phi(r) = \frac{2\pi^{\frac{1}{2}(n-1)}k}{\Gamma(\frac{1}{2}n - \frac{1}{2}) \sinh^{n-2} kr} \int_{-r}^{r} \varphi(\tau)(e^{k\tau} - e^{-kr})^{\frac{1}{2}(n-3)}(e^{kr} - e^{k\tau})^{\frac{1}{2}(n-3)} e^{k\tau}\, d\tau. \tag{5}$$

In particular, for $n = 3$ this becomes

$$\Phi(r) = \frac{2\pi k}{\sinh kr} \int_{-r}^{r} \varphi(\tau) e^{k\tau} \, d\tau. \tag{5'}$$

Thus we have obtained the integral transform from $\varphi(\tau)$, a function only of the distance from $a$ to $\omega$, to $\Phi(r)$, which depends on the distance between $a$ and $x$. This was done, as we have already said, by integrating $\varphi(\tau)$ over the set of all horospheres $\omega$ passing through $x$.

## 3. Integral Transform Associated with Horospheres in an Imaginary Lobachevskian Space[12]

### 3.1. Statement of the Problem and Preliminary Remarks

We shall now study an integral transform associated with horospheres in a three-dimensional imaginary Lobachevskian space. As was shown in Section 1.5, in such a space there exist two sets of horospheres on each of which the group of Lobachevskian motions is transitive. The first consists of horospheres of the first kind, which have dimension two and are analogous to horospheres in an ordinary Lobachevskian space. The second consists of isotropic lines.

Let $f(x)$ be a function of bounded support on an imaginary Lobachevskian space. We shall associate with it the two functions $h(\omega)$ and $\varphi(l)$. The first of these is defined on the horospheres of the first kind by the formula

$$h(\omega) = \int_{\omega} f(x) \, d\sigma, \tag{1}$$

where $d\sigma$ is the invariant measure on the horosphere $\omega$. The second of these is defined on the isotropic lines (we take the model of the imaginary Lobachevskian space on the hyperboloid whose equation is $[x, x] = -1$) by the formula

$$\varphi(l) = \int_{l} f(x) \, dl \equiv \int_{-\infty}^{\infty} f(b + t\xi) \, dt \tag{2}$$

(see Section 1.7). In order that this equation define $\varphi(l)$ uniquely, we shall assume the vectors $\xi$ and $b$ normalized in accordance with $\xi_0 = 1$

[12] This section may be read after Sections 1–3 of Chapter 6.

and $b_0 = 0$. Thus $\varphi(l)$ is a function of the vectors $\xi$ and $b$ which satisfy conditions

$$[\xi, \xi] = 0, \quad [\xi, b] = 0, \quad [b, b] = \quad 1, \quad \xi_0 = 1, \quad b_0 = 0$$

(see Section 1.3a).

The association of $f(x)$ with the pair of functions $\{h(\omega), \varphi(l)\}$ will be called *the integral transform associated with horospheres in an imaginary Lobachevskian space.* The basic problem to which we direct our attention in the present section is to derive the inversion formula for this transform. The final result giving $f(x)$ in terms of these two functions will be found at the end of Section 3.4. In Chapter VI the result will be used to expand a function on an imaginary Lobachevskian space in a series of functions transforming according to irreducible representations of the Lorentz group (in other words, to decompose a representation of the Lorentz group on an imaginary Lobachevskian space into its irreducible components).

It was shown in Section 1.5. that every horosphere $\omega$ is determined by an equation of the form $[x, \xi] = 1$ for $\xi$ on the positive sheet of the cone, that is, for $[\xi, \xi] = 0$, $\xi_0 > 0$. Therefore $h(\omega)$ may be considered a function defined on this positive cone, and we may write $h(\xi)$ instead of $h(\omega)$. By making use of the invariant measure on the horosphere defined in Section 1.7, Eq. (1) can be rewritten in the form

$$h(\xi) = \int f(x)\delta(|[x, \xi]| - 1)\, dx. \tag{3}$$

Here $dx$ is the invariant measure on the hyperboloid whose equation is $[x, x] = -1$. Thus the correspondence $f(x) \to h(\xi)$ associates with each $f(x)$ defined on this hyperboloid an $h(\xi)$ defined on the upper sheet of the $[\xi, \xi] = 0$ cone. We may ask what properties $h(\xi)$ has if $f(x)$ is infinitely differentiable and has bounded support. In answer, it is first seen immediately from Eq. (3) that $h(\xi)$ is infinitely differentiable. Second, it can be shown in exactly the same way as for a real Lobachevskian space that $h(\xi)$ vanishes in a neighborhood of the vertex of the cone. Unlike the real case, however, $h(\xi)$ does not in general have bounded support. Its asymptotic behavior is in fact given by the following.

**Proposition.** *Let $f(x)$ be an infinitely differentiable function with bounded support defined on an imaginary Lobachevskian space* (*that is, on the hyperboloid whose equation is $[x, x] = -1$*), *and let $h(\xi)$ be a function of the cone obtained from $f(x)$ by Eq.* (3). *For each $\xi$ on the positive cone,* *consider the function $h(t\xi)$. Then as $t \to +\infty$ this function has the asymptotic expansion*

$$h(t\xi) \sim \sum_{n=1}^{\infty} a_n(\xi) t^{-n}, \tag{4}$$

*where $a_n(\xi)$ is a homogeneous function of degree $-n$ in $\xi$ and is infinitely differentiable on the intersection $\Omega$ of the $[\xi, \xi] = 0$ cone and the plane whose equation is $\xi_0 = 1$. Further, this asymptotic expansion is term-by-term differentiable in $t$.*

Proof. According to Eq. (3) we have

$$h(t\xi) = \int f(x)\delta([x, t\xi] - 1)\, dx. \tag{5}$$

Now $\delta(t)$ is a homogeneous generalized function, so that this may be written in the form

$$h(t\xi) = t^{-1} \int f(x)\delta([x, \xi] - t^{-1})\, dx. \tag{6}$$

We now expand $\delta([x, \xi] - t^{-1})$ in powers of $t^{-1}$ and integrate term by term. The resulting asymptotic series is

$$h(t\xi) \sim \sum_{n=1}^{\infty} a_n(\xi) t^{-n},$$

where

$$a_n(\xi) = \frac{(-1)^{n-1}}{(n-1)!} \int f(x)\delta^{(n-1)}([x, \xi])\, dx. \tag{7}$$

These integrals converge because $f(x)$ is infinitely differentiable and has bounded support. Thus we have arrived at Eq. (4). The fact that the $a_n(\xi)$ are homogeneous and infinitely differentiable in $\xi$ on $\Omega$ also follows directly from (7) and from the bounded support and infinite differentiability of $f(x)$. To prove the term-by-term differentiability, one need only differentiate (5) with respect to $t$ and then again expand the right-hand side in powers of $t^{-1}$. This then proves the proposition.

It should be remarked that the class of $h(\xi)$ functions, obtained by the integral transform of Eq. (3) from infinitely differentiable functions $f(x)$ with bounded support, is not exactly the class of functions on the cone that is infinitely differentiable, vanishes in a neighborhood of the vertex, and has the asymptotic behavior given by Eq. (4). This is because the $h(\xi)$ functions have certain additional symmetry properties, as do the Radon transforms of functions with bounded support on a plane (see Chapter I, Section 1.6).

We shall derive the inversion formula for the integral transform associated with horospheres in an imaginary Lobachevskian space in roughly the following way. As in a real Lobachevskian space, this derivation is based on the equation

$$\int \Phi(x, a; \mu) f(x)\, dx = \int (|[a, \xi]| - 1)_+^\mu h(\xi)\, d\xi, \tag{8}$$

where $dx$ is the invariant measure in an imaginary Lobachevskian space, $d\xi$ is the invariant measure on the cone, and $\Phi(x, a; \mu)$is defined by

$$\Phi(x, a; \mu) = \int (|[a, \xi]| - 1)_+^\mu\, \delta(|[x, \xi]| - 1)\, d\xi. \tag{9}$$

This last equation is the analog of Eq. (3) in Section 2.2. In that section we derived the inversion formula by finding the residue of $\Phi(x, a; \mu)$ at $\mu = -3$, and established that this residue is equal to the $\delta$ function concentrated at $a$ to within a constant factor. In this way we obtained $f(a)$ in terms of $h(\xi)$ for a real Lobachevskian space.

Now for the imaginary Lobachevskian space we shall again find the residue of $\Phi(x, a; \mu)$ at $\mu = -3$. In this case, however, we shall obtain a different result; in addition to the $\delta$ function concentrated at $a$, another term arises, and this term differs from zero only for

$$|[a, x]| < 1,$$

or in other words for values of $x$ such that

$$r^2(a, x) < 0.$$

To put it differently, the residue of the left-hand side of (8) is, to within a constant factor, the sum of $f(x)$ evaluated at $a$ and the integral of $f(x)$ over the region given by $r^2(a, x) < 0$. In Section 3.4 we shall see that this integral can be written in terms of $\varphi(l)$, the integrals of $f(x)$ over isotropic lines. Thus in the case of an imaginary Lobachevskian space, $f(a)$ is specified not only by $h(\xi)$, but also by $\varphi(l)$.

The derivation in the case of imaginary Lobachevskian space meets with the following complications. Recall that in the real space the integral in the analog of Eq. (9) which defines $\Phi(x, a; \mu)$ converged for $\operatorname{Re} \mu > -1$, so that analytic continuation in $\mu$ could be used to define this function for general $\mu$. But in the imaginary space the integral on the right-hand side of (9) may diverge at all values of $\mu$ for certain special pairs of values of $a$ and $x$ [to be specific, for $r^2(a, x) < 0$]. In just the same way, because $h(\xi)$ does not have bounded support the integral on the right-hand side of (8) may also diverge at all values of $\mu$.[13] We therefore must somehow give meaning to these divergent integrals. Sections 3.2 and 3.5 are devoted to this problem. In them we develop the method of analytic continuation in the coordinates.

**Remark.** We saw in Section 2.2 that the real-space analog of Eq. (9) has a simple geometric meaning: it is the integral of $(|[a, \xi]| - 1)_+^\mu$ over the set of all horospheres passing through $x$ and with respect to a measure invariant under rotations about $x$. The situation is similar for Eq. (9), except that the direction vectors through $x$ now form a noncompact set, and therefore the invariant measure on this set is unbounded. Difficulties associated with integrating a function over a noncompact set of such lines arises also in other problems. They arise, for instance, in attempting to calculate the convolution of functions on a locally compact Lie group when these functions are constant on left and right cosets with respect to a noncompact subgroup; another example is in the study of the analog of spherical functions for noncompact subgroups; there are others. For this reason the method by which we shall define the integral of Eq. (9) would seem to be of greater interest than just in its application to the present problem. We may remark also that Eq. (9) may be thought of as the Radon transform of the function which is concentrated on the $[\xi, \xi] = 0$ cone and given by $(|[a, \xi]| - 1)_+^\mu$ for $\xi$ on the cone. Thus in attempting to give meaning to the integral of Eq. (9) we are dealing with problems similar to those studied in Sections 2 and 3 of Chapter I. In fact the methods developed in our present chapter may be considered supplementary to the methods of those sections of Chapter I. #

### 3.2. Regularizing Integrals by Analytic Continuation in the Coordinates

Consider the functional defined on the space of infinitely differentiable functions $f(x)$ on an imaginary Lobachevskian space by

$$(\Phi, f) = \int \Phi(x, a; \mu) f(x)\, dx, \tag{1}$$

[13] The integrand in Eq. (9) is constant on the manifolds which form the intersection of the $[\xi, \xi] = 0$ cone and each of the two-dimensional planes $[x, \xi] = 1$ and $[a, \xi] = c$. But if $r^2(a, x) < 0$, these manifolds are unbounded and the integral diverges for all $\mu$. The situation is similar in Eq. (8).

where $dx$ is the invariant measure on the imaginary Lobachevskian space. Here $\Phi(x, a; \mu)$ is a function of the points $a$ and $x$ in the space and of a complex parameter $\mu$, and is formally defined as follows. Consider the model of the imaginary Lobachevskian space on the hyperboloid whose equation is $[x, x] = -1$ and write

$$\begin{aligned}\Phi(x, a; \mu) &= \int (|[a, \xi]| - 1)_+^\mu \, \delta(|[x, \xi]| - 1) \, d\xi \\ &= \int (|[a, \xi]| - 1)_+^\mu \, \delta([x, \xi] - 1) \, d\xi \\ &\quad + \int (|[a, \xi]| - 1)_+^\mu \, \delta([x, \xi] + 1) \, d\xi \\ &\equiv \Phi_1(x, a; \mu) + \Phi_2(x, a; \mu),\end{aligned} \tag{2}$$

where $d\xi$ is the invariant measure on the $[\xi, \xi] = 0$ cone and both integrals are over the upper sheet of this cone. As has been mentioned in Section 3.1, for certain pairs of values of $a$ and $x$ the integral on the right-hand side of (2) may diverge for all values of $\mu$. We wish therefore to give it meaning for such points, or in other words to regularize the integral. To this problem we devote this section.

We turn first to the somewhat simpler problem of regularizing the integral

$$J_+(x, b; \mu) = \int [b, \xi]_+^\mu \, \delta([x, \xi] - 1) \, d\xi. \tag{3}$$

In fact the problem of Eq. (2) can be reduced to this one. Indeed, if we were to treat the integrals of Eq. (2) as though they converged, we could write

$$\begin{aligned}\Phi_1(x, a; \mu) = \int \{&([a, \xi] - 1)_+^\mu \\ &+ (-[a, \xi] - 1)_+^\mu\} \, \delta([x, \xi] - 1) \, d\xi.\end{aligned} \tag{2'}$$

Since the integrand is concentrated on the $[x, \xi] = 1$ plane, this may be written in the form

$$\begin{aligned}\Phi_1(x, a; \mu) &= \int [a - x, \xi]_+^\mu \, \delta([x, \xi] - 1) \, d\xi \\ &\quad + \int [-a - x, \xi]_+^\mu \, \delta([x, \xi] - 1) \, d\xi\end{aligned}$$

from which we arrive at

$$\Phi_1(x, a; \mu) = J_+(x, a - x; \mu) + J_+(x, -a - x; \mu). \tag{4}$$

We shall now consider Eq. (4) to be the definition of $\Phi_1(x, a; \mu)$. Then $\Phi_2(x, a; \mu)$ is defined analogously, and finally Eq. (2) may be used to define $\Phi(x, a; \mu)$. Thus the regularization of the integral of Eq. (2) is reduced to a study of the integral of Eq. (3) for all real points $b$ and $x$. We turn now to this problem.

Consider Eq. (3) first for points $b$ such that $[b, b] > 0$ and $b_0 > 0$. We shall call the set of these points the *interior of the positive cone*, and we shall denote the fact that a point $b$ lies inside the positive cone simply by $b > 0$. Now let $b > 0$ and let $\xi$ lie on the upper or positive sheet of the $[\xi, \xi] = 0$ cone, the boundary of the interior of the positive cone. Then it is easily seen that[14]

$$[b, \xi] > (b_0 - [b_1^2 + b_2^2 + b_3^2]^{\frac{1}{2}})\xi_0 > 0. \tag{5}$$

Then for $b > 0$ it follows that $J_+(x, b; \mu)$ coincides with

$$J(x, b; \mu) = \int [b, \xi]^\mu \delta([x, \xi] - 1)\, d\xi, \tag{6}$$

an integral which converges for $\operatorname{Re} \mu < -1$.

This then defines $J_+(x, b; \mu)$ for $\operatorname{Re} \mu < -1$ and for $b > 0$. We wish to extend this definition to arbitrary $b \in E_4$, and accomplish this by analytic continuation in the coordinates $b_0, b_1, b_2, b_3$ of $b$, as described in the discussion of the generalized function $P_+^\mu$ of Chapter III, Volume 1 (Section 2.3).

We introduce the four-dimensional complex linear space $Z_4$ consisting of points $z$ whose coordinates are $(z_0, z_1, z_2, z_3)$. Consider those $z = b + ic$ for which $c > 0$ (that is, for which $c$ lies in the positive cone). Let the set of all such $z$ be called the "upper half-plane" in $Z_4$. Similarly, we call the set of $z = b - ic$ such that $c > 0$ the "lower half-plane" in $Z_4$.

We now assert that $J(x, b; \mu)$ as defined by Eq. (6) for $b > 0$ can be continued analytically into the upper and lower half-planes of $Z_4$. This analytic continuation will be defined by

$$J(x, z; \mu) = \int [z, \xi]^\mu \delta([x, \xi] - 1)\, d\xi, \tag{7}$$

[14] This follows from the fact that

$$[b, \xi] = b_0\xi_0 - b_1\xi_1 - b_2\xi_2 - b_3\xi_3 > b_0\xi_0 - (b_1^2 + b_2^2 + b_3^2)^{\frac{1}{2}} \cdot (\xi_1^2 + \xi_2^2 + \xi_3^2)^{\frac{1}{2}}.$$

But since $\xi$ lies on the upper sheet of the cone,

$$\xi_0 = (\xi_1^2 + \xi_2^2 + \xi_3^2)^{\frac{1}{2}},$$

which leads directly to Eq. (5).

where raising a number to the complex power $\mu$ is understood in the sense that

$$t^{\mu} = \exp\{\mu \ln |t| + i\mu \arg t\}, \qquad -\pi < \arg t < \pi.$$

To prove this we show first that $J(x, z; \mu)$ as defined by Eq. (7) depends analytically on the coordinates of $z$ for $z$ in the upper half-plane. To see this, note that for $z = b + ic$, $c > 0$, and $\xi$ on the upper sheet of the $[\xi, \xi] = 0$ cone, we have $\mathrm{Im}[z, \xi] = [c, \xi] > 0$. Hence for $Z$ in the upper half-plane $[z, \xi]^{\mu}$ is single valued. It is easily shown also that the integral of Eq. (7) converges uniformly in $b$ within any region such that $z = b + ic$, where $c > c' > 0$.[15] Hence Eq. (7) defines an analytic function of the coordinates of $z$ in the upper half-plane.

In exactly the same way one proves that Eq. (7) also gives the analytic continuation of $J(x, b; \mu)$ into the lower half-plane.

We have thus defined $J(x, z; \mu)$ for all $z$ in the upper or lower half-planes. It is now a simple matter to define $J_{+}(x, b; \mu)$, or rather the value of the integral of Eq. (3), for any real $b$. For this purpose we write

$$\begin{aligned} J(x, b + i0; \mu) &= \lim_{c \to +0} J(x, b + ic; \mu) \\ &= \lim_{c \to +0} \int [b + ic, \xi]^{\mu}\, \delta([x, \xi] - 1)\, d\xi \end{aligned} \tag{8}$$

and

$$\begin{aligned} J(x, b - i0; \mu) &= \lim_{c \to +0} J(x, b - ic; \mu) \\ &= \lim_{c \to +0} \int [b - ic, \xi]^{\mu}\, \delta([x, \xi] - 1)\, d\xi \end{aligned} \tag{9}$$

(where $c \to +0$ denotes that $c$ approaches zero from inside the positive cone). The existence of these limits will be established in Section 3.5.

Now to define $J_{+}(x, b; \mu)$ we make use of the formula

$$t_{+}^{\mu} = \frac{e^{i\mu\pi}(t - i0)^{\mu} - e^{-i\mu\pi}(t + i0)^{\mu}}{2i \sin \mu\pi} \tag{10}$$

established for generalized functions of a single variable (see Volume 1, Chapter I, Section 3.6). Accordingly the integral $J_{+}(b, x; \mu)$ is defined by

$$J_{+}(x, b; \mu) = \frac{1}{2i \sin \mu\pi} [e^{i\mu\pi} J(x, b - i0; \mu) - e^{-i\mu\pi} J(x, b + i0; \mu)]. \tag{11}$$

[15] By $c > c'$ we mean $c - c' > 0$.

This definition holds for Re $\mu < -1$ and for arbitrary real $x$ and $b$. But it has been stated [see Eq. (4)] that

$$\Phi_1(x, a; \mu) = J_+(x, a - x; \mu) + J_+(x, -a - x; \mu). \tag{12}$$

Thus the value of the integral

$$\Phi_1(x, a; \mu) = \int (|[a, \xi]| - 1)_+^\mu \, \delta([x, \xi] - 1) \, d\xi \tag{13}$$

is defined for Re $\mu < -1$ with the aid of Eqs. (11) and (12), in which $J(x, b + i0; \mu)$ and $J(x, b - i0; \mu)$ are given by Eqs. (8) and (9). Then $\Phi_2(x, a; \mu)$ is defined similarly, and thereby $\Phi(x, a; \mu)$. As a result we arrive at the expression

$$\begin{aligned}\Phi(x, a; \mu) = \frac{1}{2i \sin \mu\pi} \lim_{c \to +0} \int \delta(|[x, \xi]| - 1) \\ \times \{e^{i\mu\pi}[([a - ic, \xi] - 1)^\mu + ([-a - ic, \xi] - 1)^\mu] \\ - e^{-i\mu\pi}[([a + ic, \xi] - 1)^\mu + ([-a + ic, \xi] - 1)^\mu]\} \, d\xi,\end{aligned} \tag{14}$$

which we take for a definition of $\Phi(x, a; \mu)$ for Re $\mu < -1$. It is easily shown thar for $c > 0$ this integral converges absolutely and uniformly in $x$.

Later we shall need an explicit expression for $\Phi(x, a; \mu)$ in terms of the distance between $a$ and $x$. We give the results here without the proof, which will be found in Section 3.5.

*Let $a$ and $x$ be points of an imaginary Lobachevskian space (realized, as usual, on the hyperboloid whose equation is* $[x, x] = -1$). *Then* $\Phi(x, a; \mu)$ *is given for* Re $\mu < -1$ *by the following formulas.*

*If* $|[a, x]| > 1$, *then*

$$\Phi(x, a; \mu) = -\frac{2\pi e^{\frac{1}{2}(\mu+1)kr}}{(\mu + 1) \cosh^{\mu+1} (\frac{1}{2}kr)} \sinh^\mu kr. \tag{15$_1$}$$

*where* $\cosh kr = |[a, x]|$.

*If* $|[a, x]| < 1$, *then*

$$\begin{aligned}\Phi(x, a; \mu) = \frac{4\pi}{(\mu + 1) \sin \mu\pi} [\cos\{\tfrac{1}{2}(\mu + 1)(\pi - k\rho)\} \cos^{-\mu-1}(\tfrac{1}{2}k\rho) \\ + \cos\{\tfrac{1}{2}(\mu + 1)k\rho\} \sin^{-\mu-1}(\tfrac{1}{2}k\rho)] \sin^\mu k\rho,\end{aligned} \tag{15$_2$}$$

*where* $\cos k\rho = |[a, x]|$.

Recall that if $x$ is such that $|[a, x]| > 1$, then the distance between $x$ and $a$ is real. If, on the other hand, $x$ is such that $|[a, x]| < 1$, the distance $r = i\rho$ from $a$ to $x$ is imaginary.

Having defined $\Phi(x, a; \mu)$ for Re $\mu < -1$, we define it for Re $\mu > -1$ by analytic continuation in $\mu$ and hence also by Eqs. ($15_1$) and ($15_2$).

Let us now finally define the generalized function $(\Phi, f)$, where $f(x)$ is an infinitely differentiable function of bounded support on an imaginary Lobachevskian space. For Re $\mu > -1$ the generalized function is given by

$$(\Phi, f) = \int \Phi(x, a; \mu) f(x)\, dx, \tag{16}$$

where $\Phi(x, a; \mu)$ is given by Eqs. ($15_1$) and ($15_2$). For fixed $f(x)$ this is an analytic function of $\mu$ and can therefore be extended by analytic continuation in $\mu$ to Re $\mu < -1$. Now recall that

$$\int \Phi(x, a; \mu) f(x)\, dx = \int (|[a, \xi]| - 1)_+^\mu h(\xi)\, d\xi \tag{17}$$

was the basic equation used in deriving the inversion formula for the integral transform associated with horospheres of an imaginary Lobachevskian space, where for Re $\mu < -1$ the function $\Phi(x, a; \mu)$ is defined by Eq. (14). Note that the integral on the right-hand side of (14) converges absolutely and uniformly in $x$. We now multiply both sides of (14) by $f(x)$ and integrate over $x$ (after changing the order of integration on the right-hand side). By using the fact that

$$\int f(x) \delta(|[x, \xi]| - 1)\, dx = h(\xi),$$

we arrive at

$$\begin{aligned}\int \Phi(x, a; \mu) f(x)\, dx = {} & \frac{1}{2i \sin \mu\pi} \lim_{c \to +0} \int h(\xi) \\ & \times \{e^{i\mu\pi}[([a - ic, \xi] - 1)^\mu + ([-a - ic, \xi] - 1)^\mu] \\ & - e^{-i\mu\pi}[([a + ic, \xi] - 1)^\mu + ([-a + ic, \xi] - 1)^\mu]\}\, d\xi \end{aligned} \tag{18}$$

(where, as before, $c \to +0$ denotes that $c$ approaches zero from inside the positive cone). If $c > 0$ and if $\xi$ lies on the $[\xi, \xi] = 0$ cone, then $[c, \xi] > 0$. Hence we may again use Eq. (10) to take the right-hand side of (18) as the definition of the integral

$$\int (|[a, \xi]| - 1)_+^\mu h(\xi)\, d\xi.$$

This proves Eq. (17) for $\Phi(x, a; \mu)$ defined by (14) and the integral on the right-hand side defined by the right-hand side of (18).

### 3.3. Derivation of the Inversion Formula

Let us now return to our original goal, namely to express $f(x)$, a function of bounded support on an imaginary Lobachevskian space, in terms of

$$h(\xi) = \int f(x)\delta(|[x, \xi]| \quad 1)\, dx$$

and

$$\varphi(l) = \int f(x)\, dl \equiv \int_{-\infty}^{\infty} f(b + t\xi)\, dt,$$

which are its integrals over horospheres of the first kind and over isotropic lines.

As already mentioned, this formula will be derived by using the relation

$$\int \Phi(x, a; \mu) f(x)\, dx = \int (|[a, \xi]| - 1)_+^{\mu} h(\xi)\, d\xi, \tag{1}$$

which was developed in Section 3.2. We shall obtain the desired inversion formula by finding the residues of both sides of Eq. (1) at $\mu = -3$.

Consider first the expression on the left-hand side, that is, let us find

$$\operatorname*{res}_{\mu=-3} \int \Phi(x, a; \mu) f(x)\, dx.$$

Equations $(15_1)$ and $(15_2)$ of Section 3.2, according to which this is a function of the distance between $a$ and $x$, imply that without loss of generality we may choose $a$ to have any given coordinates so long as $x$ is chosen at the proper distance; thus let the coordinates of $a$ be (0, 0, 0, 1). Then $[a, x] = -x_3$, and the distance $r$ between $a$ and $x$ is given by $\cosh^2 kr = x_3^2$. Now $x$ lies on the hyperboloid whose equation is $[x, x] = -1$, so that we obtain

$$\sinh^2 kr = x_0^2 - x_1^2 - x_2^2 .$$

Obviously $\sinh^2 kr > 0$ if $|x_3| > 1$, and $\sinh^2 kr < 0$ if $|x_3| < 1$. For the latter case, therefore, we may write

$$\sin^2 k\rho = -x_0^2 + x_1^2 + x_2^2 ,$$

where $r = \rho i$. We now insert these expressions for $\sinh^2 kr$ and $\sin^2 k\rho$ into Eqs. $(15_1)$ and $(15_2)$ of Section 3.2. Then with $a$ chosen as above, $\Phi(x, a; \mu)$ is given by the following two expressions.

If $| x_3 | > 1$, then

$$\Phi(x, a; \mu) = -\frac{2\pi^{\frac{1}{2}(\mu+1)kr}}{(\mu + 1)\cosh^{\mu+1}(\frac{1}{2}kr)}(x_0^2 - x_1^2 - x_2^2)_+^{\frac{1}{2}\mu}, \qquad (2_1)$$

where $\cosh kr = | x_3 |$.

If $| x_3 | < 1$, then

$$\Phi(x, a; \mu) = \frac{4\pi}{(\mu + 1)\sin \mu\pi}\{\cos[\tfrac{1}{2}(\mu + 1)(\pi - k\rho)] \cos^{-\mu-1}(\tfrac{1}{2}k\rho)$$
$$+ \cos[\tfrac{1}{2}(\mu + 1)k\rho] \sin^{-\mu-1}(\tfrac{1}{2}k\rho)\}(x_1^2 + x_2^2 - x_0^2)_+^{\frac{1}{2}\mu}, \qquad (2_2)$$

where $\cos k\rho = | x_3 |$.

The residue of the left-hand side of Eq. (1) at $\mu = -3$ is found on the basis of the following assertion (see Volume 1, Chapter III, Section 2.2). The generalized function $(x_0^2 - x_1^2 - x_2^2)_+^{\frac{1}{2}\mu}$ has a simple pole at $\mu = -3$ with residue

$$\operatorname*{res}_{\mu=-3} (x_0^2 - x_1^2 - x_2^2)_+^{\frac{1}{2}\mu} = -4\pi\delta(x_0, x_1, x_2).$$

The generalized function $(x_1^2 + x_2^2 - x_0^2)_+^{\frac{1}{2}\mu}$ has no singularity at $\mu = -3$. Further, the factors multiplying these functions in Eqs. $(2_1)$ and $(2_2)$ have the following properties. At $\mu = -3$ the factor in Eq. $(2_1)$ is continuous and equal to $\pi$ at $r = 0$, and the factor of Eq. $(2_2)$ has a simple pole with residue $-2\cos^2 k\rho$. Note also that with the special choice we have made for the coordinates of $a$, namely $(0, 0, 0, 1)$, we have

$$\cos^2 k\rho = x_3^2 = [a, x]^2$$

and

$$\sin^2 k\rho = x_1^2 + x_2^2 - x_0^2 = 1 - x_3^2 = 1 - [a, x]^2.$$

In accordance with these remarks we arrive easily at

$$\operatorname*{res}_{\mu=-3} \int \Phi(x, a; \mu) f(x)\, dx = \operatorname*{res}_{\mu=-3} \int_{|[a,x]|>1} \Phi(x, a; \mu) f(x)\, dx$$
$$+ \operatorname*{res}_{\mu=-3} \int_{|[a,x]|<1} \Phi(x, a; \mu) f(x)\, dx \qquad (3)$$
$$= -8\pi^2 f(a) - 2\int_{|[a,u]|<1} [a, x]^2(1 - [a, x]^2)^{-\frac{3}{2}} f(x)\, dx.$$

To obtain the first term we have used the fact that $f(-a) = f(a)$. In the second term the integral is understood in terms of its regularization, which is obtained by replacing the exponent $-\frac{3}{2}$ by $\frac{1}{2}\mu$ and by evaluating the analytic generalized function of $\mu$ so obtained at $\mu = -3$. Although Eq. (3) was obtained on the assumption that the coordinates of $a$ are (0, 0, 0, 1), the invariance of $\Phi(x, a; \mu)$ under simultaneous motion of $a$ and $x$ implies that the result is valid for all $a$ in our imaginary Lobachevskian space.

We now turn to the residue at $\mu = -3$ of the right-hand side of Eq. (1). Recall that the integral involved is an abbreviated notation for the right-hand side of Eq. (18) of Section 3.2. Because $\sin \mu\pi$ vanishes at $\mu = -3$, we have

$$\operatorname*{res}_{\mu=-3} \int (|[a, \xi]| - 1)_+^\mu h(\xi)\, d\xi$$

$$= (2\pi i)^{-1} \lim_{c\to+0} \int h(\xi)\{([a - ic, \xi] - 1)^{-3} + ([-a - ic, \xi] - 1)^{-3}$$

$$- ([a + ic, \xi] - 1)^{-3} - ([-a + ic, \xi] - 1)^{-3}\}\, d\xi. \tag{4}$$

It is easily shown that the integral whose limit is being taken on the right-hand side is absolutely convergent. Let us, then, take the limit under the integral sign. [We omit the detailed justification of this procedure which is related to the asymptotic behavior of $h(t\xi)$ as $t \to \infty$.] Bearing in mind that for $c$ in the positive cone and $\xi$ on its boundary, $[c, \xi] > 0$, and that

$$(t - i0)^{-3} - (t + i0)^{-3} = \pi i \delta''(t)$$

(see Volume 1, Chapter I, Section 3.6) we arrive at

$$\operatorname*{res}_{\mu=-3} \int (|[a, \xi]| - 1)_+^\mu h(\xi)\, d\xi = \tfrac{1}{2} \int h(\xi)\delta''(|[a, \xi]| - 1)\, d\xi. \tag{5}$$

Remark. This equation could also have been obtained directly in a formal way from the relation

$$\operatorname*{res}_{\mu=-3} t_+^\mu = \tfrac{1}{2}\delta''(t). \quad \#$$

We may now compare Eqs. (3) and (5) for the residues of both sides of Eq. (1). After some simple operations we arrive at

$$f(a) = -(4\pi)^{-2} \int h(\xi)\delta''(|[a, \xi]| - 1)\, d\xi$$

$$-(2\pi)^{-2} \int_{|[a,x]|<1} [a, x]^2(1 - [a, x]^2)^{-\frac{5}{2}} f(x)\, dx. \tag{6}$$

In the first term the integral is over the upper sheet of the $[\xi, \xi] = 0$ cone, and in the second it is over the region defined by $|[a, x]| < 1$ on the $[x, x] = -1$ hyperboloid. Further, the second integral is understood in the sense of its regularization.

We have now completed the second important step in obtaining the inversion formula [the first was the derivation of Eq. (17) of Section 3.2]. Let us analyze the result we have obtained.

The first term in Eq. (6) is similar to the right-hand side of the inversion formula for the case of an ordinary Lobachevskian space [see Eq. (17) of Section 2.2]. The second term involves integrating $f(x)$ over a region in which $|[a, x]| < 1$, that is, over a set of points whose distance from $a$ is imaginary. We shall show later that we can write this in terms of the integrals of $f(x)$ along isotropic lines, that is, in terms of $\varphi(l)$, thus obtaining the final form of the inversion formula.

The reason for the difference between the inversion formulas for the real and imaginary Lobachevskian spaces may be seen in the following way. At the basis of the derivation for both cases lies the calculation of the residue of $\sinh^\mu kr$ at $\mu = -3$. For the real Lobachevskian space $\sinh^\mu kr$ goes as $(x_1^2 + x_2^2 + x_3^2)^{\frac{1}{2}\mu}$, and thus at $\mu = -3$ it has a simple pole with residue proportional to the $\delta$ function. In the imaginary Lobachevskian space, however, $r$ may be real or imaginary, and therefore we must deal not only with $\sinh^\mu kr$ but also with $\sin^\mu k\rho$. The first of these functions behaves as $(x_0^2 - x_1^2 - x_2^2)_+^{\frac{1}{2}\mu}$, and thus also has a simple pole at $\mu = -3$ with residue proportional to the $\delta$ function. But $\sin^\mu k\rho$ behaves as $(x_1^2 + x_2^2 - x_0^2)_+^{\frac{1}{2}\mu}$, which is regular at $\mu = -3$. The only singularity arises from the coefficient multiplying this function. For this reason the derivations and results differ in the two cases.

**Remark.** We have not yet shown that the first term in Eq. (6), namely the integral

$$\int h(\xi)\delta''(|[a, \xi]| - 1)\, d\xi, \tag{7}$$

converges absolutely. We shall do this here. For this purpose we rewrite (7) in a form in which the integration is over $\Omega$, the sphere which is the intersection of the $[\xi, \xi] = 0$ cone with the $\xi_0 = 1$ hyperplane. Let $\xi$ lie on the $[\xi, \xi] = 0$ cone. Consider one of its generators $l$ and let $\eta$ be the point where $l$ intersects $\Omega$. We may thus write in general $\xi = q\eta$, where $q = \xi_0$. Let us take $q$ and $\eta$ to be new parameters specifying $\xi$. In these parameters the invariant measure $d\xi$ on the cone may be written $d\xi = q\,dq\, d\omega$, where $d\omega$ is the Euclidean measure on $\Omega$.

Then changing variables by writing $\xi = q\eta$ in Eq. (7), we arrive after some elementary operations at[16]

$$\int h(\xi)\delta''(|[a, \xi]| - 1)\, d\xi$$

$$= \int_{\Omega} \left\{ \frac{\partial^2 h(t^{-1}\eta)}{\partial t^2}\bigg|_{t=[a,\eta]} + \frac{\partial^2 h(t^{-1}\eta)}{\partial t^2}\bigg|_{t=-[a,\eta]} \right\} d\omega. \tag{8}$$

Thus what we need to prove now is that the integral on the right-hand side of this equation converges absolutely for every $h(\xi)$ of the form

$$h(\xi) = \int f(x)\delta(|[x, \xi]| - 1)\, dx,$$

where $f(x)$ is an infinitely differentiable function of bounded support on an imaginary Lobachevskian space. Obviously this integral can have singularities only on the set of $\eta$ such that $[a, \eta] = 0$. But it can be shown that the integrand is bounded even in a neighborhood of this set. For this purpose we recall the asymptotic expansion

$$h(t^{-1}\eta) \sim \sum_{n=1}^{\infty} a_n(\eta) t^n, \tag{9}$$

for $t \to 0$ [see Eq. (4) of Section 3.1]. In this expression the $a_n(\eta)$ are infinitely differentiable functions on $\Omega$. Now it was established in Section 3.1 that this asymptotic expansion may be differentiated term by term with respect to $t$, and we thus find that

$$\frac{\partial^2 h(t^{-1}\eta)}{\partial t^2}\bigg|_{t=[a,\eta]} \sim \sum_{n=2}^{\infty} n(n-1)a_n(\eta)[a, \eta]^{n-2}.$$

This expansion shows that the first term in the integrand of Eq. (8) is bounded in the neighborhood we are considering. Similarly, the same can be shown for the second term. Thus the right-hand side of (8) is the integral of a bounded function over a compact set and hence it converges absolutely. #

[16] In the case of an ordinary Lobachevskian space one can obtain a similar formula, namely

$$\int h(\xi)\delta''([a, \xi] - 1)\, d\xi = \int_{\Omega} \frac{\partial^2 h(t^{-1}\eta)}{\partial t^2}\bigg|_{t=[a,\eta]} d\omega.$$

### 3.4. Derivation of the Inversion Formula (Continued)

We are left still with the problem of writing the integral

$$I = \int_{|[a,x]|<1} [a, x]^2(1 - [a, x]^2)^{-\frac{3}{2}} f(x)\, dx \tag{1}$$

in terms of the integral of $f(x)$ over isotropic lines, that is, in terms of $\varphi(l)$. We now turn to this task.

Note that $[a, x]^2 (1 - [a, x]^2)^{-\frac{3}{2}}$ is constant on the planes whose equations are of the form $[a, x] = p$. We shall therefore perform the integration in Eq. (1) by first integrating $f(x)$ over the intersection of such a plane for $|p| < 1$ with the $[x, x] = -1$ hyperboloid and then integrating the result so obtained over $p$. We may thus write

$$I = \int_{-1}^{1} p^2(1 - p^2)^{-\frac{3}{2}}\, dp \int_{[a,x]=p} f(x)\, d\sigma. \tag{2}$$

Here $d\sigma$ denotes the measure on the intersection of the hyperboloid and plane, and is defined by $dx = dp\, d\sigma$.

Now the integral

$$I(p) \equiv \int_{[a,x]=p} f(x)\, d\sigma$$

can be expressed in terms of integrals over isotropic lines. This is because the intersection of the three-dimensional hyperboloid $[x, x] = -1$ with the $p$ plane is a two-dimensional hyperboloid which is single sheeted if $|p| < 1$ and may thus be decomposed (in fact in two ways) into its line generators, that is, into isotropic lines. Then $I(p)$ can be expressed in terms of $\varphi(l)$ if $|p| < 1$.[17]

Let us obtain this expression first for the special case in which $a$ has coordinates (0, 0, 0, 1). Then $[a, x] = x_3$, and hence

$$I(p) \equiv \int_{[a,x]=p} f(x)\, d\sigma = \int_{x_3=p} f(x)\, d\sigma. \tag{3}$$

Now let us use $\theta$, defined by $p = \cos\theta$, instead of $p$ as our parameter. (The geometric meaning of this parameter is discussed in Section 3.4a.)

[17] In terms of the geometry of the imaginary Lobachevskian space, $|[a, x]| < 1$ implies that the points $x$ lie at an imaginary distance from $a$, this distance being given by $r = \rho i$, where $\cos^2 k\rho = [a, x]^2$. Thus when the space is realized on a single-sheeted hyperboloid (with diametrically opposite points $x$ and $-x$ identified), two-dimensional single-sheeted hyperboloids become spheres of imaginary radius and the two-sheeted hyperboloids ($|p| > 1$) become spheres of real radius. What we have seen above is that the spheres of imaginary radius can be decomposed into isotropic lines.

We may now establish which isotropic lines will lie in the plane whose equation is $x_3 = \cos\theta$. For the model on the $[x, x] = -1$ hyperboloid, the equation of an isotropic line is $x = b + t\xi$, where

$$[b, b] = -1, \quad [b, \xi] = 0, \quad [\xi, \xi] = 0, \tag{4}$$

and $b$ and $\xi$ are normalized so that $b_0 = 0$, $\xi_0 = 1$ (see Section 1.3a). Since $l$ lies in the $x_3 = \cos\theta$ plane, the coordinates of $\xi$ will be $(1, \xi_1, \xi_2, 0)$ and those of $b$ will be $(0, b_1, b_2, \cos\theta)$. Then from Eq. (4) we obtain

$$\xi_1^2 + \xi_2^2 = 1, \quad b_1^2 + b_2^2 = \sin^2\theta, \quad b_1\xi_1 + b_2\xi_2 = 0,$$

so that

$$\xi = (1, \cos\alpha, \sin\alpha, 0) \tag{5}$$

and

$$b = (0, \pm\sin\theta\sin\alpha, \mp\sin\theta\cos\alpha, \cos\theta), \tag{6}$$

where the parameter $\alpha$ defining the isotropic line can take on values from 0 to $2\pi$. The two possible signs in Eq. (6) occur because there are two possible ways of decomposing the $[x, x] = -1$ hyperboloid into line generators. We shall use only the upper signs, thus using only a single family of isotropic lines for each $\theta$, this family covering the two-dimensional hyperboloid whose equations are $[x, x] = -1$, $x_3 = \cos\theta$.[18]

To proceed, we change variables of integration in the integral on the right-hand side of Eq. (3). Note that if $b$ and $\xi$ are given by Eqs. (5) and (6), the point $x = b + t\xi$ runs once over the whole two-dimensional hyperboloid when $\alpha$ goes from 0 to $2\pi$ and $t$ goes from $-\infty$ to $+\infty$. We shall thus take $\alpha$ and $t$ to be the new variables of integration. In terms of these variables

$$\begin{aligned} x_0 &= t, \\ x_1 &= \sin\theta\sin\alpha + t\cos\alpha, \\ x_2 &= -\sin\theta\cos\alpha + t\sin\alpha, \\ x_3 &= \cos\theta, \end{aligned} \tag{7}$$

and hence $d\sigma \equiv dx_1\, dx_2 \,|\, x_0 \,|^{-1} = dt\, d\alpha$. Then Eq. (3) becomes

$$I(\cos\theta) = \int_{x_3=\cos\theta} f(x)\, d\sigma = \int_0^{2\pi} d\alpha \int_{-\infty}^{\infty} f(b + t\xi)\, dt, \tag{8}$$

[18] Actually we do not lose the second family of generators. They are obtained when $\cos\theta$ is replaced by $-\cos\theta$, for the intersection of $[x, x] = -1$ with $x_3 = \cos\theta$ is the same as its intersection with $x_3 = -\cos\theta$ (recall that $x$ and $-x$ are identified).

where $\xi = \xi(\alpha)$ is given by Eq. (5) and $b = b(\theta, \alpha)$ is given by (6) with the upper signs. But by definition of the integral over the isotropic line $x = b + t\xi$ we have

$$\int_{-\infty}^{\infty} f(b + t\xi)\, dt = \varphi(\xi, b) \equiv \varphi(l). \tag{9}$$

Summing up, therefore, if the coordinates of $a$ are (0, 0, 0, 1), the integral of Eq. (8) may be written

$$I(\cos\theta) = \int_0^{2\pi} \varphi(\xi(\alpha), b(\theta, \alpha))\, d\alpha. \tag{10}$$

We now wish to generalize this formula to the case of arbitrary $a$. To do this we first write (10) in an invariant form. Note that the integrand actually depends only on $\xi(\alpha)$, since for fixed $\theta$ the vector $b(\theta, \alpha)$ is uniquely determined by $\xi(\alpha)$. We shall therefore write this integrand henceforth simply as $\varphi(\xi, \theta)$. Then Eq. (10) becomes

$$I(\cos\theta) = \int_0^{2\pi} \varphi(\xi(\alpha), \theta)\, d\alpha. \tag{11}$$

Now for any $\alpha$ the point $\xi(\alpha)$ lies on the intersection of the sphere $\Omega$ given by $[\xi, \xi] = 0$, $\xi_0 = 1$ with the plane whose equation is $\xi_3 = [a, \xi] = 0$. This means that we may write (11) as the integral over $\xi$ along this intersection. This can also be written as an integral over the entire sphere $\Omega$ if we use the usual $\delta$-function notation, namely

$$I(\cos\theta) = \int_\Omega \varphi(\xi, \theta)\delta([a, \xi])\, d\omega, \tag{12}$$

where $d\omega$ is the Euclidean measure of the sphere. In order to write this in invariant form we shall replace the integration over $\Omega$ by integration over an arbitrary "contour" $\Gamma$ on the $[\xi, \xi] = 0$ cone.

So far we have assumed that $\xi$ is normalized according to $\xi_0 = 1$. Let us drop this condition and now let $\xi$ be any vector on the $[\xi, \xi] = 0$ cone, while $b$ as before satisfies the conditions $[b, b] = -1$, $[b, \xi] = 0$, and $b_0 = 0$. Also as before, we define $\varphi(\xi, b)$ by

$$\varphi(\xi, b) = \int_{-\infty}^{\infty} f(b + t\xi)\, dt. \tag{13}$$

If the isotropic line $x = b + t\xi$ lies in the $x_3 = [a, x] = \cos\theta$ plane, as we have seen, $b$ is uniquely determined by $\xi$ and is independent of its normalization. Therefore we can again replace $\varphi(\xi, b)$ by $\varphi(\xi, \theta)$. This function is obviously homogeneous in $\xi$ of degree $-1$.

As a result the integrand of Eq. (12) is now defined on the entire $[\xi, \xi] = 0$ cone. It is obviously homogeneous of degree $-2$ in $\xi$. We are now able to replace the integration over $\Omega$ by integration over an arbitrary surface $\Gamma$ intersecting each generator of the cone.

We define the integral over $\Gamma$ by

$$\int_{\Gamma} \varphi(\xi, \theta)\delta([a, \xi])\, d\omega, \tag{14}$$

where

$$d\omega = |\, \xi_0 \,|^{-1}(\xi_1\, d\xi_2\, d\xi_3 - \xi_2\, d\xi_1\, d\xi_3 + \xi_3\, d\xi_1\, d\xi_2). \tag{15}$$

This integral is independent of $\Gamma$. Indeed, the integrand is homogeneous of degree $-2$ and the differential form $d\omega$ is homogeneous of degree 2. This means that the entire expression under the integral sign (that is, the integrand plus the differential form) is homogeneous of degree zero and is thus invariant when $\xi$ is replaced by $\alpha\xi$ for $\alpha > 0$. This in turn implies that (14) is invariant under arbitrary deformations of $\Gamma$.[19] It is easily seen, in particular, that if $\Gamma$ is chosen as $\Omega$, (14) reduces to (12) [for when $\xi_0 = 1$, Eq. (15) defines the Euclidean measure on the sphere].

We have thus arrived at the result

$$I(\cos\theta) \equiv \int_{[a,x]=\cos\theta} f(x)\, d\sigma = \int_{\Gamma} \varphi(\xi, \theta)\delta([a, \xi])\, d\omega, \tag{16}$$

where $\Gamma$ is any surface on the $[\xi, \xi] = 0$ cone crossing all of its generators, and the differential form $d\omega$ is defined by (15). This is the invariant form of Eq. (10), invariant in the sense that it does not depend on the choice of coordinates for $a$. To see this we need only apply (16) to the function $f(xg)$ and recall that both $[a, x]$ and $[a, \xi]$ are invariant under simultaneous motion of $a$, $x$, and $\xi$ and that the integral on the right-hand side of (16) is independent of the choice of $\Gamma$. Thus the result is valid for any $a$ in the imaginary Lobachevskian space.

Let us now return to Eq. (2), set $p = \cos\theta$, and insert the expression we have obtained for $I(\cos\theta)$. Then (2) becomes

$$I = \int_0^{\pi} \cot^2\theta\, d\theta \int_{\Gamma} \varphi(\xi, \theta)\delta([a, \xi])\, d\omega, \tag{17}$$

[19] Compare this with Volume 1, Appendix B2.5. It is useful to point out that if $\Gamma$ is defined by an equation of the form $P(\xi) = 1$, the differential form $d\omega$ satisfies the relation $d\xi = dP\, d\omega$ on $\Gamma$. Equation (14) is the *residue* of the homogeneous function $\varphi(\xi, \theta)\delta([a, \xi])$ defined on the surface whose equation is $[\xi, \xi] = 0$.

which represents the solution of the problem we have set out to investigate, namely to write Eq. (1) in terms of integrals of $f(x)$ over isotropic lines. This completes the derivation of the inversion formula, which we now summarize below.

*Let $f(x)$ be an infinitely differentiable function of bounded support on an imaginary Lobachevskian space. We form the integrals of this function over the horospheres of the first kind and over the isotropic lines; these integrals are defined in the following way (in the realization of the space on the hyperboloid $[x, x] = -1$). The integral over the horosphere of the first kind whose equation is $|[x, \xi]| = 1$ is*

$$h(\xi) = \int f(x)\delta(|[x, \xi]| - 1)\, dx, \tag{18}$$

*where $[\xi, \xi] = 0$, $\xi_0 > 0$; the integral over the isotropic line $x = b + t\xi$ is*

$$\varphi(\xi, b) = \int_{-\infty}^{\infty} f(b + t\xi)\, dt, \tag{19}$$

*where $[b, b] = -1$, $[b, \xi] = [\xi, \xi] = 0$, $b_0 = 0$.*

*Then the value of $f(x)$ at any point $a$ of the imaginary Lobachevskian space is given in terms of $h(\xi)$ and $\varphi(\xi, b)$ by the inversion formula*

$$f(a) = -(4\pi)^{-2} \int h(\xi)\delta''(|[a, \xi]| - 1)\, d\xi$$

$$+ (2\pi)^{-2} \int_0^{\pi} \cot^2\theta\, d\theta \int_{\Gamma} \varphi(\xi, \theta)\, d\omega. \tag{20}$$

Here $d\xi = |\xi_0|^{-1}\, d\xi_1\, d\xi_2\, d\xi_3$ is the invariant measure on the $[\xi, \xi] = 0$ cone; $\varphi(\xi, \theta)$ denotes the value of $\varphi(\xi, b)$ for an isotropic line $x = b + t\xi$ lying in the $[a, x] = \cos\theta$ plane (that is, such that $[a, b] = \cos\theta$).[20] The second integral in the second term is over any surface $\Gamma$ on the $[\xi, \xi] = 0$ cone that intersects all generators of the cone; the measure $d\omega$ is defined by

$$d\omega = |\xi_0|^{-1}(\xi_1\, d\xi_2\, d\xi_3 - \xi_2\, d\xi_1\, d\xi_3 + \xi_3\, d\xi_1\, d\xi_2).$$

Remark. In conclusion we point out that $h(\xi)$ and $\varphi(\xi, b)$ satisfy certain symmetry relations. The symmetry relation for $h(\xi)$ in the

[20] We choose a *single* family of generators on each of the intersections of $[a, x] = \cos\theta$ with the $[x, x] = -1$ hyperboloid. This choice was made at Eq. (6) for the case in which the coordinates of $a$ were (0, 0, 0, 1), and it is uniquely specified for other $a$ by continuity in $a$. Once $a$ is given, the generator $b + t\xi$ is uniquely determined by $\xi$ and $\theta$.

present case is similar to the case of an ordinary Lobachevskian space. It may be written in the form

$$H(a, t) = H(a, t^{-1}), \tag{21}$$

where[21]

$$H(a, t) = \int h(\xi)\delta(|[a, \xi]| - t)\, d\xi. \tag{22}$$

We shall not, however, go into this derivation here. The other symmetry relation is

$$\int_\Gamma \varphi(\xi, \theta)\delta([a, \xi])\, d\omega = \int_\Gamma \varphi(\xi, \pi - \theta)\delta([a, \xi])\, d\omega. \tag{23}$$

To prove it we need only note that the intersections of $[x, x] = -1$ with $[a, x] = \cos\theta$ and with $[a, x] = \cos(\pi - \theta) = -\cos\theta$ coincide, since $x$ and $-x$ are considered the same point on the hyperboloid. #

### 3.4a. Parallel Isotropic Lines

We wish to consider in more detail the second term in the inversion formula (20) of Section 3.4. As we have seen, the integral in this term is over the set of isotropic lines lying on all possible spheres of imaginary radius with center at $a$ (that is, on two-dimensional single-sheeted hyperboloids, in the realization on $[x, x] = -1$). This set of lines has also another geometrical description, and it is to this description that we now turn.

We first define *parallelism* of isotropic lines. We shall call two isotropic lines parallel if their direction vectors are proportional. In the two realizations of an imaginary Lobachevskian space this concept is represented in two ways. Recall (see Section 1.3a) that an isotropic line may be thought of as a two-dimensional plane in $E_{n+1}$ tangent to the $[x, x] = 0$ cone along one of its generators and that the direction vector of the line lies along this generator. Thus parallel isotropic lines are represented by

[21] Note that the situation here differs from that in the ordinary Lobachevskian space in that the integral of Eq. (22) diverges and must therefore be understood in the sense of its regularization. This may be defined by noting (see the Remark at the end of Section 3.3) that the integral

$$H_t''(a, t) = \int h(\xi)\, \delta''(|[a, \xi]| - t)\, d\xi$$

converges absolutely. Then $H(a, t)$ is uniquely determined by $H_t''(a, t)$ and the initial conditions $H(a, 1) = H'(a, 1) = 0$.

two-dimensional planes tangent to this cone along the same generator. It then becomes obvious that for the realization on the $[x, x] = -1$ hyperboloid parallel isotropic lines are parallel generators of this hyperboloid. Similarly, for the realization on the $x_0 = 1$ plane, that is, outside the unit sphere, parallel isotropic lines are lines tangent to the sphere (that is, to the absolute) at the same point.

We assert that every isotropic line lying on a sphere of imaginary radius centered at $a$ is parrallel to one of the generators of the isotropic cone with vertex at $a$ (that is, to one of the isotropic lines passing through $a$). To prove the assertion, we realize the space on the $[x, x] = -1$ hyperboloid. Then the isotropic lines are given by equations of the form $x = b + t\xi$, where $[b, b] = -1$, $[b, \xi] = [\xi, \xi] = 0$. It is easily seen that all the points of this line lie at a fixed distance from $a$ if and only if $[a, \xi] = 0$. But then $x = a + t\xi$ is also the equation of an isotropic line, and this line passes through $a$ and is parallel to $x = b + t\xi$.

The converse will also obviously hold: every isotropic line parallel to one of the generators of the isotropic cone with vertex at $a$ lies on some sphere of imaginary radius with center at $a$.

We have thus shown that the integration in Eq. (20) of Section 3.4 is over a set of isotropic lines parallel to the generators of the isotropic cone with vertex at $a$.

Furthermore, the parameter $\theta$ appearing in that equation also has a simple geometric meaning. This is because $r = \theta i/k$ is the radius of the sphere with center at $a$ on which lies the isotropic line $l \equiv l(\xi, \theta)$. In other words $r$ is the distance from $a$ to any point on $l$.

Hence the integral

$$\int_0^{\pi} \int_{\Gamma} \cot^2 \theta \varphi(\xi, \theta) \delta([a, \xi]) \, d\theta \, d\omega \tag{1}$$

may be calculated by the following sequence. Consider the isotropic cone with vertex at $a$. With each generator $l'$ of the cone we associate the set of isotropic lines parallel to it. Then $\varphi(l) \equiv \varphi(\xi, \theta)$ is integrated first over the set of these lines with weight function $\cot^2 \theta$, where $r = \theta i/k$ is the distance from $a$ to $l$. This results in a function defined on the set of generators of the isotropic cone with vertex at $a$ (or, equivalently, on the set of points of tangency of this cone with the absolute). Then expression (1) is the integral of this function over the set on which it is defined (with the appropriate measure).

Remark. To define the distance between parallel isotropic lines we prove that the distance between any two points of parallel isotropic lines is independent of the choice of points. Indeed, consider the parallel isotropic lines $x = a + t\xi$ and $y = b + t'\xi$. Because $[a, \xi] = [b, \xi] = [\xi, \xi] = 0$, we have $[x, y] = [a, b]$. Now the distance $r(x, y)$ between $x$ and $y$ is defined by $\cosh^2 kr = [x, y]^2$, so that $r(x, y) = r(a, b)$ and is thus independent of $x$ and $y$. Hence we shall define the *distance between parallel isotropic lines* as the distance between any two points on them. Let $r = \theta i/k$ be the distance between two parallel isotropic lines. Then $\theta$ is the angle between the representations of these lines when the imaginary Lobachevskian space is realized outside of the unit sphere, which may be seen as follows. Recall that parallel isotropic lines in this realization are represented by lines tangent tot the absolute at a given point. The imaginary Lobachevskian motions are then represented by the projective transformations of the three-dimensional space which map the absolute (that is, the unit sphere) into itself. It is easily shown that these are conformal mappings and thus preserve angles between lines tangent to the sphere at a given point. Further, the distance between parallel isotropic lines remains invariant under imaginary Lobachevskian motions. Thus we need only prove the assertion for a single pair of parallel isotropic lines separated by a given distance $r$. The reader will easily verify that the assertion is valid for the pair of lines tangent to the absolute at the point whose coordinates are $(1, 0, 0, 1)$ and passing, respectively, through $(1, 1, 0, 1)$ and $(1, \cos\theta, \sin\theta, 1)$. #

### 3.5. Calculation of $\Phi(x, a; \mu)$

In deriving the inversion formula we made use of the function defined formally by

$$\Phi(x, a; \mu) = \int (|[a, \xi]| - 1)_+^\mu \, \delta(|[x, \xi]| - 1) \, d\xi, \tag{1}$$

where $a$ and $x$ are points on the $[x, x] = -1$ hyperboloid, and $\xi$ is a point on the upper sheet of the $[\xi, \xi] = 0$ cone. Since for some $a$ and $x$ this integral may diverge for all $\mu$, we defined it in the following way. We write

$$\Phi(x, a; \mu) = \Phi_1(x, a; \mu) + \Phi_2(x, a; \mu),$$

where

$$\Phi_1(x, a; \mu) = J_+(x, a - x; \mu) + J_+(x, -a - x; \mu), \tag{2}$$

and, in turn,

$$J_+(x, b; \mu) = \int [b, \xi]_+^\mu \, \delta([x, \xi] - 1)\, d\xi$$

$$= \frac{1}{2i \sin \mu\pi} \lim_{c \to +0} \int \delta([x, \xi] - 1) \cdot \{e^{i\mu\pi}[b - ic, \xi]^\mu$$

$$- e^{-i\mu\pi}[b + ic, \xi]^\mu\}\, d\xi$$

$$= \frac{1}{2i \sin \mu\pi} \{e^{i\mu\pi} J(x, b - i0; \mu) - e^{-i\mu\pi} J(x, b + i0; \mu)\} \tag{3}$$

[see Section 3.2, Eqs. (4), (8), (9), and (11)]. The notation $c \to +0$ denotes that $c$ approaches zero from inside the positive cone, that is, so that $[c, c] > 0$ and $c_0 > 0$. The function $\Phi_2(x, a; \mu)$ is defined similarly. What we wish to do now is to obtain an explicit expression for $\Phi(x, a; \mu)$ in terms of the coordinates of $a$ and $x$ (an expression we have already used in Section 3.3).

Since the integrals on the right-hand side of (3) are invariant under simultaneous motion of $b$ and $x$, we have

$$\Phi(x, a; \mu) = \Phi(xg, ag; \mu).$$

But any point on the $[x, x] = -1$ hyperboloid can be moved to the point with coordinates (0, 0, 0, 1), so that we need calculate $\Phi(x, a; \mu)$ only when $x$ has these coordinates.

Now Eq. (2) implies that we actually need only calculate the integral

$$J(x, z; \mu) = \int [z, \xi]^\mu \, \delta([x, \xi] - 1)\, d\xi \tag{4}$$

for any complex point $z = (z_0, z_1, z_2, z_3)$ on the upper or lower half-planes (that is, such that either $z = b + ic$ or $z = b - ic$, where $c > 0$).[22] Further, as we have already mentioned, we may restrict our considerations to the case in which the coordinates of $x$ are (0, 0, 0, 1). Then Eq. (4) becomes

$$J(x, z; \mu) = \int [z, \xi]^\mu \, \delta(\xi_3 + 1)\, d\xi. \tag{5}$$

We shall perform this integration assuming first that $c$ lies in the positive cone, that is, that $z = b > 0$. For this case we have

$$J(x, b; \mu) = \int [b, \xi]^\mu \, \delta(\xi_3 + 1)\, d\xi. \tag{6}$$

[22] Recall that the notation $c > 0$ means that $c$ lies inside the positive cone.

But since $b > 0$ and $\xi$ lies on the positive sheet of the $[\xi, \xi] = 0$ cone, $[b, \xi] > 0$. It then follows that the integral in (6) converges for $\text{Re}\,\mu < -1$. Now for any $b > 0$ there exists a motion $g$ leaving $x = (0, 0, 0, 1)$ invariant and carrying $b$ into $bg = (P^{\frac{1}{2}}, 0, 0, b_3)$, where $P = b_0^2 - b_1^2 - b_2^2$. This motion does not alter (6), so that

$$J(x, b; \mu) = \int (P^{\frac{1}{2}}\xi_0 - b_3\xi_3)^{\mu}\delta(\xi_3 + 1)\, d\xi, \tag{7}$$

where

$$d\xi = \xi_0^{-1}\, d\xi_1\, d\xi_2\, d\xi_3 = (\xi_1^2 + \xi_2^2 + \xi_3^2)^{-\frac{1}{2}}\, d\xi_1\, d\xi_2\, d\xi_3 .$$

Then straightforward calculation shows that

$$J(x, b; \mu) = -\frac{2\pi(P^{\frac{1}{2}} + b_3)^{\mu+1}}{(\mu + 1)P^{\frac{1}{2}}} \tag{8}$$

for our special point $x$ and for the value of $P$ given. We have thus calculated $J(x, z; \mu)$ for this special $x$ and for $z$ in the positive cone. Let us now return to Eq. (5) and calculate it for $z$ on the upper half-plane, that is, for $z = b + ic$ such that $c$ lies in the positive cone.

If we define the complex number $[z, \xi]$ raised to the power $\mu$ by the expression

$$[z, \xi]^{\mu} = \exp\{\mu \ln|[z, \xi]| + i\mu \arg[z, \xi]\}, \tag{9}$$

where $-\pi < \arg\,[z, \xi] < \pi$, the integral of Eq. (5) becomes the analytic continuation of Eq. (6) of Section 3.2. It is therefore sufficient to continue expression (8) into the upper half-plane. By considering Eq. (5) for $z = bi$, $b > 0$, it is easily shown that this continuation is given by

$$J(x, z; \mu) = -\frac{2\pi}{\mu + 1}\, e^{\frac{1}{2}i\mu\pi}(-P)^{-\frac{1}{2}}\{(-P)^{\frac{1}{2}} - iz_3\}^{\mu+1}, \tag{10}$$

where $P = z_0^2 - z_1^2 - z_2^2$. [We are omitting the not very complicated but rather lengthy proof that $J(x, z; \mu)$ as defined by Eq. (10) depends analytically on $\mu$ and on the coordinates of $z$ in the upper half-plane. This proof reduces to showing that in the upper half-plane neither $-P$ nor $(-P)^{\frac{1}{2}} - iz_3$ takes on negative values.] In exactly the same way it can be shown that the analytic continuation of $J(x, b; \mu)$ into the lower half-plane is given by

$$J(x, b; \mu) = -\frac{2\pi}{\mu + 1}\, e^{-\frac{1}{2}i\mu\pi}(-P)^{-\frac{1}{2}}\{(-P)^{\frac{1}{2}} + iz_3\}^{\mu+1}, \tag{10'}$$

where again $P = z_0^2 - z_1^2 - z_2^2$.

Now for an arbitrary real point $b$ let us calculate

$$J(x, b + i0; \mu) = \lim_{c \to +0} J(x, b + ic; \mu).$$

This is done by setting $z = b + ic_\epsilon$ in Eq. (10), where $c_\epsilon = (\epsilon, 0, 0, 0)$ with $\epsilon > 0$. Then in the limit $\epsilon \to 0$ we arrive easily at

$$J(x, b + i0; \mu) = -\frac{2\pi}{\mu + 1} e^{\frac{1}{2}i\mu\pi}(-P \mp i0)^{-\frac{1}{2}}\{(-P \mp i0)^{\frac{1}{2}} - ib_3\}^{\mu+1}, \quad (11)$$

with the upper sign for $b_0 > 0$, the lower sign for $b_0 < 0$, and $P = b_0^2 - b_1^2 - b_2^2$. Similarly,

$$J(x, b - i0; \mu) = -\frac{2\pi}{\mu + 1} e^{-\frac{1}{2}i\mu\pi}(-P \pm i0)^{-\frac{1}{2}}\{(-P \pm i0)^{\frac{1}{2}} + ib_3\}^{\mu+1}. \quad (11')$$

Expressions (11) and (11′) can now be inserted into Eq. (3) for $J_+(x, b; \mu)$, which is then expressed in terms of the coordinates of $b$ and of $x = (0, 0, 0, 1)$. Then Eq. (2) can be used to find $\Phi_1(x, a; \mu)$ for this $x$, namely

$$\begin{aligned} \Phi_1(x, a; \mu) = {} & \frac{\pi i}{(\mu + 1) \sin \mu\pi} \\ & \times \{(-P \pm i0)^{-\frac{1}{2}}[e^{\frac{1}{2}\mu\pi i}\{(-P \pm i0)^{\frac{1}{2}} + i(a_3 - 1)\}^{\mu+1} \\ & - e^{-\frac{1}{2}i\mu\pi}\{(-P \pm i0)^{\frac{1}{2}} + i(a_3 + 1)\}^{\mu+1}] \\ & - (-P \mp i0)^{-\frac{1}{2}}[e^{-\frac{1}{2}i\mu\pi}\{(-P \mp i0)^{\frac{1}{2}} - i(a_3 - 1)\}^{\mu+1} \\ & - e^{\frac{1}{2}i\mu\pi}\{(-P \mp i0)^{\frac{1}{2}} - i(a_3 + 1)\}^{\mu+1}]\}, \end{aligned}$$

where $P = a_0^2 - a_1^2 - a_2^2$, and the upper sign is used when $a_0 > 0$, while the lower one is used for $a_0 < 0$. This expression can be simplified by using the relations

$$(-P \pm i0)^{\frac{1}{2}} = P_-^{\frac{1}{2}} \pm iP_+^{\frac{1}{2}} \quad (12_1)$$

and

$$(-P \pm i0)^{-\frac{1}{2}} = P_-^{-\frac{1}{2}} \mp iP_+^{-\frac{1}{2}} \quad (12_2)$$

(see Volume 1, Chapter III, Section 2.4). Recall further that diametrically opposite points of the $[x, x] = -1$ hyperboloid are identified, so that we may consider $a_3 > 0$. Also, for $P > 0$ we have

$$a_3 - 1 < P^{\frac{1}{2}} < a_3 + 1. \quad (13)$$

This is because $a$ lies on the hyperboloid, so that $P = a_0^2 - a_1^2 - a_2^2 = a_3^2 - 1$, and therefore $P > 0$ implies $a_3^2 > 1$. This together with $a_3 > 0$ implies Eq. (13).

Upon using these facts to simplify the integral, we arrive at the following result. If $x = (0, 0, 0, 1)$, then $\Phi_1(x, a; \mu)$ is given by

$$\Phi_1(x, a; \mu) = \frac{-2\pi P^{\frac{1}{2}\mu}}{(\mu + 1)} \left(1 + \frac{a_3 - 1}{P^{\frac{1}{2}}}\right)^{\mu+1} \tag{$14_1$}$$

for $P > 0$ and $a_0 > 0$. Further, $\Phi_1(x, a; \mu)$ vanishes for $P > 0$ and $a_0 < 0$, while for $P < 0$ it is given by

$$\begin{aligned} \Phi_1(x, a; \mu) = {} & \frac{\pi i(-P)^{\frac{1}{2}\mu}}{(\mu + 1) \sin \mu\pi} \\ & \times \left\{ e^{\frac{1}{2}i\mu\pi} \left[ \left(1 - \frac{i(1 - a_3)}{(-P)^{\frac{1}{2}}}\right)^{\mu+1} + \left(1 - \frac{i(1 + a_3)}{(-P)^{\frac{1}{2}}}\right)^{\mu+1} \right] \right. \\ & \left. - e^{-\frac{1}{2}i\mu\pi} \left[ \left(1 + \frac{i(1 - a_3)}{(-P)^{\frac{1}{2}}}\right)^{\mu+1} + \left(1 + \frac{i(1 + a_3)}{(-P)^{\frac{1}{2}}}\right)^{\mu+1} \right] \right\}. \end{aligned} \tag{$14_2$}$$

As before, $P = a_0^2 - a_1^2 - a_2^2$. A similar expression is obtained for $\Phi_2(x, a; \mu)$ and finally for $\Phi(x, a; \mu)$.

It is now a simple matter to find $\Phi(x, a; \mu)$ for any two points $a$ and $x$ on the $[x, x] = -1$ hyperboloid. This is because for the special $x$ we have chosen we have $[a, x] = -a_3$. To obtain a similar invariant expression for $P$ we use the fact that according to the definition of the distance between two points of an imaginary Lobachevskian space, $\cosh^2 kr = [a, x]^2$. In our case we have $\cosh^2 kr = a_3^2$. Further, on the $[x, x] = -1$ hyperboloid, $P = a_3^2 - 1$, so that in general $P = \sinh^2 kr$. When we insert these expression for $a_3$ and $P$ into Eqs. ($14_1$) and ($14_2$) we arrive at an expression for $\Phi(x, a; \mu)$ in terms of the distance $r$ between $a$ and $x$ for the special point $x$. But then the expression will not change under simultaneous motion of $a$ and $x$, and it will therefore hold for any two points of an imaginary Lobachevskian space. In this way one arrives at the expressions for $\Phi(x, a; \mu)$ given in Section 3.2.

CHAPTER VI

# HARMONIC ANALYSIS ON SPACES HOMOGENEOUS WITH RESPECT TO THE LORENTZ GROUP

## 1. Homogeneous Spaces and the Associated Representations of the Lorentz Group

### 1.1. Homogeneous Spaces

Let us start with some fundamental definitions. A group $G$ is called a *transformation group* of some space $X$ if to every element $g \in G$ there corresponds a one-to-one bicontinuous transformation of $X$. Further it is assumed that the identity transformation of $X$ corresponds to the unit element $e$ of $G$ that for every $x \in X$ and for any two elements $g_1, g_2 \in G$ we have

$$x(g_1 g_2) = (xg_1)g_2 .$$

Here $xg$ denotes the point in $X$ into which $x$ is mapped under $g$. If $G$ is continuous then one also usually requires that the mapping $x \to xg$ be continuous with respect to $g$ for all $x$.

If for every pair of points $x$, $y \in X$ there exists $g \in G$ such that $y = xg$, $G$ is called a *transitive group of motions of* $X$, and $X$ itself is said to be *homogeneous with respect to* $G$. For instance, the $n$-dimensional sphere is a space homogeneous with respect to the group of rotations of $(n+1)$-dimensional Euclidean space. The $n$-dimensional ordinary and imaginary Lobachevskian spaces considered in Chapter V are spaces homogeneous with respect to the group of linear transformations of $E_{n+1}$ with determinant one that leave the quadratic form $[x, x] = x_0^2 - x_1^2 - \cdots - x_n^2$ invariant and map each sheet of the $[x, x] = 0$ cone into itself.

### 1.2. Representations of the Lorentz Group Associated with Homogeneous Spaces

Let $X$ be a homogeneous space acted upon by the Lorentz group $G$. With $X$ we may associate a representation of $G$. This representation is constructed on some space of functions $f(x)$ defined on $X$. It is a correspondence of each $g \in G$ with some operator $T(g)$ on the function space, where $T(g)$ is defined by an equation of the form

$$T(g)f(x) = f(xg)\alpha(x, g). \tag{1}$$

Recall that the condition that the operators $T(g)$ form a representation of $G$ is that for any two elements $g_1, g_2 \in G$ we have

$$T(g_1g_2) = T(g_1)T(g_2).$$

This can be stated as a condition on the multiplier $\alpha(x, g)$, namely

$$\alpha(x, g_1g_2) = \alpha(x, g_1)\alpha(xg_1, g_2).$$

We have already discussed representations associated with certain homogeneous spaces. For instance, in Chapter III we studied operator-irreducible representations of the Lorentz group. All these representations acted on the space of functions defined on the complex projective line (see Section 1.5 of the present chapter). The operators of these representations are given by

$$T_\chi(g)f(z) = f\left(\frac{\alpha z + \gamma}{\beta z + \delta}\right)(\beta z + \delta)^{n_1-1}(\bar{\beta}\bar{z} + \bar{\delta})^{n_2-1}, \tag{2}$$

where $\chi = (n_1, n_2)$ is a pair of complex numbers defining the representation ($n_1 - n_2$ is always assumed an integer). In general a representation associated with a homogeneous space may be reducible. We shall set as our problem the decomposition of such representations into their irreducible components. In this way $f(z)$ is decomposed into functions transforming under irreducible representations of $G$ (the analog of the Fourier decomposition of a function on the line). Actually it is just this problem that was solved in Chapter IV where the homogeneous space was chosen as the Lorentz group itself, and where we obtained the Fourier transform of a function on the group.

The analogous problem will be solved here for other homogeneous spaces.

### 1.3. The Relation between Representation Theory and Integral Geometry

It is particularly important when dealing with representation theory to study integral transforms which simplify a representation. These transforms are related to going over from points of the homogeneous space $X$ to new geometrical objects in this space. Some system $Y$ of such geometrical objects (say lines or surfaces) is considered, and each $f(x)$ over $X$ is associated with a certain function $\check{f}(y)$ over $Y$ and obtained from $f(x)$ by integrating it over the geometrical object $y \in Y$. The set $Y$ is always chosen so that the correspondence $f(x) \to \check{f}(y)$ is mutually single valued. In other words, if the integral of $f(x)$ over every $y \in Y$ is known, then $f(x)$ is determined.

The role of integral geometry in representation theory lies just in this transition from the space of certain geometrical objects to the space of others. It is found that in most homogeneous spaces[1] there exist geometrical objects such that this kind of transition introduces a significant simplification into the representation. We shall call such objects *horospheres.*

The reason for this terminology is that in an ordinary Lobachevskian space these objects actually are horospheres. Since horospheres in such a space can be specified by points on the cone whose equation is

$$x_0^2 - x_1^2 - x_2^2 - x_3^2 = 0, \qquad x_0 > 0,$$

the homogeneous space of horospheres is identical with this cone. Thus the analysis of a representation associated with a Lobachevskian space reduces to an investigation of the considerably simpler representation associated with the cone.

What in general do horospheres actually represent in other cases? A general definition of such a horosphere, at any rate one that can be used for any complex semisimple Lie group, will be given in Section 1.6. For the Lorentz group this definition reduces to the following. Consider the subgroup of matrices of the form

$$\begin{Vmatrix} 1 & \zeta \\ 0 & 1 \end{Vmatrix}. \tag{1}$$

Horospheres in a homogeneous space related to the Lorentz group are those surfaces which are generated by the motions belonging to this subgroup and all surfaces into which they can be transformed.

Now consider the set of horospheres of some space $X$ homogeneous with respect to the Lorentz group. The group also will act on this set, although perhaps not transitively. It is found that the homogeneous subspaces of the space of horospheres always either are identical with the complex affine plane (defined in Section 1.5) or can be obtained from it by taking a *quotient space* [that is, by identifying certain sets of points, such as for instance all points of the form $(e^{i\varphi}z_1, e^{i\varphi}z_2)$, where $0 \leqslant \varphi < 2\pi$].

[1] This is true, in any case, in every homogeneous space related to a semisimple Lie group.

Therefore the problem of decomposing the representations of the Lorentz group associated with the space of horospheres reduces to the easily solved problem of decomposing the representations associated with the complex affine plane (a problem which will be discussed in Sections 2.1 and 2.2). Essentially the decomposition of these representations into their irreducible components reduces to the ordinary Fourier transform.

Just as simple is the problem of decomposing a representation on a space of functions over the horospheres of some other space homogeneous with respect to any simisimple Lie group. In order to return from the decomposition of representations associated with horosphere spaces to the decomposition of some initial representation associated with a given homogeneous space $X$, one must be able to write the functions $f(x)$ over $X$ in terms of their integrals over horospheres. In other words one must solve the problem of integral geometry for the horospheres of a given homogeneous space. We see that the problems of integral geometry and representation theory are closely related to classical geometry (Plücker, Klein, and others) in which, as in our present case, new homogeneous spaces are constructed of geometrical objects in some other space. In integral geometry, however, problems are treated from a more modern point of view; that is, the transition from one space to another always involves also the transformation of the functions defined on these spaces. This might be compared with the difference between classical and quantum mechanics; that is, the transformations of classical mechanics are point transformations, while those of quantum mechanics are transformations in function spaces.

Many interesting problems have already been solved in the realm of integral geometry, and these have led to new and deep integral transform formulas. These early problems of integral geometry remind one (though on a different level) of the initial stages in the development of algebraic geometry. Several such problems are discussed in the present volume.

### 1.4. Homogeneous Spaces and Associated Subgroups of Stability

We wish now to show how to describe in intrinsic terms all spaces homogeneous with respect to a group $G$.

Let $X$ be a homogeneous space. Consider some point $x_0$ fixed in $X$. With each point $x$ we shall associate the set of transformations which carry $x_0$ into $x$; let us study these sets of transformations. Consider first those transformations which carry $x_0$ into $x_0$. Clearly they form a subgroup. We shall call this the *subgroup of stability* (also called the *little group*) of $x_0$. Now let $g$ be a transformation mapping $x_0$ into $x$; then the transformations of the form $hg$, where $h$ runs through the stability subgroup of $x_0$, also map $x_0$ into $x$. It is easily shown that these are all the transformations that map $x_0$ into $x$. Thus the set of such transformations is a right coset of the stability subgroup of $x_0$. There exists therefore a correspondence between the points of the homogeneous space and the right cosets with respect to the stability subgroup of $x_0$. This correspondence is mutually single valued, and obviously the transformation $g$ on $X$ corresponds to multiplying the cosets by $g$.

*We thus arrive at the conclusion that every space $X$ homogeneous with respect to $G$ can be obtained in the following way. Some subgroup $H$ of $G$ is chosen. The points $x$ of $X$ are identified as the right cosets of $G$ with respect to $H$. The motion corresponding to $g \in G$ is defined by right multiplication of these cosets by $g$.*

We have established that each homogeneous space is defined by some subgroup of stability. Although the choice of the fixed point $x_0$ is still at our disposal, choosing some other point will lead to the same space. It is easily seen that on changing the fixed point, the subgroup of stability is changed to a conjugate subgroup. Specifically, if $H$ is the stability subgroup of $x_0$ and $x_0 g = x$, the stability subgroup of $x$ will be $g^{-1}Hg$. Consequently the spaces associated with conjugate subgroups are identical, and the classification of the homogeneous spaces reduces to the classification of all nonconjugate subgroups of $G$.

Remark. The subgroups of stability of the homogeneous spaces discussed in the present volume are not discrete. The case of discrete subgroups of stability is also of interest, and the study of representations associated with spaces that have discrete subgroups of stability leads the theory of automorphic functions [see, for instance, I. M. Gel'fand and I. I. Pyatetskii-Shapiro (18), I. M. Gel'fand and S. V. Fomin (29), R. Godement (23)]. We shall not, however, discuss this case here. #

## 1.5. Examples of Spaces Homogeneous with Respect to the Lorentz Group

In this chapter we shall consider homogeneous spaces whose group of motions is the Lorentz group, that is, the group of complex unimodular matrices in two dimensions. We start by giving some examples.

Consider the two-dimensional complex plane, that is, the space of pairs of complex numbers $(z_1, z_2)$, omitting the origin $(0, 0)$.[2]

[2] The origin is excluded in order that the space be transitive.

With every element

$$g = \begin{Vmatrix} \alpha & \beta \\ \gamma & \delta \end{Vmatrix}, \qquad \alpha\delta - \beta\gamma = 1,$$

of the Lorentz group we associate the affine transformation

$$(z_1, z_2) \to (\alpha z_1 + \gamma z_2, \beta z_1 + \delta z_2). \tag{1}$$

It is easily shown that the complex plane is then homogeneous with respect to the Lorentz group (so long as the origin is omitted). We shall call this space the *complex affine plane.*

Let us now find the stability subgroup of the complex affine plane. We choose our fixed point to be (0, 1). Obviously the transformations of the form of (1) that leave this point stationary consist of matrices of the form

$$\begin{Vmatrix} 1 & \zeta \\ 0 & 1 \end{Vmatrix},$$

and these therefore form the stability subgroup of the complex affine plane.

Now consider the space whose elements are the lines of the complex affine plane passing through the origin. Under affine transformations of the plane these lines transform into each other. Thus they also form a space homogeneous with respect to the Lorentz group, and we shall call this space the *complex projective line.* Note that each such line, whose equation is $z_1 = kz_2$, is specified by the value of $k$ (with the line $z_2 = 0$ being specified by $k = \infty$). Thus the complex projective line can also be realized as the set of all complex numbers plus the point at infinity. The motions then become the linear-fractional transformations

$$z' = \frac{\alpha z + \gamma}{\beta z + \delta}. \tag{2}$$

Let us now find the stability subgroup of the complex projective line. We choose the fixed point to be $z = 0$, and then the linear fractional transformations leaving this point fixed are of the form

$$z' = \frac{\alpha z}{\beta z + \delta}.$$

Consequently the stability subgroup of the complex projective line is the group of matrices of the form

$$\begin{Vmatrix} \alpha & \beta \\ 0 & \alpha^{-1} \end{Vmatrix}.$$

Consider further the space consisting of ordered pairs $(z_1, z_2)$, $z_1 \neq z_2$ of points on the complex projective line. With each matrix $g$ we associate the transformation

$$(z_1, z_2) \to \left(\frac{\alpha z_1 + \gamma}{\beta z_1 + \delta}, \frac{\alpha z_2 + \gamma}{\beta z_2 + \delta}\right). \tag{3}$$

This space is also homogeneous,[3] and it is easily shown that the subgroup of stability for $(0, \infty)$ chosen as the fixed point consists of the diagonal matrices

$$\begin{Vmatrix} \alpha & 0 \\ 0 & \alpha^{-1} \end{Vmatrix}.$$

Other important examples of spaces homogeneous with respect to the Lorentz group are three-dimensional Lobachevskian space, three-dimensional imaginary Lobachevskian space, and the isotropic cone. All of these were defined in Chapter III, and some of their realizations were discussed in Chapter V.

A model of Lobachevskian space is the set of all positive definite Hermitian matrices in two dimensions

$$h = \begin{Vmatrix} x_0 - x_3 & x_2 - ix_1 \\ x_2 + ix_1 & x_0 + x_3 \end{Vmatrix}$$

with determinant $x_0^2 - x_1^2 - x_2^2 - x_3^2 = 1$. Then with each matrix $g$ of the Lorentz group we associate the Lobachevskian motion which maps $h$ into the matrix

$$h' = g^*hg. \tag{4}$$

Similarly, imaginary Lobachevskian space consists of all two-dimensional Hermitian matrices $h$ with determinant $-1$, and $h$ and $-h$ are identified. The isotropic cone, finally, consists of all two-dimensional matrices $h \geqslant 0$ with determinant zero. Motions of imaginary Lobachevskian space and of the isotropic cone are defined also by Eq. (4).

Another model of these spaces is obtained by associating with each Hermitian matrix $h$ the point $x = (x_0, x_1, x_2, x_3)$ of four-dimensional Euclidean space $E_4$. We then obtain the models on the upper sheet of the two-sheeted hyperboloid, on the one-sheeted hyperboloid, and on the upper sheet of the cone, as discussed in the previous chapter. The motions in these spaces are induced by hyperbolic rotations in $E_4$, i.e., by linear

[3] If we had not required that $z_1 \neq z_2$, the space would not have been homogeneous, but would consist of two homogeneous spaces. The first of these would be the space we are discussing, and the second would be the space of pairs of the form $(z, z)$, which is isomorphic to the complex projective line.

transformations with determinant one which leave invariant the quadratic form $x_0^2 - x_1^2 - x_2^2 - x_3^2$ and map each sheet of the isotropic cone into itself.

Let us now find the stability subgroup of ordinary Lobachevskian space, using the model of Hermitian matrices. We choose the unit matrix $e$ to be the fixed point. Then the stability group of $e$ consists of matrices $g$ such that

$$g^*g = e,$$

that is, of unitary matrices. *Thus the subgroup of stability of Lobachevskian space is the special unitary group in two dimensions, which consisting of the matrices*

$$u = \begin{Vmatrix} \alpha & \beta \\ -\bar{\beta} & \bar{\alpha} \end{Vmatrix}, \qquad |\alpha|^2 + |\beta|^2 = 1.$$

We may do this also for imaginary Lobachevskian space, choosing the fixed matrix to be

$$\sigma = \begin{Vmatrix} -1 & 0 \\ 0 & 1 \end{Vmatrix}.$$

Then the stability subgroup of $\sigma$ consists of matrices $g$ such that

$$g^*\sigma g = \pm\sigma$$

(we write $\pm\sigma$ since $\sigma$ and $-\sigma$ are identified). *It follows that the subgroup of stability of imaginary Lobachevskian space is the group of matrices of the form*[4]

$$\begin{Vmatrix} \alpha & \beta \\ \bar{\beta} & \bar{\alpha} \end{Vmatrix} \quad \text{and} \quad \begin{Vmatrix} \beta & \alpha \\ -\bar{\alpha} & -\bar{\beta} \end{Vmatrix}, \qquad |\alpha|^2 - |\beta|^2 = 1.$$

Finally we turn to the cone. Let us choose the fixed point to be

$$h_0 = \begin{Vmatrix} 1 & 0 \\ 0 & 0 \end{Vmatrix}.$$

It is easily verified that the subgroup leaving $h_0$ invariant, and therefore the *stability subgroup of the cone, is the group of matrices of the form*

$$\begin{Vmatrix} e^{-i\varphi} & 0 \\ \zeta & e^{i\varphi} \end{Vmatrix}.$$

[4] It is easily verified that the connected component of the unit element in this subgroup of stability, that is, the subgroup of matrices of the form

$$\begin{Vmatrix} \alpha & \beta \\ \bar{\beta} & \bar{\alpha} \end{Vmatrix}, \qquad \text{where} \quad |\alpha|^2 - |\beta|^2 = 1,$$

is isomorphic to the subgroup of real matrices.

We give one last example of a space homogeneous with respect to the Lorentz group, a space whose elements are not points but lines. Recall that a model of imaginary Lobachevskian space is the surface of the single-sheeted hyperboloid whose equation is

$$x_0^2 - x_1^2 - x_2^2 - x_3^2 = -1.$$

Now consider the set of line generators of this surface (that is, the isotropic lines of the imaginary Lobachevskian space). It is easily verified that each such line can be carried into any other by a Lobachevskian motion. Thus the line generators of this hyperboloid themselves form a space homogeneous under the Lorentz group. In order to find the stability subgroup of this space, let us take the model of the Hermitian matrices with determinant $-1$. In this model an isotropic line is any set of matrices whose elements are linear functions of some real parameter $t$. Let the fixed point be the isotropic line

$$h_t = \begin{Vmatrix} t & 1 \\ 1 & 0 \end{Vmatrix}, \qquad -\infty < t < \infty.$$

The motions in imaginary Lobachevskian space that leave this line invariant are represented then by matrices $g$ such that for any $t$

$$g^* h_t g = \pm h_{t'}, \qquad -\infty < t' < +\infty.$$

As is easily shown, such $g$ are of the form

$$\begin{Vmatrix} \lambda & 0 \\ \zeta & \lambda^{-1} \end{Vmatrix},$$

where $\lambda \neq 0$ is either real or pure imaginary, and these matrices thereby make up the stability subgroup of the space consisting of all isotropic lines in imaginary Lobachevskian space. Since this subgroup has three real dimensions and the Lorentz group has six, it follows that the dimension of the space of isotropic lines of an imaginary Lobachevskian space is $6 - 3 = 3$.

The accompanying table summarizes the results on the above spaces homogeneous with respect to the Lorentz group.

## 1.6. Group-Theoretical Definition of Horospheres

We wish now to discuss the concept of a horosphere from the group-theoretical point of view, an approach which has several advantages over

| Name of space | Elements | Transformation induced by $y = \begin{Vmatrix} \alpha & \beta \\ \gamma & \delta \end{Vmatrix}$ | Subgroup of Stability |
|---|---|---|---|
| Complex affine plane | Ordered pairs $(z_1, z_2)$ of complex numbers, but $(z_1, z_2) \neq (0, 0)$ | $(z_1, z_2) \to (\alpha z_1 + \gamma z_2, \beta z_1 + \delta z_2)$ | Matrices of the form $\begin{Vmatrix} 1 & \zeta \\ 0 & 1 \end{Vmatrix}$ |
| Complex projective line | The complex numbers $z$, including the point at infinity | $z \to \dfrac{\alpha z + \gamma}{\beta z + \delta}$ | Matrices of the form $\begin{Vmatrix} \alpha & \beta \\ 0 & \alpha^{-1} \end{Vmatrix}$ |
| Point pairs on the complex projective line | Ordered pairs $(z_1, z_2)$ for $z_1, z_2$ on the complex projective line and $z_1 \neq z_2$ | $(z_1, z_2) \to \left(\dfrac{\alpha z_1 + \gamma}{\beta z_1 + \delta}, \dfrac{\alpha z_2 + \gamma}{\beta z_2 + \delta}\right)$ | Matrices of the form $\begin{Vmatrix} \alpha & 0 \\ 0 & \alpha^{-1} \end{Vmatrix}$ |
| Lobachevskian 3-space[a] | Positive definite unimodular Hermitian matrices $h$ | $h \to g^*hg$ | Unitary unimodular matrices $\begin{Vmatrix} \alpha & \beta \\ -\bar{\beta} & \bar{\alpha} \end{Vmatrix}$, $\lvert \alpha \rvert^2 + \lvert \beta \rvert^2 = 1$ |
| Imaginary Lobachevskian 3-space[a] | Hermitian matrices $h$ of determinant $-1$, with $h$ and $-h$ identified | $h \to g^*hg$ | Matrices of the forms $\begin{Vmatrix} \alpha & \beta \\ \bar{\beta} & \bar{\alpha} \end{Vmatrix}$ and $\begin{Vmatrix} \beta & \alpha \\ -\bar{\alpha} & -\bar{\beta} \end{Vmatrix}$, $\lvert \alpha \rvert^2 - \lvert \beta \rvert^2 = 1$ |
| The cone[a] | Hermitian matrices $h$ of determinant zero, with $h$ and $-h$ identified | $h \to g^*hg$ | Matrices of the form $\begin{Vmatrix} e^{-i\varphi} & 0 \\ \zeta & e^{i\varphi} \end{Vmatrix}$ |
| Isotropic lines of imaginary Lobachevskian 3-space | Sets $l$ of Hermitian matrices of determinant $-1$ whose elements are linear functions of a real parameter $t$ | $l \to g^*lg$ | Matrices of the form $\begin{Vmatrix} \lambda & 0 \\ \zeta & \lambda^{-1} \end{Vmatrix}$, where $\lambda$ is real or pure imaginary |

[a] These spaces can also be realized as surfaces in a four-dimensional linear space. If $[x, x] = x_0^2 - x_1^2 - x_2^2 - x_3^2$, then Lobachevskian 3-space is the positive sheet of the $[x, x] = 1$ hyperboloid, imaginary Lobachevskian 3-space is the $[x, x] = -1$ hyperboloid with diametrically ... The motions are then the hyperbolic rotations of $E_4$.

the geometrical. These advantages are related to the fact that although the geometrical approach can be used to define horospheres in a natural way in homogeneous spaces with a Riemannian (or pseudo-Riemannian) metric, the group-theoretical definition is applicable also for other homogeneous spaces.[5]

We shall illustrate the group-theoretical definition of a horosphere on the group of two-dimensional complex unimodular matrices. (As was shown in Chapter III, this group is isomorphic to the Lorentz group.) Let $X$ be a space homogeneous under this group, and we shall define horospheres in $X$ so that if $X$ is Lobachevskian or imaginary Lobachevskian space the definition will coincide with that of Chapter V.

A horosphere in a space $X$ homogeneous under the group of complex two-dimensional unimodular matrices is the orbit of any point $x \in X$ under the subgroup $Z$ of the matrices

$$\begin{Vmatrix} 1 & \zeta \\ 0 & 1 \end{Vmatrix}$$

or under any subgroup conjugate to it. This definition implies that every homogeneous family of horospheres has a stationary subgroup which is either $Z$ or some conjugate subgroup. Thus such a family is a homogeneous space of a very simple kind; namely, its structure is such that it is isomorphic to the complex affine line or one of its quotient spaces.

By this definition a horosphere $\omega$ is given by some point $x \in X$ and some element $g$ of the Lorentz group; it consists of all points of the form $xg^{-1}\zeta g$, where $\zeta$ runs over $Z$. If we write $y = xg^{-1}$, all points on $\omega$ are of the form $y\zeta g$. In the future we shall denote the elements of horospheres in this way.

**Assertion.** *This group-theoretical definition of horospheres for space homogeneous under the Lorentz group coincides with the geometrical definition of Chapter V when X is three-dimensional Lobachevskian or imaginary Lobachevskian space.* We shall prove this assertion separately for Lobachevskian and imaginary Lobachevskian spaces.

[5] It is true, nevertheless, that the geometric definition (say, based on orthogonal trajectories of sets of parallel geodesics) can be extended to many inhomogeneous spaces, at least to Riemannian spaces of varying curvature. Moreover, the corresponding problems of integral geometry are not without interest. It would be of profit, for instance, to solve the analog of the problem dealt with in the second section of Chapter V for a two-dimensional Riemannian manifold of varying curvature such that the curvature is bounded both above and below. Similar problems in higher dimensions are undoubtedly also of interest.

Proof (*Lobachevskian Space*). Let us use the Hermitian-matrix model of Lobachevskian space, and consider the horosphere $\omega$ (group-theoretical definition) defined by the Hermitian matrix $y$ and the identity element of the Lorentz group. Then the horosphere consists of points $x$ of the form $x = y\zeta$, where $\zeta$ runs over $Z$. In this model, recall, the motions are transformations of the form $y \to g^*yg$, so that $\omega$ consists of Hermitian matrices of the form

$$x = \begin{Vmatrix} 1 & 0 \\ \zeta & 1 \end{Vmatrix} \begin{Vmatrix} y_{11} & y_{12} \\ y_{21} & y_{22} \end{Vmatrix} \begin{Vmatrix} 1 & \zeta \\ 0 & 1 \end{Vmatrix}.$$

It is easily verified that all these matrices have the same matrix element in the upper left-hand corner, and we shall denote it by $x_{11} = C$. Now this equation can also be written down for the model on the $[x, x] = 1$ hyperboloid. Let us use the same symbol $x$ to denote the matrix

$$x = \begin{Vmatrix} x_{11} & x_{12} \\ x_{21} & x_{22} \end{Vmatrix}.$$

and the point $x = (x_0, x_1, x_2, x_3)$ which corresponds to it on the hyperboloid. Then $x_0 - x_3 = x_{11}$. Thus in the second model all the points on $\omega$ have coordinates such that $x_0 - x_3 = C$, or in other words such that $[x, \xi] = C$, where $\xi = (1, 0, 0, 1)$. But $\xi$ lies on the $[\xi, \xi] = 0$ cone, and therefore $[x, \xi] = C$ is the equation of a horosphere in the geometrical definition.

We have shown that if $\omega$ is a group-theoretically defined horosphere associated with the identity element of the Lorentz group, then it is also a horosphere according to the geometrical definition. But every horosphere associated with any other element $g$ of the Lorentz group can be obtained by Lobachevskian motion from a horosphere associated with the identity. Therefore all horospheres defined group theoretically in three-dimensional Lobachevskian space are also horospheres as defined geometrically. The converse is implied immediately by homogeneity of the set of horospheres in the geometrical definition.

In conclusion, therefore, in Lobachevskian 3-space the grouptheoretical and geometrical (Chapter V) definitions of horospheres coincide.

Proof (*Imaginary Lobachevskian Space*). Proceeding as in the above case, it is easily verified that horospheres of the form $y\zeta$, where $y$ is a Hermitian matrix such that $y_{11} \neq 0$, are horospheres of the first kind in the geometric definition. For $y_{11} = 0$ the situation is somewhat different. Consider therefore an Hermitian matrix of the form

$$y = \begin{Vmatrix} 0 & y_{12} \\ y_{21} & y_{22} \end{Vmatrix}.$$

Writing $x = \zeta^* y \zeta$. for $\zeta \in Z$, we obtain the new matrix

$$\begin{Vmatrix} 0 & x_{12} \\ x_{21} & x_{22} \end{Vmatrix} = \begin{Vmatrix} 1 & 0 \\ \bar{\zeta} & 1 \end{Vmatrix} \begin{Vmatrix} 0 & y_{12} \\ y_{21} & y_{22} \end{Vmatrix} \begin{Vmatrix} 1 & \zeta \\ 0 & 1 \end{Vmatrix},$$

which has the property that $x_{12} = y_{12}$. Thus the horosphere $\omega$ associated with some point $y$ such that $y_{11} = 0$ and with the unit element of the Lorentz group consists of matrices $x$ such that $x_{11} = 0$ and $x_{12} = C$ (where $|C| = 1$).

Let us now turn the model on the $[x, x] = -1$ hyperboloid. Again we use the same symbol $x$ to denote the Hermitian matrix and the point $x = (x_0, x_1, x_2, x_3)$ corresponding to it; then $x_{12} = x_2 - ix_1$. Consequently $x_2 - ix_1 = C$, or $x_1 = C_1$ and $x_2 = C_2$. Thus our horosphere is represented on the $[x, x] = -1$ hyperboloid by the line generator whose equations are $x_0 = x_3$, $x_1 = C_1$, $x_2 = C_2$, $C_1^2 + C_2^2 = 1$. In other words it corresponds to an isotropic line in imaginary Lobachevskian space.

Since every horosphere of the form $y\zeta g$, where $y_{11} = 0$, can be obtained by Lobachevskian motion from a horosphere of the form $y\zeta$, all these horospheres correspond to isotropic lines. This shows, incidentally, that the horospheres (group-theoretical definition) of imaginary Lobachevskian space decompose into two classes. In the first class are the horospheres associated with matrices $y$ such that $y_{11} \neq 0$; each such horosphere is a two-dimensional manifold and corresponds to a horosphere of the first kind in the geometric definition. The second class consists of horospheres associated with $y$ such that $y_{11} = 0$, and as we have seen these correspond to isotropic lines.[6]

In conclusion, therefore, we see that for imaginary Lobachevskian 3-space the group-theoretical and geometrical (Chapter V) definitions of horospheres coincide.

Before proceeding we want finally to study the horospheres (in the group-theoretical definition) of the point pairs on the complex projective line, i.e., of ordered pairs $(z_1, z_2)$, $z_1 \neq z_2$, of points on the complex projective line. Consider first the horospheres $\omega_0 = \{yz\}$, where $y = (y_1, y_2)$ is some element in this space and where $z$ runs over the subgroup $Z'$ of matrices of the form

$$\begin{Vmatrix} 1 & 0 \\ z & 1 \end{Vmatrix}.$$

[6] It now becomes clear why we did not call the isotropic plane $[x, \xi] = 0$ a horosphere of the second kind, but rather the lines of which it is composed.

Note that this subgroup is conjugate to $Z$, since

$$\begin{Vmatrix} 1 & 0 \\ z & 1 \end{Vmatrix} = \begin{Vmatrix} 0 & -1 \\ 1 & 0 \end{Vmatrix} \cdot \begin{Vmatrix} 1 & -z \\ 0 & 1 \end{Vmatrix} \cdot \begin{Vmatrix} 0 & 1 \\ -1 & 0 \end{Vmatrix}.$$

In this space the Lobachevskian motions are given by

$$(z_1, z_2) \to (z_1', z_2') = \left( \frac{\alpha z_1 + \gamma}{\beta z_1 + \delta}, \frac{\alpha z_2 + \gamma}{\beta z_2 + \delta} \right), \tag{1}$$

so that $\omega_0$ consists of points of the form $(z_1, z_2) = (y_1 + z, y_2 + z)$. The equation of this horosphere may consequently be written

$$z_1 - z_2 = C, \tag{2}$$

and $z_1 \neq z_2$ implies that $C \neq 0$.

Now consider an arbitrary horosphere $\omega = \{yzg\}$, where $g$ is any element of the Lorentz group. Clearly this horosphere is obtained from $\omega_0 = \{yz\}$ by the motion $g$. To find its equation, therefore, we use Eq. (1) with $z_1$ and $z_2$ satisfying Eq. (2). Then after some simple algebra (and dropping the primes) we find that $\omega$ is given by

$$z_1 - z_2 = (a - bz_1)(a - bz_2), \tag{3}$$

where $a$ and $b$ are complex numbers, not both equal to zero, depending on the matrix elements of $g$ and on $C$.

Since each horosphere is given by Eq. (3), it is uniquely defined by a pair of complex numbers $(a, b)$, except that $(a, b)$ and $(-a, -b)$ define the same horosphere. Now let us see how $(a, b)$ changes under a motion $g$, that is, under a transformation of the kind given by Eq. (1). By replacing $z_1$ and $z_2$ in (3) by their expressions in terms of $z_1'$ and $z_2'$ and dropping the primes, we arrive at

$$z_1 - z_2 = [(\alpha a + \gamma b) - (\beta a + \delta b)z_1] \cdot [(\alpha a + \gamma b) - (\beta a + \delta b)z_2].$$

Thus under the motion $g$ the parameters $(a, b)$ are transformed to $(a', b')$ given by

$$a' = \alpha a + \gamma b, \qquad b' = \beta a + \delta b, \tag{4}$$

which is the same as the transformation $g$ induces on the complex affine plane.

*We have thus shown that the space of horospheres in the space of point pairs on the complex projective line coincides with the complex projective plane with diametrically opposite points identified.*

The group-theoretical definition of horospheres we have given here is easily extended to the group $G$ of complex unimodular matrices of any dimension. In particular, let $X$ be a space homogeneous under $G$. Then a horosphere in $X$ is any set of points of the form $y\zeta g$ where $y \in X$, $g \in G$, and $\zeta$ runs over the subgroup $G$ of matrices of the form

$$\begin{Vmatrix} 1 & \zeta_{12} & \cdots & \zeta_{1n} \\ 0 & 1 & \cdots & \zeta_{2n} \\ & & \cdots & \\ 0 & 0 & \cdots & 1 \end{Vmatrix}. \tag{5}$$

This includes as degenerate cases subgroups of matrices in which some of the $\zeta_{ij}$ vanish and each matrix is composed of submatrices in blocks of unit matrices along the main diagonal, null matrices below it, and arbitrary matrices above it.

To extend the definition further to any complex semisimple Lie group $G$, rather than the subgroup of matrices of the form (5), one must choose any maximal nilpotent[7] subgroup of $G$. [Equation (5) defines the maximal nilpotent subgroup of $G$, the group of complex unimodular matrices in $n$ dimension.] The subgroups consisting of submatrices as described below Eq. (5) (degenerate case) are replaced by certain nonmaximal nilpotent subgroups whose description in terms of root vectors will not be given here.

Finally, we extend the notion horospheres to any Lie group $G$. Let $g(t)$ be a one-parameter subgroup of $G$. We define the *horospherical subgroup* associated with $g(t)$ as the set of elements $\zeta \in G$ such that $\lim_{t\to+\infty} g(-t)\zeta g(t) = e$, where $e$ is the unit element. We then call horospheres in a homogeneous space the trajectories generated by horospherical subgroups. For a complex semisimple Lie group the horospherical subgroups are all the maximal nilpotent subgroups as well as certain nonmaximal ones (degenerate case). Thus this definition coincides with the previous one for complex semisimple Lie groups.

## 1.7. Fourier Integral Expansions of Functions on Homogeneous Spaces

The principal aim of this chapter is to decompose certain spaces of functions over homogeneous spaces into irreducible subspaces invariant

[7] Let $G$ be some group. Consider the sets $G_k$, $k = 0, 1, \ldots, n, \ldots$ in $G$ defined inductively as follows. Let $G_0 = e$, where $e$ is the identity element of $G$. When $G_k$ has been defined, $G_{k+1}$ is defined as the set of elements $s \in G$ such that for any $t \in G$ the element $s^{-1}t^{-1}st \in G_k$. A group $G$ is called *nilpotent* if there exists an $n$ such that $G = G_n$.

under the group. Such a decomposition is analogous to the decomposition of the space of functions $f(x)$ over the line into one-dimensional subspaces invariant under translation. For this simple case the decomposition is given by the Fourier integral, and therefore the decompositions we obtain in this chapter are the analogs of the Fourier integral on homogeneous spaces. For the present we wish to state the results that will be obtained later.

Consider the functions on the complex affine plane. With each function $f(z_1, z_2)$ with square integrable modulus we associate the homogeneous function $F(z_1, z_2; \rho, n)$ defined by

$$F(z_1, z_2; \rho, n) = \frac{i}{2} \int f(\lambda z_1, \lambda z_2) \lambda^{-n_1} \bar{\lambda}^{-n_2} \, d\lambda \, d\bar{\lambda}, \tag{1}$$

where $n_1 = \frac{1}{2}(n + i\rho)$ and $n_2 = \frac{1}{2}(-n + i\rho)$ (here $n$ is an integer and $\rho$ is a real number). Under the transformations of the form

$$T(g) f(z_1, z_2) = f(\alpha z_1 + \gamma z_2, \beta z_1 + \delta z_2)$$

$F$ transforms according to a unitary irreducible representation of the principal series

$$T_\chi(g) F(z_1, z_2; \rho, n) = F(\alpha z_1 + \gamma z \ , \beta z_1 + \delta z_2; \rho, n), \tag{2}$$

where $\chi = (n_1, n_2)$. The function $f(z_1, z_2)$ is given in terms of its "Fourier components"[8] $F(z_1, z_2; \rho, n)$ by

$$f(z_1, z_2) = (2\pi)^{-2} \sum_{n=-\infty}^{\infty} \int_{-\infty}^{\infty} F(z_1, z_2; \rho, n) \, d\rho. \tag{3}$$

Further, the analog of Plancherel's theorem will be found valid [see Section 2.1, Eq. (10)].

Now consider the functions $f(x)$ over the cone whose equation is $[x, x] = 0$. In this case the Fourier components are given by

$$F(x; \rho) = \int_0^{\infty} f(tx) t^{-\frac{1}{2}i\rho} \, dt, \tag{4}$$

and the Fourier expansion in these components by

$$f(x) = (4\pi)^{-1} \int_{-\infty}^{\infty} F(x; \rho) \, d\rho. \tag{5}$$

[8] The functions $F(z_1, z_2; \rho, n)$ are analogous to the individual $a_n e^{in\varphi}$, that is, to each entire term of the Fourier series. We therefore call them the Fourier components of $f(z_1, z_2)$.

Under the transformation $f(x) \to f(xg)$ of a function over the cone, its Fourier components transform according to

$$T_\chi(g)F(x; \rho) = F(xg; \rho), \tag{6}$$

which are representations equivalent to irreducible unitary representations of the principal series for which $\chi = (\frac{1}{2}i\rho, -\frac{1}{2}i\rho)$.

Now consider functions $f(x)$ defined on real Lobachevskian space. To write out the Fourier integral expansion for such functions, we turn to the model on the $[x, x] = 1$ hyperboloid. Then the Fourier components of $f(x)$ are the functions $F(\xi; \rho)$ on the $[\xi, \xi] = 0$ cone defined by

$$F(\xi; \rho) = \int f(y)[y, \xi]^{\frac{1}{2}i\rho-1}\, dy. \tag{7}$$

These functions are homogeneous in $\xi$ of degree $\frac{1}{2}i\rho - 1$. Under the transformation $f(x) \to f(xg)$ the Fourier components transform according to

$$T_\chi(g)F(\xi, \rho) = F(\xi g; \rho), \tag{8}$$

that is, according to representations equivalent to irreducible unitary representations of the principal series with $\chi = (\frac{1}{2}i\rho, \frac{1}{2}i\rho)$. Further, $f(x)$ is given in terms of its Fourier components by

$$f(x) = (4\pi)^{-3} \int_0^\infty \rho^2\, d\rho \int F(\xi; \rho)\delta([x, \xi] - 1)\, d\xi. \tag{9}$$

It is remarkable that Eq. (7) can be written in another form analogous to the formula for the Fourier transform of a function $f(x)$ on $n$-dimensional Euclidean space. In fact since $[y, \xi] = e^{k\,\tau(y,\xi)}$, where $\tau(y, \xi)$ is the distance from $y$ to the horosphere whose equation is $[x, \xi] = 1$ (see Chapter V, Section 2.2), we may write

$$F(\xi; \rho) = \int f(y) \exp[(\tfrac{1}{2}i\rho - 1)k\,\tau(y, \xi)]\, dy. \tag{10}$$

The Fourier components of $f(x)$ in $n$-dimensional Euclidean space can be written in the form

$$F(\xi) = \int f(y) e^{i|\xi|\tau(y,\xi)}\, dy, \tag{11}$$

where

$$\xi = (\xi_1, ..., \xi_n), \qquad (x, \xi) = x_1\xi_1 + \cdots + x_n\xi_n,$$
$$|\xi| = (\xi_1^2 + \cdots + \xi_n^2)^{\frac{1}{2}},$$

and $\tau(y, \xi)$ is the distance from $y$ to the plane whose equation is $(x, \xi) = 0$. The resemblance between Eqs. (10) and (11) is obvious [see also Section 3.2, Eq. (12)].

Finally, consider imaginary Lobachevskian space. In this space there exist two kinds of Fourier components. When the space is realized on the $[x, x] = -1$ hyperboloid, the Fourier components of the first kind are functions over the $[\xi, \xi] = 0$ cone. They are then given by

$$F(\xi; \rho) = \int f(y)|[y, \xi]|^{\frac{1}{2}i\rho-1}\, dy, \tag{12}$$

where $dy$ is the invariant measure on the imaginary Lobachevskian space. The Fourier components of the second kind are functions over the set of line generators $l$ of the $[x, x] = -1$ hyperboloid. They are given by

$$F(l; 2n) = \int f(y)\delta([y, \xi])e^{-2in\theta}\, dy. \tag{13}$$

Here $\xi$ is the direction vector of $l$ normalized so that $\xi = (1, \xi_1, \xi_2, \xi_3)$, and $\theta = -ikr$, where $r$ is the distance in the imaginary Lobachevskian space from $y$ to the isotropic line $l$. Under the transformation $f(x) \rightarrow f(xg)$ the Fourier components of the first kind transform according to

$$T_\chi(g)F(\xi; \rho) = F(\xi g; \rho), \tag{14}$$

that is, under representations equivalent to irreducible unitary representations of the Lorentz group for which $\chi = (\frac{1}{2}i\rho, \frac{1}{2}i\rho)$. The Fourier components of the second kind, however, transform under representations equivalent to $T_\chi(g)$ with $\chi = (2n, 2n)$. A function $f(x)$ is given in terms of its Fourier components by

$$f(x) = \tfrac{1}{2}(4\pi)^{-3} \int_0^\infty \rho^2\, d\rho \int_\Omega F(\eta; \rho)|[x, \eta]|^{-\frac{1}{2}i\rho-1}\, d\sigma$$

$$+ 4\pi^{-2} \sum_{n=1}^{\infty} n \int_\Omega F(l; 2n)e^{2in\theta}\, \delta([x, \eta])\, d\sigma,$$

for which the notation is defined in Sections 4.4 and 4.5. We remark that it is possible also in the case of the imaginary Lobachevskian space to rewrite the Fourier transform formulas in a form analogous to those for a Euclidean space.

## 2. Representations of the Lorentz Group Associated with the Complex Affine Plane and with the Cone, and Their Irreducible Components

### 2.1. Unitary Representations of the Lorentz Group Associated with the Complex Affine Plane

It has already been remarked in Section 1.3 that the decomposition of any representation of the Lorentz group $G$ derived from a homogeneous space can be reduced to the decomposition of the representation derived from the complex affine plane or one of its quotient spaces obtained by identifying certain points (see Section 1.3). Let us therefore first show how to decompose the representation associated with the complex affine plane.

For simplicity consider the following unitary representation of $G$. The representation is on the Hilbert space $H$ of functions $f(z_1, z_2)$ of two complex variables such that the integral

$$\|f\|^2 = \left(\frac{i}{2}\right)^2 \int |f(z_1, z_2)|^2 \, dz_1 \, d\bar{z}_1 \, dz_2 \, d\bar{z}_2 \tag{1}$$

converges. It consists of associating with each element

$$g = \begin{Vmatrix} \alpha & \beta \\ \gamma & \delta \end{Vmatrix}$$

of the Lorentz group the operator $T(g)$ defined by

$$T(g)f(z_1, z_2) = f(\alpha z_1 + \gamma z_2, \beta z_1 + \delta z_2). \tag{2}$$

It is obvious that $T(g_1 g_2) = T(g_1)T(g_2)$, that is, that $T(g)$ is indeed a representation of $G$. This representation is unitary because the transformation

$$(z_1, z_2) \to (\alpha z_1 + \gamma z_2, \beta z_1 + \delta z_2), \qquad \alpha\delta - \beta\gamma = 1,$$

leaves invariant the volume element of the complex affine plane.

Now the irreducible representations of the Lorentz group are given by the same equation,

$$T_\chi(g)f(z_1, z_2) = f(\alpha z_1 + \gamma z_2, \beta z_1 + \delta z_2), \tag{3}$$

but they act on one of the spaces of homogeneous functions that we have called $D_\chi$. Thus the decomposition of $T(g)$ defined by Eq. (2) reduces to the expansion of $f(z_1, z_2) \in H$ in homogeneous functions.

Let us therefore turn to the problem of finding these homogeneous components of $f(z_1, z_2)$. Let $f(z_1, z_2)$ be a function of bounded support on the affine plane, vanishing in a neighborhood of the origin. With $f(z_1, z_2)$ we associate the homogeneous function $F(z_1, z_2; \chi)$, where $\chi = (n_1, n_2)$ is a pair of complex numbers whose difference is an integer. This function is defined by the equation

$$F(z_1, z_2; \chi) = \frac{i}{2} \int f(\lambda z_1, \lambda z_2) \lambda^{-n_1} \bar{\lambda}^{-n_2} \, d\lambda \, d\bar{\lambda}. \tag{4}$$

From the assumptions we have made concerning the properties of $f$, the integral of Eq. (4) obviously converges for all $n_1, n_2$. We shall call $F$ the *Mellin transform* of $f(z_1, z_2)$. Then from the definition it follows immediately that $F$ has the following properties.

*Property 1.* $F$ is homogeneous in $z_1, z_2$ of degree $(n_1 - 1, n_2 - 1)$. In other words for any $a \neq 0$ we have

$$F(az_1, az_2; \chi) = a^{n_1 - 1} \bar{a}^{n_2 - 1} F(z_1, z_2; \chi). \tag{5}$$

*Property 2.* The transformation

$$T(g) f(z_1, z_2) = f(\alpha z_1 + \gamma z_2, \beta z_1 + \delta z_2)$$

on $f(z_1, z_2)$ induces the transformation

$$T_\chi(g) F(z_1, z_2; \chi) = F(\alpha z_1 + \gamma z_2, \beta z_1 + \delta z_2; \chi) \tag{6}$$

on its Mellin transform.

Let us obtain a formula for expanding $f(z_1, z_2)$ in terms of $F(z_1, z_2; \chi)$. The expansion will contain $F$ only for $\chi = (\frac{1}{2}n + \frac{1}{2}i\rho, -\frac{1}{2}n + \frac{1}{2}i\rho)$, where $n$ is an integer and $\rho$ is any real number. Recall that these values of $\chi$ correspond to irreducible unitary representation of $G$ belonging to the principal series. In order to obtain the desired expansion we first reduce the Mellin transform to the ordinary Fourier transform. For this purpose we write

$$\lambda = e^{\sigma + i\tau}, \qquad n_1 = \tfrac{1}{2}(n + i\rho), \qquad n_2 = \tfrac{1}{2}(-n + i\rho)$$

in Eq. (4). Then

$$\begin{aligned} F(z_1, z_2; \rho, n) &\equiv F(z_1, z_2; \chi) \\ &= \int_{-\infty}^{\infty} \int_0^{2\pi} f(e^{\sigma + i\tau} z_1, e^{\sigma + i\tau} z_2) e^{-i(\rho\sigma + n\tau) + 2\sigma} \, d\tau \, d\sigma, \end{aligned} \tag{7}$$

which we see is the Fourier transform (in $\sigma$ and $\tau$) of the function

$$e^{2\sigma} f(e^{\sigma + i\tau} z_1, e^{\sigma + i\tau} z_2).$$

Then by taking the inverse Fourier transform we obtain

$$e^{2\sigma}f(e^{\sigma+i\tau}z_1, e^{\sigma+i\tau}z_2) = (2\pi)^{-2} \sum_{n=-\infty}^{\infty} \int_{-\infty}^{\infty} F(z_1, z_2; \rho, n) e^{i(\rho\sigma+n\tau)}\, d\rho.$$

Finally, we write $\sigma = \tau = 0$ and then it is seen that $f(z_1, z_2)$ is given in terms of its Mellin transform by

$$f(z_1, z_2) = (2\pi)^{-2} \sum_{n=-\infty}^{\infty} \int_{-\infty}^{\infty} F(z_1, z_2; \rho, n)\, d\rho. \tag{8}$$

Again we remark that only those $F(z_1, z_2; \chi)$ that transform according to the representations of the principal series enter into this expression.

We now assert that Eqs. (4) and (8) give the decomposition of a unitary representation into its irreducible components. For this purpose we must obtain the analog of Plancherel's theorem, or in other words we must show that with a suitably defined scalar product in the space of the $F(z_1, z_2; \rho, n)$ and with a suitable measure $d\mu(\rho, n)$ we may write

$$\|f\|^2 = \int \|F(z_1, z_2; \rho, n)\|_\chi^2\, d\mu(\rho, n). \tag{9}$$

To prove this we write down the usual Plancherel theorem for the Fourier transform of Eq. (7). This yields

$$\int_0^{2\pi} \int_{-\infty}^{\infty} |f(e^{\sigma+i\tau}z, e^{\sigma+i\tau})|^2 e^{4\sigma}\, d\sigma\, d\tau$$
$$= (2\pi)^{-2} \sum_{n=-\infty}^{\infty} \int_{-\infty}^{\infty} |F(z, 1; \rho, n)|^2\, d\rho.$$

Now integrate both sides with respect to $z$ and $\bar{z}$ and write $e^{\sigma+i\tau} z = z_1$, $e^{\sigma+i\tau} = z_2$. We then arrive at

$$\left(\frac{i}{2}\right)^2 \int |f(z_1, z_2)|^2\, dz_1\, d\bar{z}_1\, dz_2\, d\bar{z}_2$$
$$= (2\pi)^{-2} \sum_{n=-\infty}^{\infty} \frac{i}{2} \int_{-\infty}^{\infty} d\rho \int |F(z, 1; \rho, n)|^2\, dz\, d\bar{z}, \tag{10}$$

which is the desired analog of Plancherel's theorem. To put this in the form of Eq. (9) we introduce the norm

$$\|F(z_1, z_2; \rho, n)\|_\chi^2 = \frac{i}{2} \int |F(z, 1; \rho, n)|^2\, dz\, d\bar{z} \tag{11}$$

in the space of the $F$ functions and define integration with respect to $d\mu(\rho, n)$ by

$$\int G(\rho, n)\, d\mu(\rho, n) = (2\pi)^{-2} \sum_{n=-\infty}^{\infty} \int_{-\infty}^{\infty} G(\rho, n)\, d\rho. \tag{12}$$

Then by using (1) for the left-hand side of Eq. (10), we arrive at Eq. (9).

This shows that the mapping $f(z_1, z_2) \to F(z_1, z_2; \rho, n)$ is isometric with respect to the norm of Eq. (1) and the norm

$$\| F \|^2 = (2\pi)^{-2} \sum_{n=-\infty}^{\infty} \frac{i}{2} \int_{-\infty}^{\infty} d\rho \int |F(z, 1; \rho, n)|^2\, dz\, d\bar{z}. \tag{13}$$

We have so far defined this mapping only for $f(z_1, z_2)$ of bounded support and vanishing in a neighborhood of the origin. Since these functions form an everywhere dense set in $H$, however, the mapping can be extended to all of $H$.

This shows that Eqs. (4) and (8) give the decomposition of a unitary representation $T(g)$ of the Lorentz group into its irreducible components. The only irreducible unitary representations entering into this decomposition are those of the principal series. Note that each of the representations of the principal series enters twice, since $T_\chi$ is equivalent to $T_{-\chi}$, where $-\chi = (-n_1, -n_2)$.

## 2.2. Unitary Representation of the Lorentz Group Associated with the Cone

We shall now study the representation of the Lorentz group associated with another homogeneous space we have discussed, namely the set of points on the positive cone

$$[x, x] \equiv x_0^2 - x_1^2 - x_2^2 - x_3^2 = 0, \qquad x_0 > 0, \tag{1}$$

in four-dimensional real Euclidean space $E_4$. The motions of this space are induced by linear transformations with determinant one on $E_4$ that preserve the quadratic form $[x, x]$ and map each sheet of the cone into itself. This is the space we have called *the cone*.

We wish to show that the cone can be obtained as a quotient space of the complex affine plane. Specifically, if all points of the form $(e^{i\varphi} z_1, e^{i\varphi} z_2)$ of the complex affine plane are identified, where $0 \leqslant \varphi < 2\pi$, one obtains a homogeneous space isomorphic to the cone.

The proof goes as follows. With every point $x \in E_4$ we associate the Hermitian matrix

$$h = \begin{Vmatrix} x_0 - x_3 & x_2 - ix_1 \\ x_2 + ix_1 & x_0 + x_3 \end{Vmatrix}.$$

Then the points on the $[x, x] = 0$ cone will correspond to nonnegative definite Hermitian matrices with determinant zero. As we know, under this correspondence the motions of the cone as defined after Eq. (1) induce transformation carrying the Hermitian matrix $h$ into $g^*hg$, where

$$g = \begin{Vmatrix} \alpha & \beta \\ \gamma & \delta \end{Vmatrix}.$$

Now let us associate with each point $z = (z_1, z_2)$ of the complex affine plane the point on the cone whose matrix is

$$h_z = \begin{Vmatrix} \bar{z}_1 & 0 \\ \bar{z}_2 & 0 \end{Vmatrix} \begin{Vmatrix} z_1 & z_2 \\ 0 & 0 \end{Vmatrix} = \begin{Vmatrix} z_1\bar{z}_1 & \bar{z}_1 z_2 \\ z_1\bar{z}_2 & z_2\bar{z}_2 \end{Vmatrix}.$$

Obviously every point on the cone can be obtained in this way from a point of the complex affine plane. Note also that under the transformation

$$(z_1, z_2) \to (\alpha z_1 + \gamma z_2, \beta z_1 + \delta z_2)$$

of the complex affine plane, $h_z$ is transformed into $g^*h_zg$. Thus the motions of the affine plane can be associated with motions of the cone.

Let us see which points of the plane are mapped in this way onto the same point of the cone. Let $(z_1, z_2)$ and $(z_1', z_2')$ be mapped onto the same point of the cone. Then

$$|z_1|^2 = |z_1'|^2, \qquad |z_2|^2 = |z_2'|^2, \qquad \bar{z}_1 z_2 = \bar{z}_1' z_2', \qquad z_1\bar{z}_2 = z_1'\bar{z}_2'.$$

This implies that $z_1' = e^{i\varphi}z_1$ and $z_2' = e^{i\varphi}z_2$. Thus all points of the form $(e^{i\varphi}z_1, e^{i\varphi}z_2)$, and only such points, are mapped onto a given point of the cone, which proves the original assertion.

We now wish to construct a unitary representation of the Lorentz group on the functions $f(x)$ over the cone for which the integral

$$\|f\|^2 = \int |f(x)|^2 \, dx \tag{2}$$

converges, where $dx$ is the invariant measure on the cone (see Chapter V, Section 1.6). This representation is obtained by associating each element $g$ of the Lorentz group with the operator defined by

$$T(g)f(x) = f(xg), \tag{3}$$

where as usual $xg \in E_4$ is the point into which $x$ is mapped by the motion $g$.[9] This representation is unitary, for $dx$ is invariant so that

$$\| T(g)f \|^2 = \int |f(xg)|^2 \, dx = \int |f(x)|^2 \, dx = \| f \|^2.$$

Our problem is now to decompose this unitary representation into its irreducible components. Since the cone is a quotient space of the affine plane, the problem reduces to the one already solved of decomposing the representations defined on the affine plane. We may therefore proceed as follows.

With each function $f(x)$ on the cone is associated a function $f(z_1, z_2)$ on the complex plane with the property that

$$f(e^{i\varphi} z_1, e^{i\varphi} z_2) = f(z_1, z_2). \tag{4}$$

Thus we may use the Fourier integral expansion of $f(z_1, z_2)$ that we obtained in Section 2.1, namely

$$f(z_1, z_2) = (2\pi)^{-2} \sum_{n=-\infty}^{\infty} \int_{-\infty}^{\infty} F(z_1, z_2; \rho, n) \, d\rho, \tag{5}$$

where the Fourier transform is defined by

$$\begin{gathered} F(z_1, z_2; \rho, n) = \frac{i}{2} \int f(\lambda z_1, \lambda z_2) \lambda^{-n_1} \bar{\lambda}^{-n_2} \, d\lambda \, d\bar{\lambda}, \\ n_1 = \tfrac{1}{2}(n + i\rho), \qquad n_2 = \tfrac{1}{2}(-n + i\rho). \end{gathered} \tag{6}$$

Equation (4) now implies that $F$ fails to vanish only if $n = 0$. Then Eq. (6) becomes

$$F(z_1, z_2; \rho, 0) = 2\pi \int_0^{\infty} f(tz_1, tz_2) t^{1-i\rho} \, dt. \tag{7}$$

Remark. The $F(z_1, z_2; \rho, 0)$ functions form a space remarkable in that it contains a vector invariant under those operators of the representation that correspond to unitary matrices (that is, invariant under the maximal compact subgroup of the Lorentz group). In fact, $(|z_1|^2 + |z_2|^2)^{\frac{1}{2}i\rho - 1}$ is such a vector. When the carrier space of an irreducible representation of a semisimple Lie group contains a vector invariant under the maximal compact subgroup, the representation is usually said to be of class I. #

To summarize the above discussion, let $f(z_1, z_2)$ be a function over the

[9] The explicit form of the transformation from $x$ to $xg$ is given in Chapter III, Section 1.1.

complex affine plane satisfying Eq. (4). Then its Fourier integral expansion is of the form

$$f(z_1, z_2) = (2\pi)^{-2} \int_{-\infty}^{\infty} F(z_1, z_2; \rho)\, d\rho, \tag{8}$$

where

$$F(z_1, z_2; \rho) = 2\pi \int_0^{\infty} f(tz_1, tz_2) t^{1-i\rho}\, dt. \tag{9}$$

Then together with the results of Section 2.1 this implies the analog of Plancherel's theorem

$$\left(\frac{i}{2}\right)^2 \int |f(z_1, z_2)|^2\, dz_1\, d\bar{z}_1\, dz_2\, d\bar{z}_2$$
$$= (2\pi)^{-2} \left(\frac{i}{2}\right) \int_{-\infty}^{\infty} d\rho \int |F(z, 1; \rho)|^2\, dz\, d\bar{z}. \tag{10}$$

By returning from the points $z_1$, $z_2$ of the complex affine plane to the points on the cone, we arrive finally at the following result.

Consider the representation

$$T(g)f(x) = f(xg) \tag{11}$$

of the Lorentz group, where $f(x)$ may be any function on the cone for which the integral

$$\|f\|^2 = \int |f(x)|^2\, dx \tag{12}$$

converges. The irreducible components of this representation are of class I and belong to the principal series, having weights $\chi = (\frac{1}{2}i\rho, \frac{1}{2}i\rho)$.

The Fourier components of $f(x)$, given by

$$F(x; \rho) = \int_0^{\infty} f(tx) t^{-\frac{1}{2}i\rho}\, dt, \tag{13}$$

are homogeneous functions on the cone, of degree $\frac{1}{2}i\rho - 1$; in other words

$$F(ax; \rho) = a^{\frac{1}{2}i\rho - 1} F(x; \rho). \tag{14}$$

Under the transformation carrying $f(x)$ into $f(xg)$, the function $F(x; \rho)$ transforms according to

$$T_\chi(g)F(x; \rho) = F(xg; \rho), \tag{15}$$

that is, according to a representation equivalent to an irreducible unitary representation of the principal series for which $\chi = (\frac{1}{2}i\rho, \frac{1}{2}i\rho)$. Further, $f(x)$ is given in terms of its Fourier transform by

$$f(x) = (4\pi)^{-1} \int_{-\infty}^{\infty} F(x; \rho)\, d\rho. \tag{16}$$

The analog of Plancherel's theorem

$$\int |f(x)|^2 \, dx = (4\pi)^{-1} \int_{-\infty}^{\infty} \int_{\Omega} |F(\omega; \rho)|^2 \, d\omega \, d\rho \tag{17}$$

holds valid. Here the integral on the right is over the sphere $\Omega$ which is the intersection of the $[x, x] = 0$ cone with the $x_0 = 1$ plane, and $d\omega$ is the Euclidean measure on this sphere.

We note in conclusion that the irreducible representation of Eq. (15) can be restricted to the functions defined on $\Omega$. This is because the homogeneous functions $F(x; \rho)$ are uniquely determined by their values on $\Omega$. Let us obtain an explicit expression for the representation defined in this way. Let $\omega$ be a point on $\Omega$. By $\omega g$ we denote the point on the cone into which $\omega$ is mapped under the motion $g$, and by $\omega_g$ the point on $\Omega$ which lies on the same generator of the cone as does $\omega g$. The homogeneity of $F(x; \rho)$ implies that

$$F(\omega g; \rho) = (\omega_0')^{\frac{1}{2}i\rho - 1} F(\omega_g; \rho),$$

where $\omega_0'$ is the zeroth coordinate of $\omega g$. Consequently we may rewrite Eq. (15) in the form

$$T_\chi(g) F(\omega; \rho) = (\omega_0')^{\frac{1}{2}i\rho - 1} F(\omega_g; \rho).$$

This is the desired explicit expression. Equation (3) of Chapter III, Section 1.1 (which given the linear transformation in $E_4$ in terms of the matrix elements of $g$) can now be used to obtain expressions for the coordinates of $\omega_g$ and the coefficient $(\omega_0')^{\frac{1}{2}i\rho - 1}$. This is left as an exercise for the reader.

## 3. Decomposition of the Representation of the Lorentz Group Associated with Lobachevskian Space

### 3.1. Representation of the Lorentz Group Associated with Lobachevskian Space

Let $X$ be Lobachevskian 3-space; we associate with it a unitary representation of the Lorentz group of the following form. Let $H$ be the space of functions $f(x)$ over $X$ such that the integral

$$\|f\|^2 = \int |f(x)|^2 \, dx \tag{1}$$

converges (where $dx$ is the invariant measure on $X$). We then define the representation $T(g)$ in the usual way by the equation

$$T(g)f(x) = f(xg). \tag{2}$$

It is obvious that $T(g)$ is then a unitary representation of the Lorentz group. Our problem is again to decompose it into its irreducible components.

Remark. This problem can be attacked in many different ways. One of these involves reducing the representation to the regular representation. Recall that the regular representation of $G$ is defined on the functions $f(g)$ on $G$ such that $\int |f(g)|^2 \, dg < +\infty$. Specifically, the representation consists of the translation operators on these functions:

$$T(g_0)f(g) = f(gg_0). \tag{3}$$

Since the stability subgroups of the points $x \in X$ are isomorphic to the subgroup $U$ of unitary matrices, $X$ can be realized as the space of cosets $G/U$ of $G$ with respect to $U$. In this way every $f(x)$ on $X$ can be thought of as some $f(g)$ on $G$ such that $f(g)$ is constant on these cosets. Since $U$ is compact, $f(g)$ is square integrable on the group if $f(x)$ is square integrable on $X$. Consequently the carrier space of the representation $T(g)$ is an invariant subspace of the regular representation. We may now use the decomposition of the regular representation of the Lorentz group into its irreducible components (Chapter IV) to decompose the representation on the functions over $X$. This method is, however, rather inconvenient in that it cannot be extended to imaginary Lobachevskian space, in which the subgroups of stability are no longer compact. Therefore we shall decompose this representation by using the more general method based on horospheres discussed in Section 1.7. #

## 3.2. Decomposition by the Horosphere Method

We now proceed as discussed above to use the method of horospheres to decompose the representation of the Lorentz group associated with Lobachevskian space. Later we shall use the same method for imaginary Lobachevskian space.

With $f(x)$ defined on Lobachevskian space $X$ we associate its integral

$$h(\omega) = \int_{\omega} f(x) \, d\sigma \tag{1}$$

over a horosphere $\omega$, where $d\sigma$ is the invariant measure on $\omega$. In other words, in this way we associate with $f(x)$ a function $h(\omega)$ defined on the set of horospheres. Under the transformation carrying $f(x)$ into $f(xg)$ the function $h(\omega)$ is transformed into a function we shall call $h_g(\omega)$ defined by Eq. (1) with $f(x)$ replaced by $f(xg)$. Since $d\sigma$ is invariant under Lobachevskian motions, we have (see Chapter V, Section 1.7)

$$h_g(\omega) = \int_{\omega} f(xg)\, d\sigma = \int_{\omega g} f(x)\, d\sigma_g = h(\omega g),$$

where $\omega g$ is the image of $\omega$ under the motion $g$, and $d\sigma_g$ is the measure on $\omega g$.

Thus the operators

$$T(g)f(x) = f(xg) \tag{2}$$

on the $f(x)$ functions correspond to the operators

$$Q(g)h(\omega) = h(\omega g) \tag{3}$$

on the $h(\omega)$ functions. Then the problem of decomposing the representation associated with Lobachevskian space reduces essentially to the following two problems.

1. To obtain the inversion formula giving $f(x)$ in terms of $h(\omega)$.

2. To decompose the representation of Eq. (3) into its irreducible components.

The first of these is the problem of integral geometry which we solved in Section 2 of Chapter V. The second has in fact also been solved, as is seen from the following. In Chapter V, Section 2.1 it was established that $h(\omega)$ can be though of as a function $h(\xi)$ defined on the upper sheet of $[\xi, \xi] = 0$ cone. (This requires writing the equation of each horosphere $\omega$ in the form $[x, \xi] = 1$.) The motion $\omega \to \omega g$ then corresponds to the motion $\xi \to \xi g$ on the cone, and therefore the $Q(g)$ can be written as operators on the cone in the form

$$Q(g)h(\xi) = h(\xi g). \tag{4}$$

Consequently the decomposition of (3) reduces to the decomposition of the representations of $G$ on the cone (or, equivalently, on the complex affine plane in which points are identified if they differ only by a common factor of the form $e^{i\varphi}$, where $0 \leqslant \varphi < 2\pi$). This problem was solved in Section 2.

Remark. Note, however, that Eq. (4) defines a representation which is not quite equivalent to the representation associated with the cone (see Section 2.2). This is because the transforms $h(\xi)$ of the $f(x)$ functions satisfy the additional symmetry condition

$$\int h(\xi)\delta([a, \xi] - t)\, d\xi = \int h(\xi)\delta([a, \xi] - t^{-1})\, d\xi \tag{5}$$

established in Section 2.2 of Chapter V. Thus Eq. (4) defines a representation not on all the functions $h(\xi)$ on the cone, but only on those satisfying this condition. #

We now proceed to derive the formulas for the decomposition of the representation associated with Lobachevskian space $X$. First we find the Fourier transform of a function $f(x)$ defined on $X$. For this purpose we write

$$h(\xi) = \int f(x)\delta([x, \xi] - 1)\, dx. \tag{6}$$

As was shown in Section 2.2, the Fourier transform of $h(\xi)$ is given by

$$F(\xi; \rho) = \int_0^\infty h(t\xi)t^{-\frac{1}{2}i\rho}\, dt. \tag{7}$$

Inserting Eq. (6), we obtain

$$F(\xi; \rho) = \int_0^\infty \int f(x)\delta([x, t\xi] - 1)t^{-\frac{1}{2}i\rho}\, dx\, dt. \tag{8}$$

Recall that $F(\xi; \rho)$ is a homogeneous function on the cone. Under the transformation carrying $f(x)$ into $f(xg)$ these $F$ functions transform according to the irreducible representation

$$T_\chi(g)F(\xi; \rho) = F(\xi g; \rho),$$

where $\chi = (\frac{1}{2}i\rho, \frac{1}{2}i\rho)$.

The expression for $F(\xi; \rho)$ can be simplified. For this purpose we change the order of integration and use the obvious relation

$$\int_0^\infty t^{-\frac{1}{2}\rho}\delta([x, t\xi] - 1)\, dt = [x, \xi]^{\frac{1}{2}i\rho-1}.$$

Then Eq. (8) becomes

$$F(\xi; \rho) = \int f(x)[x, \xi]^{\frac{1}{2}i\rho-1}\, dx. \tag{9}$$

There exists another form of the expression for $F(\xi; \rho)$ which is somewhat reminiscent of the Fourier transform in Euclidean space. To obtain it, we recall that $[x, \xi] = e^{k\tau(x,\xi)}$ where $\tau(x, \xi)$ is the distance from $x$ to the horosphere whose equation is $[x, \xi] = 1$. Thus Eq. (9) can be written

$$F(\xi; \rho) = \int f(x) \exp[(\tfrac{1}{2}i\rho - 1)k\tau(x, \xi)]\, dx. \tag{10}$$

Remark. Fourier transforms in Euclidean space, usually written in the form

$$F(\xi) = \int f(x) e^{i(x,\xi)}\, dx, \tag{11}$$

can be put in analogous form. Indeed, we have

$$(x, \xi) = x_1\xi_1 + \cdots + x_n\xi_n = |\,\xi\,|\,\tau(x, \xi),$$

where $|\,\xi\,| = (\xi_1^2 + \cdots + \xi_n^2)^{\frac{1}{2}}$, and $\tau(x, \xi)$ is the distance from $x$ to the plane whose equation is $(x, \xi) = 0$. Thus this Fourier transform can be written

$$F(\xi) = \int f(x) e^{i|\xi|\tau(x,\xi)}\, dx. \tag{12}$$

The similarity of Eqs. (10) and (12) is obvious. #

Let us now find the symmetry conditions imposed on $F(\xi; \rho)$ by Eq. (5). Note that Eq. (5) can be written in the form

$$\int h(t\xi)\delta([a, \xi] - 1)\, d\xi = t^{-2} \int h(t^{-1}\xi)\delta([a, \xi] - 1)\, d\xi. \tag{13}$$

By multiplying both sides of this equation by $t^{-\frac{1}{2}i\rho}$, integrating over $t$ from 0 to $\infty$, and using Eq. (8), we arrive at

$$\int F(\xi; \rho)\delta([a, \xi] - 1)\, d\xi = \int F(\xi; -\rho)\delta([a, \xi] - 1)\, d\xi. \tag{14}$$

This is the symmetry condition that must be satisfied by $F(\xi; \rho)$; in it $a$ is an arbitrary point in $X$. The existence of this condition is related to the equivalence of the representations $T_\chi(g)$ and $T_{-\chi}(g)$, where $-\chi = (-\frac{1}{2}i\rho, -\frac{1}{2}i\rho)$.

Proceeding, we now express $f(x)$ in terms of $F$. Recall first that in Section 2.2 of the present chapter we found that

$$h(\xi) = (4\pi)^{-1} \int_{-\infty}^{\infty} F(\xi; \rho)\, d\rho, \tag{15}$$

and in Section 2.2 of Chapter V we found that

$$f(x) = -2(4\pi)^{-2} \int h(\xi)\delta''([x, \xi] - 1)\, d\xi. \tag{16}$$

By combining these two expressions we arrive at

$$f(x) = -2(4\pi)^{-3} \int_{-\infty}^{\infty} \int F(\xi; \rho)\delta''([x, \xi] - 1)\, d\xi\, d\rho, \tag{17}$$

which is the desired result. This equation can, however, be simplified. For this purpose we write it in the form

$$f(x) = -2(4\pi)^{-3} \int \int_{-\infty}^{\infty} \int_{-\infty}^{\infty} F(\xi; \rho)\delta(t - 1)\delta''([x, \xi] - t)\, d\rho\, dt\, d\xi,$$

which we integrate by parts with respect to $t$, then writing $\xi = t\eta$. Now recall that $f$ is homogeneous in $\xi$ of degree $\frac{1}{2}i\rho - 1$ and that $\delta(x)$ is homogeneous of degree $-1$; then after some simple operations we arrive at

$$f(x) = (4\pi)^{-3} \int_{0}^{\infty} \rho^2\, d\rho \int F(\xi; \rho)\delta([x, \xi] - 1)\, d\xi \tag{18}$$

[derived with the aid of Eq. (14)].

Now the integral over $\xi$ in (18) is actually taken over the intersection of the $[\xi, \xi] = 0$ cone with the plane whose equation is $[x, \xi] = 1$. We assert that it can also be written as the integral over any surface $\Gamma$ on the $[\xi, \xi] = 0$ cone which intersects each of the generators exactly once. More specifically,

$$\int F(\xi; \rho)\delta([x, \xi] - 1)\, d\xi = \int_{\Gamma} F(\xi; \rho)[x, \xi]^{-\frac{1}{2}i\rho - 1}\, d\omega. \tag{19}$$

[Here $d\omega$ is defined by $d\xi = dP d\omega$, where $P(\xi) = 1$ is the equation of $\Gamma$.] This is because the integrand on the right-hand side of (19) is homogeneous of degree $-2$, so that the integral does not depend on the choice of $\Gamma$ (see the discussion in Section 3.4 of Chapter V). If, in particular, we choose $\Gamma$ to be the intersection of the cone and the plane whose equation is $[x, \xi] = 1$, we obtain exactly the left-hand side of (19). Thus Eq. (18) may be written

$$f(x) = (4\pi)^{-3} \int_{0}^{\infty} \rho^2\, d\rho \int_{\Gamma} F(\xi; \rho)[x, \xi]^{-\frac{1}{2}i\rho - 1}\, d\omega, \tag{20}$$

where $\Gamma$ is any "contour" on the $[\xi, \xi] = 0$ cone. When $\Gamma$ is chosen to be the sphere $\Omega$ defined by $[\xi, \xi] = 0$, $\xi_0 = 1$, the measure $d\omega$ becomes the Euclidean measure on $\Omega$. To summarize the results of this section, we have found that the Fourier transform of a function $f(x)$ defined on Lobachevskian space $X$ is given by Eq. (9), and that $f(x)$ can be expressed in terms of its Fourier transform by Eq. (20).

### 3.3. The Analog of Plancherel's Theorem for Lobachevskian Space

We have thus derived Fourier transform formulas for functions on Lobachevskian space. If we wish further to show that these formulas will yield a decomposition of the representation of the Lorentz group into its irreducible components we must prove the analog of Plancherel's theorem. We shall do this here in a relatively nonrigorous way, without justifying change in the order of integration.

Consider the integral

$$\int |f(x)|^2 \, dx = \int f(x)\bar{f}(x) \, dx \tag{1}$$

and replace $\bar{f}(x)$ by its Fourier integral expansion for Lobachevskian space, namely by using Eq. (20) of Section 3.2. After inverting the order of integration we arrive at

$$\int |f(x)|^2 \, dx = (4\pi)^{-3} \int_0^\infty \rho^2 \int_\Gamma \bar{F}(\xi; \rho) \int [x, \xi]^{\frac{1}{2}i\rho-1} f(x) \, dx \, d\omega \, d\rho.$$

Now the desired analog of Plancherel's theorem is obtained simply by using Eq. (9) of Section 3.2. The result is

$$\int |f(x)|^2 \, dx = (4\pi)^{-3} \int_0^\infty \rho^2 \, d\rho \int_\Gamma |F(\xi; \rho)|^2 \, d\omega, \tag{2}$$

where $\Gamma$ is any "contour" on the $[\xi, \xi] = 0$ cone, and $d\omega$ is the corresponding differential form.

This equation can also be written in the form

$$\int |f(x)|^2 \, dx = \int \| F \|_\rho^2 \, d\mu, \tag{3}$$

where

$$\| F \|_\rho^2 = \int_\Gamma |F(\xi; \rho)|^2 \, d\omega \tag{4}$$

and $d\mu = (4\pi)^{-3}\rho^2\, d\rho$. In other words the mapping $f(x) \to F(\xi;\rho)$ is isometric with respect to the norms

$$\|f\|^2 = \int |f(x)|^2\, dx \tag{5}$$

and

$$\|F\|^2 = \int \|F\|_\rho^2\, d\mu. \tag{6}$$

This in turn implies that this mapping decomposes $H$ (the space of functions on Lobachevskian space with finite norm $\|f\|$) into a direct integral of spaces $H_\rho$ each of which consists of functions $F(\xi;\rho)$ for which the norm $\|F\|_\rho$ is finite. The transformation $T(g)$ on $f(x)$, that is, $T(g)f(x) = f(xg)$, then induces under this mapping the transformation of $F(\xi;\rho)$ by the irreducible unitary representation of the Lorentz group $T_\chi(g)$, with $\chi = (\frac{1}{2}i\rho, \frac{1}{2}i\rho)$.

One may ask the following further question. What are all the $F(\xi;\rho)$ functions which correspond to square integrable $f(x)$. It can be shown that these will include functions with all possible *positive* real values of $\rho$ for which the integral of Eq. (6) converges. If, however, one considers all $F(\xi;\rho)$ functions defined by Eq. (15) of Section 3.2 for *all* real values of $\rho$, the square integrable $f(x)$ functions will correspond to all those for which the integral

$$\int_{-\infty}^{\infty} \rho^2\, d\rho \int_\Gamma |F(\xi;\rho)|^2\, d\omega$$

converges and such that Eq. (14) of Section 3.2 is satisfied.

We wish now to state the result of Sections 3.1–3.3.

*The representation*

$$T(g)f(x) = f(xg)$$

*of the Lorentz group on functions defined on Lobachevskian space can be decomposed into irreducible representations of the principal series, of class I, that is, into representations whose weight is* $\chi = (\frac{1}{2}i\rho, \frac{1}{2}i\rho)$, $0 \leqslant \rho < \infty$. *Each of these representations enters with multiplicity one.*

*The Fourier integral representation of a square integrable function* $f(x)$ *is*

$$f(x) = (4\pi)^{-3}\int_0^{\infty} \rho^2\, d\rho \int_\Gamma F(\xi;\rho)[x, \xi]^{-\frac{1}{2}i\rho-1}\, d\omega, \tag{7}$$

*where* $\Gamma$ *is any contour on the* $[\xi, \xi] = 0$ *cone and* $d\omega$ *is the differential form defined by* $d\xi = dP\, d\omega$, *where* $P(\xi) = 1$ *is the equation of* $\Gamma$. *Here*

$$F(\xi;\rho) = \int f(x)[x, \xi]^{\frac{1}{2}i\rho-1}\, dx. \tag{8}$$

*Further, the analog of Plancherel's theorem*

$$\int |f(x)|^2\, dx = (4\pi)^{-3} \int_0^{\infty} \rho^2\, d\rho \int_{\Gamma} |F(\xi; \rho)|^2\, d\omega \tag{9}$$

*is valid. The mapping* $f(x) \to F(\xi; \rho)$ *defined by Eq.* (8) *is an isometric mapping of all* $f(x)$ *such that*

$$\|f\|^2 \equiv \int |f(x)|^2\, dx < \infty$$

*onto the space of all* $F(\xi; \rho)$, $0 \leqslant \rho < \infty$, *such that*

$$\|F\|^2 \equiv (4\pi)^{-3} \int_0^{\infty} \rho^2\, d\rho \int_{\Gamma} |F(\xi; \rho)|^2\, d\omega < \infty.$$

*Treated as a function on the entire real* $\rho$ *line,* $F(\xi; \rho)$ *satisfies the symmetry relation*

$$\int F(\xi; \rho)\delta([a, \xi] - 1)\, d\xi = \int F(\xi; -\rho)\delta([a, \xi] - 1)\, d\xi. \tag{10}$$

## 4. Decomposition of the Representation of the Lorentz Group Associated with Imaginary Lobachevskian Space

### 4.1. Representation of the Lorentz Group Associated with Imaginary Lobachevskian Space

The representation of the Lorentz group associated with imaginary Lobachevskian 3-space can be constructed in exactly the same way as that associated with real Lobachevskian space. Consider the Hilbert space $H$ of function $f(x)$ on imaginary Lobachevskian space $X$ such that

$$\|f\|^2 = \int |f(x)|^2\, dx < +\infty \tag{1}$$

(here $dx$ is the invariant measure on $X$). As before, we define the representation $T(g)$ by the translation equation

$$T(g)f(x) = f(xg). \tag{2}$$

The invariance of $dx$ implies that $T(g)$ is a unitary representation of the Lorentz group.

This representation cannot be decomposed into its irreducible components by the method described in Section 3.1, that is, by going over to the regular representation of the group. This is because the stability subgroup of a point in imaginary Lobachevskian space is noncompact, and therefore square integrable functions on $X$ correspond to functions on $G$ that are not square integrable. The integral transform method involving horospheres, however, can be extended to the present case without any essential changes.

Imaginary Lobachevskian space differs, however, from ordinary Lobachevskian space in that there are two homogeneous manifolds of horospheres. One of these, as we saw in Section 1.5 of Chapter V, consists of the horospheres of the first kind, which are of dimension two, and the other consists of isotropic lines. Correspondingly, Eq. (2) is associated with two representations on horospheres, first on those of the first kind and second on the isotropic lines.

## 4.2. Decomposition of the Representation Associated with Horospheres of the First Kind

In accordance with the general method described in Section 1.3, with every infinitely differentiable function $f(x)$ with bounded support on imaginary Lobachevskian space $X$ we associate the function $h(\omega)$ on the set of horospheres of the first kind which is defined by

$$h(\omega) = \int_{\omega} f(x)\, d\sigma, \tag{1}$$

where $d\sigma$ is the invariant measure on the horosphere $\omega$.

The operator $Q(g)$ induced on the $h(\omega)$ functions by the translation operator

$$T(g)f(x) = f(xg) \tag{2}$$

of $f(x)$ can be found as follows. By definition $f(xg)$ corresponds to the function

$$h_g(\omega) = \int_{\omega} f(xg)\, d\sigma.$$

Since measures on horospheres of the first kind remain invariant under the motion $x \to xg$ (see Chapter V, Section 1.7), we have

$$h_g(\omega) = \int_{\omega g} f(x)\, d\sigma = h(\omega g).$$

Thus the desired operator $Q(g)$ is given by

$$Q(g)h(\omega) = h(\omega g). \tag{3}$$

Obviously the $Q(g)$ operators form a representation of the Lorentz group.

Having associated with $T(g)$ a representation $Q(g)$ on functions defined on the set of horospheres of the first kind, we shall decompose $Q(g)$ into its irreducible components. This decomposition, as we shall now show, reduces to the decomposition of the representation of the Lorentz group associated with the cone, a problem we solved in Section 2.2. Indeed, consider the model on the $[x, x] = -1$ hyperboloid. As was shown in Chapter V, Section 1.5, in this model the horospheres of the first kind are given by equations of the form $|[x, \xi]| = 1$, where $\xi$ is a point on the upper sheet of the $[\xi. \xi] = 0$ cone. Thus $h(\omega)$ can be treated as a function on this cone, and we shall write $h(\xi)$ instead of $h(\omega)$. Now it is obvious that the equation of the transformed horosphere $\omega g$ is $|[x, \xi g]| = 1$, so that $Q(g)$ may be defined by

$$Q(g)h(\xi) = h(\xi g). \tag{3'}$$

This establishes the assertion that the decomposition of the representation associated with horospheres of the first kind reduces to the decomposition of representations associated with the cone.

Recall that the Fourier components of functions on the cone are given by

$$F(\xi; \rho) = \int_0^\infty h(t\xi)t^{-\frac{1}{2}i\rho}\, dt. \tag{4}$$

These functions are homogeneous in $\xi$ of degree $\frac{1}{2}i\rho - 1$, and when $h(\xi)$ is transformed into $h(\xi g)$ they transform according to irreducible representations of the Lorentz group equivalent to $T_\chi(g)$ of the principal series with $\chi = (\frac{1}{2}i\rho, \frac{1}{2}i\rho)$. The $h(\xi)$ functions, in turn, are given in terms of $F(\xi; \rho)$ by the integral

$$h(\xi) = (4\pi)^{-1} \int_{-\infty}^\infty F(\xi; \rho)\, d\rho. \tag{5}$$

Thus to decompose the representation of Eq. (3) we write it in terms of functions $h(\xi)$ defined on the cone, where $\xi$ is the point on the cone corresponding to the horosphere $\omega$. Then we express $h(\xi)$ in terms of its Fourier component $F(\xi; \rho)$ in accordance with Eq. (4). Conversely, $h(\xi)$ [and thus also $h(\omega)$] is given in terms of $F(\xi; \rho)$ by Eq. (5). Then in decomposing the representations of the Lorentz group associated with horospheres of the first kind in imaginary Lobachevskian space we obtain the representations $T_\chi(g)$ with $\chi = (\frac{1}{2}i\rho, \frac{1}{2}i\rho)$, namely representations of class I.

Remark. In general the integral of Eq. (4) diverges, and it must be understood in the sense of its regularization, which is defined as follows. Write $t = e^\tau$ in Eq. (4) so that it becomes

$$F(\xi; \rho) = \int_{-\infty}^{\infty} e^\tau h(e^\tau \xi) e^{-\frac{1}{2}i\rho\tau}\, d\tau.$$

It was shown in Section 3.1 of Chapter V that $h(\xi)$, if it is the transform of an infinitely differentiable function of bounded support on imaginary Lobachevskian space, vanishes in a neighborhood of the vertex of the cone and possesses the asymptotic expansion

$$h(t\xi) \sim \sum_{n=1}^{\infty} a_n(\xi) t^{-n}$$

as $t \to +\infty$. Consequently $e^\tau h(e^\tau \xi)$ vanishes for all $\tau$ less than some positive number $N$ and has a finite limit $a_1(\xi)$ as $\tau \to +\infty$. This means that

$$F(\xi; \rho) = \int_{-\infty}^{\infty} [e^\tau h(e^\tau \xi) - \theta(\tau) a_1(\xi)] e^{-\frac{1}{2}i\rho\tau}\, d\tau + a_1(\xi) \int_0^{\infty} e^{-\frac{1}{2}i\rho\tau}\, d\tau,$$

where $\theta(\tau) = 1$ for $\tau > 0$, and $\theta(\tau) = 0$ for $\tau < 0$. The first term in this equation converges, and the second is given by

$$\int_0^{\infty} e^{-\frac{1}{2}i\rho\tau}\, d\tau = -2i\rho^{-1} + 2\pi\delta(\rho)$$

(see Volume 1, page 360, Entry 23 in the table). This defines the Fourier component $F(\xi; \rho)$ of any $h(\xi)$ that is the transform of an infinitely differentiable $f(x)$ of bounded support on an imaginary Lobachevskian space. With this definition of $F(\xi; \rho)$, Eq. (5) remains valid although $h(\xi)$ may not be of bounded support. #

### 4.3. Decomposition of the Representation Associated with Isotropic Lines

We have now constructed the representation of the Lorentz group associated with horospheres of the first kind and have decomposed it into its irreducible components. Although in the case of real Lobachevskian space this was all that was needed, it will no longer suffice, since to find $f(x)$ we must know not only its integrals over horospheres of the first kind but also along isotropic lines (see Section 3.4 of Chapter V).

We must thus study in addition to the representation $Q(g)h(\omega) = h(\omega g)$ the representation associated with isotropic lines.

Let us therefore construct the representation associated with the isotropic lines of an imaginary Lobachevskian space. Let $f(x)$ be an infinitely differentiable function of bounded support on imaginary Lobachevskian space. With it we associate the function $\varphi(l)$ defined on the set of isotropic lines by the integral of $f(x)$ along each line $l$:

$$\varphi(l) = \int_l f(x)\, dl. \tag{1}$$

Recall that the integral along $l$ is defined by

$$\int_l f(x)\, dl = \int_{-\infty}^{\infty} f(b + t\xi)\, dt, \tag{2}$$

where $b$ is some point on $l$ and $\xi$ is the normalized direction vector of the line (normalized, that is, according to $\xi_0 = 1$).[10]

Let us start by studying the transformation of $\varphi(l)$ induced by the transformation $f(x) \to f(xg)$ on the original function. Let $\varphi_g(l)$ denote the function obtained by integrating $f(xg)$, namely

$$\varphi_g(l) = \int_l f(xg)\, dl = \int_{-\infty}^{\infty} f(bg + t\xi g)\, dt.$$

Obviously the set of points $bg + t\xi g$, $-\infty < t < \infty$, is the isotropic line $lg$ into which $l$ is carried under the motion $g$. If $\xi_g$ is the normalized direction vector of $lg$, we may write

$$\int_{-\infty}^{\infty} f(bg + t\xi g)\, dt = \beta^{-1}(l, g) \int_{-\infty}^{\infty} f(bg + t\xi_g)\, dt = \beta^{-1}(l, g)\varphi(lg),$$

where $\beta(l, g)$ is the ratio of the components of $\xi g$ and $\xi_g$. This shows that the translation operator $T(g)$ defined by $T(g)f(x) = f(xg)$ corresponds to the operator $R(g)$ defined by

$$R(g)\varphi(l) = \beta^{-1}(l, g)\varphi(lg) \tag{3}$$

on the $\varphi(l)$ functions on the isotropic lines. Recall that the direction vectors of the isotropic lines are normalized so that their zeroth coordinates are equal to one; hence $\beta(l, g)$ is equal to the zeroth coordinate of $\xi g$. Note also that in this normalization $\beta(l, g)$ has the same value for any two parallel isotropic lines.[11] Finally, it is obvious that the $R(g)$

[10] Unless stated otherwise, we shall assume that our imaginary Lobachevskian space is realized on the hyperboloid whose equation is $[x, x] = -1$.

[11] Parallel isotropic lines are discussed in Section 3.4a of Chapter V.

operators also form a representation of the Lorentz group. We shall say that this is the representation associated with isotropic lines in the imaginary Lobachevskian space.

Let us now decompose this representation into its irreducible components. To do this we first obtain the Fourier component of $\varphi(l)$. Consider the isotropic line $l$ and the set of all isotropic lines parallel to it. As was shown in Section 3.4a of Chapter V, when the imaginary Lobachevskian space is realized as the exterior of the unit sphere, these line are all the lines tangent to this sphere at the same point as $l$. Further, the distance $r$ from $l$ to any other line $l_1$ of this set is $i\theta/k$, where $\theta$ is the oriented angle between the two lines. This means that each line of our set is uniquely determined by a number $\theta$ and that $\theta$ and $\theta + \pi$ denote the same line. Thus $\varphi(l_1)$ is a function of $l$ and $\theta$, and we write $\varphi(l_1) \equiv \varphi(l, \theta)$ and bear in mind that $\varphi(l, \theta + \pi) = \varphi(l, \theta)$.

Now to expand the representation of Eq. (3) into its irreducible components we obtain the Fourier series in $\theta$ for $\varphi(l, \theta)$. Since this function has period $\pi$ in $\theta$, the series may be written in the form

$$\varphi(l, \theta) = \pi^{-1} \sum_{n=-\infty}^{\infty} F(l; 2n) e^{2in\theta}, \tag{4}$$

where

$$F(l; 2n) = \int_0^{\pi} \varphi(l, \theta) e^{-2in\theta}\, d\theta. \tag{5}$$

We shall call the $F(l; 2n)$ the Fourier components of $\varphi(l)$ (a terminology which will be justified below). By writing $\theta = 0$ in (4) and using the fact that $\varphi(l, 0) \equiv \varphi(l)$, we arrive at

$$\varphi(l) = \pi^{-1} \sum_{n=-\infty}^{\infty} F(l; 2n), \tag{6}$$

which is the analog of Eq. (5) of Section 2.2, and gives $\varphi(l)$ in terms of its Fourier components $F(l; 2n)$.

We must now show that these functions are indeed the Fourier components of $\varphi(l)$, that is, that when $\varphi(l)$ is transformed according to Eq. (3) the $F(l; 2n)$ transform according to irreducible representations of the Lorentz group.

Note first that the $F(l; 2n)$ have the following property: If the distance between two parallel isotropic lines $l_1$ and $l_2$ is $r = i\theta/k$, then

$$F(l_2; 2n) = e^{2in\theta} F(l_1; 2n). \tag{7}$$

This follows directly from Eq. (5) and the periodicity of $\varphi(l, \theta)$ in $\theta$. Proceeding, we wish to find the form of the operators acting on the $F(l; 2n)$ when $\varphi(l)$ is transformed according to Eq. (3). These operators will be given by

$$\begin{aligned} R_{2n}(g)F(l; 2n) &= \int_0^{\pi} R(g)\varphi(l, \theta)e^{-2in\theta}\, d\theta \\ &= \int_0^{\pi} \beta^{-1}(l, \theta; g)\varphi(lg, \theta)e^{-2in\theta}\, d\theta, \end{aligned} \tag{8}$$

where $\beta^{-1}(l, \theta; g) \equiv \beta^{-1}(l_1, g)$, where $l_1$ is the isotropic line parallel to $l$ and separated from it by the distance $r = i\theta/k$. Now the motion $g$ maps lines into lines parallel to themselves, so that $\beta^{-1}(l, g)$ has the same value for all lines parallel to each other, and therefore Eq. (8) can be written

$$\begin{aligned} R_{2n}(g)F(l; 2n) &= \beta^{-1}(l, g) \int_0^{\pi} \varphi(lg, \theta)e^{-2in\theta}\, d\theta \\ &= \beta^{-1}(l, g)F(lg; 2n). \end{aligned} \tag{9}$$

We thus find that when $\varphi(l)$ is transformed according to Eq. (3), its Fourier components transform according to Eq. (9), or simply

$$R_{2n}(g)F(l; 2n) = \beta^{-1}(l, g)F(lg; 2n). \tag{10}$$

These operators $R_{2n}(g)$ form a representation of the Lorentz group on the functions over the set of isotropic lines. We assert that this representation is equivalent to the irreducible unitary representation $T_{\chi}(g)$ with $\chi = (2n, -2n)$.

**Proof.** The $T_{\chi}(g)$ representations operate on spaces of homogeneous functions $f(z_1, z_2)$ of the complex variables $z_1, z_2$ (Chapter III, Section 2.4.). In order, therefore, to establish the equivalence of $T_{\chi}(g)$ and $R_{2n}(g)$ we must first find the relation between isotropic lines and pairs of complex numbers. For this purpose we associate with each pair of complex numbers $(z_1, z_2) \neq (0, 0)$ an isotropic line $l$ in the following way. Each such pair can be associated with the unitary matrix

$$U(z_1, z_2) = \begin{Vmatrix} u & v \\ -\bar{v} & \bar{u} \end{Vmatrix},$$

where $u = cz_1$, $v = cz_2$, and $c = (|z_1|^2 + |z_2|^2)^{-\frac{1}{2}}$. Moreover, we saw Section 1.5 that to each complex unimodular matrix in two dimensions, and in particular to such a unitary matrix, there corresponds a motion of the imaginary Lobachevskian space. Now consider the fixed isotropic line $l_0$ given by $x = b^0 + t\xi^0$, where $b^0 = (0, 0, 1, 0)$ and $\xi^0 = (1, 0, 0, -1)$. Then with each pair $(z_1, z_2)$ we associate the isotropic line $l \equiv l(z_1, z_2)$ obtained from $l_0$ by the motion corresponding to U. It is easily shows that in this way we obtain all isotropic lines (because the motions corresponding to unitary matrices form a transitive group on the isotropic lines).

We now wish to express in terms of $z_1$ and $z_2$ the components of the direction vector $\xi$ and the coordinates of the fixed point $b$ on the isotropic line $l(z_1, z_2)$ whose equation we write $x = b + t\xi$. This may be done by seeing what happens to $\xi^0$ and $b^0$ under the motion corresponding to $U(z_1, z_2)$. Calculation yields

$$\xi_1 = i(\bar{u}v - u\bar{v}), \qquad \xi_2 = \bar{u}v + u\bar{v}, \qquad \xi_3 = |v|^2 - |u|^2,$$

$$b_1 = \frac{i}{2}(\bar{u}^2 + \bar{v}^2 - u^2 - v^2), \qquad b_2 = \tfrac{1}{2}(u^2 + v^2 + \bar{u}^2 + \bar{v}^2),$$

$$b_3 = uv + \bar{u}\bar{v}.$$

These equations are explicit expressions for the isotropic line in terms of the pair of numbers $(z_1, z_2) \neq (0, 0)$ with which it is associated. This association has the following properties.

*Property 1.* $l(z_1, z_2) = l(z_1', z_2')$ if and only if $z_1' = \lambda z_1$ and $z_2' = \lambda z_2$, where $\lambda \neq 0$ is a real or pure imaginary number.

*Property 2.* $l(z_1, z_2)$ is parallel to $l(e^{\frac{1}{2}i\theta} z_1, e^{\frac{1}{2}i\theta} z_2)$. The distance between these two lines is $i\theta/k$.

*Property 3.* The transformation

$$(z_1, z_2) \rightarrow (\alpha z_1 + \gamma z_2, \beta z_1 + \delta z_2) \tag{11}$$

in the complex affine plane induces the transformation $l \rightarrow lg$ on the set of isotropic lines where $g$ is the motion associated with

$$\begin{Vmatrix} \alpha & \beta \\ \gamma & \delta \end{Vmatrix}, \qquad \alpha\delta - \beta\gamma = 1.$$

The proof of these assertions is left as an exercise for the reader.

From all of this it is evident that functions over the set of isotropic lines can be thought of as functions $f(z_1, z_2)$ over the complex affine plane such that

$$f(\lambda z_1, \lambda z_2) = f(z_1, z_2) \tag{12}$$

for any real or imaginary number $\lambda \neq 0$. Let us see in particular what the properties are of the functions of $z_1$ and $z_2$ that correspond to the functions $F(l;\, 2n)$. These functions satisfy Eq. (7) for parallel lines $l_1$ and $l_2$. Then from Property 2 it follows that if $f(z_1, z_2)$ is the function on the complex affine plane corresponding to $F(l;\, 2n)$, then

$$f(e^{\frac{1}{2}i\theta}z_1\,,\, e^{\frac{1}{2}i\theta}z_2) = e^{2in\theta}f(z_1\,,\, z_2). \tag{13}$$

Equations (12) and (13) imply that for any complex number $\alpha \neq 0$,

$$f(\alpha z_1\,,\, \alpha z_2) = \alpha^{2n}\bar{\alpha}^{-2n}f(z_1\,,\, z_2). \tag{14}$$

In other words $f(z_1\,,\, z_2)$ is a homogeneous function of degree $(2n,\, -2n)$.

Now let us write Eq. (10) in terms of $f(z_1\,,\, z_2)$. It is clear from Property 3 that the function

$$f(\alpha z_1 + \gamma z_2\,,\, \beta z_1 + \delta z_2) \tag{15}$$

corresponds to $F(lg;\, 2n)$, so that we need only find an expression for $\beta(l, g)$ in terms of $z_1$, $z_2$, and the matrix elements of $g$. For this purpose we recall that $\beta(l, g)$ is equal to the zeroth component of $\xi g$ where $\xi$ is the direction vector of $l$ normalized according to $\xi_0 = 1$. Now we have already obtained an expression for the components of $\xi$ in terms of $z_1$ and $z_2$, so that it is a simple matter to obtain the result

$$\beta(l, g) = \frac{|\,\alpha z_1 + \gamma z_2\,|^2 + |\,\beta z_1 + \delta z_2\,|^2}{|\,z_1\,|^2 + |\,z_2\,|^2}\,. \tag{16}$$

Consequently Eqs. (15) and (16) show that the $R_{2n}(g)$ operators act on the $f(z_1\,,\, z_2)$ functions [which have the homogeneity indicated in (14)] according to

$$R_{2n}(g)f(z_1\,,\, z_2) = \frac{|\,z_1\,|^2 + |\,z_2\,|^2}{|\,\alpha z_1 + \gamma z_2\,|^2 + |\,\beta z_1 + \delta z_2\,|^2} \times f(\alpha z_1 + \gamma z_2\,,\, \beta z_1 + \delta z_2). \tag{17}$$

We wish to transform this to the usual form

$$T_\chi(g)f(z_1\,,\, z_2) = f(\alpha z_1 + \gamma z_2\,,\, \beta z_1 + \delta z_2)$$

for a representation of the Lorentz group. For this all we need do is consider functions of the form

$$f_1(z_1\,,\, z_2) = \frac{f(z_1\,,\, z_2)}{|\,z_1\,|^2 + |\,z_2\,|^2}\,.$$

Then according to (14) these functions have the property that

$$f_1(\alpha z_1\,,\, \alpha z_2) = \alpha^{2n-1}\bar{\alpha}^{-2n-1}f_1(z_1\,,\, z_2), \tag{18}$$

that is, that they belong to $D_\chi$ with $\chi = (2n, -2n)$ (Chapter III, Section 2.2). Further, the operator $R_{2n}(g)$ defined in Eq. (17) corresponds to the operator

$$T_\chi(g) f_1(z_1, z_2) = f_1(\alpha z_1 + \gamma z_2, \beta z_1 + \delta z_2)$$

on the functions of $D_\chi$. This completes the proof that the representation of Eq. (10) is equivalent to $T_\chi(g)$ with $\chi = (2n, -2n)$. In particular, it shows that $R_{2n}(g)$ is an irreducible representation.

To summarize, we have proven the following result.

The representation

$$R(g)\varphi(l) = \beta^{-1}(l, g)\varphi(lg) \tag{19}$$

acting on functions over the isotropic lines [this representation obtained from the representation $T(g)f(x) = f(xg)$ acting on the functions over imaginary Lobachevskian space] is decomposed into its irreducible components in the following way.

The Fourier components of $\varphi(l)$ are given by

$$F(l; 2n) = \int_0^\pi \varphi(l, \theta) e^{-2in\theta}\, d\theta, \tag{20}$$

where $\varphi(l, \theta)$ is the value of $\varphi$ on the isotropic line parallel to and separated by the distance $r = i\theta/k$ from $l$. These Fourier components have the property that

$$F(l_2; 2n) = e^{2in\theta} F(l_1, 2n), \tag{21}$$

where $l_1$ and $l_2$ are parallel isotropic lines separated by the distance $i\theta/k$. When $\varphi(l)$ transforms according to (19), its Fourier components transform according to

$$R_{2n}(g) F(l; 2n) = \beta^{-1}(l, g) F(lg; 2n). \tag{22}$$

The representation $R_{2n}(g)$ of the Lorentz group defined by this equation is equivalent to the irreducible unitary representation $T_\chi(g)$ for which $\chi = (2n, -2n)$.

### 4.4. Decomposition of the Representation Associated with Imaginary Lobachevskian Space

Let us now go on to solve the problem of this section, namely to decompose into its irreducible components the representation

$$T(g) f(x) = f(xg) \tag{1}$$

of the Lorentz group associated with imaginary Lobachevskian space. With this representation we have associated the two representations

$$Q(g)h(\xi) = h(\xi g) \tag{2}$$

and

$$R(g)\varphi(l) = \beta^{-1}(l, g)\varphi(lg) \tag{3}$$

on horospheres of the first kind and on isotropic lines, respectively, and we have decomposed these into their irreducible components.

In order to do the same for the representation of Eq. (1) we express $f(x)$ in terms of the Fourier components of $h(\xi)$ and $\varphi(l)$. Let us start with $h(\xi)$. It was shown in Section 4.2 that the Fourier components of $h(\xi)$ are given by

$$F(\xi; \rho) = \int_0^\infty h(t\xi)t^{-\frac{1}{2}i\rho}\, dt. \tag{4}$$

Further, $h(\xi)$ is the integral of $f(x)$ over the horosphere $\omega$ whose equation is $|[x, \xi]| = 1$, and is given in terms of $f(x)$ by

$$h(\xi) = \int f(x)\delta(|[x, \xi]| - 1)\, dx, \tag{5}$$

where $dx$ is the invariant measure on the imaginary Lobachevskian space (see Chapter V, Section 3.1). By inserting (5) into (4) we arrive at

$$F(\xi; \rho) = \int_0^\infty t^{-\frac{1}{2}i\rho}\, dt \int f(x)\delta(|[x, t\xi]| - 1)\, dx. \tag{6}$$

This equation can be simplified by interchanging the order of integration. From the fact that

$$\int_0^\infty t^{-\frac{1}{2}i\rho}\delta(|[x, t\xi]| - 1)\, dt = |[x, \xi]|^{\frac{1}{2}i\rho - 1},$$

we arrive at

$$F(\xi; \rho) = \int f(x)|[x, \xi]|^{\frac{1}{2}i\rho-1}\, dx, \tag{7}$$

which is the analog of Eq. (6) of Section 3.2. As in the case of the ordinary Lobachevskian space, this can also be written in the form

$$F(\xi; \rho) = \int f(x)e^{(\frac{1}{2}i\rho-1)k\tau(x,\xi)}\, dx, \tag{8}$$

where $\tau(x, \xi)$ is the distance from $x$ to $\omega$.

We now express the Fourier components $F(l; 2n)$ for the representation associated with isotropic lines in terms of $f(x)$. As was shown in Section 4.3, these Fourier components can be written in terms of $\varphi(l)$ in the form

$$F(l; 2n) = \int_0^\pi \varphi(l, \theta) e^{-2in\theta} \, d\theta, \tag{9}$$

where $\varphi(l, \theta)$ is the value of $\varphi$ on the isotropic line $l_\theta$ parallel to and separated by a distance $r = i\theta/k$ from $l$. We have in addition

$$\varphi(l) \equiv \varphi(\xi, b) = \int_{-\infty}^{\infty} f(b + t\xi) \, dt, \tag{10}$$

where $\xi$ is the normalized direction vector of $l$, and $b$ is a point on $l$. Inserting (10) into (9), we obtain

$$F(l; 2n) = \int_0^\pi e^{-2in\theta} \, d\theta \int_{-\infty}^{\infty} f(b_\theta + t\xi_\theta) \, dt, \tag{11}$$

where $\xi_\theta$ and $b_\theta$ are the normalized direction vector and a point of $l_\theta$. Since the normalized direction vectors of parallel isotropic lines are the same, we may write

$$F(l; 2n) = \int_0^\pi e^{-2in\theta} \, d\theta \int_{-\infty}^{\infty} f(b_\theta + t\xi) \, dt, \tag{11$'$}$$

which is the expression for $F(l; 2n)$ in terms of $f(x)$. This formula can also be written as an integral over the entire $[x, x] = -1$ hyperboloid. To do this we note that when $\theta$ varies from 0 to $\pi$ and $t$ from $-\infty$ to $\infty$, the point $b_\theta + t\xi$ runs through the set of all points of the imaginary Lobachevskian space that lie on isotropic lines parallel to $l$. It is easily verified that this set is the intersection of the $[x, x] = -1$ hyperboloid and the plane whose equation is $[x, \xi] = 0$, where $\xi$ is the direction vector of $l$. Having shown that $F(l; 2n)$ is actually an integral of $f(x)$ over this intersection, we wish to show that the measure $dt \, d\theta$ with respect to which this integral is taken in Eq. (11′) can be written in the form

$$dt \, d\theta = \delta([x, \xi]) \, dx, \tag{12}$$

where $dx$ is the invariant measure on imaginary Lobachevskian space $X$. This is so because both sides of Eq. (12) are invariant under simultaneous rotation of $x$ and $\xi$ about the $x_0$ axis and because such a rotation can be found which will carry any generator $l$ of the hyperboloid into the generator that passes through the point (0, 1, 0, 0) and has direction vector $\xi = (1, 0, 0, 1)$. With this remark, Eq. (12) becomes

$$dt\, d\theta = \delta(x_0 - x_3) \frac{dx_0\, dx_1\, dx_3}{|x_2|}. \tag{12'}$$

In order to prove (12), then, we shall prove (12′). But for this we need only note that the coordinates of a point on the line $b_\theta + t\xi$ can be written

$$x_0 = t, \qquad x_1 = \cos\theta, \qquad x_2 = \sin\theta, \qquad x_3 = t,$$

and then use this to rewrite the right-hand side of (12′) in terms of $t$ and $\theta$. Thus (12′) and therefore also (12) are valid, and Eq. (11′) can be written

$$F(l; 2n) = \int f(x) e^{-2in\theta} \delta([x, \xi])\, dx, \tag{13}$$

which is the analog of Eq. (7) for $F(\xi; \rho)$.

Similarly, it is a simple matter to obtain the analog of Eq. (8) for $F(l; 2n)$. For this purpose we write $\theta i = k\tau(x, l)$, where $\tau(x, l)$ is the distance from the isotropic line to the point $x$ (or, equivalently, the distance from $x$ to any point $y$ on $l$). Thus (13) may be written

$$F(l; 2n) = \int f(x) e^{-2n\tau(l,x)} \delta([x, \xi])\, dx, \tag{14}$$

which is the desired analog of Eq. (8).[12]

From now on it will be convenient to deal with unnormalized direction vectors $\xi$. We shall then introduce the function $F(\xi, b; 2n)$, which is defined by Eq. (13) except that now $\xi$ is *any* direction vector of the line $l$ whose equation is $x = b + t\xi$ and now $\cos\theta = [x, b]$. Like the function $\varphi(\xi, b)$ introduced in Section 3.4 of Chapter V, $F(\xi, b; 2n)$ is a homogeneous function of $\xi$ of degree $-1$.

We have now obtained expressions for the Fourier components $F(\xi; \rho)$ and $F(l; 2n)$ of a function $f(x)$ defined on imaginary Lobachevskian space $X$. We now wish to derive expressions for $f(x)$ in terms of its Fourier components, which we do by using the inversion formula of Chapter V, Section 3.4, namely

$$f(a) = -(4\pi)^{-2} \int h(\xi) \delta''(|[a, \xi]| - 1)\, d\xi$$
$$-(2\pi)^{-2} \int_0^\pi \cot^2\theta\, d\theta \int_\Gamma \varphi(\xi, \theta) \delta([a, \xi])\, d\omega. \tag{15}$$

[12] Recall that $\tau(x, l)$ is defined only if $x$ and $l$ lie in an isotropic plane; otherwise the distance from $x$ to some point $y$ of $l$ depends on $y$. These considerations are not, however, relevant in the present case, since the $\delta$ function in the integrand differs from zero only on the plane whose equation is $[x, \xi] = 0$, and on this plane $\tau(x, l)$ is defined.

Here $d\xi$ is the invariant measure on the $[\xi, \xi] = 0$ cone, $\Gamma$ is an arbitrary surface on this cone that intersects each of its generators, and $d\omega$ is a differential form defined by $d\xi = dP\, d\omega$, where $P(\xi) = 1$ is the equation of $\Gamma$. Further, $\varphi(\xi, \theta)$ is the value of $\varphi(\xi, b)$ on the isotropic line whose equation is $x = b + t\xi$ and whose distance from $a$ is $r = i\theta/k$.

We may now insert the expressions

$$h(\xi) = (4\pi)^{-1} \int_{-\infty}^{\infty} F(\xi; \rho)\, d\rho \tag{16}$$

and

$$\varphi(\xi, b) = \pi^{-1} \sum_{n=-\infty}^{\infty} F(\xi, b; 2n) \tag{17}$$

into (15) [see Eq. (5) of Section 4.2 and Eq. (6) of Section 4.3], arriving at the following expression for $f(a)$ in terms of $F(\xi; \rho)$ and $F(\xi, b; 2n)$:

$$f(a) = -(4\pi)^{-3} \int_{-\infty}^{\infty} \int F(\xi; \rho)\delta''(|[a, \xi]| - 1)\, d\xi\, d\rho$$

$$-2(2\pi)^{-3} \int_0^{\pi} \cot^2 \theta\, d\theta \int_\Gamma \sum_{n=-\infty}^{\infty} F(\xi, \theta; 2n)\delta([a, \xi])\, d\omega. \tag{18}$$

Here we have written $F(\xi, \theta; 2n)$ instead of $F(\xi, b; 2n)$ for this function on the isotropic line $l = l(\xi, \theta)$ with direction vector $\xi$ at a distance $r = i\theta/k$ from $a$.

This expression for $f(a)$ can be simplified by using the homogeneity of $F(\xi; \rho)$ and $F(l; 2n)$. Proceeding exactly as in Section 3.2, we find that the first term $J_1$ of Eq. (18) can be written in the form

$$J_1 = 2(8\pi)^{-3} \int_{-\infty}^{\infty} \rho(\rho + 2i)\, d\rho \int_\Gamma F(\xi; \rho)|[a, \xi]|^{-\frac{1}{2}i\rho-1}\, d\omega. \tag{19}$$

It can also be shown just as in the case of the ordinary Lobachevskian space that the $F(\xi; \rho)$ functions have the symmetry property

$$\int F(\xi; \rho)\delta(|[a, \xi]| - 1)\, d\xi = \int F(\xi; -\rho)\delta(|[a, \xi]| - 1)\, d\xi. \tag{20}$$

Since $F(\xi; \rho)$ is homogeneous in $\xi$ of degree $\frac{1}{2}i\rho - 1$, this symmetry property can also be written in the form

$$\int_\Gamma F(\xi; \rho)|[a, \xi]|^{-\frac{1}{2}i\rho-1}\, d\omega = \int_\Gamma F(\xi; -\rho)|[a, \xi]|^{\frac{1}{2}i\rho-1}\, d\omega, \tag{20'}$$

where $\Gamma$ and $d\omega$ have the same meaning as in Eq. (15). Then inserting this into (19) we find that

$$J_1 = \tfrac{1}{2}(4\pi)^{-2} \int_0^{\infty} \rho^2 \, d\rho \int_{\Gamma} F(\xi; \rho) |[a, \xi]|^{-\frac{1}{2}i\rho - 1} \, d\omega. \tag{21}$$

Let us now turn to the second term of Eq. (18), namely

$$J_2 = -2(2\pi)^{-3} \int_0^{\pi} \cot^2 \theta \, d\theta \int_{\Gamma} \sum_{n=-\infty}^{\infty} F(\xi, \theta; 2n) \delta([a, \xi]) \, d\omega. \tag{22}$$

Recall that

$$F(l_2; 2n) = e^{2in\theta} F(l_1; 2n),$$

if $l_1$ and $l_2$ are parallel isotropic lines separated by the distance $i\theta/k$. Consequently we may write

$$F(\xi, \theta; 2n) = e^{2in\theta} F(\xi, a; 2n). \tag{23}$$

When this is inserted into (22), it becomes

$$J_2 = -2(2\pi)^{-3} \int_0^{\pi} \cot^2 \theta \, d\theta \int_{\Gamma} \sum_{n=-\infty}^{\infty} e^{2in\theta} F(\xi, a; 2n) \delta([a, \xi]) \, d\omega, \tag{24}$$

which is an expression for $J_2$ in terms of the values of $F(l; 2n)$ on the set of isotropic lines passing through the point $a$. When the order of integration is changed in (24)(as previously, we shall not go into the proof of the validity of this procedure), it becomes

$$J_2 = -2(2\pi)^{-3} \sum_{n=-\infty}^{\infty} \int_0^{\pi} \cos^2 \theta \sin^{-2} \theta \, e^{2in\theta} \, d\theta$$
$$\times \int_{\Gamma} F(\xi, a; 2n) \delta([a, \xi]) \, d\omega. \tag{24'}$$

Now the symmetry relation

$$\int_{\Gamma} \varphi(\xi, \theta) \delta([a, \xi]) \, d\omega = \int_{\Gamma} \varphi(\xi, \pi - \theta) \delta([a, \xi]) \, d\omega$$

for the $\varphi(\xi, \theta)$ functions (see Chapter V, section 4.4) implies that

$$\int_{\Gamma} F(\xi, a; 2n) \delta([a, \xi]) \, d\omega = \int_{\Gamma} F(\xi, a; -2n) \delta([a, \xi]) \, d\omega. \tag{25}$$

We insert this in (24′) to obtain

$$J_2 = -\tfrac{1}{2}\pi^{-3} \sum_{n=-\infty}^{\infty} \alpha_n \int_\Gamma F(\xi, a; 2n)\delta([a, \xi])\, d\omega, \tag{24″}$$

where

$$\alpha_n = \int_0^{\frac{1}{2}\pi} \cos^2\theta \sin^{-2}\theta \cos 2n\theta\, d\theta. \tag{26}$$

This defining integral for the $\alpha_n$ diverges, and we therefore understand it in the sense of its regularization, that is the value at $\lambda = -2$ of the integral

$$\int_0^{\frac{1}{2}\pi} \cos^2\theta \sin^\lambda\theta \cos 2n\theta\, d\theta.$$

By using the formula[13]

$$\int_0^{\pi} \sin^q x \cos px\, dx = \frac{\pi q}{2^{q-2}(q^2 - p^2)} \frac{\cos \frac{1}{2}p\pi}{B(\frac{1}{2}q + \frac{1}{2}p, \frac{1}{2}q - \frac{1}{2}p)},$$

we find that

$$\alpha_n = -4\pi \mid n \mid.$$

Thus the second term in Eq. (18) may be written

$$J_2 = 4\pi^{-2} \sum_{n=1}^{\infty} n \int_\Gamma F(\xi, a; 2n)\delta([a, \xi])\, d\omega. \tag{27}$$

We now insert Eqs. (21) and (27) into (18). Then we have

$$f(a) = \tfrac{1}{2}(4\pi)^{-3} \int_0^\infty \rho^2\, d\rho \int_\Gamma F(\xi; \rho) |[a, \xi]|^{-\frac{1}{2}i\rho - 1}\, d\omega$$
$$+ 4\pi^{-2} \sum_{n=1}^{\infty} n \int_\Gamma F(\xi, a; 2n)\delta([a, \xi])\, d\omega. \tag{28}$$

Equation (23) can be used to replace the isotropic line whose equation is $x = a + t\xi$ by an arbitrary isotropic line with equation $x = b + t\xi$ if the integrand is multiplied by $e^{2in\theta}$, where $\cos\theta = [a, b]$ (or in other

[13] I. M. Ryshik and I. S. Gradstein, "Tables of Series, Products, and Integrals," VEB Deutscher Verlag der Wissenschaften, Berlin, 1963. Formula **3.454**(4), p. 162.

words where $i\theta/k$ is the distance from $a$ to this new line). Thus we find we arrive at

$$f(a) = \tfrac{1}{2}(4\pi)^{-3} \int_0^\infty \rho^2 \, d\rho \int_\Gamma F(\xi; \rho) |[a, \xi]|^{-\frac{1}{2}\rho - 1} \, d\omega$$
$$+ 4\pi^{-2} \sum_{n=1}^\infty n \int_\Gamma F(\xi, b; 2n) e^{2in\theta} \delta([a, \xi]) \, d\omega. \tag{28'}$$

*Summary.* Let $f(x)$ be an infinitely differentiable function with bounded support over imaginary Lobachevskian space $X$. This function has two kinds of Fourier components. The first kind $F(\xi; \rho)$ is a continuous function of $\rho$, $0 \leqslant \rho < \infty$, and is given in terms of $f(x)$ by

$$F(\xi; \rho) = \int f(x) |[x, \xi]|^{\frac{1}{2}i\rho - 1} \, dx. \tag{29}$$

The second kind of Fourier component $F(\xi, b; 2n)$ depends on the discrete variable $n$ (and is therefore actually a set of Fourier components) and is given in terms of $f(x)$ by

$$F(\xi, b; 2n) = \int f(x) e^{-2in\theta} \delta([x, \xi]) \, dx, \tag{30}$$

where $i\theta/k$ is the distance from $x$ to the isotropic line whose equation is $y = b + t\xi$. If the direction vector $\xi$ is normalized according to $\xi_0 = 1$, we write $F(l; 2n)$ rather than $F(\xi, b; 2n)$. These functions have the following properties. $F(\xi; \rho)$ is a function on the cone $[\xi, \xi] = 0$, $\xi_0 > 0$, homogeneous of degree $\frac{1}{2}i\rho - 1$. $F(l; 2n)$ is a function on the set of isotropic lines with the property that

$$F(l_2; 2n) = e^{2in\theta} F(l_1; 2n) \tag{31}$$

for parallel isotropic lines of $l_1$ and $l_2$ separated by the distance $i\theta/k$.

When $f(x)$ is acted on by a representation of the Lorentz group of the form

$$T(g) f(x) = f(xg), \tag{32}$$

$F(\xi; \rho)$ for fixed $\rho$ is acted upon by the irreducible unitary representation

$$Q(g) F(\xi; \rho) = F(\xi g; \rho), \tag{33}$$

equivalent to $T_\chi(g)$ with $\chi = (\frac{1}{2}i\rho, \frac{1}{2}i\rho)$, which is in the principal series. Under the same transformation of $f(x)$, $F(l; 2n)$ for fixed $n$ transforms under the irreducible unitary representation

$$R_{2n}(g) F(l; 2n) = \beta^{-1}(l, g) F(lg; 2n), \tag{34}$$

equivalent to $T_\chi(g)$ with $\chi = (2n, -2n)$ (see Section 4.3). Finally, $f(x)$ is given in terms of $F(\xi; \rho)$ and $F(\xi, b; 2n)$ by Eq. (28′).

Remark. The first and second terms in Eq. (28′) are actually quite similar. This is because the space of horospheres of the first kind is related to the space of isotropic lines. Specifically, they are fiber spaces with the same basis, the two-dimensional sphere $\Omega$ (which can, of course, be replaced by any surface $\Gamma$ homeomorphic to it). A fiber in the space of horospheres of the first kind is a ray (that is, the multiplicative group of positive real numbers). In the space of isotropic lines, a fiber is a circle (that is, the multiplicative group of complex numbers $z$ such that $|z| = 1$). It is natural to call both these spaces "cones". It should be noted that they are both quotient spaces of the complex affine plane. First, if we identify all points of the form $(e^{i\theta}z_1, e^{i\theta}z_2)$, where $0 \leqslant \theta < 2\pi$, we obtain the space of horospheres of the first kind. If, on the other hand, we identify all points of the form $(\lambda z_1, \lambda z_2)$, where $\lambda \neq 0$ runs through the set of all real and pure imaginary numbers, we obtain the set of isotropic lines. #

## 4.5. The Analog of Plancherel's Theorem for Imaginary Lobachevskian Space

As a last step in our discussion of imaginary Lobachevskian space, we wish to establish the analog of Plancherel's theorem for the Fourier transform. As in the case of real Lobachevskian space, the proof we give will not be rigorous, for we shall not justify changing the order of integration.

In the present case the analog of Plancherel's theorem is

$$\int |f(x)|^2\, dx = \tfrac{1}{2}(4\pi)^{-3} \int_0^\infty \rho^2\, d\rho \int_\Gamma |F(\xi; \rho)|^2\, d\omega + 4\pi^{-2} \sum_{n=1}^{\infty} n \int_\Gamma |F(\xi; 2n)|^2\, d\omega. \tag{1}$$

As before, $\Gamma$ is an arbitrary surface on the $[\xi, \xi] = 0$ cone intersecting each of its generators, $d\omega$ is a differential form defined by $d\xi = dP\, d\omega$, where $P(\xi) = 1$ is the equation of $\Gamma$, and $F(\xi; 2n)$ is the value of $F(\xi, b; 2n)$ for one of the isotropic lines $l$ with direction vector $\xi$. [Since all of these values differ only by a factor of the form $e^{2in\theta}$, the absolute value $|F(\xi; 2n)|$ is independent of the choice of isotropic line with direction vector $\xi$.]

**Proof.** Consider the integral

$$\int |f(x)|^2\, dx = \int f(x)\bar{f}(x)\, dx$$

and replace $\bar{f}(x)$ by its expression as given in Eq. (28′) of Section 4.4. We then have

$$\int |f(x)|^2\, dx = \tfrac{1}{2}(4\pi)^{-3} \int f(x)\, dx \int_0^\infty \rho^2\, d\rho \int_\Gamma \bar{F}(\xi;\rho)|[x,\xi]|^{\frac{1}{2}i\rho-1}\, d\omega$$

$$+ 4\pi^{-2} \int f(x)\, dx \left[\sum_{n=1}^{\infty} n \int_\Gamma \bar{F}(\xi, b; 2n) e^{-2in\theta}\delta([x,\xi])\, d\omega\right]. \quad (2)$$

Consider the first term $I_1$ on the right-hand side of this equation. According to Eq. (29) of Section 4.4,

$$\int f(x)|[x,\xi]|^{\frac{1}{2}i\rho-1}\, dx = F(\xi;\rho),$$

so that by changing the order of integration, we find that

$$I_1 = \tfrac{1}{2}(4\pi)^{-3} \int_0^\infty \rho^2\, d\rho \int_\Gamma |F(\xi;\rho)|^2\, d\omega.$$

Now consider the second term, $I_2$. In this term also we change the order of integration and summation. Then

$$I_2 = 4\pi^{-2} \sum_{n=1}^{\infty} n \int_\Gamma \bar{F}(\xi, b; 2n)\, d\omega \int f(x) e^{-2in\theta}\delta([x,\xi])\, dx,$$

where $\theta$ is the distance from $x$ to the line $l$. Now according to Eq. (30) of Section 4.4

$$\int f(x) e^{-2in\theta}\delta([x,\xi])\, dx = F(\xi, b; 2n),$$

so that

$$I_2 = 4\pi^{-2} \sum_{n=1}^{\infty} n \int_\Gamma |F(\xi, b; 2n)|^2\, d\omega.$$

When these expressions for $I_1$ and $I_2$ are inserted into Eq. (2), we arrive at Eq. (1), which is thus proved.

This result may be stated in the following way. Let $H$ be the Hilbert space of functions $f(x)$ over imaginary Lobachevskian space $X$ such that the integral

$$\|f\|^2 = \int |f(x)|^2\, dx \quad (3)$$

converges, and let $H_1$ be the space of pairs of functions $\{F(\xi; \rho); F(\xi, b; 2n)\}$ for which

$$\|F\|^2 = \tfrac{1}{2}(4\pi)^{-3} \int_0^{\infty} \rho^2 \, d\rho \int_{\Gamma} |F(\xi; \rho)|^2 \, d\omega + 4\pi^{-2} \sum_{n=1}^{\infty} n \int_{\Gamma} |F(\xi; 2n)|^2 \, d\omega \tag{4}$$

converges.[14]

Then the mapping

$$f(x) \to \{F(\xi; \rho); F(\xi, b; 2n)\},$$

defined by

$$F(\xi; \rho) = \int f(x) |[\xi, x]|^{\frac{1}{2}i\rho - 1} \, dx,$$

$$F(\xi, b; 2n) = \int f(x) e^{-2in\theta} \delta([x, \xi]) \, dx,$$

is an isometric imbedding of $H$ in $H_1$. It can be shown that if we allow only positive values of $\rho$ and $n$, the image of $H$ under this mapping consists of *all* pairs of functions $\{F(\xi; \rho); F(\xi, b; 2n)\}$ for which Eq. (4) is finite. If, on the other hand, we allow negative as well as positive $\rho$ and $n$, the functions $F(\xi; \rho)$ and $F(\xi, b; 2n)$ must have the (symmetry) properties that for any $x \in X$

$$\int_{\Gamma} F(\xi; \rho) |[x, \xi]|^{-\frac{1}{2}i\rho - 1} \, d\omega = \int_{\Gamma} F(\xi; -\rho) |[x, \xi]|^{\frac{1}{2}i\rho - 1} \, d\omega$$

and

$$\int_{\Gamma} F(\xi; 2n) \, \delta([x, \xi]) \, d\omega = \int_{\Gamma} F(\xi; -2n) \, \delta([x, \xi]) \, d\omega.$$

## 4.6. Integral Transform Associated with Planes in Lobachevskian Space

In this chapter we have been obtaining Fourier expansions of functions over ordinary and imaginary Lobachevskian space. At the basis of our development lay the integral transform associating functions over such a space with functions on the space of horospheres. From the point of

[14] Here $F(\xi; 2n)$ denotes the value of $F(l; 2n)$ on any one of the isotropic lines $l$ with direction vector $\xi$.

view of integral geometry, it is interesting also to consider integral transforms associating such functions with functions on the space of planes. We shall not, however, discuss these in this volume, but shall merely mention some ways in which these functions are of interest.

Consider first ordinary Lobachevskian space $X$. As has already been mentioned in Chapter V, Section 1.4, a plane in such a space is given by an equation of the form $[x, \xi] = 0$, where $x = (x_0, x_1, ..., x_n)$ is the set of homogeneous coordinates of a point in the space, and $[\xi, \xi] < 0$. (If $[\xi, \xi] \geqslant 0$, then $[x, \xi] > 0$ for every $x \in X$.) It is easily verified that this set of planes is a homogeneous space and is identical with imaginary Lobachevskian space. Thus by integrating a function $f$ on Lobachevskian space over planes we associate with it a function $\varphi$ on imaginary Lobachevskian space. It should be mentioned, however, that $\varphi$ cannot be an "arbitrary" function on the imaginary Lobachevskian space, since as we have seen the expansion of a function on imaginary Lobachevskian space involves more irreducible representations than in ordinary Lobachevskian space. There exists an inversion formula expressing $f$ in terms of $\varphi$. (This formula is derived by the same method as the one used in Chapter V in dealing with integral transforms associated with horospheres.)

Now consider imaginary Lobachevskian space. By analogy with the case of ordinary Lobachevskian space, we call planes in this space surfaces given by equations of the form $[x, \xi] = 0$. Unlike the previous case, $\xi$ may now be any vector. The set of planes of the imaginary Lobachevskian space can consequently be broken up into three families: planes of the first kind, for which $[\xi, \xi] > 0$; planes of the second kind, for which $[\xi, \xi] = 0$; and planes of the third kind, for which $[\xi, \xi] < 0$. It is easily shown that each of these families is homogeneous, and that the set of planes of the first kind is identical with ordinary Lobachevskian space, the set of planes of the second kind is identical with the absolute, and the set of planes of the third kind is identical with imaginary Lobachevskian space. It is also easily shown that the planes of the second and third kind stratify into isotropic lines and that the planes of the first kind are compact manifolds.

Consider the planes of the first kind. By integrating a function $f$ on imaginary Lobachevskian space over planes of the first kind, we associate with $f$ a function $\varphi$ on ordinary Lobachevskian space The mapping $f \rightarrow \varphi$ is not one-to-one (since otherwise the $\varphi$ functions could be expanded in terms of the same irreducible representations as the $f$ functions, which we know is not true). Consequently if we wish to know $f$ uniquely it is not sufficient to know its integrals over planes of the first kind. It can be shown, however, that if the integrals of $f$ over isotropic lines are also known, then $f$ is uniquely determined. The inversion formula can be obtained in the same way as in Chapter V for the integral transforms associated with horospheres. It is noteworthy that this inversion formula can be used to reduce the problem of finding the Fourier integral expansion for a function on imaginary Lobachevskian space to the simpler problem of finding the Fourier integral expansion of a function on ordinary Lobachevskian space and a function on the space of isotropic lines of imaginary Lobachevskian space.

## 5. Integral Geometry and Harmonic Analysis on the Point Pairs on the Complex Projective Line

Consider the space of ordered pairs $(z_1, z_2)$, $z_1 \neq z_2$, which we have called the point pairs on the complex projective line. With each complex matrix

$$g = \begin{Vmatrix} \alpha & \beta \\ \gamma & \delta \end{Vmatrix}, \qquad \alpha\delta - \beta\gamma = 1,$$

we associate the transformation

$$(z_1, z_2) \to \left(\frac{\alpha z_1 + \gamma}{\beta z_1 + \delta}, \frac{\alpha z_2 + \gamma}{\beta z_2 + \delta}\right) \tag{1}$$

on this space, and this space is homogeneous with respect to these transformations.

We wish to associate with this space a unitary representation of the Lorentz group. This representation will be constructed in the space of functions $f(z_1, z_2)$ such that

$$\| f \|^2 = \left(\frac{i}{2}\right)^2 \int | f(z_1, z_2)|^2 \, dz_1 \, d\bar{z}_1 \, dz_2 \, d\bar{z}_2 < \infty. \tag{2}$$

Specifically, the operator of the representation is defined by

$$T(g) f(z_1, z_2) = f\left(\frac{\alpha z_1 + \gamma}{\beta z_1 + \delta}, \frac{\alpha z_2 + \gamma}{\beta z_2 + \delta}\right) \times (\beta z_1 + \delta)^{n_1'-1} (\bar{\beta}\bar{z}_1 + \bar{\delta})^{n_2'-1} (\beta z_2 + \delta)^{n_1''-1} (\bar{\beta}\bar{z}_2 + \bar{\delta})^{n_2''-1}, \tag{3}$$

wher $n_1' = -\bar{n}_2'$, $n_1'' = -\bar{n}_2''$. This representation is called the *Kronecker product* of the two irreducible unitary representations (of the principal series)

$$T_1(g) f(z) = f\left(\frac{\alpha z + \gamma}{\beta z + \delta}\right) (\beta z + \delta)^{n_1'-1} (\bar{\beta}\bar{z} + \bar{\delta})^{n_2'-1}$$

and

$$T_2(g)f(z) = f\left(\frac{\alpha z + \gamma}{\beta z + \delta}\right)(\beta z + \delta)^{n_1''-1}(\bar{\beta}\bar{z} + \bar{\delta})^{n_2''-1}.$$

Our problem is to expand $f(z_1, z_2)$ in a Fourier integral, i.e., to decompose the representation of Eq. (3) into its irreducible components. We shall show here how to solve this problem by using the methods of integral geometry. In so doing we shall, however, merely state the results (whose detailed proof will be found in Gel'fand and Graev (9)], where the analogous problem was solved for an arbitrary complex semisimple group).

We start by formulating the problem of integral geometry for the point pairs on the complex projective line. Consider the horospheres of this space, that is, manifolds generated by motions belonging to the subgroup of matrices of the form

$$\begin{Vmatrix} 1 & \zeta \\ 0 & 1 \end{Vmatrix}$$

or some subgroup conjugate to it. Our problem is then to find $f(z_1, z_2)$ if we know its integrals over all possible horospheres. (The problem will be stated more accurately somewhat later.)

As was shown in Section 1.6, the horospheres of this space are given by equations of the form

$$z_1 - z_2 = (a - bz_1)(a - bz_2), \tag{4}$$

where $(a, b) \neq (0, 0)$. Under the transformation of Eq. (1), the parameters $a$ and $b$ defining a given horosphere $\omega$ transform according to

$$a' = \alpha a + \gamma b, \qquad b' = \beta a + \delta b.$$

In other words, the set of these horospheres is a homogeneous space. This space is identical with the complex affine plane in which diametrically opposite points $(a, b)$ and $(-a, -b)$ are identified (for these points define the same horosphere). The Fourier integral expansion of a function on the space of horospheres then becomes rather elementary.

Let us define the integral over the horosphere given by Eq. (4). Let $(z_1^0, z_2^0)$ be any element in our space of point pairs, and let us assume that $z_1^0 \neq 0$ and $z_2^0 \neq 0$. Then if $b \neq 0$ the horosphere can be given by the parametric equations

$$z_1 = \frac{u}{v} + \frac{z_1^0}{v^2(z_1^0\zeta + 1)}, \qquad z_2 = \frac{u}{v} + \frac{z_2^0}{v^2(z_2^0\zeta + 1)}, \tag{5}$$

where $\zeta$ is a complex parameter and

$$u = s^{\frac{1}{2}}a, \qquad v = s^{\frac{1}{2}}b,$$

where we write $s = z_1^0 z_2^0 (z_1^0 - z_2^0)^{-1}$, and $s^{\frac{1}{2}}$ is either one of the square roots. Let $f(z_1, z_2)$ be an infinitely differentiable function concentrated on a sufficiently small neighborhood of $(z_1^0, z_2^0)$.[15] We define the integral of $f(z_1, z_2)$ over the horosphere of Eq. (5) by

$$\varphi(u, v) = \frac{i}{2} \int f\left(\frac{u}{v} + \frac{z_1^0}{v^2(\zeta z_1^0 + 1)}, \frac{u}{v} + \frac{z_2^0}{v^2(\zeta z_2^0 + 1)}\right)$$
$$\times\, a(v[\zeta z_1^0 + 1], v[\zeta z_2^0 + 1])\, d\zeta\, d\bar{\zeta}, \tag{6}$$

where we have written

$$a(z_1, z_2) = z_1^{n_1'-1} \bar{z}_1^{n_2'-1} z_2^{n_1''-1} \bar{z}_2^{n_2''-1}. \tag{7}$$

It can be shown that with this definition of the integral, $\varphi(u, v)$ has the following property. When $f(z_1, z_2)$ is transformed by the operator of the representation $T(g)$ according to Eq. (3), $\varphi(u, v)$ is transformed according to

$$T(g)\varphi(u, v) = \varphi(u\alpha + v\gamma, u\beta + v\delta). \tag{8}$$

In other words the mapping $f(z_1, z_2) \to \varphi(u, v)$ carries the representation on the $f(z_1, z_2)$ functions into the representation of Eq. (8)

Now the problem of integral geometry is to find $f(z_1, z_2)$ when $\varphi(u, v)$ is known. The solution of this problem, which we present without proof in a form entirely sufficient for purposes of harmonic analysis [that is, in the form of an expression for $f(z_1^0, z_2^0)$ in terms of $\varphi(u, v)$], is

$$f(z_1^0, z_2^0) = \frac{i}{2} \int |w_1 w_2|\, a^{-\frac{1}{2}}(w_1 w_2^{-1}, w_2 w_1^{-1})$$
$$\times \frac{\partial^2}{\partial\lambda\, \partial\bar{\lambda}} \{|\lambda|^2 \varphi(\lambda[w_1 w_2]^{\frac{1}{2}} z, \lambda[w_1 w_2]^{\frac{1}{2}})\}_{\lambda=1}\, dz\, d\bar{z}. \tag{9}$$

where $w_j = z_j^0 (z_j^0 - z)^{-1}$, $j = 1, 2$ and $(w_1 w_2)^{\frac{1}{2}}$ is either of the square roots.

Remark. Equation (9) can be used also to obtain an expression for $f$ at any point $(z_1, z_2)$. This is done by applying Eq. (9) to $T(g)f$ and bearing in mind that the latter function corresponds to

$$T(g)\varphi(u, v) = \varphi(u\alpha + v\gamma, u\beta + v\delta). \quad \#$$

[15] This assumption ensures that all the relevant integrals converge.

We can now proceed to solve the basic problem, namely to obtain the Fourier integral expansion of $f(z_1, z_2)$. Consider the Fourier transform of $\varphi(u, v)$ (which we have called its Mellin transform in Section 2.1, where it was written in homogeneous coordinates). We have

$$\psi(z; \chi) \equiv \psi(z; \rho, n) = \alpha^{-1}(\chi)\tfrac{1}{2}i \int \varphi(\lambda z, \lambda)\lambda^{-n_1}\bar{\lambda}^{-n_2}\, d\lambda\, d\bar{\lambda}, \tag{10}$$

where

$$n_1 = \tfrac{1}{2}(n + i\rho), \qquad n_2 = \tfrac{1}{2}(-n + i\rho)$$

(recall that $n$ is an integer and that $\rho$ is any real number). The Fourier integral expansion of $\varphi(u, v)$ is

$$\varphi(u, v) = (2\pi)^{-2} \int \alpha(\chi)\psi(u/v; \chi)v^{n_1-1}\bar{v}^{n_2-1}\, d\chi, \tag{11}$$

where the integral over $\chi$ is understood as integration over $\rho$ and summation over $n$. For convenience in writing Eqs. (10) and (11) we have introduced the normalizing factor $\alpha(\chi)$ defined by

$$\alpha(\chi) = (z_1^0)^{\frac{1}{2}(n_1'-n_1''-n_1-1)}(\bar{z}_1^0)^{\frac{1}{2}(n_2'-n_2''-n_2-1)}(z_2^0)^{\frac{1}{2}(n_1''-n_1'-n_1-1)}$$
$$\times (\bar{z}_2^0)^{\frac{1}{2}(n_2''-n_2'-n_2-1)}(z_2^0 - z_1^0)^{\frac{1}{2}(n_1'+n_1''+n_1-1)}(\bar{z}_2^0 - \bar{z}_1^0)^{\frac{1}{2}(n_2'+n_2''+n_2-1)}. \tag{12}$$

Note that as defined by Eq. (6), $\varphi(u, v)$ has the property that

$$\varphi(-u, -v) = a(-1, -1)\varphi(u, v) = (-1)^{(n_1'-n_2')+(n_1''-n_2'')}\varphi(u, v).$$

This means that Eq. (11) involves $\psi(z; \chi) \equiv \psi(z; \rho, n)$ either only with even $n$ (when $n_1' - n_2'$ and $n_1'' - n_2''$ are of the same parity and therefore $\varphi$ is an even function) or only with odd $n$(when $n_1' - n_2'$ and $n_1'' - n_2''$ are of opposite parity and therefore $\varphi$ is an odd function).

By inserting the expression for $\varphi(u, v)$ in terms of $f(z_1, z_2)$ into Eq. (10), we obtain an expression for $\psi(z; \chi)$ in terms of the latter function. The result is easy to obtain and is found to be

$$\psi(z; \chi) = \tfrac{1}{4}\left(\frac{i}{2}\right)^2 \int A(z_1 - z, z_2 - z; \chi)f(z_1, z_2)\, dz_1\, d\bar{z}_1\, dz_2\, d\bar{z}_2, \tag{13}$$

where we have written

$$A(z, z'; \chi) = z^{\frac{1}{2}(n_1''-n_1'+n_1-1)}\bar{z}^{\frac{1}{2}(n_2''-n_2'+n_2-1)}z'^{\frac{1}{2}(n_1'-n_1''+n_1-1)}$$
$$\times \bar{z}'^{\frac{1}{2}(n_2'-n_2''+n_2-1)}(z' - z)^{\frac{1}{2}(-n_1'-n_1''-n_1-1)}(\bar{z}' - \bar{z})^{\frac{1}{2}(-n_2'-n_2''-n_2-1)}. \tag{14}$$

It is seen that the appropriate choice of the normalizing factor $\alpha(\chi)$ makes $\psi(z;\chi)$ independent of the choice of $(z_1^0, z_2^0)$. What remains now is to obtain an expression for $f(z_1, z_2)$ in terms of $\psi(z;\chi)$, which would then be its Fourier integral expansion. This can be done by substituting Eq. (11) into the expression for $\varphi(u, v)$ in Eq. (9). Then elementary operations lead to

$$f(z_1^0, z_2^0) = -(2\pi)^{-2}(\tfrac{1}{2}i) \int n_1 n_2 \bar{A}(z_1^0 - z, z_2^0 - z; \chi)\psi(z;\chi)\, dz\, d\bar{z}\, d\chi, \quad (15)$$

where the kernel $A(z_1^0 - z, z_2^0 - z; \chi)$ is given in (14).

Equations (13) and (15) can be used also to obtain the following analog of Plancherel's theorem for $f(z_1, z_2)$ and its Fourier transform[16]:

$$\left(\frac{i}{2}\right)^2 \int |f(z_1, z_2)|^2\, dz_1\, d\bar{z}_1\, dz_2\, d\bar{z}_2$$
$$= -\pi^{-2}\left(\frac{i}{2}\right)^2 \int n_1 n_2 \,|\,\psi(z;\chi)|^2\, dz\, d\bar{z}\, d\chi.$$

**Remark.** This analog of Plancherel's theorem implies that the mapping $f(z_1, z_2) \to \psi(z;\chi)$ is isometric with respect to the norms

$$\|f\|^2 = \left(\frac{i}{2}\right)^2 \int |f(z_1, z_2)|^2\, dz_1\, d\bar{z}_1\, dz_2\, d\bar{z}_2$$

and

$$\|\psi\|^2 = -\pi^{-2}\left(\frac{i}{2}\right) \int n_1 n_2 \,|\,\psi(z;\chi)|^2\, dz\, d\bar{z}\, d\chi.$$

So far we have defined this mapping only for $f(z_1, z_2)$ with bounded support. Because the mapping is isometric, however, it can be extended to all $f(z_1, z_2)$ such that $\|f\| < \infty$. An interesting problem, incidentally, is to find those $\psi(z;\chi)$ functions which are Fourier transforms of square integrable $f(z_1, z_2)$; however, we shall not go into it here. #

[16] Recall that $n_1 n_2 = |\,n_1\,|^2 = \frac{1}{4}(n^2 + \rho^2)$.

CHAPTER VII

# REPRESENTATIONS OF THE GROUP OF REAL UNIMODULAR MATRICES IN TWO DIMENSIONS

We turn in this chapter to an outline of the representation theory of the group of real matrices with determinant 1 in two dimensions. We shall restrict our considerations to the construction of the irreducible representations and to a study of their properties, omitting questions of harmonic analysis. Thus the content of Chapter VII is similar to that of Chapter III, where the analogous problems were treated for complex matrices.

The group of real matrices has several interesting properties that differ considerably from these of the group of complex matrices. The most important are a more complicated structure of the representations at integer points and the existence of representations acting on analytic function spaces. These properties will be described in more detail in Sections 2.1–2.4, which summarize the results of this chapter.

As in Chapter III, we shall proceed by means of investigating invariant bilinear functionals. In a certain sense this treatment is simpler than it was in the complex case, since we shall be dealing with functions of a real instead of a complex variable.

## 1. Representations of the Real Unimodular Matrices in Two Dimensions Acting on Homogeneous Functions of Two Real Variables

### 1.1. The $D_\chi$ Spaces of Homogeneous Functions

In this section we shall discuss the representations of the group $G$ of real matrices

$$g = \begin{Vmatrix} \alpha & \beta \\ \gamma & \delta \end{Vmatrix}, \qquad \alpha\delta - \beta\gamma = 1.$$

In analogy with the group of complex matrices, these representations will be constructed on spaces of homogeneous functions $f(x_1, x_2)$, but now functions of real rather than complex variables. A function $f(x_1, x_2)$ of two real variables $(x_1, x_2) \neq (0, 0)$ is called *homogeneous of degree* $s$, where $s$ may be any complex number, if for every real number $a > 0$ we have

$$f(ax_1, ax_2) = a^s f(x_1, x_2).$$

Further, $f(x_1, x_2)$ is said to have *even parity* or to be *even* if

$$f(-x_1, -x_2) = f(x_1, x_2),$$

and to have *odd parity* or to be *odd* if

$$f(-x_1, -x_2) = -f(x_1, x_2).$$

Consider all possible pairs of numbers $\chi = (s, \epsilon)$, where $s$ is any complex number and $\epsilon = 0$ or 1. With each such pair of numbers we associate a function space $D_\chi$ consisting of functions $f(x_1, x_2)$ with the following properties.

*1.* Every $f(x_1, x_2) \in D_\chi$ is homogeneous of degree[1] $s - 1$ and has even parity if $\epsilon = 0$ or odd parity if $\epsilon = 1$ (we shall sometimes also call $\epsilon$ the parity). In other words for any real $a \neq 0$ we have

$$f(ax_1, ax_2) = | a |^{s-1} \operatorname{sgn}^\epsilon a \, f(x_1, x_2).$$

*2.* Every $f(x_1, x_2) \in D_\chi$ is infinitely differentiable in $x_1$ and $x_2$ everywhere except at the origin (0, 0).

We topologize $D_\chi$ in the following way. A sequence of functions $\{f_n\}$ in $D_\chi$ is said to *converge to zero* if on every bounded closed set not containing the origin the $f_n$ converge uniformly to zero together with all their derivatives. It can be shown that $D_\chi$ is complete with respect to this topology.

It is possible also to realize $D_\chi$ in other ways. Consider, for instance, a curve $l$ in the plane that intersects every straight line passing through the origin. The homogeneity and parity conditions then determine $f(x_1, x_2)$ from its values on $l$. Thus $D_\chi$ may be considered not the space of homogeneous functions of given parity on the plane, but as functions on the line. It should be noted, however, that these functions are not entirely arbitrary on the curve $l$, for if some line $a_1x_1 + a_2x_2 = 0$ intersects $l$ at several points, the values of the function at these points must be in a ratio independent of the choice of function.

[1] The choice of $s - 1$ rather than $s$ will be found convenient later.

### 1.2. Two Useful Realizations of $D_\chi$

Later we shall find the following two realizations of $D_\chi$ particular by useful.

Consider the line $x_2 = 1$ in the $(x_1, x_2)$ plane. Obviously each homogeneous function $f(x_1, x_2)$ of given parity $\epsilon$ is uniquely determined by its values on this line. Thus with each $f(x_1, x_2) \in D_\chi$ we may associate the function of a single variable

$$\varphi(x) = f(x, 1). \tag{1}$$

Obviously $f(x_1, x_2)$ is given in terms of $\varphi(x)$ by

$$f(x_1, x_2) = | x_2 |^{s-1} \operatorname{sgn}^\epsilon x_2 \, \varphi(x_1/x_2). \tag{2}$$

In this way $D_\chi$ is realized as the set of functions of a single variable $x$. It is easy to discover which functions $\varphi(x)$ may belong to $D_\chi$. In fact they must have the following properties.

1. Every $\varphi(x)$ must possess all derivatives.

2. If $\varphi(x) \in D_\chi$, then its "inversion" $\varphi(x) = | x |^{s-1} \operatorname{sgn}^\epsilon x \, \hat{\varphi}(-1/x)$ must also possess all derivatives.

This second condition characterizes the behavior of $\varphi(x)$ for large $| x |$. In particular it implies that asymptotically as $| x | \to \infty$ its behavior is given by

$$\varphi(x) \sim C \, | x |^{s-1} \operatorname{sgn}^\epsilon x.$$

The second realization of $D_\chi$ is obtained by considering the circle whose equation is $x_1^2 + x_2^2 = 1$. Every $f(x_1, x_2) \in D_\chi$ is then uniquely determined by its values on this circle. Consequently $D_\chi$ can be realized as the space of infinitely differentiable functions $\varphi(\theta)$ on the circle (here $\theta$ is the polar angle). These functions must satisfy the subsidiary condition

$$\varphi(\theta + \pi) = (-1)^\epsilon \varphi(\theta)$$

arising from their parity. The function $f(x_1, x_2)$ is given in terms of its $\varphi(\theta)$ by

$$f(r \cos \theta, r \sin \theta) = r^{s-1} \varphi(\theta).$$

### 1.3. Representation of $G$ on $D_\chi$

Let $G$ be the group of real unimodular matrices in two dimensions. Then each matrix

$$g = \begin{Vmatrix} \alpha & \beta \\ \gamma & \delta \end{Vmatrix}, \qquad \alpha\delta - \beta\gamma = 1$$

in $G$ defines the linear transformation

$$x_1' = \alpha x_1 + \gamma x_2, \qquad x_2' = \beta x_1 + \delta x_2$$

of the $(x_1, x_2)$ plane, which induces the transformation

$$f(x_1, x_2) \to f(\alpha x_1 + \gamma x_2, \beta x_1 + \delta x_2)$$

in $D_\chi$, for it is obvious that $f(x_1, x_2) \in D_\chi$ implies that

$$f(\alpha x_1 + \gamma x_2, \beta x_1 + \delta x_2) \in D_\chi .$$

Thus this is a linear transformation of $D_\chi$, and we shall denote it by $T_\chi(g)$; it is defined by

$$T_\chi(g) f(x_1, x_2) = f(\alpha x_1 + \gamma x_2, \beta x_1 + \delta x_2) \tag{1}$$

for every $f \in D_\chi$. It is easily verified that $T_\chi(g)$ is continuous with respect to the topology in $D_\chi$. Further, it is continuous also with respect to $g$. In other words, given a sequence of matrices $g_m$ such that $\lim_{m\to\infty} g_m = g$, then

$$\lim_{m\to\infty} T_\chi(g_m) f = T_\chi(g) f$$

for every $f \in D_\chi$. Finally, it is also easily verified that if $g_1, g_2 \in G$, then

$$T_\chi(g_1 g_2) = T_\chi(g_1) T_\chi(g_2).$$

Consequently, $T_\chi(g)$ is a representation of $G$. Summarizing, *with every pair of numbers $\chi = (s, \epsilon)$, where $s$ is a complex number and $\epsilon = 0$ or $1$, we associate a representation $T_\chi(g)$ of $G$ defined on the space $D_\chi$ of infinitely differentiable functions $f(x_1, x_2)$, homogeneous of degree $s - 1$ and of given parity (even for $\epsilon = 0$ and odd for $\epsilon = 1$). The representation is defined by Eq.* (1).

## 1.4. The $T_\chi(g)$ Operators in Other Realizations of $D_\chi$

It was pointed out in Section 1.2 that $D_\chi$ can also be realized as the space of infinitely differentiable functions $\varphi(x)$ of a single variable, and that

$$f(x_1, x_2) = |x_2|^{s-1} \operatorname{sgn}^\epsilon x_2 \, \varphi(x_1/x_2).$$

This equation can be used to find the form of $T_\chi(g)$ in this realization of $D_\chi$. Proceeding exactly as in Chapter II, we arrive at

$$T_\chi(g)\varphi(x) = |\beta x + \delta|^{s-1} \operatorname{sgn}^\epsilon (\beta x + \delta)\, \varphi\left(\frac{\alpha x + \gamma}{\beta x + \delta}\right).$$

If $D_\chi$ is realized as the space of all infinitely differentiable functions $\varphi(\theta)$ on the circle (satisfying the symmetry condition due to parity), then

$$\varphi(\theta) = f(\cos\theta, \sin\theta).$$

Again, this equation can be used to find the form of $T_\chi(g)$ in this realization of $D_\chi$. The result is

$$T_\chi(g)\varphi(\theta) = \varphi(\theta')\rho^{s-1}(\theta, g),$$

where

$$\rho^2(\theta, g) = (\alpha\cos\theta + \gamma\sin\theta)^2 + (\beta\cos\theta + \delta\sin\theta)^2,$$

and $\theta'$ is given by

$$\cos\theta' = \frac{\alpha\cos\theta + \gamma\sin\theta}{\rho(\theta, g)}, \qquad \sin\theta' = \frac{\beta\cos\theta + \delta\sin\theta}{\rho(\theta, g)}.$$

**Remark.** It is not difficult also to write the general form of $T_\chi(g)$ when $D_\chi$ is realized as the space of functions on some curve $l$. We assume that $l$ intersects every line $a_1x_1 + a_2x_2 = 0$. Then to every pair of numbers $(x_1, x_2) \neq (0, 0)$ there corresponds a point $M(\omega_1, \omega_2)$ on $l$ and a number $\rho(x_1, x_2)$ such that

$$x_1 = \rho(x_1, x_2)\omega_1, \qquad x_2 = \rho(x_1, x_2)\omega_2.$$

Proceeding in a simple way one finds that the formula for the representation is

$$T_\chi(g)f(\omega_1, \omega_2) = f\left(\frac{\alpha\omega_1 + \gamma\omega_2}{\rho_g(\omega_1, \omega_2)}, \frac{\beta\omega_1 + \delta\omega_2}{\rho_g(\omega_1, \omega_2)}\right)$$
$$\times |\rho_g(\omega_1, \omega_2)|^{s-1} \operatorname{sgn}^\epsilon \rho_g(\omega_1, \omega_2),$$

where $\rho_g(\omega_1, \omega_2) = \rho(\alpha\omega_1 + \gamma\omega_2, \beta\omega_1 + \delta\omega_2)$. #

### 1.5. The Dual Representations

Let $D'_\chi$ be the space dual to $D_\chi$, that is, the space of linear functionals $F$ on $D_\chi$. Then we can define a new representation $\tilde{T}_\chi(g)$ of $G$ on $D'_\chi$ according to

$$(\tilde{T}_\chi(g)F, \varphi) = (F, T_\chi^{-1}(g)\varphi).$$

Here $(F, \varphi)$ denotes, as usual, the value of the functional $F$ on the function $\varphi$. Direct calculation shows that $\tilde{T}_\chi(g_1g_2) = \tilde{T}_\chi(g_1)\,\tilde{T}_\chi(g_2)$ for all $g_1, g_2 \in G$, so that this is indeed a representation.

We shall call $\tilde{T}_\chi(g)$ the representation *dual* to $T_\chi(g)$. We assert that $\tilde{T}_\chi(g)$ is an extension of $T_{-\chi}(g)$, where $\chi = (s, \epsilon)$ and $-\chi = (-s, \epsilon)$. This means that $D_{-\chi}$ can be imbedded in $D'_\chi$ and that on $D_{-\chi}$ the operators $\tilde{T}_\chi(g)$ coincide with $T_{-\chi}(g)$.

For the proof, note that every function $\psi(x) \in D_{-\chi}$ defines a linear functional in $D_\chi$ in accordance with[2]

$$(\psi, \varphi) = \int_{-\infty}^{+\infty} \psi(x)\varphi(x)\, dx, \qquad \varphi \in D_\chi .$$

Thus $D_{-\chi}$ is included in $D'_\chi$. Further, direct calculation shows that

$$(T_{-\chi}(g)\psi, \varphi) = (\psi, T_\chi^{-1}(g)\varphi).$$

Consequently

$$(T_{-\chi}(g)\psi, \varphi) = (\tilde{T}_\chi(g)\psi, \varphi),$$

and therefore $\tilde{T}_\chi(g)\psi = T_{-\chi}(g)\psi$, which completes the proof.

## 2. Summary of the Basic Results concerning Representations on $D_\chi$

We shall now summarize the results on the representations of $G$ on the spaces we have called $D_\chi$. These results will be obtained later in Sections 4.1–5.7.

In many respects the representations of the group of real matrices are similar to the representations of the group of complex matrices studied in Chapter III. But they are different enough to be in some sense simpler and in another sense more complicated than in the complex case, just as the real numbers are in some sense simpler and in another sense more complicated than the complex numbers. For this reason and in spite of the fact that they are so similar to the results of Chapter III, we shall describe the results for the real case in the same detail. One may thus follow Chapter VII without having read Chapter III.

### 2.1. Irreducibility of Representations on $D_\chi$

It will be shown in Section 4 that every continuous operator on $D_\chi$ that commutes with the operators of a representation is a multiple of the unit

[2] That the integral converges is easily shown by using the asymptotic behavior of $\varphi(x)$ and $\psi(x)$ for large $|x|$ (see Section 1.2).

operator. In Chapter III we called this property *operator irreducibility* and we remarked that operator irreducibility is not necessarily true irreducibility, for a representation may be operator irreducible although it has invariant subspaces. In Section 3.1 of Chapter III we gave stronger definitions of irreducibility. It can be shown that the representation in $D_\chi$ for noninteger $\chi$ is irreducible in this stronger sense, but we shall not go into the proof here.

A special role is played by the representation $D_\chi$ for integer points $\chi = (s, \epsilon)$, that is, when $s$ is an integer having the same parity as $\epsilon + 1$. For these representations the operators are given by[3]

$$T_s(g)f(x) = f\left(\frac{\alpha x + \gamma}{\beta x + \delta}\right)(\beta x + \delta)^{s-1}. \tag{1}$$

For reasons that will become clear later we shall call these *analytic representations*. We shall see directly that for these integer points $D_s$ has invariant subspaces. Unlike the complex case, however, there are three such subspaces rather than one.

Consider first the case of nonnegative integer $s$. Obviously $D_s$ contains the invariant subspace $E_s$ consisting of polynomials of degree $s - 1$ and lower,[4] for it is seen that the $T_s(g)$ transform such polynomials into other polynomials of degree $s - 1$ and lower. The dimension of this space is clearly $s$. Since $E_s$ is an invariant subspace, the representation of $G$ can also be realized on the factor space $D_s/E_s$, that is, on the space of functions in $D_s$ defined only up to a polynomial of degree $s - 1$ or lower.

We shall show later that there exist also two other invariant subspaces in $D_s$, which we shall call $D_s^+$ and $D_s^-$. The first of these consists of functions which are the limits of functions analytic in the upper half-plane (up to polynomials of degree $s - 1$ and lower). The second is defined similarly with relation to the lower half-plane. We shall see also that $D_s^+ \cap D_s^- = E_s$ and that $D_s^+ \cup D_s^- = D_s$. Thus $D_s/E_s$ is the direct sum of $D_s^+/E_s$ and $D_s^-/E_s$. It can be shown also that these last two factor spaces have no invariant subspaces.

Now let us turn to $D_{-s}$, where $s$ is a positive integer. On such a space the representation is defined by

$$T_{-s}(g)f(x) = f\left(\frac{\alpha x + \gamma}{\beta x + \delta}\right)(\beta x + \delta)^{-s-1}. \tag{2}$$

[3] These representations are uniquely determined by $s$, and we shall often denote them by $T_s(g)$ and the spaces on which they act by $D_s$.

[4] We set $E_s = 0$ for $s = 0$.

Consider the subspace $F_{-s}$ of $D_{-s}$, which consists of functions $f(x)$ with vanishing moments

$$b_k = \int_{-\infty}^{\infty} x^k f(x)\, dx = 0,$$

where $k = 0, 1, ..., s - 1$. It is easily shown that when $f(x)$ is transformed according to Eq. (2) these $b_k$ transform among themselves and therefore $F_{-s}$ is an invariant subspace. The representation of $G$ can therefore also be realized on the factor space $D_{-s}/F_{-s}$. Obviously the elements of this factor space are uniquely determined by the $b_k$, $k = 0, 1, ..., s - 1$. Thus $D_{-s}/F_{-s}$ is finite dimensional of dimension $s$.

We shall see in Section 4 that $D_{-s}$ has two other invariant subspaces, namely $F_{-s}^+$, the space of functions which are the limits of functions analytic in the upper half-plane, and $F_{-s}^-$, defined similarly with respect to the lower half-plane. It can also be shown that $F_{-s}^+$ and $F_{-s}^-$ contain no invariant subspaces. These spaces do not intersect, and their direct sum is $F_{-s}$.

## 2.2. Equivalence of Representations on $D_\chi$ and the Role of Integer Points

The concept of the equivalence of two representations was introduced in Section 3.2 of Chapter III. We shall show in Section 4.2 that two representations $D_{\chi_1}$ and $D_{\chi_2}$ are equivalent if and only if one of the following two conditions is fulfilled.

*Case 1.* $\chi_1 = \chi_2$.

*Case 2.* $\chi_1 = (s, \epsilon)$, $\chi_2 = (-s, \epsilon)$, and $\chi_1$ is not an integer point (that is, $s$ is not an integer having the same parity as $\epsilon + 1$).

It will be shown also that if $A$ maps $D_{\chi_1}$ onto $D_{\chi_2}$ so that

$$AT_{\chi_1}(g) = T_{\chi_2}(g)A, \tag{1}$$

i. e., if $A$ is an intertwining operator for $T_{\chi_1}$ and $T_{\chi_2}$, then in each of these cases $A$ is uniquely determined up to a multiplicative factor. In the trivial Case 1, that is, if $D_{\chi_1}$ and $D_{\chi_2}$ coincide, $A$ is a multiple of the unit operator. In Case 2 it is given by

$$A\varphi(x) = \frac{1}{\Gamma(-s)} \int_{-\infty}^{+\infty} | x_1 - x |^{-s-1} \operatorname{sgn}^\epsilon(x_1 - x)\, \varphi(x_1)\, dx_1, \tag{2}$$

where the integral is understood in the sense of its regularization. For nonnegative $s$, in particular, $\epsilon$ has the same parity as $s$, and $A$ is given up to a multiplicative factor by

$$A\varphi(x) = \varphi^{(s)}(x).$$

**Remark.** The operator $B$ which is the inverse of $A$ is given, up to a multiplicative factor, by replacing $s$ in (2) by $-s$:

$$B\varphi(x) = \frac{1}{\Gamma(s)} \int_{-\infty}^{+\infty} | x_1 - x |^{s-1} \operatorname{sgn}^{\epsilon}(x_1 - x) \, \varphi(x_1) \, dx_1 \, . \quad \# \tag{2'}$$

Once we have agreed to call essentially different only inequivalent representations, we may speak of the set of all inequivalent representations of $G$. The elements of this set will actually be classes of equivalent representations. Since each $T_{\chi}(g)$ representation is given by $s - 1$ and by $\epsilon$, we may introduce the concept of the "Riemann surface" of representations. The points of this Riemann surface are given by a complex number $s$ and by $\epsilon = 0$ or 1. Thus the universal covering space of this representation space is a pair of complex planes of the variable $s$. We shall call a branch point of this Riemann surface a point such that each of its neighborhoods contains equivalent representations. The order of the branch point is defined in the natural way. In view of what has been said already, the only branch points on the Riemann surface are those with $s = 0$ ($\epsilon = 0, 1$). These branch points both have order 2.

### 2.3. The Problem of Equivalence at Integer Points

Let us now turn to the representations

$$T_s f(x) = f\left(\frac{\alpha x + \gamma}{\beta x + \delta}\right) (\beta x + \delta)^{s-1},$$

$$T_{-s} f(x) = f\left(\frac{\alpha x + \gamma}{\beta x + \delta}\right) (\beta x + \delta)^{-s-1}$$

for $s \neq 0$ an integer. To be specific we shall assume $s$ positive. In Section 4.2 we shall see that there exists an operator $A$ which maps $D_{-s}$ into $D_s$, intertwining with the representations. This operator is defined uniquely (up to a constant multiplicative factor) and may be written

$$A\varphi(x) = \int_{-\infty}^{\infty} (x_1 - x)^{s-1} \varphi(x_1) \, dx_1 \, .$$

Obviously $A$ does not map $D_{-s}$ into $D_s$ isomorphically, since it maps every $\varphi(x) \in D_{-s}$ into a polynomial of degree $s - 1$ or less. Thus the representations on $D_s$ and $D_{-s}$ are not equivalent. The kernel of the mapping $A$ is $F_{-s}$, so that $A$ maps $D_{-s}/F_{-s}$ isomorphically onto $E_s$ (see Section 2.1).

Now consider intertwining linear mappings in the other direction, i.e., from $D_s$ into $D_{-s}$. In this case the situation is somewhat more interesting. It will be shown that there exist two linearly independent linear operators $A_+$ and $A_-$ mapping $D_s$ into $D_{-s}$ and intertwining with the representations. These operators are

$$A_+\varphi(x) = \frac{1}{2\pi i} \int_{-\infty}^{\infty} \frac{\varphi^{(s)}(x_1)\, dx_1}{x_1 - x - i0},$$

$$A_-\varphi(x) = -\frac{1}{2\pi i} \int_{-\infty}^{\infty} \frac{\varphi^{(s)}(x_1)\, dx_1}{x_1 - x + i0}.$$

It will be shown also that the kernel of $A_+$ is $D_s^-$ and that its range is $F_{-s}^+$, so that it maps $D_s/D_s^-$ isomorphically onto $F_{-s}^+$. Similarly, $A_-$ maps $D_s/D_s^+$ isomorphically onto $F_{-s}^-$.

We shall see also that there exist no pairs of representations other than those already discussed in Sections 2.2 and 2.3 that possess intertwining operators.

Remark. In Section 3.3 of Chapter III we called two irreducible representations *closest relatives* if they were induced by a given operator irreducible representation, one of them on an invariant subspace and the other in the corresponding factor space. We have also called two irreducible representations *related* if they could be connected by a finite chain of closest relatives. It can be shown that the finite-dimensional representation on $E_s$ is a closest relative of the representations on $F_{-s}^+$ and $F_{-s}^-$. Thus the representations on these two subspaces of $D_{-s}$ are related. It can be shown that they are not, however, closest relatives, and that there exist no other such relations between the irreducible representations of $G$. #

## 2.4. Unitary Representations

In Sections 5.1 and 5.2 we shall find the conditions under which there exists in the carrier space of a representation a Hermitian positive definite functional $(\varphi, \psi)$ invariant under the representation. When such a functional exists it can be thought of as a scalar product, and when the space is completed with respect to this scalar product, the representa- becomes a unitary representation of $G$ on a Hilbert space.

It will be shown that a scalar product can be introduced in $D_\chi$, $\chi = (s, \epsilon)$, in the following two cases.

*Case 1.* $s = i\rho$ is imaginary, and $\epsilon = 0$ or 1. Then the scalar product is of the form

$$(\varphi, \psi) = \int_{-\infty}^{+\infty} \varphi(x)\bar{\psi}(x)\, dx.$$

We shall call a representation on $D_\chi$, with $\chi = (i\rho, \epsilon)$, a representation of the principal series.

*Case 2.* $\epsilon = 0$, and $s \neq 0$ is a real number such that $-1 < s < 1$. For this case the scalar product is

$$(\varphi, \psi) = \frac{1}{\Gamma(-s)} \int_{-\infty}^{\infty} \int_{-\infty}^{\infty} | x_1 - x_2 |^{-s-1} \varphi(x_1)\bar{\psi}(x_2)\, dx_1\, dx_2 .$$

A representation on such a $D_\chi$ will be called a representation of the supplementary series.

It is remarkable that these two cases do not include all the irreducible unitary representations of $G$. There exist also unitary representations acting on invariant subspaces of $D_{-s}$ for nonnegative integers $s$. Recall that $D_{-s}$ is the space on which the representation is given by

$$T_{-s}(g)f(x) = f\left(\frac{\alpha x + \gamma}{\beta x + \delta}\right) (\beta x + \delta)^{-s-1},$$

and that it contains the two invariant subspaces $F_{-s}^{+}$ and $F_{-s}^{-}$ consisting, respectively, of the limits of functions analytic in the upper half-plane and the limits of functions analytic in the lower half-plane. We shall see that it is not possible to introduce a scalar product on all of $D_{-s}$, but that it is possible to introduce one on each of these two invariant subspaces. In exhibiting these scalar products, it is convenient to realize $F_{-s}^{+}$, for instance, not as a space of functions on the real axis which are the limits of certain analytic functions, but as the space of functions $\varphi(z)$ analytic in the upper half-plane. Then the scalar product in $F_{-s}^{+}$ is given by

$$(\varphi, \psi) = \frac{1}{\Gamma(s)} \frac{i}{2} \int \varphi(z)\bar{\psi}(z)(\operatorname{Im} z)^{s-1}\, dz\, d\bar{z},$$

where the integral is taken over the upper half-plane. A similar formula gives the scalar product in $F_{-s}^{-}$. We shall call the representations on $F_{-s}^{+}$ and $F_{-s}^{-}$ for nonnegative integers $s$ representations of the *discrete series.*

## 3. Invariant Bilinear Functionals

The results we have been summarizing in the preceding few paragraphs on the representations of $G$ will all be obtained by a single method involving the use of invariant bilinear functionals. Specifically, we shall start by solving the following problem. Consider two representations $T_{\chi_1}(g)$ and $T_{\chi_2}(g)$ of $G$. We may then ask when these representations possess an invariant bilinear functional Recall that a bilinear functional $B(\varphi, \psi)$, for $\varphi \in D_{\chi_1}$ and $\psi \in D_{\chi_2}$, is called invariant if

$$B(T_{\chi_1}(g)\varphi, T_{\lambda_2}(g)\psi) = B(\varphi, \psi)$$

for every $g \in G$.

This problem will be solved in the next few sections. The solution will proceed essentially in the same way as the solution to the analogous problem in Chapter III; that is, we shall consider not all matrices of $G$, but just the following three:[5]

$$g_1 = \begin{Vmatrix} 1 & 0 \\ x_0 & 1 \end{Vmatrix}, \qquad g_2 = \begin{Vmatrix} \alpha & 0 \\ 0 & \alpha^{-1} \end{Vmatrix}, \qquad g_3 = \begin{Vmatrix} 0 & 1 \\ -1 & 0 \end{Vmatrix}.$$

It can be shown that every matrix of $G$ can be written as a product of such matrices, so that in determining whether a bilinear functional is invariant it is sufficient to consider the operators corresponding to just these three forms.

## 3.1. Invariance under Translation and Dilation

Consider two representations of $G$,

$$T_{\chi_1}(g)\varphi(x) = |\beta x + \delta|^{s_1-1} \operatorname{sgn}^{\epsilon_1}(\beta x + \delta)\, \varphi\left(\frac{\alpha x + \gamma}{\beta x + \delta}\right) \tag{1}$$

and

$$T_{\chi_2}(g)\psi(x) = |\beta x + \delta|^{s_2-1} \operatorname{sgn}^{\epsilon_2}(\beta x + \delta)\, \psi\left(\frac{\alpha x + \gamma}{\beta x + \delta}\right), \tag{1'}$$

acting on $D_{\chi_1}$ and $D_{\chi_2}$. We want to find a bilinear functional $B(\varphi, \psi)$ invariant with respect to these representations for $g$ in one of the two forms

$$g = \begin{Vmatrix} 1 & 0 \\ x_0 & 1 \end{Vmatrix} \quad \text{and} \quad g = \begin{Vmatrix} \alpha & 0 \\ 0 & \alpha^{-1} \end{Vmatrix},$$

which we shall call, respectively, the translation form and the dilation form. Further, we shall restrict our considerations temporarily to

[5] These matrices give rise to the translations $x \to x + x_0$, the dilations $x \to \alpha^2 x$, and the inversions $x \to -1/x$.

infinitely differentiable functions $\varphi(x)$ and $\psi(x)$ with bounded support. The representations of matrices of the translation form are

$$T_{\chi_1}(g)\varphi(x) = \varphi(x + x_0),$$

$$T_{\chi_2}(g)\psi(x) = \psi(x + x_0).$$

Thus for this form the invariance of the bilinear functional becomes

$$B(\varphi, \psi) = B(\varphi(x + x_0), \psi(x + x_0)).$$

We now make use of the following result. Every bilinear functional $B(\varphi, \psi)$ invariant with respect to translations on the space $K$ of infinitely differentiable functions of bounded support is of the form

$$B(\varphi, \psi) = (B_0 , \omega). \tag{2}$$

Here $B_0$ is a linear functional on $K$, and $\omega(x)$ is the function defined by[6]

$$\omega(x) = \int \varphi(x_1)\psi(x + x_1)\, dx_1 . \tag{3}$$

Thus we know the general form of a bilinear functional invariant under operators corresponding to matrices of the translation form.

We wish further that $B(\varphi, \psi)$ be invariant under $T_\chi(g)$ for $g$ in the dilation form. These operators are given by

$$T_{\chi_1}(g)\varphi(x) = |\alpha|^{-s_1+1} \operatorname{sgn}^{\epsilon_1}\alpha\, \varphi(\alpha^2 x)$$

and

$$T_{\chi_2}(g)\psi(x) = |\alpha|^{-s_2+1} \operatorname{sgn}^{\epsilon_2}\alpha\, \psi(\alpha^2 x).$$

The condition that $B(\varphi, \psi)$ be invariant undert these operators may consequently be written

$$B(\varphi, \psi) = |\alpha|^{-s_1-s_2+2} \operatorname{sgn}^{\epsilon_1+\epsilon_2}\alpha\, B(\varphi(\alpha^2 x), \psi(\alpha^2 x)). \tag{4}$$

Note first that *this requiries that $T_{\chi_1}(g)$ and $T_{\chi_2}(g)$ have the same parity.* Indeed, if we set $\alpha = -1$ in (4), we obtain

$$B(\varphi, \psi) = (-1)^{\epsilon_1+\epsilon_2} B(\varphi, \psi),$$

which implies that $\epsilon_1 = \epsilon_2$ .

[6] See Chapter III, Section 4.2, the discussion after Eq. (4).

We now substitute $(B_0, \omega)$ for $B(\varphi, \psi)$ in (4). By bearing in mind that

$$\int \varphi(\alpha^2 x_1)\psi(\alpha^2[x + x_1])\, dx_1$$
$$= \alpha^{-2} \int \varphi(x_1)\psi(\alpha^2 x + x_1)\, dx_1 = \alpha^{-2}\omega(\alpha^2 x),$$

we arrive at

$$(B_0, \omega) = |\alpha|^{-s_1-s_2}(B_0, \omega(\alpha^2 x)).$$

Now let $\alpha > 0$ and replace $\alpha$ by $\alpha^{\frac{1}{2}}$. Then this equation becomes

$$(B_0, \omega) = \alpha^{-\frac{1}{2}(s_1+s_2)}(B_0, \omega(\alpha x)), \tag{5}$$

which shows that $B_0$ is a homogeneous generalized function of degree $\lambda = -\frac{1}{2}(s_1 + s_2) - 1$.

For the reader's benefit we recall some of the basic properties of homogeneous generalized functions of a single variable.[7] For every complex number $\lambda$ there exist, up to a constant factor, one even and one odd homogeneous generalized function of degree $\lambda$, and every other homogeneous generalized function of this degree is a linear combination of these. They are defined as follows. For $\lambda \neq -1, -2, \ldots$, the even generalized function $|x|^\lambda$ is given by

$$(|x|^\lambda, \varphi) = \int_{-\infty}^{+\infty} |x|^\lambda \varphi(x)\, dx,$$

and the odd generalized function $|x|^\lambda \operatorname{sgn} x$ is given by

$$(|x|^\lambda \operatorname{sgn} x, \varphi) = \int_{-\infty}^{\infty} |x|^\lambda \operatorname{sgn} x\, \varphi(x)\, dx.$$

Both of these integrals converge for $\operatorname{Re}\lambda > -1$. For $\operatorname{Re}\lambda < -1$ they are understood in the sense of their regularizations, which is to say that they are given by analytic continuation in $\lambda$. For $\lambda = -k$, where $k$ is a positive integer, there exist two linearly independent homogeneous generalized functions of degree $\lambda$; these are $\delta^{(k-1)}(x)$ and $x^{-k}$. One of these is even and the other is odd.

We have thus established that an invariant bilinear functional $B(\varphi, \psi)$ can exist only for representations $T_{\chi_1}(g)$ and $T_{\chi_2}(g)$ of the same parity. When it exists, this functional is given for each pair of functions $\varphi$ and $\psi$ of bounded support by the expression

$$B(\varphi, \psi) = \int B_0(x) \int \varphi(x + x_1)\psi(x)\, dx_1\, dx,$$

[7] See Volume 1, pages 80 and 81.

where $B_0(x)$ is a generalized homogeneous function of degree $-\frac{1}{2}(s_1 + s_2) - 1$. In general $B_0(x)$ is of one of the two following forms. If $\frac{1}{2}(s_1 + s_2) \neq 0, 1, 2, \ldots$, then

$$B_0(x) = C_1 \mid x \mid^{-\frac{1}{2}(s_1+s_2)-1} + C_2 \mid x \mid^{-\frac{1}{2}(s_1+s_2)-1} \operatorname{sgn} x. \tag{6}$$

If $\frac{1}{2}(s_1 + s_2) = 0, 1, 2, \ldots$, then

$$B_0(x) = C_1 \delta^{(\frac{1}{2}s_1+\frac{1}{2}s_2)}(x) + C_2 x^{-\frac{1}{2}(s_1+s_2)-1}. \tag{7}$$

### 3.2. Necessary and Sufficient Conditions for the Existence of an Invariant Bilinear Functional

Let us now find the values of $s_1$ and $s_2$ for which an invariant bilinear functional may exist. Assume that such a functional exists, so that according to Section 3.1 it is given for $\varphi$ and $\psi$ by

$$B(\varphi, \psi) = \int B_0(x_1 - x_2)\varphi(x_1)\psi(x_2)\, dx_1\, dx_2\,,$$

where $B_0(x)$ is a homogeneous generalized function of degree $-\frac{1}{2}(s_1 + s_2) - 1$. Let us now use, in addition to the requirement of invariance under translation and dilation, the requirement that the functional be invariant under inversion. Assume, therefore, that $g_0$ is given by

$$g_0 = \begin{Vmatrix} 0 & 1 \\ -1 & 0 \end{Vmatrix},$$

so that

$$T_{\chi_1}(g_0)\varphi(x) = \mid x \mid^{s_1-1} \operatorname{sgn}^{\epsilon} x\, \varphi(-1/x), \tag{1}$$

$$T_{\chi_2}(g_0)\psi(x) = \mid x \mid^{s_2-1} \operatorname{sgn}^{\epsilon} x\, \psi(-1/x). \tag{1'}$$

Now since these transformations will in general transform functions of bounded support into functions whose support may not be bounded, we shall have to place additional restrictions on $\varphi(x)$ and $\psi(x)$. Let us require, therefore, that each of these functions vanish in some neighborhood of $x = 0$. In this way $T_{\chi_1}(g_0)\varphi$ and $T_{\chi_2}(g_0)\psi$ will have bounded support.

Now the condition that the bilinear functional be invariant under these operators becomes

$$\begin{aligned} &\int B_0(x_1 - x_2)\varphi(x_1)\psi(x_2)\, dx_1\, dx_2 \\ &\quad = \int B_0(x_1 - x_2) \mid x_1 \mid^{s_1-1} \mid x_2 \mid^{s_2-1} \operatorname{sgn}^{\epsilon} (x_1 x_2) \\ &\qquad \times \varphi(-1/x_1)\psi(-1/x_2)\, dx_1\, dx_2\,. \end{aligned} \tag{2}$$

Now let us write $x_1 = -1/x_1'$ and $x_2 = -1/x_2'$. Then this equation becomes

$$\int\int B_0(x_1 - x_2)\varphi(x_1)\psi(x_2)\, dx_1\, dx_2$$
$$= \int B_0\left(\frac{x_1 - x_2}{x_1x_2}\right) | x_1 |^{-s_1-1} | x_2 |^{-s_2-1} \operatorname{sgn}^{\epsilon}(x_1x_2)$$
$$\times \varphi(x_1)\, \psi(x_2)\, dx_1\, dx_2\,, \tag{2'}$$

We wish to find now the values of $s_1$ and $s_2$ for which this is possible.

Consider first the case in which $\frac{1}{2}(s_1 + s_2) \neq 0, 1, \ldots$. Then $B_0(x)$ is of the first form given by Eq. (6) of Section 3.1. We shall write[8]

$$B_0(x) = | x |^{-\frac{1}{2}(s_1+s_2)-1} \operatorname{sgn}^{\nu} x, \qquad \nu = 1, 0.$$

In order to ensure that the integrals in (2′) converge in the ordinary sense, we must make some further assumptions on $\varphi(x)$ and $\psi(x)$. Specifically, we assume that these functions have nonintersecting supports, that is, that there exist on the line closed sets $A_1$ and $A_2$ containing no points in common and such that $\varphi(x) = 0$ outside of $A_1$ and $\psi(x) = 0$ outside of $A_2$. Then the integrals on both sides of (2′) converge, and we obtain

$$\int | x_1 - x_2 |^{-\frac{1}{2}(s_1+s_2)-1} \operatorname{sgn}^{\nu}(x_1 - x_2)\, \varphi(x_1)\, \psi(x_2)\, dx_1\, dx_2$$
$$= \int | x_1 - x_2 |^{-\frac{1}{2}(s_1+s_2)-1} \operatorname{sgn}^{\nu}(x_1 - x_2)\, | x_1 |^{\frac{1}{2}(s_2-s_1)}\, | x_2 |^{\frac{1}{2}(s_1-s_2)}$$
$$\times \operatorname{sgn}^{\epsilon-\nu}(x_1x_2)\, \varphi(x_1)\psi(x_2)\, dx_1\, dx_2\,.$$

Clearly this can be true only if $\nu = \epsilon$ and $s_1 = s_2$. We conclude, therefore, that *if $T_{\chi_1}(g)$ and $T_{\chi_2}(g)$ are representations of $G$ of the same parity and with indices $s_1$ and $s_2$, respectively, such that $\frac{1}{2}(s_1 + s_2) \neq 0, 1, \ldots,$ then a bilinear functional invariant with respect to these representations can exist only if $s_1 = s_2$. When it exists, this functional is unique up to a constant multiple. For functions with bounded support it is given by*

$$B(\varphi, \psi) = \int | x_1 - x_2 |^{-s_1-1} \operatorname{sgn}^{\epsilon}(x_1 - x_2)\, \varphi(x_1)\psi(x_2)\, dx_1\, dx_2\,. \tag{3}$$

For Re $s_1 > 0$, the integral is understood in the sense of its regularization.

Let us now turn to the case in which $\frac{1}{2}(s_1 + s_2) = 0, 1, \ldots$. In this case $B_0(x)$ is given by Eq. (7) of Sections 3.1. Let us assume first that

$$B_0(x) = \delta^{(\frac{1}{2}s_1+\frac{1}{2}s)}(x).$$

[8] This might not seem to be a sufficiently general form, but it can be shown that either $C_1$ or $C_2$ must be zero in the expression for $B_0(x)$ given in Eq. (6) of Section 3.1 if $B_0(x)$ is to be invariant.

Then Eq. (2′) becomes

$$\int \delta^{(\frac{1}{2}s_1+\frac{1}{2}s_2)}(x_1 - x_2)\varphi(x_1)\psi(x_2)\,dx_1\,dx_2$$
$$= \int \delta^{(\frac{1}{2}s_1+\frac{1}{2}s_2)}\left(\frac{x_1 - x_2}{x_1x_2}\right) | x_1 |^{-s_1-1} | x_2 |^{-s_2-1} \operatorname{sgn}^\epsilon (x_1x_2)$$
$$\times \varphi(x_1)\psi(x_2)\,dx_1\,dx_2\,. \tag{4}$$

This equation can be simplified by using the fact that $\varphi(x_1)\psi(x_2) = 0$ in a neighborhood of the lines $x_1 = 0$ and $x_2 = 0$. Thus a power of $x_1x_2$ can be carried outside the $\delta$ function.[9] Then (4′) becomes

$$\int \delta^{(\frac{1}{2}s_1+\frac{1}{2}s_2)}(x_1 - x_2)\varphi(x_1)\psi(x_2)\,dx_1\,dx_2$$
$$= \int \delta^{(\frac{1}{2}s_1+\frac{1}{2}s_2)}(x_1 - x_2)| x_1 |^{-s_1} | x_2 |^{-s_2}(x_1x_2)^{\frac{1}{2}(s_1+s_2)} \operatorname{sgn}^\epsilon (x_1x_2)$$
$$\times \varphi(x_1)\psi(x_2)\,dx_1\,dx_2\,,$$

so that

$$\int \varphi^{(\frac{1}{2}s_1+\frac{1}{2}s_2)}(x)\psi(x)\,dx = \int [| x |^{\frac{1}{2}(s_2-s_1)}\varphi(x)]^{(\frac{1}{2}s_1+\frac{1}{2}s_2)} | x |^{\frac{1}{2}(s_1-s_2)}\psi(x)\,dx.$$

It follows that

$$\varphi^{(\frac{1}{2}s_1+\frac{1}{2}s_2)}(x) = | x |^{\frac{1}{2}(s_1-s_2)}[| x |^{\frac{1}{2}(s_2-s_1)}\varphi(x)]^{(\frac{1}{2}s_1+\frac{1}{2}s_2)}.$$

This equation is obviously valid for $s_1 = s_2$. If, however, $s_1 \neq s_2$, then $s_1 = -s_2$, since otherwise differentiation would yield more than one term on the right-hand side.

*Thus if* $\frac{1}{2}(s_1 + s_2) = 0, 1, \ldots$ *and the bilinear functional is given by the kernel* $\delta^{(\frac{1}{2}s_1+\frac{1}{2}s_2)}(x_1 - x_2)$, *the functional can be invariant only in two cases:*

*Case 1.* $s_1 = s_2$.

*Case 2.* $s_1 = -s_2$.

Let us now assume[10] that

$$B_0(x) = x^{-\frac{1}{2}(s_1+s_2)-1}.$$

[9] See the analogous discussion following Eq. (4) of Chapter III, Section 4.5.

[10] It might seem that we ought to treat the general case of a linear combination of the $\delta$-function kernel we have just been discussing plus the one we are now starting to discuss. It can be shown, however, that nothing new is added in this way.

Then again assuming that $\varphi(x)$ and $\psi(x)$ have nonintersecting supports, we have from Eq. (2′)

$$\int (x_1 - x_2)^{-\frac{1}{2}(s_1+s_2)-1}\varphi(x_1)\psi(x_2)\, dx_1\, dx_2$$

$$= \int (x_1 - x_2)^{-\frac{1}{2}(s_1+s_2)-1} \mid x_1 \mid^{-s_1-1} \mid x_2 \mid^{-s_2-1}(x_1x_2)^{\frac{1}{2}(s_1+s_2)+1}$$

$$\times \operatorname{sgn}^{\epsilon}(x_1x_2)\, \varphi(x_1)\psi(x_2)\, dx_1\, dx_2\,. \tag{5}$$

Obviously this can be a true statement only if $s_1 = s_2$ (recall that $s_1$ and $s_2$ are nonnegative integers) and $s_1 + 1$ is of the same parity as $\epsilon$. We have found such conditions for $s_1$ and $s_2$ before, when treating the functional whose kernel is $\delta^{(\frac{1}{2}s_1+\frac{1}{2}s_2)}(x_1 - x_2)$.

Note that the representation operators in the present case are given by

$$T_\chi(g)\varphi(x) = (\beta x + \delta)^{s_1-1}\varphi\left(\frac{\alpha x + \gamma}{\beta x + \delta}\right),$$

$$T_\chi(g)\psi(x) = (\beta x + \delta)^{s_2-1}\psi\left(\frac{\alpha x + \gamma}{\beta x + \delta}\right),$$

where $s_1$ and $s_2$ are integers. We call such representations *analytic representations*.[11] It is seen that analytic representations may in general possess two different bilinear functionals. This is a peculiarity of the real group. In the complex group, as we have seen in Chapter III, the invariant bilinear functional on $D_\chi$ is always uniquely defined up to a constant multiple.

Thus we have found the necessary conditions for the existence of bilinear functionals invariant under $T_{\chi_1}(g)$ and $T_{\chi_2}(g)$ for $\frac{1}{2}(s_1 + s_2) = 0, 1, \ldots$ . Such functionals can exists only if the representations have the same parity, that is, if $\epsilon_1 = \epsilon_2 = \epsilon$ and if either $s_1 = s_2$ or $s_1 = -s_2$ . When it exists, the bilinear functional is unique up to a constant factor, except for a special case. If $T_{\chi_1}(g)$ and $T_{\chi_2}(g)$ are analytic representations such that $s_1 = s_2 = 0, 1, \ldots$, there may exist two independent invariant bilinear functionals.

[11] As was mentioned in Section 2.1, the spaces on which these representations act contain invariant subspaces. Specifically, for $s_1 = 0, -1, -2, \ldots$ there exist two invariant subspaces, one of which consists of functions which are the limits of functions analytic in the upper half-plane, and the other consists of functions obtained similarly from the lower half-plane. The situation is similar for $s_1 = 1, 2, \ldots$ . This is why these representations are called analytic.

It can be shown that these necessary conditions are also sufficient conditions for the existence of invariant bilinear functionals. The proof is almost identical with that for the complex group (see Chapter III), so we shall omit it and merely summarize the final results.

Consider two representations of the real group

$$T_{\chi_1}(g)\varphi(x) = |\beta x + \delta|^{s_1 - 1} \operatorname{sgn}^{\epsilon_1}(\beta x + \delta)\, \varphi\left(\frac{\alpha x + \gamma}{\beta x + \delta}\right)$$

and

$$T_{\chi_2}(g)\psi(x) = |\beta x + \delta|^{s_2 - 1} \operatorname{sgn}^{\epsilon_2}(\beta x + \delta)\, \psi\left(\frac{\alpha x + \gamma}{\beta x + \delta}\right).$$

These representations will have an invariant bilinear functional if and only if they are of the same parity (that is, $\epsilon_1 = \epsilon_2$) and if either $s_1 = s_2$ or $s_1 = -s_2$. The desired functional $B(\varphi, \psi)$ is defined as follows.

For $s_1 = -s_2$

$$B(\varphi, \psi) = \int_{-\infty}^{\infty} \varphi(x)\psi(x)\, dx. \tag{6}$$

For $s_1 = s_2 \neq 0, 1, \ldots$, $B(\varphi, \psi)$ is defined for functions $\varphi(x)$ and $\psi(x)$ with bounded support by

$$B(\varphi, \psi) = \int_{-\infty}^{\infty} |x_1 - x_2|^{-s_1 - 1} \operatorname{sgn}^{\epsilon_1}(x_1 - x_2)\, \varphi(x_1)\psi(x_2)\, dx_1\, dx_2 \tag{7}$$

(understood in the sense of its regularization for $\operatorname{Re} s_1 > 0$). The definition is extended to arbitrary functions $\varphi$, $\psi$ by linearity and invariance (see Chapter III, Section 4.4).

For $s_1 = s_2 = 0, 1, \ldots$, but nonanalytic representations,[12]

$$B(\varphi, \psi) = \int_{-\infty}^{+\infty} \varphi^{(s_1)}(x)\psi(x)\, dx. \tag{8}$$

In the special case in which $s_1 = s_2 = 0, 1, \ldots$ and the representations are analytic, in addition to this last functional there exists

$$B(\varphi, \psi) = \int_{-\infty}^{+\infty} (x_1 - x_2)^{-s_1 - 1}\varphi(x_1)\psi(x_2)\, dx_1\, dx_2 \tag{9}$$

(understood in the sense of its regularization).

[12] It is easily shown that this integral converges for arbitrary $\varphi(x)$ and $\psi(x)$.

### 3.3. Degenerate Invariant Bilinear Functionals for Analytic Representations

Let us consider in more detail the question of invariant bilinear functionals for analytic representations of $G$, that is, for representations of the form

$$T_s(g)\varphi(x) = (\beta x + \delta)^{s-1}\varphi\left(\frac{\alpha x + \gamma}{\beta x + \delta}\right),$$

where $s$ is an integer.

Consider first the case in which $s$ is a nonnegative integer. Then for functions with bounded support we have

$$B(\varphi, \psi) = \int B_0(x_1 - x_2)\varphi(x_1)\psi(x_2)\, dx_1\, dx_2\,,$$

where $B_0(x)$ may be any homogeneous generalized function of degree $-s-1$. Thus there exist two linearly independent invariant bilinear functionals. Let us try to obtain formulas defining these for *arbitrary* functions in $D_s$. We choose $(x - i0)^{-s-1}$ and $(x + i0)^{-s-1}$ as our linearly independent homogeneous generalized functions of degree $-s-1$,[13] and study the functionals corresponding to them. For $\varphi(x)$ and $\psi(x)$ with bounded support, these functionals are defined by

$$B_+(\varphi, \psi) = \int (x_1 - x_2 - i0)^{-s-1}\varphi(x_1)\psi(x_2)\, dx_1\, dx_2\,, \tag{1}$$

$$B_-(\varphi, \psi) = \int (x_1 - x_2 + i0)^{-s-1}\varphi(x_1)\psi(x_2)\, dx_1\, dx_2\,. \tag{1'}$$

Let us rewrite these equations so that they will remain meaningful for arbitrary functions in $D_s$. We do this by associating with every $\varphi(x)$ with bounded support the two functions

$$\varphi_+(x) = \frac{1}{2\pi i}\int \frac{\varphi(x_1)}{x_1 - x - i0}\, dx_1\,, \tag{2}$$

$$\varphi_-(x) = -\frac{1}{2\pi i}\int \frac{\varphi(x_1)}{x_1 - x + i0}\, dx_1\,. \tag{2'}$$

These functions are the boundary values of functions analytic on the upper and lower half-planes, respectively, and

$$\varphi(x) = \varphi_+(x) + \varphi_-(x).$$

[13] See Volume 1, pages 59 and 60.

It is then clear that $B_+$ and $B_-$ can be written in the forms[14]

$$B_+(\varphi, \psi) = \int \varphi_+^{(s)}(x)\psi(x)\,dx, \tag{3}$$

$$B_-(\varphi, \psi) = \int \varphi_-^{(s)}(x)\psi(x)\,dx. \tag{3'}$$

These integrals are known to converge if $\varphi(x)$ and $\psi(x)$ have bounded support. We assert that *they converge also for arbitrary $\varphi(x)$ and $\psi(x)$ in $D_s$ and hence define invariant bilinear functionals on all of $D_s$.*

Proof. We first show that $\varphi_+^{(s)}(x) \equiv [\varphi^{(s)}(x)]_+$ and $\varphi_-^{(s)}(x) \equiv [\varphi^{(s)}(x)]_-$ exist for every $\varphi(x) \in D_s$. Indeed, we have the asymptotic expansion

$$\varphi(x) \sim \sum_{k=-\infty}^{s-1} a_k x^k \tag{4}$$

as $|x| \to \infty$. By differentiation, then,

$$\varphi^{(s)}(x) \sim \sum_{k=-\infty}^{-s-1} b_k x^k$$

and therefore for large $|x|$

$$\varphi^{(s)}(x) \sim Cx^{-s-1};$$

hence this function can be used to define $\varphi_+^{(s)}(x)$ and $\varphi_-^{(s)}(x)$, which therefore exist.

Further, as is easily verified, asymptotically for large $|x|$ these functions behave according to

$$\varphi_+^{(s)}(x) \sim C_1 x^{-s-1}, \qquad \varphi_-^{(s)}(x) \sim C_2 x^{-s-1}. \tag{5}$$

In the same limit, on the other hand,

$$\psi(x) \sim Cx^{s-1},$$

and these results together imply the convergence of the integrals in (3) and (3′), which thus are invariant bilinear functionals defined on all of $D_s$ for an analytic representation.

From this result we can deduce an interesting structural property of $D_s$. Note that $B_+(\varphi, \psi)$ and $B_-(\varphi, \psi)$ are degenerate functionals, in that there exists a subspace $D_s^- \subset D_s$ of $\varphi(x)$ functions such that $B_+(\varphi, \psi) = 0$ for every $\psi \in D_s$. In fact, Eq. (3) shows that $D_s^-$ contains

[14] We have dropped the constant factors.

all $\varphi(x)$ functions such that $\varphi_+^{(s)}(x) = 0$, or in other words such that $\varphi^{(s)}(x) = \varphi_-^{(s)}(x)$. Obviously this condition means that $\varphi(x)$ is, to within a polynomial of degree $s - 1$, the boundary value of a function analytic on the lower half-plane. Further, $D_s^-$ is invariant, as follows immediately from the invariance of $B_+(\varphi, \psi)$.

In a very similar way, $B_-(\varphi, \psi)$ is degenerate on a subspace $D_s^+ \subset D_s$ of $\varphi(x)$ functions which, up to a polynomial of degree $s - 1$, are boundary values of functions analytic on the upper half-plane.

Finally, the intersection of $D_s^+$ and $D_s^-$ is the finite-dimensional subspace $E_s$ of all polynomials of degree $s - 1$ and less. Summarizing, the carrier space $D_s$, $s = 1, 2, \ldots$, of an analytic representation contains three invariant subspaces, one finite dimensional and two infinite dimensional.

Let us now turn to negative integers $s$. In this case there exists to within a constant multiple only the one invariant bilinear functional

$$B(\varphi, \psi) = \int (x_1 - x_2)^{-s-1} \varphi(x_1) \psi(x_2) \, dx_1 \, dx_2 \, . \tag{6}$$

This integral converges for all functions in $D_s$, and thereby defines a functional on the entire space. Obviously $B(\varphi, \psi)$ is degenerate on the subspace $F_s$ of $\varphi(x)$ function for which

$$\int x^k \varphi(x) \, dx = 0 \qquad \text{for} \quad k = 0, \ldots, -s - 1.$$

Further, $F_s$ is an invariant subspace. We shall see later that it is the direct sum of two other invariant subspaces $F_s^+$ and $F_s^-$, so that in this case also $D_s$ contains three invariant subspaces.

## 3.4. Conditionally Invariant Bilinear Functionals

Having shown that the bilinear functionals invariant under analytic representations of $G$ are degenerate on certain invariant subspaces, we now assert that there exist on these subspaces new invariant bilinear functionals. We shall deal with the various cases separately.

An analytic representation with index $s = -1, -2, \ldots$ possesses the invariant functional given in Eq. (6) of Section 3.3, which is degenerate on $F_s \subset D_s$. There exist, however, two bilinear functionals invariant on $F_s$, whose kernels are the associated homogeneous generalized functions

$$(x_1 - x_2)^{-s-1} \ln(x_1 - x_2 - i0) \qquad \text{and} \qquad (x_1 - x_2)^{-s-1} \ln(x_1 - x_2 + i0).$$

The invariance of these functionals,

$$B_+(\varphi, \psi) = \int (x_1 - x_2)^{-s-1} \ln(x_1 - x_2 - i0)\varphi(x_1)\psi(x_2)\, dx_1\, dx_2$$

and

$$B_-(\varphi, \psi) = \int (x_1 - x_2)^{-s-1} \ln(x_1 - x_2 + i0)\varphi(x_1)\psi(x_2)\, dx_1\, dx_2\,,$$

is easily verified by direct calculation. Note that $B_+$ is degenerate on $F_s^-$, the space of functions in $F_s$ which are the boundary values of functions analytic on the lower half-plane. Similarly $B_-$ is degenerate on $F_s^+$. It will be shown in Section 4.3 that $F_s = F_s^+ \oplus F_s^-$. Then $B_+$ is nondegenerate on $F_s^+$, and $B_-$ on $F_s^-$.

We now turn to analytic representations for $s = 1, 2, \ldots$. Then the two existing linearly independent invariant bilinear functionals on $D_s$ are degenerate on the subspace $E_s$ consisting of polynomials of degree $s - 1$ and lower. We wish to find a bilinear functional invariant on $E_s$. Let $C(\varphi, \psi)$ be such a functional. It is clearly uniquely defined by the constants

$$c_{kl} = C(x^k, x^l), \qquad k, l = 0, \ldots, s - 1.$$

We assert that $c_{kl} = 0$ for $k + l \neq s - 1$. Applying a matrix of the dilation form,

$$g = \begin{Vmatrix} \alpha & 0 \\ 0 & \alpha^{-1} \end{Vmatrix},$$

we arrive at

$$c_{kl} = \alpha^{2(k+l)-2(s-1)} c_{kl}\,,$$

which yields the desired result. Thus we need to find only the constants $c_{k,s-1-k}$. Applying a matrix of the translation form,

$$\begin{Vmatrix} 1 & 0 \\ x_0 & 1 \end{Vmatrix},$$

we arrive at

$$0 = C(x^k, x^{s-k}) = C([x + x_0]^k, [x + x_0]^{s-k}).$$

By expanding in powers of $x_0$ on the right-hand side, we obtain

$$k c_{k-1,s-k} + (s - k) c_{k,s-1-k} = 0,$$

which yields

$$c_{k,s-1-k} = (-1)^k \frac{k!(s - 1 - k)!}{(s - 1)!}$$

(we have set $c_{0,s-1} = 1$). Thus for the polynomials

$$\varphi(x) = \sum_{k=0}^{s-1} a_k x^k \quad \text{and} \quad \psi(x) = \sum_{k=0}^{s-1} b_k x^k$$

we obtain the bilinear functional

$$C(\varphi, \psi) = \sum_{k=0}^{s-1} (-1)^k \frac{k!(s-1-k)!}{(s-1)!} a_k b_{s-1-k} \,. \tag{1}$$

It is easily verified that $C(\varphi, \psi)$ is indeed invariant.

## 4. Equivalence of Two Representations

We now go on to find the conditions under which two representations $T_{\chi_1}(g)$ and $T_{\chi_2}(g)$ of the real group $G$ are equivalent.

### 4.1. Intertwining Operators

Let

$$T_{\chi_1}(g)\varphi(x) = \varphi\left(\frac{\alpha x + \gamma}{\beta x + \delta}\right) | \beta x + \delta |^{s_1 - 1} \operatorname{sgn}^{\epsilon_1} (\beta x + \delta)$$

and

$$T_{\chi_2}(g)\psi(x) = \psi\left(\frac{\alpha x + \gamma}{\beta x + \delta}\right) | \beta x + \delta |^{s_2 - 1} \operatorname{sgn}^{\epsilon_2} (\beta x + \delta)$$

be two representations of $G$ on $D_{\chi_1}$ and $D_{\chi_2}$, respectively. Let us first find the conditions under which there exists a continuous mapping of $A$ of $D_{\chi_1}$ into $D_{\chi_2}$ such that

$$AT_{\chi_1}(g) = T_{\chi_2}(g)A.$$

Recall that the representations are called equivalent if $A$ is a one-to-one continuous mapping of $D_{\chi_1}$ onto $D_{\chi_2}$.

We start by establishing a relation between such *intertwining operators* $A$ and invariant bilinear forms. Let $A$ be a continuous mapping of $D_{\chi_1}$ into $D_{\chi_2}$, and let $(\psi, \varphi)$ be an invariant bilinear form on $D_{-\chi_2}$ and $D_{\chi_2}$, where $-\chi_2 = (-s_2, \epsilon_2)$. Then we found in Section 3.2 that $(\psi, \varphi)$ exists and is given by

$$(\psi, \varphi) = \int_{\infty}^{+\infty} \psi(x)\varphi(x)\, dx, \qquad \psi \in D_{-\chi_2}, \quad \varphi \in D_{\chi_2} \,.$$

We now associate with $A$ the bilinear functional $B(\varphi, \psi)$ on $D_{\chi_1}$ and $D_{-\chi_2}$ defined by

$$B(\varphi, \psi) = (\psi, A\varphi) = \int_{-\infty}^{+\infty} \psi(x) A\varphi(x)\, dx, \qquad \varphi \in D_{\chi_1}, \quad \psi \in D_{-\chi_2}.$$

We may then assert that $A$ *intertwines with* $T_{\chi_1}(g)$ and $T_{\chi_2}(g)$ if and only if $B(\varphi, \psi)$ exists and is invariant under $T_{\chi_1}(g)$ and $T_{-\chi_2}(g)$.[15]

We have already established the conditions under which invariant bilinear functionals exist. Then by substituting $-\chi_2 = (-s_2, \epsilon_2)$ for $\chi_2 = (s_2, \epsilon_2)$ in these conditions, we obtain the following conditions for the existence of an intertwining operator $A$.

*Two representation* $T_{\chi_1}(g)$ *and* $T_{\chi_2}(g)$, *where* $\chi_1 = (s_1, \epsilon_1)$ *and* $\chi_2 = (s_2, \epsilon_2)$, *possess an intertwining operator* $A \not\equiv 0$ *which maps* $D_{\chi_1}$ *continuously into* $D_{\chi_2}$, *i.e., one such that*

$$AT_{\chi_1}(g) = T_{\chi_2}(g)A,$$

*if and only if* $\epsilon = \epsilon_1 = \epsilon_2$ *and either* $s_1 = s_2$ *or* $s_1 = -s_2$.

Let us now obtain expressions for such $A$ operators. Consider first the case in which $s_1 = s_2$, that is, $T_{\chi_1}(g) = T_{\chi_2}(g)$ (we write $\chi = \chi_1 = \chi_2$). Then every bilinear functional invariant under $T_\chi(g)$ and $T_{-\chi}(g)$ is of the form

$$B(\varphi, \psi) = \lambda \int_{-\infty}^{+\infty} \psi(x)\varphi(x)\, dx.$$

By comparing this with our previous expression for $B(\varphi, \psi)$ in terms of $A$, we conclude that *every operator on* $D_\chi$ *that commutes with* $T_\chi(g)$ *is a multiple of the unit operator. Thus all the* $T_\chi(g)$ *representations are operator irreducible.*

Now consider the case $s_1 = -s_2 = s$. Let us assume first that $T_\chi(g)$ is *not* an analytic representation with $s = 0, 1, \ldots$. Then every bilinear functional invariant under $T_\chi(g)$ and $T_{-\chi}(g)$ may be written either in the form

$$B(\varphi, \psi) = \lambda \int_{-\infty}^{+\infty} | x_1 - x_2 |^{-s-1} \operatorname{sgn}^\epsilon(x_1 - x_2)\, \varphi(x_1)\psi(x_2)\, dx_1\, dx_2,$$

for $s \neq 0, 1, \ldots$, or in the form

$$B(\varphi, \psi) = \lambda \int_{-\infty}^{+\infty} \varphi^{(s)}(x)\psi(x)\, dx, \tag{1}$$

for $s = 0, 1, \ldots$.

[15] The proof of this assertion is the same as for the case of the complex group.

We thus conclude that *an operator $A$ intertwining with $T_\chi(g)$ and $T_{-\chi}(g)$ (except for the case of analytic representations with $s = 0, 1, 2, \ldots$) may be given as follows:*

*If $s \neq 0, 1, 2, \ldots$, then*

$$A\varphi(x) = \int_{-\infty}^{+\infty} | x_1 - x |^{-s-1} \operatorname{sgn}^\epsilon(x_1 - x)\, \varphi(x_1)\, dx_1 \tag{2}$$

*(understood in the sense of its regularization for* Re $s > 0$*).*

*If $s = 0, 1, 2, \ldots$ [but $T_\chi(g)$ is not an analytic representation], then*

$$A\varphi(x) = \varphi^{(s)}(x). \tag{3}$$

These formulas can be unified in the form

$$A\varphi(x) = \frac{1}{2\Gamma(-s)} \int_{-\infty}^{+\infty} | x_1 - x |^{-s-1} \operatorname{sgn}^\epsilon(x_1 - x)\, \varphi(x_1)\, dx_1 . \tag{4}$$

All that remains is to consider the analytic representation with $s = 0, 1, 2, \ldots$, that is, $T_\chi(g)$ and $T_{-\chi}(g)$ given for such $s$ by

$$\begin{aligned} T_\chi(g)\varphi(x) &= \varphi\left(\frac{\alpha x + \gamma}{\beta x + \delta}\right)(\beta x + \delta)^{s-1}, \\ T_{-\chi}(g)\psi(x) &= \psi\left(\frac{\alpha x + \gamma}{\beta x + \delta}\right)(\beta x + \delta)^{-s-1}. \end{aligned} \tag{5}$$

For this case any invariant bilinear functional on $D_\chi \equiv D_s$ is of the form, as was shown in Section 3.2,

$$B(\varphi, \psi) = \int_{-\infty}^{+\infty} [\lambda_1 \varphi_+^{(s)}(x) + \lambda_2 \varphi_-^{(s)}(x)]\psi(x)\, dx,$$

where

$$\varphi_+^{(s)}(x) = \frac{1}{2\pi i} \int \frac{\varphi^{(s)}(x_1)}{x_1 - x - i0}\, dx_1 ,$$

$$\varphi_-^{(s)}(x) = -\frac{1}{2\pi i} \int \frac{\varphi^{(s)}(x_1)}{x_1 - x + i0}\, dx_1 ,$$

and $\lambda_1$ and $\lambda_2$ are arbitrary constants. We then conclude that *any operator $A$ intertwining with the analytic representations of Eq.* (5) *must be of the form*

$$A\varphi(x) = \frac{\lambda_1}{2\pi i} \int_{-\infty}^{+\infty} \frac{\varphi^{(s)}(x_1)}{x_1 - x - i0}\, dx_1 - \frac{\lambda_2}{2\pi i} \int_{-\infty}^{\infty} \frac{\varphi^{(s)}(x_1)}{x_1 - x + i0}\, dx_1 ,$$

*where $\lambda_1$ and $\lambda_2$ are arbitrary constants.*

### 4.2. Equivalence of Two Representations

We now find the necessary and sufficient conditions for $T_{\chi_1}(g)$ and $T_{\chi_2}(g)$ to be equivalent representations. In Section 4.1 we obtained necessary and sufficient conditions for $T_{\chi_1}(g)$ and $T_{\chi_2}(g)$ to possess an intertwining operator. These were found to be that the representations have the same parity and that $s_1$ and $s_2$ may differ only, if at all, in sign. In order that the representations be equivalent, it is further necessary that the intertwining operator $A$ be a one-to-one continuous mapping of $D_{\chi_1}$ onto the entire space $D_{\chi_2}$.

To proceed, consider the two representations

$$T_{\chi_1}(g)\varphi(x) = \varphi\left(\frac{\alpha x + \gamma}{\beta x + \delta}\right) | \beta x + \delta |^{s-1} \operatorname{sgn}^{\epsilon} (\beta x + \delta) \tag{1}$$

and

$$T_{\chi_2}(g)\psi(x) = \psi\left(\frac{\alpha x + \gamma}{\beta x + \delta}\right) | \beta x + \delta |^{-s-1} \operatorname{sgn}^{\epsilon} (\beta x + \delta) \tag{1'}$$

possessing an intertwining operator. We shall show that *except for the special case in which* $T_{\chi_1}(g)$ *and* $T_{\chi_2}(g)$ *are analytic representations, they are equivalent.*

**Proof.** By assumption there exists a continuous mapping $A$ of $D_{\chi_1}$ into $D_{\chi_2}$ such that

$$AT_{\chi_1}(g) = T_{\chi_2}(g)A. \tag{2}$$

From the results of Section 4.1 there also exists a continuous mapping $A_1$ of $D_{\chi_2}$ into $D_{\chi_1}$ such that

$$T_{\chi_1}(g)A_1 = A_1T_{\chi_2}(g). \tag{2'}$$

Consider the mapping $A_1A$ of $D_{\chi_1}$ into itself. Equations (2) and (2′) imply that

$$A_1AT_{\chi_1}(g) = T_{\chi_1}(g)A_1A,$$

and since $T_{\chi_1}(g)$ is operator irreducible, it follows that

$$A_1A = \lambda_1 E,$$

where $E$ is the unit operator.

We assert further that $\lambda_1 \neq 0$. For the proof let us assume first that $s$ is not an integer, so that $A$ and $A_1$ can be written

$$A\varphi(x) = \frac{1}{\Gamma(-s)} \int |x - x_1|^{-s-1} \operatorname{sgn}^\epsilon(x - x_1)\, \varphi(x_1)\, dx_1 ,$$

$$A_1\varphi(x) = \frac{1}{\Gamma(s)} \int |x - x_1|^{s-1} \operatorname{sgn}^\epsilon(x - x_1)\, \varphi(x_1)\, dx_1 .$$

We shall consider only functions $\varphi(x)$ which are what we have called *rapidly decreasing* (that is, rapidly decreasing with all of their derivatives) and such that their Fourier transforms $\tilde{\varphi}(\sigma) = F[\varphi(x)]$ vanish in a neighborhood of $\sigma = 0$. Let us now take the Fourier transforms of the expressions for $A$ and $A_1$. Then the convolutions become products of Fourier transforms, and we must find the Fourier transforms of $|x|^{-s-1} \operatorname{sgn}^\epsilon x$ and $|x|^{s-1} \operatorname{sgn}^\epsilon x$. These can be calculated from the formulas[16]

$$F\left[\frac{|x|^\lambda}{\Gamma(\lambda+1)}\right] = -2 \sin \frac{\lambda\pi}{2} |\sigma|^{-\lambda-1},$$

$$F\left[\frac{|x|^\lambda \operatorname{sgn} x}{\Gamma(\lambda+1)}\right] = 2i \cos \frac{\lambda\pi}{2} |\sigma|^{-\lambda-1} \operatorname{sgn} \sigma.$$

It thus follows that $F[A_1A\varphi(x)] = \lambda_1 F[\varphi(x)]$, where $\lambda_1 = 4\cos^2 \frac{1}{2}s\pi$ when $\epsilon = 0$ and $s \neq \pm 1, \pm 3, ...,$ and $\lambda_1 = 4 \sin^2 \frac{1}{2}s\pi$ when $\epsilon = 1$ and $s \neq 0, \pm 2, \pm 4, ...$ . Note that $\lambda_1 \neq 0$ in both cases.

Thus $A_1A$ is a nonvanishing multiple of the unit operator, and in the same way it can be shown that

$$AA_1 = \lambda_2 E,$$

where $\lambda_2 \neq 0$. It is then obvious that $A$ and $A_1$ are each one-to-one continuous mappings *onto* the entire space. This completes the proof that the representations of (1) and (1′) are equivalent.

Now consider the special case in which $T_{\chi_1}(g)$ and $T_{\chi_2}(g)$ are analytic representations, so that

$$T_{\chi_1}(g)\varphi(x) = \varphi\left(\frac{\alpha x + \gamma}{\beta x + \delta}\right) (\beta x + \delta)^{s-1}, \tag{3}$$

$$T_{\chi_2}(g)\psi(x) = \psi\left(\frac{\alpha x + \gamma}{\beta x + \delta}\right) (\beta x + \delta)^{-s-1}, \tag{3'}$$

[16] See Volume 1, page 359.

where $s$ is an integer. We assert that (*for* $s \neq 0$) *these representations are inequivalent.* In order to be definite, let us assume that $s > 0$. Then every operator $A$ mapping $D_{\chi_1}$ onto $D_{\chi_2}$ so that

$$AT_{\chi_1}(g) = T_{\chi_2}(g)A$$

is of the form

$$A\varphi(x) = \frac{\lambda_1}{2\pi i}\int_{-\infty}^{+\infty}\frac{\varphi^{(s)}(x_1)\,dx_1}{x_1 - x - i0} - \frac{\lambda_2}{2\pi i}\int_{-\infty}^{+\infty}\frac{\varphi^{(s)}(x_1)\,dx_1}{x_1 - x + i0}.$$

Obviously this operator is degenerate on the subspace $E_s$ of polynomials of degree $s - 1$ and lower, so that $A$ is not one-to-one. This shows that the analytic representations (3) and (3′) are inequivalent.

*Thus we have shown that* $T_{\chi_1}(g)$ *and* $T_{\chi_2}(g)$, $\chi_1 \neq \chi_2$, *are equivalent representations if and only if the following three conditions are fulfilled*:

(a) *They have the same parity.*

(b) $s_1 = -s_2$.

(c) *They are not analytic.*

### 4.3. Partially Equivalent Representations

We have seen in Section 4.2 that the analytic representations

$$T_s(g)\varphi(x) = \varphi\left(\frac{\alpha x + \gamma}{\beta x + \delta}\right)(\beta x + \delta)^{s-1}$$

and

$$T_{-s}(g)\psi(x) = \psi\left(\frac{\alpha x + \gamma}{\beta x + \delta}\right)(\beta x + \delta)^{-s-1},$$

where $s \neq 0$ is an integer,[17] possess an intertwining operator but are nevertheless inequivalent. Let us consider in somewhat more detail the relation between these representations. We shall assume, to be specific, that $s > 0$.

We saw in Section 2.1 that $T_s(g)$ and $T_{-s}(g)$ are not subspace irreducible, in fact that $D_s$ contains a finite-dimensional subspace $E_s$ consisting of polynomials of the form

$$\varphi(x) = \sum_{k=0}^{s-1} a_k x^k,$$

[17] Recall that we denote by $T_s(g)$ the analytic representation $T_\chi(g)$, $\chi = (s, \epsilon)$, and by $D_s$ the corresponding space.

and that $D_{-s}$ contains the invariant subspace $F_{-s}$ consisting of all functions $\varphi(x)$ such that

$$\int_{-\infty}^{+\infty} x^k \psi(x)\, dx = 0 \qquad \text{for} \quad k = 0, \ldots, s-1.$$

Thus $T_s(g)$ and $T_{-s}(g)$ yield two representations each of $G$, namely those on the invariant subspaces and those on the corresponding factor spaces. We then assert the following.

**Assertion 1.** *The restriction of $T_s(g)$ to $E_s$ is equivalent to the representation induced by $T_{-s}(g)$ on $D_{-s}/F_{-s}$.*

**Assertion 2.** *The representation induced by $T_s(g)$ on $D_s/E_s$ is equivalent to the restriction of $T_{-s}(g)$ to $F_{-s}$.*

Proof. Let $A_1(s)$ be a mapping of $D_{-s}$ into $D_s$ such that

$$A_1 T_{-s}(g) = T_s(g) A_1 . \tag{1}$$

Then $A_1$ is of the form

$$A_1 \varphi(x) = \int_{-\infty}^{+\infty} (x_1 - x)^{s-1} \varphi(x_1)\, dx_1 .$$

It is clear that the kernel of $A_1$ is $F_{-s} \subset D_{-s}$. Further, every $\varphi(x) \in D_{-s}$ is mapped by $A_1$ into a polynomial of degree $s - 1$ or lower, and obviously all such polynomials can be obtained in this way. Consequently $A_1$ is a continuous one-to-one mapping of $D_{-s}/F_{-s}$ onto $E_s$. Then the equivalence of the representations on these two spaces follows immediately from Eq. (1).

Let us now turn to Assertion 2. We have seen that the operator $A$ defined by

$$A\varphi(x) = \varphi^{(s)}(x)$$

maps $D_s$ into $D_{-s}$ so that

$$A T_s(g) = T_{-s}(g) A. \tag{2}$$

Again we wish to find the kernel and image under $A$. Obviously the kernel is $E_s$. Further, $A$ maps $D_s$ into $F_{-s}$, as we shall now show. Assume first that $\varphi(x)$ has bounded support. Integration by parts then yields

$$\int_{-\infty}^{+\infty} x^k \varphi^{(s)}(x)\, dx = 0 \qquad \text{for} \quad k = 0, 1, \ldots, s-1.$$

Thus if $\varphi(x)$ has bounded support, it is mapped by $A$ into $F_{-s}$. Since, further, $F_{-s}$ is invariant, it must according to Eq. (2) contain the image of every function of the form $T_s(g)\varphi(x)$, where $\varphi(x)$ has bounded support. But every functions $\varphi(x) \in D_s$ can be written as a linear combination of functions $T_s(g_k)\varphi_k(x)$, where the $\varphi_k(x)$ all have bounded support. Thus every function $\varphi(x) \in D_s$ is mapped by $A$ into $F_{-s}$.

We must now show that all of $F_{-s}$ can be obtained in this way. Since every function in $F_{-s}$ is the sum of a function with bounded support plus the inversion (see Section 1.2) of such a function, both belonging to $F_{-s}$, and since further $F_{-s}$ is invariant, it is sufficient to show that every function with bounded support in $F_{-s}$ can be obtained in this way from some function in $D_s$. Thus let $\varphi(x)$ be a function with bounded support in $F_{-s}$. Consider the integral operator

$$B\psi(x) = \int_{-\infty}^{x} \psi(t)\, dt.$$

Since $\int_{-\infty}^{+\infty} x^k \psi(x)\, dx = 0$ for $k = 0, 1, ..., s - 1$, all the functions $B\psi(x), ..., B^s\psi(x)$ also have bounded support. Now by applying the operator $A = d^s/dx^s$ to $B^s\psi(x)$, we obviously obtain $\psi(x)$, and therefore every function $\psi(x) \in F_{-s}$ is the mapping of some functions in $B_s$.

We have shown that the kernel of the mapping $A = d^s/dx^s$ is $E_s$, and that the image of $D_s$ under $A$ is $F_{-s} \subset D_{-s}$. Therefore $A$ is a continuous one-to-one mapping of $D_s/E_s$ onto $F_{-s}$. Since, further, $AT_s(g) = T_{-s}(g)A$, the representations on $D_s/E_s$ and on $F_{-s}$ are equivalent.

We now make the following assertion: *The representations of $G$ on $D_s/E_s$ and on $F_{-s}$ are reducible. More specifically, each of these subspaces is the direct sum of two invariant subspaces.*

**Proof.** Consider the operators $A_+$ and $A_-$ defined by

$$A_+\varphi(x) = \frac{1}{2\pi i} \int_{-\infty}^{+\infty} \frac{\varphi^{(s)}(x_1)\, dx_1}{x_1 - x - i0}$$

and

$$A_-\varphi(x) = -\frac{1}{2\pi i} \int_{-\infty}^{+\infty} \frac{\varphi^{(s)}(x_1)\, dx_1}{x_1 - x + i0},$$

whose sum is $A$. As was shown in Section 4.1, these map $D_s$ into $D_{-s}$ and intertwine with the representations, i.e.,

$$A_+T_s(g) = T_{-s}(g)A_+ \qquad \text{and} \qquad A_-T_s(g) = T_{-s}(g)A_-.$$

Every other intertwining operator for $T_s(g)$ and $T_{-s}(g)$ is a linear combination of $A_+$ and $A_-$. Let $D_s^-$ be the kernel of the mapping $A_+$, and $D_s^+$ be the kernel of $A_-$. Then clearly $D_s^-$ and $D_s^+$ are invariant subspaces. From the definitions of $A_+$ and $A_-$ it follows immediately that the intersection of $D_s^-$ and $D_s^+$ is $E_s$. Moreover, the sum of $D_s^-$ and $D_s^+$ is the entire space $D_s$. Indeed, let $\varphi(x)$ be any function in $D_s$. Then for large $|x|$ it may be expanded in the asymptotic series

$$\varphi(x) \sim \sum_{k=-\infty}^{s-1} a_k x^k.$$

Without loss of generality we may assume that the coefficients of the nonnegative powers of $x$ in this series vanish [this can be achieved by subtracting from $\varphi(x)$ a polynomial of degree $s-1$, which belongs to the intersection of $D_s^+$ and $D_s^-$]. Then $\varphi(x)$ is square integrable and can consequently be written in the form

$$\varphi(x) = \varphi_+(x) + \varphi_-(x),$$

where $\varphi_+(x)$ is the boundary value of some function analytic in the upper half-plane and $\varphi_-(x)$ is the boundary value of some function analytic in the lower half-plane. It is easily shown that these functions also belong to $D_s$, and it is further obvious that $A_+\varphi_-(x) = 0$ and $A_-\varphi_+(x) = 0$, so that $\varphi_-(x) \in D_s^-$ and $\varphi_+(x) \in D_s^+$. This shows that $D_s$ is the sum of $D_s^-$ and $D_s^+$.

The above implies that $D_s/E_s$ is a direct sum of the form

$$D_s/E_s = D_s^+/E_s + D_s^-/E_s.$$

Now let $F_{-s}^+$ and $F_{-s}^-$ be the images of $D_s^+$ and $D_s^-$ under $A_+$ and $A_-$, respectively. Obviously these subspaces are invariant under $T_{-s}(g)$ and their intersection is empty. Their sum is $F_{-s}$.

$F_{-s}^+$ and $F_{-s}^-$ can also be defined intrinsically. Namely, $F_{-s}^+$ is the subspace of all functions in $D_{-s}$ which are the boundary values of functions analytic in the upper half-plane; similarly $F_{-s}^-$ is the subspace of functions in $D_{-s}$ which are the boundary values of functions analytic in the lower half-plane. We prove this assertion as follows. Recall that $F_{-s}^+$ is the subspace of all functions in $F_{-s}$ that are the boundary values of functions in the upper half-plane. Thus we need only show that every function $\varphi(x) \in D_{-s}$ that is the boundary value of a function analytic in the upper half-plane belongs to $F_{-s}$, that is, has the property that

$$\int_{-\infty}^{+\infty} x^k \varphi(x)\, dx = 0 \qquad \text{for} \quad k = 0, 1, \ldots, s-1.$$

Let $\tilde{\varphi}(\xi)$ be the Fourier transform of $\varphi(x)$. Because of the large-$|x|$ behavior of $\varphi(x)$, this Fourier transform has $s - 1$ continuous derivatives, and $i^k \tilde{\varphi}^{(k)}(0) = \int_{-\infty}^{+\infty} x^k \varphi(x)\, dx$. But we know further that $\tilde{\varphi}(\xi) = 0$ for $\xi > 0$, for we have assumed that $\varphi(x)$ is the boundary value or some functions analytic in the upper half-plane. Consequently $\varphi(x) \in F_{-s}$, as asserted.

Finally, let us summarize the results which are obtained in this section. For the analytic representations

$$T_s(g)\varphi(x) = \varphi\left(\frac{\alpha x + \gamma}{\beta x + \delta}\right)(\beta x + \delta)^{s-1}$$

and

$$T_{-s}(g)\psi(x) = \psi\left(\frac{\alpha x + \gamma}{\beta x + \delta}\right)(\beta x + \delta)^{-s-1}$$

acting on $D_s$ and $D_{-s}$, $s = 1, 2, \ldots$, we have established the following.

$D_s$ contains three invariant subspaces:

1. $E_s$, the space of all polynomials of degree $s - 1$ and less;
2. $D_s^+$, the space of all $\varphi(x)$ such that $A_-\varphi(x) = 0$;
3. $D_s^-$, the space of all $\varphi(x)$ such that $A_+\varphi(x) = 0$.

Here $A_+$ and $A_-$ are mappings of $D_s$ into $D_{-s}$ defined by

$$A_+\varphi(x) = \frac{1}{2\pi i}\int_{-\infty}^{+\infty} \frac{\varphi^{(s)}(x_1)\, dx_1}{x_1 - x - i0},$$

$$A_-\varphi(x) = -\frac{1}{2\pi i}\int_{-\infty}^{+\infty} \frac{\varphi^{(s)}(x_1)\, dx_1}{x_1 - x + i0}.$$

The intersection of $D_s^+$ and $D_s^-$ is $E_s$, and their sum is the entire space $D_s$. (It is also possible to characterize $D_s^+$ and $D_s^-$ as the subspaces of $D_s$ which, up to polynomials of degree $s - 1$, and lower, are boundary values of functions analytic in the upper and lower half-planes, respectively.)

$D_{-s}$ contains three invariant subspaces:

1. $F_{-s}$, the space of all $\varphi(x)$ such that $\int_{-\infty}^{+\infty} x^k \varphi(x)\, dx = 0$ for $k = 0, 1, \ldots, s - 1$;

2. $F_{-s}^+$, the subspace of functions that are boundary values of functions analytic in the upper half-plane;

3. $F_{-s}^-$, the subspace of functions that are boundary values of functions analytic in the lower half-plane.

Further, $F_{-s}$ is the direct sum of $F_{-s}^+$ and $F_{-s}^-$.[18]

[18] $D_0$ has an even simpler structure: it is the direct sum of the two invariant subspaces $D_0^+ = F_0^+$ and $D_0^- = F_0^-$.

We have also established the equivalence of the representations in the following pairs of subspaces:

1. $E_s$ and $D_{-s}/F_{-s}$. The latter is mapped onto the former by the operator

$$A\varphi(x) = \int_{-\infty}^{+\infty} (x_1 - x)^{s-1} \varphi(x_1)\, dx_1 .$$

2. $D_s/E_s$ and $F_{-s}$. The former is mapped onto the latter by the differential operator $d^s/dx^s$.

3. $D_s^+/E_s$ and $F_{-s}^+$ (or $D_s^-/E_s$ and $F_{-s}^-$). The former is mapped onto the latter by $A_+$ (or $A_-$).

Remark. It can be shown that the inverses of $A_+$ and $A_-$, which we shall denote by $B_+$ and $B_-$, are given by

$$B_+\varphi(x) = c_1 \int_{-\infty}^{+\infty} (x_1 - x)^{s-1} \ln(x_1 - x - i0)\varphi(x_1)\, dx_1 ,$$

$$B_-\varphi(x) = c_2 \int_{-\infty}^{+\infty} (x_1 - x)^{s-1} \ln(x_1 - x + i0)\varphi(x_1)\, dx_1 . \quad \#$$

## 4.4. Other Models of $F_s^+$ and $F_s^-$

In Section 4.3 we studied the representations of $G$ on $F_s^+$ and $F_s^-$ for negative integer $s$. Recall that these spaces were defined as follows. $F_s^+$ consists of functions $\varphi(x)$ such that both $\varphi(x)$ and its inversion $\hat{\varphi}(x) = x^{s-1}\varphi(-1/x)$ are infinitely differentiable functions, and $\varphi(x)$ is the boundary value of a function $\varphi(z) = \varphi(x + iy)$ analytic in the upper half-plane $\operatorname{Im} z > 0$. The definition of $F_s^-$ is analogous.

In both of these spaces the representation of $G$ is defined by

$$T_s(g)\varphi(x) = \varphi\left(\frac{\alpha x + \gamma}{\beta x + \delta}\right) (\beta x + \delta)^{s-1}.$$

It would be natural, on the other hand, to deal not with the functions $\varphi(x)$ on the line, but with the functions $\varphi(z) = \varphi(x + iy)$, analytic in the upper half-plane, whose boundary values they are. When we do so, we obtain another realization of $F_s^+$ and $F_s^-$.

Let us thus assume $F_s^+$ to be the space of functions analytic in the half-plane $\operatorname{Im} z > 0$. Then the representation of $G$ on $F_s^+$ is given by

$$T_s(g)\varphi(z) = \varphi\left(\frac{\alpha z + \gamma}{\beta z + \delta}\right) (\beta z + \delta)^{s-1}.$$

It should be emphasized that $F_s^+$ does not consist of *all* functions analytic in the upper half-plane, but only of those satisfying certain subsidiary conditions. Specifically, $\varphi(z)$ must be infinitely differentiable together with its inversion

$$\hat{\varphi}(z) = z^{s-1}\varphi(-1/z)$$

on the closed upper half-plane.

Remark. It is sometimes also useful to take as a model for $F_s^+$ the space of functions analytic within the unit circle. This realization is obtained by replacing $\varphi(z)$ by

$$\varphi_1(w) = (1 - w)^{s-1}\varphi\left(i\,\frac{1+w}{1-w}\right),$$

which is analytic in the unit circle, and by considering $F_s^+$ to be the space of these $\varphi_1(w)$ functions. It can then be shown that $\varphi_1(w) \in F_s^+$ if and only if in addition to being analytic in the unit circle it is infinitely differentiable on the closed unit circle. The representation of $G$ on this space is most conveniently described in terms of parameters other than the matrix elements of $g$. Let

$$a = \tfrac{1}{2}[(\alpha + \delta) + i(\gamma - \beta)], \qquad b = \tfrac{1}{2}[(\alpha - \delta) - i(\gamma + \beta)].$$

Then the representation is given by[19]

$$T_s(g)\varphi(w) = \varphi\left(\frac{aw + b}{\bar{b}w + \bar{a}}\right)(\bar{b}w + \bar{a})^{s-1}.$$

The derivation of this formula is left as an exercise to the reader. #

## 5. Unitary Representations of $G$

We now go on to find the conditions under which it is possible to define on the carrier space of the representation an inner product which is invariant under the representation. Recall that an inner or scalar product is a positive definite Hermitian functional.

### 5.1. Existence of an Invariant Hermitian Functional

Let us first find the conditions under which there exists in $D_\chi$ a Hermitian functional $(\varphi, \psi)$, not necessarily positive definite, which is invariant in the sense that

$$(\varphi, \psi) = (T_\chi(g)\varphi, T_\chi(g)\psi).$$

[19] As is well known, the transformation $w' = (aw + b)/(\bar{b}w + \bar{a})$ transforms the unit circle $|w| < 1$ into itself.

This may be done by establishing a relation between invariant Hermitian functionals and the invariant bilinear functionals studied in Section 2. Let us then assume that there exists a Hermitian functional $(\varphi, \psi)$ on $D_\chi$, $\chi = (s, \epsilon)$. We may associate with it the bilinear functional

$$B(\varphi, \psi) = (\varphi, \bar{\psi}), \tag{1}$$

defined on $D_\chi$ and $D_{\bar{\chi}}$, where $\bar{\chi} = (\bar{s}, \epsilon)$.

Obviously $(\varphi, \psi)$ is invariant if and only if $B(\varphi, \psi)$ is also invariant. Now we have found the conditions under which there may exist an invariant bilinear functional on $D_{\chi_1}$ and $D_{\chi_2}$. By setting $\chi_1 = \chi$ and $\chi_2 = \bar{\chi}$, we arrive at the following condition for the existence of an invariant Hermitian functional.

*An invariant Hermitian functional $(\varphi, \psi)$ on $D_\chi$, $\chi = (s, \epsilon)$, will exist if and only if $s$ satisfies one of the two following conditions:*

*Case 1.* $s = \bar{s}$.

*Case 2.* $s = -\bar{s}$.

*In other words, $s$ must be either real or pure imaginary.*

From our previous results we can also immediately obtain an expression for the invariant Hermitian functional when it exists. We start with Case 2.

Let $s = i\rho$, $\rho \neq 0$. Then

$$(\varphi, \psi) = \lambda \int_{-\infty}^{+\infty} \varphi(x)\bar{\psi}(x)\, dx. \tag{2}$$

Now we take Case 1. Let $s$ be real, but $s \neq 0, 1, \ldots$. Then

$$(\varphi, \psi) = \lambda \int_{-\infty}^{\infty} |x_1 - x_2|^{-s-1} \operatorname{sgn}^\epsilon(x_1 - x_2)\, \varphi(x_1)\bar{\psi}(x_2)\, dx_1\, dx_2, \tag{3}$$

where $\epsilon$ is the parity of the representation (for $s > 0$ the integral is to be understood in the sense of its regularization).

Now let $s$ be a nonnegative integer. Then (except for the special case in which $T_\chi(g)$ is an analytic representation)[20]

$$(\varphi, \psi) = \lambda \int_{-\infty}^{\infty} \varphi^{(s)}(x)\bar{\psi}(x)\, dx. \tag{4}$$

[20] Note that Eqs. (3) and (4) can be unified into the single formula

$$(\varphi, \psi) = \lambda \frac{1}{\Gamma(-s)} \int_{-\infty}^{\infty} |x_1 - x_2|^{-s-1} \operatorname{sgn}^\epsilon(x_1 - x_2)\, \varphi(x_1)\, \psi(x_2)\, dx_1\, dx_2 .$$

In the special case in which $T_\chi(g)$ is an analytic representation and $s = 0, 1, \ldots$, the invariant Hermitian functional depends on two arbitrary constants. For this case we obtain

$$(\varphi, \psi) = \int_{-\infty}^{+\infty} [\lambda_1 \varphi_+^{(s)}(x) + \lambda_2 \varphi_-^{(s)}(x)] \bar{\psi}(x)\, dx, \tag{5}$$

where

$$\varphi_+^{(s)}(x) = \frac{1}{2\pi i} \int_{-\infty}^{+\infty} \frac{\varphi^{(s)}(x_1)\, dx_1}{x_1 - x - i0},$$

$$\varphi_-^{(s)}(x) = -\frac{1}{2\pi i} \int_{-\infty}^{+\infty} \frac{\varphi^{(s)}(x_1)\, dx_1}{x_1 - x + i0},$$

and $\lambda_1$, $\lambda_2$ are arbitrary constants.

### 5.2. Positive Definite Invariant Hermitian Functionals (Nonanalytic Representations)

Let us now find the additional conditions under which the invariant Hermitian form for given $s$ and $\epsilon$ is positive definite. As we have seen, if $s$ is pure imaginary, then

$$(\varphi, \psi) = \int_{-\infty}^{+\infty} \varphi(x) \bar{\psi}(x)\, dx. \tag{1}$$

This is obviously positive definite, so that we need only consider the case of real $s$. We restrict ourselves also in this section to the case of nonanalytic representations, and shall deal with that special case separately in the next section. Note that for our purposes, therefore, $T_\chi(g)$ and $T_{-\chi}(g)$ are equivalent, and we may therefore assume that $s < 0$.

Proceeding, then, with $s < 0$, our invariant Hermitian functional is defined by the convergent integral

$$(\varphi, \psi) = i^\epsilon \int_{-\infty}^{+\infty} | x_1 - x_2 |^{-s-1} \operatorname{sgn}^\epsilon(x_1 - x_2)\, \varphi(x_1) \bar{\psi}(x_2)\, dx_1\, dx_2 . \tag{2}$$

Let us write this in homogeneous coordinates, taking the model of $D_s$ as a space of homogeneous functions. This expression then becomes

$$(\varphi, \psi) = i^\epsilon \int | x_1 x_2' - x_2 x_1' |^{-s-1} \operatorname{sgn}^\epsilon (x_1 x_2' - x_2 x_1') \times \varphi(x_1, x_2) \bar{\psi}(x_1', x_2')\, d\omega\, d\omega'. \tag{2'}$$

The integral here is taken along a contour $\omega$ in the $(x_1, x_2)$ plane that crosses every line passing through the origin, and a similar contour $\omega'$ in the $(x_1', x_2')$ plane. The differential forms are

$$d\omega = x_1\, dx_2 - x_2\, dx_1; \qquad d\omega' = x_1'\, dx_2' - x_2'\, dx_1'.$$

Actually Eq. (2) is the special case of (2′) in which the contours are chosen as the lines whose equations are $x_2 = 1$ and $x_2' = 1$. We wish now to find the conditions on $s$ and $\epsilon$ for this functional to be of definite sign. It is convenient for this purpose to choose as our contours the circles $x_1 = \cos\theta$, $x_2 = \sin\theta$, and $x_1' = \cos\theta'$, $x_2' = \sin\theta'$. Then if we denote $\varphi(x_1, x_2)$ simply by $\varphi(\theta)$ along this curve, we obtain

$$(\varphi, \psi) = i^\epsilon \int_0^{2\pi} \int_0^{2\pi} |\sin(\theta - \theta')|^{-s-1} \operatorname{sgn}^\epsilon \sin(\theta - \theta')\, \varphi(\theta)\bar{\psi}(\theta')\, d\theta\, d\theta'.$$

Let the Fourier series for $\varphi(\theta)$ and $\psi(\theta)$ be

$$\varphi(\theta) = \sum_{k=-\infty}^{+\infty} a_k e^{ik\theta}, \qquad \psi(\theta) = \sum_{k=-\infty}^{+\infty} b_k e^{ik\theta}.$$

Now recall that by definition of $D_\chi$ these series contain only even $k$ for $\epsilon = 0$ and only odd $k$ for $\epsilon = 1$. Thus $(\varphi, \psi)$ can be written as a Hermitian form in $a_k$ and $b_k$, namely

$$(\varphi, \psi) = 4\pi^2 \sum_{k=-\infty}^{+\infty} \lambda_{-k} a_k \bar{b}_k, \tag{3}$$

where the $\lambda_k$ are the Fourier coefficients of $i^\epsilon |\sin\theta|^{-s-1} \operatorname{sgn}^\epsilon \sin\theta$. We mention again that if $\epsilon = 0$ this series contains only even indices, and if $\epsilon = 1$ only odd indices. We shall call this the *canonical form* of our Hermitian functional.

Now in order to calculate the conditions under which $(\varphi, \psi)$ is of definite sign we need only calculate the $\lambda_k$. These coefficients are given by the following formulas. If $\epsilon = 0$, then

$$\lambda_{2k} = \pi^{-1} \int_0^\pi \sin^{-s-1}\theta \cos 2k\theta\, d\theta = \frac{2^{s+1}\Gamma(-s)(-1)^k}{\Gamma(\frac{1}{2} - \frac{1}{2}s + k)\Gamma(\frac{1}{2} - \frac{1}{2}s - k)}. \tag{4}$$

If $\epsilon = 1$,[21]

$$\lambda_{2k+1} = \pi^{-1} \int_0^\pi \sin^{-s-1}\theta \sin(2k+1)\theta\, d\theta = \frac{2^{s+1}\Gamma(-s)(-1)^k}{\Gamma(-\frac{1}{2}s - k)\Gamma(-\frac{1}{2}s + k + 1)}. \tag{4'}$$

[21] See Ryshik, I. M., and Grastein, I. S., "Tables of Series, Products, and Integrals," formula 3.454, p. 162, V.E.B. Deutscher Verlag der Wissenschaften, Berlin, 1963.

It is seen first that if the representation is not analytic (that is, if $s \neq -n$, where $n$ is a positive integer of parity opposite from $\epsilon$), the coefficients of the canonical form do not vanish. Thus except for these special cases the invariant Hermitian functional is nondegenerate. Second, we may go on to determine the signs of the $\lambda_k$. For $\epsilon = 0$ we have

$$\operatorname{sgn} \lambda_{2k} = (-1)^k \operatorname{sgn} [\Gamma(\tfrac{1}{2} - \tfrac{1}{2}s + k)\Gamma(\tfrac{1}{2} - \tfrac{1}{2}s - k)].$$

This means that $\operatorname{sgn} \lambda_{2k} = \operatorname{sgn} \lambda_{-2k}$ and that for $| k | > -\frac{1}{2}(s + 1)$ the $\lambda_{2k}$ all have the same sign, while for $| k | < -\frac{1}{2}(s - 1)$ the signs alternate. This means that if $\chi = (s, 0)$, where $s < 0$ but $s \neq -1, -3, \ldots$ (that is, $T_\chi(g)$ is not an analytic representation), the invariant Hermitian functional

$$(\varphi, \psi) = \int | x_1 - x_2 |^{-s-1} \varphi(x_1)\bar{\psi}(x_2)\, dx_1\, dx_2$$

is positive definite only for $0 > s > -1$. In the interval

$$-4n - 1 > s > -4n - 3$$

this functional has $2n + 1$ positive coefficients in the canonical form, and the rest are negative. In the interval $-4n + 1 > s > -4n - 1$ it has $2n$ negative coefficients in the canonical form, and the rest are positive.

Now let $\chi = (s, 1)$ where $s < 0$ but $s \neq -2, -4, \ldots$. Then as is seen easily from (4′),

$$\operatorname{sgn} \lambda_{2k+1} = -\operatorname{sgn} \lambda_{-2k+1}$$

for sufficiently large $| k |$ [specifically, for $k$ such that $\operatorname{sgn}(-\frac{1}{2}s - k) = -\operatorname{sgn}(-\frac{1}{2}s + k)$]. Thus for $\chi = (s, 1)$ the representation of $T_\chi(g)$ possesses a Hermitian functional whose canonical form has an infinite number of positive and an infinite number of negative coefficients. We have thus established the following result. *Nonanalytic representations possess a positive definite Hermitian form only in the following two cases.*

*Case 1. $s$ is pure imaginary. We shall call these the representations of the principal series.*

*Case 2. Representations with $\epsilon = 0$ and real $s \neq 0$, where $|s| < 1$. We shall call these the representations of the supplementary series.* The invariant positive definite Hermitian functional is given in these cases in the following way.

For the principal series

$$(\varphi, \psi) = \int_{-\infty}^{+\infty} \varphi(x)\bar{\psi}(x)\, dx.$$

For the supplementary series with $s < 0$

$$(\varphi, \psi) = \int_{-\infty}^{+\infty} | x_1 - x_2 |^{-s-1} \varphi(x_1)\bar{\psi}(x_2) \, dx_1 \, dx_2$$

(when $s > 0$ the integral must be multiplied by $-1$).

## 5.3. Invariant Hermitian Functionals for Analytic Representations

We now turn to the special case in which the representation is analytic. Recall that this means that we may write

$$T_s(g)\varphi(x) = \varphi\left(\frac{\alpha x + \gamma}{\beta x + \delta}\right) (\beta x + \delta)^{s-1},$$

where $s$ is an integer. Consider first the case of nonnegative $s$. Then as was shown in Section 5.1, every invariant Hermitian functional on $D_s$ is a linear combination of

$$(\varphi, \psi)_+ = i^{-s} \int_{-\infty}^{+\infty} \varphi_+^{(s)}(x)\bar{\psi}(x) \, dx$$

and

$$(\varphi, \psi)_- = i^s \int_{-\infty}^{+\infty} \varphi_-^{(s)}(x)\bar{\psi}(x) \, dx.$$

where

$$\varphi_+^{(s)}(x) = \frac{1}{2\pi i} \int_{-\infty}^{+\infty} \frac{\varphi^{(s)}(x_1) \, dx_1}{x_1 - x - i0},$$

$$\varphi_-^{(s)}(x) = -\frac{1}{2\pi i} \int_{-\infty}^{+\infty} \frac{\varphi^{(s)}(x_1) \, dx_1}{x_1 - x + i0}.$$

Each of these functionals is degenerate, as we have seen. Specifically, $(\varphi, \psi)_+$ vanishes on $D_s^-$, the subspace of functions $\varphi(x)$ such that $\varphi_+^{(s)}(x) = 0$ [that is, $(\varphi, \psi)_+ = 0$ for every $\psi \in D_s$ and $\varphi \in D_s^-$]. Similarly, $(\varphi, \psi)_-$ vanishes on $D_s^+$. We have already dealt with $D_s^-$ and $D_s^+$ in Section 4.3, where we established that their intersection is the subspace $E_s$ of polynomials of degree $s - 1$ and lower and that their sum is the entire space $D_s$. It then follows that

$$D_s/E_s = D_s^+/E_s + D_s^-/E_s.$$

It was shown further that $D_s^+$ and $D_s^-$ are invariant subspaces.

Consider $(\varphi, \psi)_+$ as a Hermitian functional on $D_s^+/E_s$. Obviously on this space the functional is not degenerate. We may show further that on this space it is positive definite. We prove this by going over to the Fourier transform

$$\tilde{\varphi}(\xi) = \int_{-\infty}^{+\infty} \varphi(x) e^{i\xi x}\, dx. \tag{1}$$

**Remark.** In general this integral does not converge, since for large $|x|$ the $\varphi(x)$ functions may be expressed by an asymptotic series of the form

$$\varphi(x) \sim \sum_{k=-\infty}^{s-1} a_k x^k.$$

Since, however, every such function is defined only up to a polynomial of degree $s - 1$, we may assume that this series begins with the term $a_{-1}x^{-1}$, so that $\varphi(x)$ is square integrable. Then the integral is defined in the sense of convergence in the mean. #

Recall that the Fourier transform of $\varphi^{(s)}(x)$ is $(-i)^s \xi^s \tilde{\varphi}(\xi)$. Then it is easily shown that[22]

$$F[\varphi_+^{(s)}(x)] = i^s \xi_-^s \tilde{\varphi}(\xi).$$

Consequently Plancherel's theorem implies that

$$(\varphi, \psi)_+ \equiv i^{-s} \int_{-\infty}^{+\infty} \varphi_+^{(s)}(x) \bar{\psi}(x)\, dx = (2\pi)^{-1} \int_{-\infty}^{0} |\xi|^s \tilde{\varphi}(\xi) \bar{\tilde{\psi}}(\xi)\, d\xi. \tag{2}$$

Note that since $\varphi(x)$ is the boundary value of a function analytic in the upper half-plane, its Fourier transform is a function concentrated on $-\infty < \xi < 0$.

From Eq. (2) it follows immediately that $(\varphi, \psi)_+$ is positive definite on $D_s^+/E_s$. It can be shown similarly that $(\varphi, \psi)_-$ is positive definite on $D_s^-/E_s$. This functional is given in terms of the Fourier transforms of $\varphi(x)$ and $\psi(x)$ by

$$(\varphi, \psi) = (2\pi)^{-1} \int_0^{\infty} \xi^s \tilde{\varphi}(\xi) \bar{\tilde{\psi}}(\xi)\, d\xi.$$

We have thus established the existence of positive definite invariant Hermitian functionals on $D_s^+/E_s$ and $D_s^-/E_s$ for $s = 0, 1, \ldots$.

[22] See Volume 1, page 360. By definition, $\xi_-^s = |\xi|^s$ for $\xi < 0$ and $\xi_-^s = 0$ for $\xi > 0$.

Now let us turn to analytic representations with $s = -1, -2, \ldots$. For this case the invariant Hermitian functional on $D_s$ is given by

$$(\varphi, \psi) = \int (x_1 - x_2)^{-s-1}\varphi(x_1)\bar{\psi}(x_2)\, dx_1\, dx_2\,.$$

This functional obviously vanishes on $F_s$, the subspace of $\varphi(x)$ functions such that

$$\int_{-\infty}^{+\infty} x^k\varphi(x)\, dx = 0, \qquad k = 0, 1, \ldots, s-1.$$

As we saw in Section 4.3, $F_s$ is the direct sum of two invariant subspaces, namely $F_s^+$, consisting of functions which are the boundary values of functions analytic in the upper half-plane, and $F_s^-$, defined similarly with respect to the lower half-plane. We assert now that there exists a positive definite Hermitian form on each of $F_s^+$ and $F_s^-$. This follows immediately from the fact that the representations of $F_s^+$ and $F_s^-$ are equivalent to representations on $D_{-s}^+/E_{-s}$ and $D_{-s}^-/E_{-s}$ for $s = -1, -2, \ldots$ (see Section 4.3) and from the existence of positive definite invariant Hermitian forms on the latter subspaces. We present without proof the expressions for these forms on $F_s^+$ and $F_s^-$ for negative integers $s$ ($s = 0$ has already been treated).

*The positive definite invariant Hermitian functional on $F_s^+$ is given by*

$$(\varphi, \psi)_+ = \int (x_1 - x_2)^{-s-1} \ln(x_1 - x_2 - i0)\varphi(x_1)\bar{\psi}(x_2)\, dx_1\, dx_2\,. \tag{3}$$

*The positive definite invariant Hermitian functional on $F_s^-$ is given by*

$$(\varphi, \psi)_- = \int (x_1 - x_2)^{-s-1} \ln(x_1 - x_2 + i0)\varphi(x_1)\bar{\psi}(x_2)\, dx_1\, dx_2\,. \tag{3'}$$

Remark. These expressions can be obtained directly from the expressions for the invariant forms in $D_{-s}^+/E_{-s}$ and $D_{-s}^-/E_{-s}$, since the operators $A_+$ and $A_-$ which map these spaces onto $F_s^+$ and $F_s^-$ are known. In any case, it is relatively simple to verify that the functionals defined by (3) and (3′) are invariant and positive definite.[23] #

We shall call the representations of $G$ on $F_s^+ \cong D_{-s}^+/E_{-s}$ and $F_s^- \cong D_{-s}^-/E_{-s}$ (where $s = 0, -1, -2, \ldots$) *representations of the discrete series.*

[23] Actually (3) and (3′) define functionals of the entire space $D_s$. On $D_s$, however, they are neither invariant nor positive definite.

### 5.4. Invariant Positive Definite Hermitian Functionals on the Analytic Function Spaces $F_s^+$ and $F_s^-$

Recall that we have several realizations of $F_s^+$ and $F_s^-$. Let us obtain the expressions for the invariant Hermitian functionals in these realizations.

We turn first to the model in which $F_s^+$ is the space of functions $\varphi(z)$ analytic in the half-plane $\operatorname{Im} z > 0$ and infinitely differentiable together with its inversion $\hat{\varphi}(z) = z^{s-1}\varphi(-1/z)$ on the closed half-plane $\operatorname{Im} z \leqslant 0$. The representation in this realization is defined by

$$T_s(g)\varphi(z) = \varphi\left(\frac{\alpha z + \gamma}{\beta z + \delta}\right)(\beta z + \delta)^{s-1},$$

where $s = 0, -1, -2, \ldots$. It is possible, of course, to obtain the expression for the invariant Hermitian functional for this model from the expression we have obtained for the other model (Section 5.3). It is easier, however, to calculate it directly.

Thus for $s < 0$ we attempt to find an invariant functional of the form

$$(\varphi, \psi) = \frac{i}{2}\int_{\operatorname{Im} z>0} \varphi(z)\bar{\psi}(z)\omega(z)\, dz\, d\bar{z}, \tag{1}$$

where $\omega(z)$ is some positive function yet to be defined. We now apply $T_s(g)$ to $\varphi(z)$ and $\psi(z)$, obtaining

$$(T_s(g)\varphi, T_s(g)\psi) = \frac{i}{2}\int_{\operatorname{Im} z>0} \varphi\left(\frac{\alpha z + \gamma}{\beta z + \delta}\right)\bar{\psi}\left(\frac{\alpha z + \gamma}{\beta z + \delta}\right) | \beta z + \delta |^{2s-2}\omega(z)\, dz\, d\bar{z}.$$

Now set $(\alpha z + \gamma)/(\beta z + \delta) = w$, converting the integral to the form

$$(T_s(g)\varphi, T_s(g)\psi) = \frac{i}{2}\int_{\operatorname{Im} z>0} \varphi(w)\bar{\psi}(w)| \alpha - \beta w |^{-2s-2}\omega\left(\frac{\delta w - \gamma}{-\beta w + \alpha}\right) dw\, d\bar{w}.$$

Thus invariance requires that

$$\frac{i}{2}\int_{\operatorname{Im} z>0} \varphi(z)\bar{\psi}(z)\omega(z)\, dz\, d\bar{z}$$

$$= \frac{i}{2}\int_{\operatorname{Im} z>0} \varphi(z)\bar{\psi}(z)| \alpha - \beta z |^{-2s-2}\omega\left(\frac{\delta z - \gamma}{-\beta z + \alpha}\right) dz\, d\bar{z}. \tag{2}$$

Consequently $(\varphi, \psi)$ is invariant if and only if $\omega(z)$ satisfies the condition

$$\omega\left(\frac{\delta z - \gamma}{-\beta z + \alpha}\right) = \omega(z)| \alpha - \beta z |^{2s+2}. \tag{3}$$

This determines $\omega(z)$ to within normalization. We set $z = i$ in Eq. (3) and write $\omega(i) = 1$. Then

$$\omega\left(\frac{\delta i - \gamma}{-\beta i + \alpha}\right) = |\alpha - \beta i|^{2s+2} = |\alpha^2 + \beta^2|^{s+1}$$

We now separate the argument of $\omega$ into its real and imaginary parts, obtaining

$$\omega\left(-\frac{\alpha\gamma + \beta\delta}{\alpha^2 + \beta^2} + \frac{i}{\alpha^2 + \beta^2}\right) = |\alpha^2 + \beta^2|^{s+1},$$

so that

$$\omega(z) = (\operatorname{Im} z)^{-s-1}.$$

It is easily verified that the $\omega(z)$ so obtained satisfies (3).

Summarizing, *the invariant Hermitian functional on* $F_s^+$ *realized as functions analytic on the upper half-plane* is given, for $s = 0, -1, -2, \ldots$, by[24]

$$(\varphi, \psi) = \frac{1}{\Gamma(-s)} \frac{i}{2} \int_{\operatorname{Im} z > 0} \varphi(z)\bar{\psi}(z)(\operatorname{Im} z)^{-s-1}\, dz\, d\bar{z}. \tag{4}$$

Recall that another model of $F_s^+$ is the space of functions $\varphi_1(w)$ analytic in the circle $|w| < 1$. The transition from the functions $\varphi(z)$ analytic in the upper half-plane to these new functions is obtained by

$$\varphi_1(w) = (1 - w)^{s-1}\varphi\left(i\frac{1 + w}{1 - w}\right).$$

In this realization $\varphi_1(w) \in F_s^+$ if and only if $\varphi_1(w)$ is analytic inside the circle $|w| < 1$ and is infinitely differentiable on its boundary. In particular, every $\varphi_1(w)$ is bounded in the unit circle. The representation in this realization is given by

$$T_s(g)\varphi(w) = \varphi\left(\frac{aw + b}{\bar{b}w + \bar{a}}\right)(\bar{b}w + \bar{a})^{s-1}, \tag{5}$$

[24] The factor $1/\Gamma(-s)$ is introduced so that Eq. (4) will remain valid for $s = 0$. Recalling that $[y_+^{-s-1}/\Gamma(-s)]_{s=0} = \delta(y)$, we find that for $s = 0$

$$(\varphi, \psi) = \int_{-\infty}^{+\infty} \varphi(x)\,\bar{\psi}(x)\, dx.$$

That this functional is invariant was seen in Section 5.3.

where $a$ and $b$ are given in terms of the matrix elements of

$$g = \begin{Vmatrix} \alpha & \beta \\ \gamma & \delta \end{Vmatrix}$$

by

$$a = \tfrac{1}{2}[(\alpha + \delta) + i(\gamma - \beta)], \qquad b = \tfrac{1}{2}[(\alpha - \delta) - i(\gamma + \beta)].$$

To obtain the form of the Hermitian functional in this realization, we write $z = i(1 + w)/(1 - w)$ in Eq. (4). Then

$$(\varphi, \psi) = \frac{1}{\Gamma(-s)} \frac{i}{2} \int_{|w|<1} \varphi(w)\bar{\psi}(w)(1 - w\bar{w})^{-s-1}\, dw\, d\bar{w}. \tag{6}$$

Note that since $\varphi(w)$ is bounded in the circle, $(\varphi, \varphi) < \infty$ for every $\varphi(w) \in F_s^+$.

Equation (6) can also be written in the form of a series which may be obtained by using the fact that $1, w, w^2, \ldots$ form a complete orthogonal set with respect to the inner product on $F_s^+$ defined by (6). For the elements of this set we have

$$(w^k, w^k) = \frac{1}{\Gamma(-s)} \frac{i}{2} \int_{|w|<1} |w|^{2k}(1 - |w|^2)^{-s-1}\, dw\, d\bar{w} = \pi \frac{k!}{(k - s)!}.$$

Thus if we write

$$\varphi(w) = \sum_{k=0}^{\infty} a_k w^k, \qquad \psi(w) = \sum_{k=0}^{\infty} b_k w^k,$$

the functional becomes

$$(\varphi, \psi) = \pi \sum_{k=0}^{\infty} \frac{k!}{(k - s)!} a_k \bar{b}_k .$$

This shows that if we complete $F_s^+$ with respect to the norm $\| \varphi \| = (\varphi, \varphi)^{\frac{1}{2}}$, we obtain the space of all functions analytic in $| w | < 1$, such that $(\varphi, \varphi) < \infty$.

## 5.5. Unitary Representations of $G$ by Operators on Hilbert Space

In Sections 5.2 and 5.3 we found the conditions under which $B$ or, in the case of analytic representations, one of its invariant subspaces possesses a positive definite Hermitian functional $(\varphi, \psi)$ invariant under $T_\chi(g)$, that is, such that

$$(\varphi, \psi) = (T_\chi(g)\varphi, T_\chi(g)\psi).$$

Recall that such a functional exists in three cases: when $s$ is pure imaginary and $\epsilon = 0$ or 1, when $s \neq 0$ lies in the interval $-1 < s < 1$ and $\epsilon = 0$, and when $s = 0, -1, \ldots$ and $s + 1$ is of the same parity as $\epsilon$. (In the last case $D_\chi$ contains two invariant subspaces each of which possesses its own invariant positive definite Hermitian form.)

Now in each of these cases the Hermitian functional can be thought of as an inner product on the corresponding space. If this space is then completed with respect to the norm

$$\| \varphi \|^2 = (\varphi, \varphi),$$

one obtains a Hilbert space $H$ in which the initial space forms an everywhere dense set. Then $T_\chi(g)$ can be continued uniquely to the unitary operators on the entire space $H$. Let us denote this continuation, or rather these operators, also by $T_\chi(g)$.

Now it is clear that this continuation also has the group property

$$T_\chi(g_1 g_2) = T_\chi(g_1) T_\chi(g_2)$$

and therefore is also a representation of $G$. Consequently to every representation $T_\chi(g)$ that possesses a positive definite Hermitian functional there corresponds a representation of $G$ by unitary operators on Hilbert space. We then make the following assertion. *In this correspondence equivalent representations correspond to unitary equivalent representations, and inequivalent representations correspond to inequivalent ones.*

**Proof.** We shall first discuss the nonsingular case in which the representations do not belong to the discrete series. The discrete series will be discussed separately in Section 5.6.

Consider two equivalent representations $T_{\chi_1}(g)$ and $T_{\chi_2}(g)$ on $D_{\chi_1}$ and $D_{\chi_2}$, and assume that they possess invariant positive definite functionals. Equivalence means that there exists a one-to-one bicontinuous mapping $A$ of $D_{\chi_1}$ on $D_{\chi_2}$ that intertwines with the representations, i.e., such that $AT_{\chi_1}(g) = T_{\chi_2}(g)A$. Let $(\varphi, \psi)_1$ and $(\varphi, \psi)_2$ be the invariant Hermitian functionals on $D_{\chi_1}$ and $D_{\chi_2}$, respectively. Since these functions are defined uniquely to within a multiplicative factor, it follows that

$$(\varphi, \psi)_1 = c(A\varphi, A\psi)_2\,, \qquad c > 0.$$

Therefore when properly normalized $A$ is an isometric mapping of $D_{\chi_1}$ onto $D_{\chi_2}$, and it can then be continued to an isometric mapping of $H_1$ (the Hilbert space obtained by completing $D_{\chi_1}$) onto $H_2$ (the completion of $D_{\chi_2}$). Obviously for the extension of $A$ and the extensions of the representations the relation $AT_{\chi_1}(g) = T_{\chi_2}(g)A$ will remain valid. Thus the representations on $H_1$ and $H_2$ obtained by extending the equivalent representations $T_{\chi_1}(g)$ and $T_{\chi_2}(g)$ are unitary equivalent.

Conversely, assume that $T_{\chi_1}(g)$ and $T_{\chi_2}(g)$ induce unitary equivalent representations on the Hilbert spaces $H_1$ and $H_2$. Then there exists an isometric mapping $\hat{A}$ of $H_1$ onto $H_2$ such that $\hat{A}T_{\chi_1}(g) = T_{\chi_2}(g)\hat{A}$. Let $(\varphi, \psi)$ be the invariant inner product on $H_2$. We introduce the Hermitian bilinear functional on $H_1$ and $H_2$ defined by

$$B(\varphi, \psi) = (\hat{A}\varphi, \psi).$$

This functional is invariant under $T_{\chi_1}(g)$ and $T_{\chi_2}(g)$. In particular, $B(\varphi, \psi)$ is also an invariant Hermitian functional for the pair of spaces $D_{\chi_1}$ and $D_{\chi_2}$. But such a functional can exist, as we have seen, only if the representations have the same parity and either $s_1 = -\bar{s}_2$ or $s_1 = \bar{s}_2$. This implies that either $s_1 = s_2$ or $s_1 = -s_2$ (since in the representations we are now considering $s_1$ and $s_2$ are either real or pure imaginary). Thus the representations on $D_{\chi_1}$ and $D_{\chi_2}$ are equivalent.

We now wish to classify the unitary representations of $G$.

*Representations of the principal (continuous) series.* These representations are on the Hilbert space of functions $\varphi(x)$ on the line such that

$$(\varphi, \varphi) = \int_{-\infty}^{+\infty} \varphi(x)\bar{\varphi}(x)\, dx < \infty.$$

The representation is defined by

$$T_\chi(g)\varphi(x) = \varphi\left(\frac{\alpha x + \gamma}{\beta x + \delta}\right) | \beta x + \delta |^{i\rho - 1} \operatorname{sgn}^\epsilon (\beta x + \delta).$$

Here $\rho$ can take on any real value and $\epsilon = 0$ or 1, and $\chi = (i\rho, \epsilon)$.

We have shown that two representations of the principal series are equivalent if and only if they have the same parity and $\rho$ is the same for both or differs only in sign.

*Representations of the supplementary series.* These representations are defined by a real parameter $s \neq 0$ in the interval $-1 < s < 1$. For each $s$ the representation is on the space of functions $\varphi(x)$ on the line with the inner product defined by

$$(\varphi, \psi) = \frac{1}{\Gamma(-s)} \int | x_1 - x_2 |^{-s-1} \varphi(x_1)\bar{\psi}(x_2)\, dx_1\, dx_2$$

(for positive $s$ the integral is understood in the sense of its regularization). The representation itself is defined by

$$T_\chi(g)\varphi(x) = \varphi\left(\frac{\alpha x + \gamma}{\beta x + \delta}\right) | \beta x + \delta |^{s-1},$$

where $\chi = (s, 0)$.

We have shown that two representations of the supplementary series are equivalent if and only if $s$ is the same for both or differs only by sign.

*Representations of the discrete series.* These representations are given by integer $s = 0, -1, 2, \ldots$ . To each $s$ there correspond two representations. The first is on the space of functions analytic in the half-plane $\mathrm{Im}\, z > 0$ such that

$$(\varphi, \psi) = \frac{1}{\Gamma(-s)} \frac{i}{2} \int_{\mathrm{Im}\, z>0} \varphi(z)\bar{\psi}(z)(\mathrm{Im}\, z)^{-s-1}\, dz\, d\bar{z} < \infty.$$

The second representation is on the functions analytic in the half-plane $\mathrm{Im}\, z < 0$ such that

$$(\varphi, \psi) = \frac{1}{\Gamma(-s)} \frac{i}{2} \int_{\mathrm{Im}\, z<0} \varphi(z)\bar{\psi}(z)| \mathrm{Im}\, z |^{-s-1}\, dz\, d\bar{z} < \infty.$$

The representation itself is given in both cases by

$$T_s(g)\varphi(z) = \varphi\left(\frac{\alpha z + \gamma}{\beta z + \delta}\right) (\beta z + \delta)^{s-1}.$$

It will be shown in the next section that the representations of the discrete series are pairwise inequivalent.

## 5.6. Inequivalence of the Representations of the Discrete Series

Consider the operators $T_s(g)$ corresponding to matrices of the form

$$g = \begin{Vmatrix} \cos\theta & -\sin\theta \\ \sin\theta & \cos\theta \end{Vmatrix}.$$

We wish to find the eigenfunctions and eigenvalues (the spectrum) of these operators. We shall see that the spectra corresponding to different representations of the discrete series are different, which then implies that different representations of the discrete series are inequivalent.

We treat first the representations on the space of functions analytic in the upper half-plane. For convenience we go over to the realization of these representations in which we deal with functions analytic within $|w| < 1$, so that in accordance with Section 5.4 the inner product is given by

$$(\varphi, \psi) = \frac{1}{\Gamma(-s)} \frac{i}{2} \int_{|w|<1} \varphi(w)\bar{\psi}(w)(1 - w\bar{w})^{-s-1} \, dw \, d\bar{w},$$

and the representation is defined by

$$T_s(g)\varphi(w) = \varphi\left(\frac{aw + b}{\bar{b}w + \bar{a}}\right)(\bar{b}w + \bar{a})^{s-1}$$

(recall that $s = 0, -1, -2, \ldots$). Here $a$ and $b$ are given in terms of the matrix elemts of $g$ by

$$a = \tfrac{1}{2}[(\alpha + \delta) + i(\gamma - \beta)], \qquad b = \tfrac{1}{2}[(\alpha - \delta) - i(\gamma + \beta)].$$

For the special type of $g$ we are considering, in particular, $a = e^{i\theta}$ and $b = 0$, so that the representation of such a $g$ is defined by

$$T_s(g)\varphi(w) = \varphi(e^{2i\theta}w)e^{-i(s-1)\theta}.$$

Obviously the eigenfunctions of $T_s(g)$ are $1, w, \ldots, w^k, \ldots$. They belong to the eigenvalues[25]

$$e^{-i(s-1)\theta}, e^{-i(s-3)\theta}, \ldots, e^{-i(s-2k-1)\theta}, \ldots .$$

Note that all of the eigenvalues of $T_s(g)$ have multiplicity 1, or that the spectrum of $T_s(g)$ is simple.

In complete analogy we find that when $T_s(g)$ is realized on the space of functions analytic in the lower half-plane its eigenvalues are

$$e^{i(s-1)\theta}, e^{i(s-3)\theta}, \ldots, e^{i(s-2k-1)\theta}, \ldots .$$

It is thus seen that for different representations of the discrete series these $T_s(g)$ operators are different. Consequently the representations themselves are pairwise inequivalent.

### 5.7. Subspace Irreducibility of the Unitary Representations

We shall now show that *the unitary representations of the principal, supplementary, and discrete series are subspace irreducible, which means that there exist no proper invariant subspaces under these representations.*

[25] These eigenvalues of $T_s(g)$ for the special form of $g$ are often called the weights of the representation.

**Proof.** We shall restrict ourselves to the discrete series, since the proofs for the principal and supplementary series are exactly the same as in the complex case, which was treated in Section 6.6 of Chapter III. Let us, then, realize our representation on the space of functions analytic within the unit circle. As we have seen,

$$T_s(g)\varphi(w) = \varphi\left(\frac{aw+b}{\bar{b}w+\bar{a}}\right)(\bar{b}w+\bar{a})^{s-1}.$$

Assume that the representation is subspace reducible. Then the carrier space of the representation is the direct sum of a pair of mutually orthogonal invariant subspaces $H_1$ and $H_2$. We turn again to the functions $1, w, w^2, \ldots$. These functions are the eigenfunctions of the $T_s(g)$ operators corresponding to

$$g = \begin{Vmatrix} \cos\theta & -\sin\theta \\ \sin\theta & \cos\theta \end{Vmatrix}.$$

and they belong to different eigenvalues of $T_s(g)$.

Now the $w^n$ form an orthogonal basis. Note that each of these functions belongs to either $H_1$ or $H_2$, for if this were not so there would exist some $w^n$ such that

$$w^n = h_1 + h_2, \qquad h_1 \neq 0, h_2 \neq 0,$$

where $h_1 \in H_1$ and $h_2 \in H_2$. It is easily seen that $h_1$ and $h_2$ must also be eigenfunctions of $T_s(g)$ belonging to the same eigenvalue as $w^n$. But this is impossible, for as we have seen $T_s(g)$ has a simple spectrum. To be specific let us assume that $\varphi(w) \equiv 1$ belongs to $H_2$. This means that there is no constant term in any power series expansion of any function $\psi(w) \in H_1$, and therefore that every such function vanishes at $w = 0$. Now if $H_1$ is assumed nonempty, it contains at least one of the powers of $w$, say $w^k$. By applying $T_s(g)$ to $w^k$, where

$$g = \begin{Vmatrix} \lambda^{-1} & 0 \\ 0 & \lambda \end{Vmatrix},$$

we again obtain a function in $H_1$ which must therefore vanish at $w = 0$. Yet

$$T_s(g)w^k = \frac{(aw+b)^k}{(bw+a)^{k-s+1}},$$

where

$$a = \tfrac{1}{2}(\lambda + \lambda^{-1}), \qquad b = \tfrac{1}{2}(\lambda - \lambda^{-1}).$$

Evidently $[T_s(g)w^k]_{w=0} \neq 0$, so that by assuming that the representation of the discrete series is subspace irreducible we are led to a contradiction.

# NOTES AND REFERENCES TO THE LITERATURE

## Chapter I, Section 1

The problem of expressing a function in terms of its integrals over planes has been treated by several authors [Radon (38), John (25, 26), Khachaturov (29), Kostelyanets and Reshetnyak (32)]. The solution in terms of generalized functions is presented in Vol. 1. The given derivation of the inversion formula is due to Gel'fand and Graev, and is presented here for the first time. Plancherel's theorem was obtained by Yu. Reshetnyak. The analog of the Paley-Wiener theorem is due to Gel'fand and Graev (14). The asymptotic behavior of the Fourier transform of characteristic functions is due to John (26).

## Chapter I, Sections 2 and 3

These results are due to Gel'fand and Graev. They are partially described in Ref. (14) and partially published here for the first time. The generalized hypergeometric function is a concept due to Gel'fand.

## Chapter II, Section 1

These results are due to Gel'fand and Graev (15). Special cases were obtained earlier: lines intersecting a curve of second order by Gel'fand (6), and lines intersecting an arbitrary curve by Kirillov (30).

## Chapter II, Section 2

In essence the problem treated here was solved by Gel'fand and Naimark (17). In the geometric form in which it is here presented (for the first time) it is due to Gel'fand and Graev.

## Chapter II, Section 3

These results are due to Gel'fand and Graev. In part they are given in Ref. (14) and in part published here for the first time.

## Chapter III

The results of this chapter (for unitary representations of the Lorentz group) were obtained by Gel'fand and Naimark and appear in Ref. (17). The method, based on a study of bilinear forms, is due to Bruhat (4), who did not consider, however, integer points. The results concerning representations at integer points, published for the first time, are due to Gel'fand and Vilenkin, who are also responsible for the concept of the $D_\chi$ spaces. Tensor irreducibility was first enunciated by Gel'fand.

## Chapter IV

The results of Sections 1–3 are due to Gel'fand and Naimark (17), while the method of presentation is due to the present authors. Section 5, the Paley-Wiener theorem, is due to Gel'fand (7) and Zhelobenko (46). These results were later generalized to semisimple complex groups by Gel'fand and Graev (10). The Paley-Wiener theorem for the real Lorentz group was proven earlier (in another form) by Ehrenpreis and Mautner (5).

The present state of the theory of infinite-dimensional representations of groups is discussed in the interesting review paper by Mackey (47).

## Chapter V

This chapter is based on a paper by Gel'fand and Graev (11) which makes use of matrix realizations of the spaces involved. The geometric presentation given here is due to the present authors. A similar problem for planes in Lobachevskian and Riemannian spaces was studied by Helgason (24) and Semyanistyi (41, 42).

## Chapter VI

The general definition of a horosphere in a homogeneous space acted on by a complex semisimple group is given by Gel'fand and Graev in Ref. (9), where the relation between representation theory and integral geometry is also clarified. The concept of a horosphere is closely related to that of the boundary of a symmetric space, introduced by Karpelevich (28). The results of Section 2–4 were first published in the Gel'fand and Graev paper. The geometric treatment given here is due to the present authors. The results of Section 5 are due to Gel'fand and Graev (9) [in Ref. (9) they were obtained for an arbitrary complex semisimple group]. The same results were obtained earlier in a different way by Naimark (37). Recently Vilenkin has established the relation between the method of horospheres and the Fock-Miller transformation.

### Chapter VII

The results of this chapter are due essentially to Bargmann (1). In the form given here, they are due to the present authors and are published here for the first time. The behavior of representations at integer points is due to the present authors and is published here for the first time. The question of harmonic analysis on the group of real matrices is discussed in interesting papers by Ehrenpreis and Mautner (5), Kunze and Stein (33), and Pukánszky (48).

## BIBLIOGRAPHY


*1.* V. Bargmann, Representations of the Lorentz group, *Ann. Math.* **48**, 568 (1947).

*2.* W. Blaschke, "Vorlesungen über Integralgeometrie." Teubner, 1936–1937.

*3.* S. Bochner, "Vorlesungen über Fouriersche Integrale." Akad. Verlagsges. Leipzig, 1932.

*4.* F. Bruhat, Sur les représentations induites des groupes de Lie, *Bull. Soc. Math. France* **84**, 97 (1956).

*5.* L. Ehrenpreis and F. Mautner, Some properties of the Fourier transform on semi-simple Lie groups. I, *Ann. Math.* **61**, 406 (1955); II, *Trans. Am. Math. Soc.* **84**, 1 (1957).

*6.* I. M. Gel'fand, Integral geometry and its relation to representation theory (in Russian), *Uspekhi Mat. Nauk* **15**, 155 (1960).

*7.* I. M. Gel'fand, On the structure of the ring of rapidly decreasing functions on a Lie group (in Russian), *Doklady Akad. Nauk S.S.S.R.* **124**, 19 (1959).

*8.* I. M. Gel'fand and S. V. Fomin, Geodesic flows on manifolds of constant curvature (in Russian), *Uspekhi Mat. Nauk* **7**, 118 (1952). English translation: *Am. Math. Soc. Translations*, Ser. 2, Vol. 1.

*9.* I. M. Gel'fand and M. I. Graev, Geometry of homogeneous spaces, representations of groups in homogeneous spaces, and related questions of integral geometry (in Russian), *Communs. (Trudy) Moscow Math. Soc.* **8**, 321 (1959). English translation: *Am. Math. Soc. Translations*, Ser. 2, Vol. 37.

*10.* I. M. Gel'fand and M. I. Graev, Fourier transforms of rapidly decreasing functions on complex semisimple Lie groups (in Russian), *Doklady Akad. Nauk S.S.S.R.* **131**, 496 (1960).
*11.* I. M. Gel'fand and M. I. Graev, Application of the method of horospheres to the spectral analysis of functions in ordinary and in imaginary Lobachevskian spaces (in Russian). *Communs. (Trudy) Moscow Math. Soc.* **11**, 243 (1962).
*12.* I. M. Gel'fand and M. I. Graev, On a general method for decomposing the regular representation of a Lie group into irreducible representations (in Russian), *Doklady Akad. Nauk S.S.S.R.* **92**, 221 (1953).
*13.* I. M. Gel'fand and M. I. Graev, Analog of the Plancherel formula for the classical groups (in Russian), *Communs. (Trudy) Moscow Math. Soc.* **4**, 375 (1955). English translation: *Am. Math. Soc. Translations*, Ser. 2, Vol. 9.
*14.* I. M. Gel'fand and M. I. Graev, Integrals of test functions and generalized functions over hyperplanes (in Russian), *Doklady Akad. Nauk S.S.S.R.* **135**, 1307 (1960).
*15.* I. M. Gel'fand and M. I. Graev, Integral transforms associated with complexes of lines in a complex affine space (in Russian), *Doklady Akad. Nauk S.S.S.R.*. **138**, 1266 (1961).
*16.* I. M. Gel'fand, R. A. Minlos, and Z. Ya. Shapiro, "Representations of the Rotation and Lorentz Groups and Their Applications." Oxford, New York, 1963.
*17.* I. M. Gel'fand and M. A. Naimark, Unitary representations of the Lorentz group (in Russian), *Izvestia (Bulletin) Akad. Nauk. S.S.S.R. (ser. mat.)* **11**, 411 (1947).
*18.* I. M. Gel'fand and I. I. Pyatetskii-Shapiro, Theory of representations and theory of automorphic functions (in Russian), *Uspekhi Mat. Nauk* **14**, 171 (1959). English translation: *Am. Math. Soc. Translations*, Ser. 2, Vol. 26.
*19.* I. M. Gel'fand and Z. Ya. Shapiro, Homogeneous functions and their applications (in Russian), *Uspekhi Mat. Nauk* **10**, 118 (1952). English translation: Homogeneous functions and their extensions, *Am. Math. Soc. Translations*, Ser. 2, Vol. 8.
*20.* I. M. Gel'fand and G. E. Shilov, "Generalized Functions," Vol. 1. Academic, New York, 1964.
*21.* I. M. Gel'fand and G. E. Shilov, "Generalized Functions," Vol. 2 ("Spaces of test functions and generalized functions") (in Russian). Fizmatgiz, Moscow, 1958. English translation in preparation.
*22.* I. M. Gel'fand and N. Ya. Vilenkin, "Generalized Functions," Vol. 4. Academic, New York, 1964.
*23.* R. Godement, "Seminaire H. Cartan École Normale Supérieure, 1957–1958." Paris, 1958.
*24.* S. Helgason, Differential operators on homogeneous spaces, *Acta. Math.* **102**, 239 (1959).
*25.* F. John, Bestimmung einer Funktion aus ihren Integralen über gewisse Männigfaltigkeiten, *Math. Ann.* **100**, 488 (1934).
*26.* F. John, Abhängigkeiten zwischen den Flächenintegralen einer stetigen Funktion, *Math. Ann.* **111**, 541 (1935).
*27.* F. John, "Plane Waves and Spherical Means, Applied to Partial Differential Equations." Wiley (Interscience), New York, 1955.
*28.* F. I. Karpelevich, Geodesic lines and harmonic functions on symmetric spaces (in Russian), *Doklady Akad. Nauk S.S.S.R.* **124**, 1199 (1959).
*29.* A. A. Khachaturov, Determination of the value of a measure on a region in a space of $n$ dimensions from its values for all half-spaces (in Russian), *Uspekhi Mat. Nauk* **9**, 205 (1954).

*30.* A. A. Kirillov, On a certain problem of I. M. Gel'fand (in Russian), *Doklady Akad. Nauk S.S.S.R.* **137**, 276 (1961).

*31.* F. Klein, "Vorlesungen über Höhere Geometrie." Springer, Berlin, 1926.

*32.* P..O. Kostelyanets and Yu. G. Reshetnyak, Determination of an additive function from its values on a half-space (in Russian), *Uspehki Mat. Nauk* **9**, 135 (1954).

*33.* R. A. Kunze and E. M. Stein, Uniformly bounded representations and harmonic analysis of the 2 × 2 real unimodular group, *Am. J. Math.* **82**, 1 (1960).

*34.* J. Leray, Les solutions élémentaires d'une équation aux dérivées partielles à coefficients constants, *Compt. rend. acad. sci.* **234**, 1112 (1952). See also this author's book "Hyperbolic Differential Equations." New York, 1955.

*35.* J. Leray, Le calcul différentiel et intégral sur une variété analytique complexe (Problème de Cauchy, III), *Bull. Soc. Math. France* **87**, 81 (1959).

*36.* M. A. Naimark, "Linear Representations of the Lorentz Group." Macmillan, (Permagon), New York, 1964.

*37.* M. A. Naimark, Decomposition of a tensor product of irreducible representations of the proper Lorentz group into irreducible representations (in Russian), *Communs. (Trudy) Moscow Math. Soc.* **8**, 121 (1959). English translation (together with two subsequent articles by the same author on the same subject): *Am. Math. Soc. Translations*, Ser. 2, Vol. 36.

*38.* J. Radon, Über die Bestimmung von Funktionen durch ihre Integralwärte längs gewisser Männigfaltigkeiten, *Ber. Verh. Sächs. Akad.* **69**, 262 (1917).

*39.* L. A. Santalo Sors, "Introduction to Integral Geometry." Hermann, Paris, 1953.

*40.* A. Selberg, Harmonic analysis and discontinuous groups in weakly symmetric Riemannian spaces with applications to Dirichlet series, *J. Indian Math. Soc. (N.S.)* **20**, 47 (1956).

*41.* V. I. Semyanistyi, On some integral transforms in Euclidean space (in Russian), *Doklady Akad. Nauk S.S.S.R.* **134**, 536 (1960).

*42.* V. I. Semyanistyi, Homogeneous functions and some problems of integral geometry in spaces of constant curvature (in Russian), *Doklady Akad. Nauk S.S.S.R.* **136**, 288 (1961).

*43.* E. Titchmarsh, "The Theory of Functions." Oxford Univ. Press, London and New York, 1950.

*44.* A. Weil. "L'intégration dans les Groupes Topologiques et ses Applications." Paris, 1940.

*45.* H. Weyl, "The Classical Groups." Princeton Univ. Press, Princeton, New Jersey, 1946.

*46.* D. P. Zhelobenko, Structure of the group ring of the Lorentz group (in Russian), *Doklady Akad. Nauk S.S.S.R.* **126**, 482 (1959).

*47.* G. W. Mackey, Infinite dimensional group representations. Colloquium Lecture given at Stillwater, Oklahoma, Aug. 29-Sept. 1 at the 66th Summer Meeting of the American Mathematical Society, 1961.

*48.* L. Pukánszky, On Kronecker products of irreducible representations of the 2 × 2 real unimodular group, part I, *Trans. Am. Math. Soc.* **100**, 116 (1961).

# Index

In this index the group of complex unimodular matrices in two dimensions is referred to as $G$. When no specific group is referred to, $G$ is often understood. The analogous real group is referred to as "the real unimodular group."

# Index of Radon Transforms of Particular Functions




7 8 9 0 1 2 3 4 5



QUEEN MARY COLLEGE LIBRARY